Intelligence Integration in Distributed Knowledge Management

Dariusz Król
Wroclaw University of Technology, Poland

Ngoc Thanh Nguyen
Wroclaw University of Technology, Poland

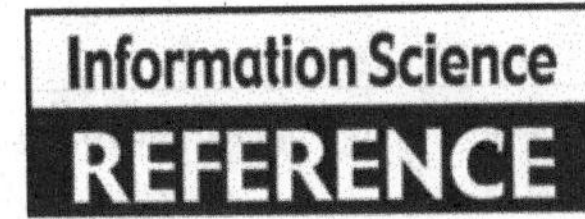

INFORMATION SCIENCE REFERENCE
Hershey · New York

Director of Editorial Content: Kristin Klinger
Managing Development Editor: Kristin M. Roth
Senior Managing Editor: Jennifer Neidig
Managing Editor: Jamie Snavely
Assistant Managing Editor: Carole Coulson
Copy Editor: Lanette Ehrhardt
Typesetter: Jeff Ash
Cover Design: Lisa Tosheff
Printed at: Yurchak Printing Inc.

Published in the United States of America by
Information Science Reference (an imprint of IGI Global)
701 E. Chocolate Avenue, Suite 200
Hershey PA 17033
Tel: 717-533-8845
Fax: 717-533-8661
E-mail: cust@igi-global.com
Web site: http://www.igi-global.com

and in the United Kingdom by
Information Science Reference (an imprint of IGI Global)
3 Henrietta Street
Covent Garden
London WC2E 8LU
Tel: 44 20 7240 0856
Fax: 44 20 7379 0609
Web site: http://www.eurospanbookstore.com

Library of Congress Cataloging-in-Publication Data

Intelligence integration in distributed knowledge management / Dariusz Krol and Ngoc Thanh Nguyen, editors.

p. cm.

Includes bibliographical references and index.

Summary: "This book covers a broad range of intelligence integration approaches in distributed knowledge systems, from Web-based systems through multi-agent and grid systems, ontology management to fuzzy approaches"--Provided by publisher.

ISBN 978-1-59904-576-4 (hardcover) -- ISBN 978-1-59904-578-8 (ebook)

1. Expert systems (Computer science) 2. Intelligent agents (Computer software) 3. Electronic data processing--Distributed processing. I. Krol, Dariusz. II. Nguyên, Ngoc Thanh.

QA76.76.E95I53475 2009

006.3--dc22

2008016377

British Cataloguing in Publication Data
A Cataloguing in Publication record for this book is available from the British Library.

Table of Contents

Preface xiv

Section I
Advanced Methods for Integration

Chapter I
Logical Inference Based on Incomplete and/or Fuzzy Ontologies 1
Juliusz L. Kulikowski, Polish Academy of Sciences, Poland

Chapter II
Using Logic Programming and XML Technologies for Data Extraction from Web Pages 17
Amelia Bădică, University of Craiova, Romania
Costin Bădică, University of Craiova, Romania
Elvira Popescu, University of Craiova, Romania

Chapter III
A Formal Analysis of Virtual Enterprise Creation and Operation 48
Andreas Jacobsson, Blekinge Institute of Technology, Sweden
Paul Davidsson, Blekinge Institute of Technology, Sweden

Chapter IV
Application of Uncertain Variables to Knowledge-Based Resource Distribution 63
Donat Orski, Wroclaw University of Technology, Poland

Chapter V
A Methodology of Design for Virtual Environments 85
Clive Fencott, University of Teesside, UK

Chapter VI
An Ontological Representation of Competencies as Codified Knowledge 104
Salvador Sanchez-Alonso, University of Alcalá, Spain
Dirk Frosch-Wilke, University of Applied Sciences, Germany

Section II
Integration Aspects for Agent Systems

Chapter VII
Aspects of Openness in Multi-Agent Systems: Coordinating the Autonomy in Agent Societies 119
Marcos De Oliveira, University of Otago, New Zealand
Martin Purvis, University of Otago, New Zealand

Chapter VIII
How Can We Trust Agents in Multi-Agent Environments? Techniques and Challenges 132
Kostas Kolomvatsos, National and Kapodistrian University of Athens, Greece
Stathes Hadjiefthymiades, National and Kapodistrian University of Athens, Greece

Chapter IX
The Concept of Autonomy in Distributed Computation and Multi-Agent Systems 154
Mariusz Nowostawski, University of Otago, New Zealand

Chapter X
An Agent-Based Library Management System Using RFID Technology 171
Maryam Purvis, University of Otago, New Zealand
Toktam Ebadi, University of Otago, New Zealand
Bastin Tony Roy Savarimuthu, University of Otago, New Zealand

Chapter XI
Mechanisms to Restrict Exploitation and Improve Societal Performance in Multi-Agent Systems 182
Sharmila Savarimuthu, University of Otago, New Zealand
Martin Purvis, University of Otago, New Zealand
Maryam Purvis, University of Otago, New Zealand
Mariusz Nowostawski, University of Otago, New Zealand

Chapter XII
Norm Emergence in Multi-Agent Societies 195
Bastin Tony Roy Savarimuthu, University of Otago, New Zealand
Maryam Purvis, University of Otago, New Zealand
Stephen Cranefield, University of Otago, New Zealand

Chapter XIII
Multi-Agent Systems Engineering: An Overview and Case Study 207
Scott A. DeLoach, Kansas State University, USA
Madhukar Kamar, Software Engineer, USA

Section III
Fuzzy-Based and Other Methods for Integration

Chapter XIV
Modeling, Analysing, and Control of Agents Behaviour 226
František Čapkovič, Institute of Informatics, Slovak Academy of Sciences, Slovak Republic

Chapter XV
Using Fuzzy Segmentation for Colour Image Enhancement of Computed Tomography Perfusion Images 253
Martin Tabakov, Wrocław University of Technology, Poland

Chapter XVI
Fuzzy Mediation in Shared Control and Online Learning 263
Giovanni Vincenti, Research and Development at Gruppo Vincenti, Italy
Goran Trajkovski, Algoco eLearning Consulting, USA

Chapter XVII
Utilizing Past Web for Knowledge Discovery 286
Adam Jatowt, Kyoto University, Japan
Yukiko Kawai, Kyoto Sangyo University, Japan
Katsumi Tanaka, Kyoto University, Japan

Chapter XVIII
Example-Based Framework for Propagation of Tasks in Distributed Environments 305
Dariusz Król, Wrocław University of Technology, Poland

Chapter XIX
Survey on the Application of Economic and Market Theory for Grid Computing 316
Xia Xie, Huazhong University of Science and Technology, China
Jin Huang, Huazhong University of Science and Technology, China
Song Wu, Huazhong University of Science and Technology, China
Hai Jin, Huazhong University of Science and Technology, China
Melvin Koh, Asia Pacific Science & Technology Center, Sun Microsystems, Singapore
Jie Song, Asia Pacific Science & Technology Center, Sun Microsystems, Singapore
Simon See, Asia Pacific Science & Technology Center, Sun Microsystems, Singapore

Compilation of References 335

About the Contributors 358

Index 364

Detailed Table of Contents

Preface xiv

Section I
Advanced Methods for Integration

Chapter I
Logical Inference Based on Incomplete and/or Fuzzy Ontologies 1
Juliusz L. Kulikowski, Polish Academy of Sciences, Poland

In this chapter, a concept of using incomplete or fuzzy ontologies in decision making is presented. A definition of ontology and of ontological models is given, as well as their formal representation by taxonomic trees, bi-partite graphs, multigraphs, relations, super-relations and hyper-relations. The definitions of the corresponding mathematical notions are also given. Then, the concept of ontologies representing incomplete or uncertain domain knowledge is presented. This concept is illustrated by an example of decision making in medicine. The aim of this chapter is to give an outlook on the possibility of ontological models extension in order to use them as an effective and universal form of domain knowledge representation in computer systems supporting decision making in various application areas.

Chapter II
Using Logic Programming and XML Technologies for Data Extraction from Web Pages 17
Amelia Bădică, University of Craiova, Romania
Costin Bădică, University of Craiova, Romania
Elvira Popescu, University of Craiova, Romania

The Web is designed as a major information provider for the human consumer. However, information published on the Web is difficult to understand and reuse by a machine. In this chapter, we show how well established intelligent techniques based on logic programming and inductive learning combined with more recent XML technologies might help to improve the efficiency of the task of data extraction from Web pages. Our work can be seen as a necessary step of the more general problem of Web data management and integration.

Chapter III
A Formal Analysis of Virtual Enterprise Creation and Operation 48
Andreas Jacobsson, Blekinge Institute of Technology, Sweden
Paul Davidsson, Blekinge Institute of Technology, Sweden

This chapter introduces a formal model of virtual enterprises, as well as an analysis of their creation and operation. It is argued that virtual enterprises offer a promising approach to promote both innovations and collaboration between companies. A framework of integrated ICT-tools, called Plug and Play Business, which support innovators in turning their ideas into businesses by dynamically forming virtual enterprises, is also formally specified. Furthermore, issues regarding the implementation of this framework are discussed and some useful technologies are identified.

Chapter IV
Application of Uncertain Variables to Knowledge-Based Resource Distribution 63
Donat Orski, Wroclaw University of Technology, Poland

The chapter concerns a class of systems composed of operations performed with the use of resources allocated to them. In such operation systems, each operation is characterized by its execution time depending on the amount of a resource allocated to the operation. The decision problem consists in distributing a limited amount of a resource among operations in an optimal way, that is, in finding an optimal resource allocation. Classical mathematical models of operation systems are widely used in computer supported projects or production management, allowing optimal decision making in deterministic, well-investigated environments. In the knowledge-based approach considered in this chapter, the execution time of each operation is described in a nondeterministic way, by an inequality containing an unknown parameter, and all the unknown parameters are assumed to be values of uncertain variables characterized by experts. Mathematical models comprising such two-level uncertainty are useful in designing knowledge-based decision support systems for uncertain environments. The purpose of this chapter is to present a review of problems and algorithms developed in recent years, and to show new results, possible extensions and challenges, thus providing a description of a state-of-the-art in the field of resource distribution based on the uncertain variables.

Chapter V
A Methodology of Design for Virtual Environments.. 85
Clive Fencott, University of Teesside, UK

This chapter undertakes a methodological study of virtual environments (VEs), a specific subset of interactive systems. It takes as a central theme the tension between the engineering and aesthetic notions of VE design. First of all method is defined in terms of underlying model, language, process model, and heuristics. The underlying model is characterized as an integration of Interaction Machines and Semiotics with the intention to make the design tension work to the designer's benefit rather than trying to eliminate it. The language is then developed as a juxtaposition of UML and the integration of a range of semiotics-based theories. This leads to a discussion of a process model and the activities that comprise it. The intention throughout is not to build a particular VE design method, but to investigate the methodological concerns and constraints such a method should address.

Chapter VI
An Ontological Representation of Competencies as Codified Knowledge 104
Salvador Sanchez-Alonso, University of Alcalá, Spain
Dirk Frosch-Wilke, University of Applied Sciences, Germany

In current organizations, the models of knowledge creation include specific processes and elements that drive the production of knowledge aimed at satisfying organizational objectives. The knowledge life cycle (KLC) model of the Knowledge Management Consortium International (KMCI) provides a comprehensive framework for situating competencies as part of the organizational context. Recent work on the use of ontologies for the explicit description of competency-related terms and relations can be used as the basis for a study on the ontological representation of competencies as codified knowledge, situating those definitions in the KMCI lifecycle model. In this chapter, we discuss the similarities between the life cycle of knowledge management (KM) and the processes in which competencies are identified and assessed. The concept of competency, as well as the standard definitions for this term that coexist nowadays, will then be connected to existing KLC models in order to provide a more comprehensive framework for competency management in a wider KM framework. This paper also depicts the framework's integration into the KLC of the KMCI in the form of ontological definitions.

Section II
Integration Aspects for Agent Systems

Chapter VII
Aspects of Openness in Multi-Agent Systems: Coordinating the Autonomy
in Agent Societies 119
Marcos De Oliveira, University of Otago, New Zealand
Martin Purvis, University of Otago, New Zealand

In the distributed multi-agent systems discussed in this chapter, heterogeneous autonomous agents interoperate in order to achieve their goals. In such environments, agents can be embedded in diverse contexts and interact with agents of various types and behaviours. Mechanisms are needed for coordinating these multi-agent interactions, and so far they have included tools for the support of conversation protocols and tools for the establishment and management of agent groups and electronic institutions. In this chapter, we explore the necessity of dealing with openness in multi-agent systems and its relation with the agent's autonomy. We stress the importance to build coordination mechanisms capable of managing complex agent societies composed by autonomous agents and introduce our institutional environment approach, which includes the use of commitments and normative spaces. It is based on a metaphor in which agents may join an open system at any time, but they must obey regulations in order to maintain a suitable reputation, that reflects its degree of cooperation with other agents in the group, and make them a more desired partner for others. Coloured Petri Nets are used to formalize a workflow in the institutional environment defining a normative space that guides the agents during interactions in the conversation space.

Chapter VIII
How Can We Trust Agents in Multi-Agent Environments? Techniques and Challenges.................. 132
Kostas Kolomvatsos, National and Kapodistrian University of Athens, Greece
Stathes Hadjiefthymiades, National and Kapodistrian University of Athens, Greece

The field of Multi-agent systems (MAS) has been an active area for many years due to the importance that agents have to many disciplines of research in computer science. MAS are open and dynamic systems where a number of autonomous software components, called agents, communicate and cooperate in order to achieve their goals. In such systems, trust plays an important role. There must be a way for an agent to make sure that it can trust another entity, which is a potential partner. Without trust, agents cannot cooperate effectively and without cooperation they cannot fulfill their goals. Many times, trust is based on reputation. It is an indication that we may trust someone. This important research area is investigated in this book chapter. We discuss main issues concerning reputation and trust in MAS. We present research efforts and give formalizations useful for understanding the two concepts.

Chapter IX
The Concept of Autonomy in Distributed Computation and Multi-Agent Systems.......................... 154
Mariusz Nowostawski, University of Otago, New Zealand

The concept of autonomy is one of the central concepts in distributed computational systems, and in multi-agent systems in particular. With diverse implications in philosophy, social sciences and the theory of computation, autonomy is a rather complicated and somewhat vague notion. Most researchers do not discuss the details of this concept, but rather assume a general, common-sense understanding of autonomy in the context of computational multi-agent systems. In this chapter, we will review the existing definitions and formalisms related to the notion of autonomy. We re-introduce two concepts: relative autonomy and absolute autonomy. We argue that even though the concept of absolute autonomy does not make sense in computational settings, it is useful if treated as an assumed property of computational units. For example, the concept of autonomous agents facilitates more flexible and robust architectures. We adopt and discuss a new formalism based on results from the study of massively parallel multi-agent systems in the context of Evolvable Virtual Machines. We also present the architecture for building such architectures based on our multi-agent system KEA, where we use an extended notion of dynamic and flexibly linking. We augment our work with theoretical results from chemical abstract machine algebra for concurrent and asynchronous information processing systems. We argue that for open distributed systems, entities must be connected by multiple computational dependencies and a system as a whole must be subjected to influence from external sources. However, the exact linkages are not directly known to the computational entities themselves. This provides a useful notion and the necessary means to establish an autonomy in such open distributed systems.

Chapter X
An Agent-Based Library Management System Using RFID Technology... 171
Maryam Purvis, University of Otago, New Zealand
Toktam Ebadi, University of Otago, New Zealand
Bastin Tony Roy Savarimuthu, University of Otago, New Zealand

The objective of this research is to describe a mechanism to provide an improved library management system using RFID and agent technologies. One of the major issues in large libraries is to track misplaced items. By moving from conventional technologies such as barcode-based systems to RFID-based systems and using software agents that continuously monitor and track the items in the library, we believe an effective library system can be designed. Due to constant monitoring, the up-to-date location information of the library items can be easily obtained.

Chapter XI
Mechanisms to Restrict Exploitation and Improve Societal Performance
in Multi-Agent Systems 182
Sharmila Savarimuthu, University of Otago, New Zealand
Martin Purvis, University of Otago, New Zealand
Maryam Purvis, University of Otago, New Zealand
Mariusz Nowostawski, University of Otago, New Zealand

Societies are made of different kinds of agents, some cooperative and uncooperative. Uncooperative agents tend to reduce the overall performance of the society, due to exploitation practices. In the real world, it is not possible to decimate all the uncooperative agents; thus the objective of this research is to design and implement mechanisms that will improve the overall benefit of the society without excluding uncooperative agents. The mechanisms that we have designed include referrals and resource restrictions. A referral scheme is used to identify and distinguish noncooperators and cooperators. Resource restriction mechanisms are used to restrict noncooperators from selfish resource utilization. Experimental results are presented describing how these mechanisms operate.

Chapter XII
Norm Emergence in Multi-Agent Societies.. 195
Bastin Tony Roy Savarimuthu, University of Otago, New Zealand
Maryam Purvis, University of Otago, New Zealand
Stephen Cranefield, University of Otago, New Zealand

Norms are shared expectations of behaviours that exist in human societies. Norms help societies by increasing the predictability of individual behaviours and by improving cooperation and collaboration among members. Norms have been of interest to multi-agent system researchers, as software agents intend to follow certain norms. But, owing to their autonomy, agents sometimes violate norms, which needs monitoring. In order to build robust MAS that are norm compliant and systems that evolve and adapt norms dynamically, the study of norms is crucial. Our objective in this chapter is to propose a mechanism for norm emergence in artificial agent societies and provide experimental results. We also study the role of autonomy and visibility threshold of an agent in the context of norm emergence.

Chapter XIII
Multi-Agent Systems Engineering: An Overview and Case Study 207
Scott A. DeLoach, Kansas State University, USA
Madhukar Kamar, Software Engineer, USA

This chapter provides an overview of the Multi-agent Systems Engineering (MaSE) methodology for analyzing and designing multi-agent systems. MaSE consists of two main phases that result in the creation of a set of complementary models that get successively closer to implementation. MaSE has been used to design systems ranging from a heterogeneous database integration system to a biologically based, computer virus-immune system to cooperative robotics systems. The authors also provide a case study of an actual system developed using MaSE in an effort to help demonstrate the practical aspects of developing systems using MaSE.

Section III
Fuzzy-Based and Other Methods for Integration

Chapter XIV
Modeling, Analysing, and Control of Agents Behaviour... 226
František Čapkovič, Institute of Informatics, Slovak Academy of Sciences, Slovak Republic

An alternative approach to modeling and analysis of agents' behaviour is presented in this chapter. The agents and agent systems are understood here to be discrete-event systems (DES). The approach is based on the place/transition Petri nets (P/T PN) that yield both the suitable graphical or mathematical description of DES and the applicable means for testing the DES properties as well as for the synthesis of the agents' behaviour. The reachability graph (RG) of the P/T PN-based model of the agent system and the space of feasible states are found. The RG adjacency matrix helps to form an auxiliary hypermodel in the space of the feasible states. State trajectories representing the actual interaction processes among agents are computed by means of the mutual intersection of both the straight-lined reachability tree (developed from a given initial state toward a prescribed terminal one) and the backtracking reachability tree (developed from the desired terminal state toward the initial one; however, oriented toward the terminal state). Control interferences are obtained on the base of the most suitable trajectory chosen from the set of feasible ones.

Chapter XV
Using Fuzzy Segmentation for Colour Image Enhancement of Computed
Tomography Perfusion Images... 253
Martin Tabakov, Wrocław University of Technology, Poland

This chapter presents a methodology for an image enhancement process of computed tomography perfusion images by means of partition generated with appropriately defined fuzzy relation. The proposed image processing is used to improve the radiological analysis of the brain perfusion. Colour image segmentation is a process of dividing the pixels of an image in several homogenously- coloured and topologically connected groups, called regions. As the concept of homogeneity in a colour space is imprecise, a measure of dependency between the elements of such a space is introduced. The proposed measure is based on a pixel metric defined in the HSV colour space. By this measure a fuzzy similarity relation is defined, which next is used to introduce a clustering method that generates a partition, and so a segmentation. The achieved segmentation results are used to enhance the considered computed tomography perfusion images with the purpose of improving the corresponding radiological recognition.

Chapter XVI
Fuzzy Mediation in Shared Control and Online Learning... 263
Giovanni Vincenti, Research and Development at Gruppo Vincenti, Italy
Goran Trajkovski, Algoco eLearning Consulting, USA

This chapter presents an innovative approach to the field of information fusion. Fuzzy mediation differentiates itself from other algorithms, as this approach is dynamic in nature. The experiments reported in this work analyze the interaction of two distinct controllers as they try to maneuver an artificial agent through a path. Fuzzy mediation functions as a fusion engine to integrate the two inputs to produce a single output. Results show that fuzzy mediation is a valid method to mediate between two distinct controllers. The work reported in this chapter lays the foundation for the creation of an effective tool that uses positive feedback systems instead of negative ones to train human and nonhuman agents in the performance of control tasks.

Chapter XVII
Utilizing Past Web for Knowledge Discovery.. 286
Adam Jatowt, Kyoto University, Japan
Yukiko Kawai, Kyoto Sangyo University, Japan
Katsumi Tanaka, Kyoto University, Japan

The Web is a useful data source for knowledge extraction, as it provides diverse content virtually on any possible topic. Hence, a lot of research has been recently done for improving mining in the Web. However, relatively little research has been done taking directly into account the temporal aspects of the Web. In this chapter, we analyze data stored in Web archives, which preserve content of the Web, and investigate the methodology required for successful knowledge discovery from this data. We call the collection of such Web archives past Web; a temporal structure composed of the past copies of Web pages. First, we discuss the character of the data and explain some concepts related to utilizing the past Web, such as data collection, analysis and processing. Next, we introduce examples of two applications, temporal summarization and a browser for the past Web.

Chapter XVIII
Example-Based Framework for Propagation of Tasks in Distributed Environments......................... 305
Dariusz Król, Wrocław University of Technology, Poland

In this chapter, we propose a generic framework in C# to distribute and compute tasks defined by users. Unlike the more popular models such as middleware technologies, our multinode framework is task-oriented desktop grid. In contrast with earlier proposals, our work provides simple architecture to define, distribute and compute applications. The results confirm and quantify the usefulness of such ad-hoc grids. Although significant additional experiments are needed to fully characterize the framework, the simplicity of how they work in tandem with the user is the most important advantage of our current proposal. The last section points out conclusions and future trends in distributed environments.

Chapter XIX
Survey on the Application of Economic and Market Theory for Grid Computing 316
Xia Xie, Huazhong University of Science and Technology, China
Jin Huang, Huazhong University of Science and Technology, China
Song Wu, Huazhong University of Science and Technology, China
Hai Jin, Huazhong University of Science and Technology, China
Melvin Koh, Asia Pacific Science & Technology Center, Sun Microsystems, Singapore
Jie Song, Asia Pacific Science & Technology Center, Sun Microsystems, Singapore
Simon See, Asia Pacific Science & Technology Center, Sun Microsystems, Singapore

In this chapter, we present a survey on some of the commercial players in the Grid industry, existing research done in the area of market-based Grid technology and some of the concepts of dynamic pricing model that we have investigated. In recent years, it has been observed that commercial companies are slowly shifting from owning their own IT assets in the form of computers, software and so forth, to purchasing services from utility providers. Technological advances, especially in the area of Grid computing, have been the main catalyst for this trend. The utility model may not be the most effective model and the price still needs to be determined at the point of usage. In general, market-based approaches are more efficient in resource allocations, as it depends on price adjustment to accommodate fluctuations in the supply and demand. Therefore, determining the price is vital to the overall success of the market.

Compilation of References 335

About the Contributors 358

Index 364

Preface

Rapid advances and wide availability have caused knowledge management to permeate the lives of people from all walks of life. The development of the distributed knowledge technologies has extended the reach of computer intelligence to almost everyone.

In our book, intelligence integration can be understood in two aspects. The first is referred to as methods for integration of human intelligence useful for management and social sciences. The second aspect is related to integration methods for intelligent computer systems such as agent systems, Web-based systems, ad hoc systems and so forth. The subject of this edited book is focused on the second aspect. It covers a broad range of intelligence integration approaches in distributed knowledge systems, from Web-based systems through multi-agent and grid systems, and ontology management to fuzzy approaches. It presents cutting edge research in knowledge management in the first decade of the 21st century. The new directions include integration of computational intelligence, distributed computing and data mining.

In order to achieve the goals of better knowledge integration in the field of distributed environment that collect modern approaches from artificial intelligence, computer communication, and information systems, several issues need to be addressed. These issues can be summarized by new computing ideas for, among other things:

- Advanced data analysis, including Web mining and knowledge discovery;
- Coordination, collaboration, cooperation and other related dynamic mechanisms;
- Data, code, signal and behavior propagation strategy;
- Data migration and metadata evolution;
- Decision analysis, optimization and control;
- E-learning algorithms and architectures;
- Error detection and communication methods;
- Robust grid computing and multi-agent systems;
- Information processing using intelligent and hybrid systems;
- Integrity maintenance in open systems; and
- Representation, elicitation and processing of uncertain, imprecise and incomplete knowledge.

The research reported in this book is focused first and foremost on the above topics. The approach followed to explain these topics is intentionally broad and exploratory.

This volume is focused on topics worthy of interest due to their significant advances. From the submissions, the editors have selected 19 of the most interesting chapters for publication. These chapters have been divided into three parts: *Advanced Methods for Integration*, *Integration Aspects for Agent Systems*, and *Fuzzy-based and other Methods for Integration.*

The first section, *Advanced Methods for Integration,* consists of six chapters.

It starts with the chapter of J.L. Kulikowski, which gives an outlook on the possibility of ontological models extension serving to effective and universal domain knowledge representation in computer systems supporting decision making in various application areas. It is given a definition of ontology and of ontological models as well as their formal representation by taxonomic trees, bi-partite graphs, multi-graphs, relations, super-relations and hyper-relations. The definitions of the corresponding mathematical notions are also given. Then, the concept of ontologies representing incomplete or uncertain domain knowledge is presented. This concept is illustrated by an example of decision making in medicine.

The second chapter is by A. Bădică et al., and discusses data extraction from Web pages. The Web is designed as a major information provider for the human consumer. However, information published on the Web is difficult to understand and reuse by a machine. In this chapter, the authors show how well established intelligent techniques based on logic programming and inductive learning combined with more recent XML technologies might help to improve the efficiency of the task of data extraction from Web pages. Their work can be seen as a necessary step of the more general problem of Web data management and integration.

In the third chapter, A. Jacobsson and P. Davidsson introduce a formal model of virtual enterprises as well as an analysis of their creation and operation. It is argued that virtual enterprises offer a promising approach to promote both innovations and collaboration between companies. A framework of integrated ICT-tools, called Plug and Play Business, which support innovators in turning their ideas into businesses by dynamically forming virtual enterprises, is also formally specified. Furthermore, issues regarding the implementation of this framework are discussed and some useful technologies are identified.

The fourth chapter, by D. Orski, concerns a class of systems composed of operations performed with the use of resources allocated to them. In such operation systems, each operation is characterized by its execution time depending on the amount of a resource allocated to the operation. The decision problem consists in distributing a limited amount of a resource among operations in an optimal way, that is, in finding an optimal resource allocation. In the knowledge-based approach considered in this chapter, the execution time of each operation is described in a nondeterministic way, by an inequality containing an unknown parameter, and all the unknown parameters are assumed to be values of uncertain variables characterized by experts.

In the fifth chapter, C. Fencott undertakes a methodological study of virtual environments, a specific subset of interactive systems. The underlying model is characterized as an integration of interaction machines and semiotics with the intention to make the design tension work to the designer's benefit rather than trying to eliminate it. The language is then developed as a juxtaposition of UML and the integration of a range of semiotics-based theories. This leads to a discussion of a process model and the activities that comprise it. The intention throughout is not to build a particular design method, but to investigate the methodological concerns and constraints such a method should address.

In the last chapter of the first section, S. Sanchez-Alonso and D. Frosch-Wilke discuss the similarities between the life cycle of knowledge management and the processes in which competencies are identified and assessed. This chapter also presents the framework's integration into the knowledge life cycle of the knowledge management consortium international in the form of ontological definitions. It includes a brief discussion on some current definitions of the term competency and details the most interesting efforts in the standardization of competency definitions. At the end, it provides a preliminary mapping of competency-related concepts to terms in upper ontologies.

The second section of this book refers to *Integration Aspects for Agent Systems* and consists of seven chapters.

The first chapter, by M. Oliveira and M. Purvis is about some interesting aspects of coordinating and integrating the autonomy in agent societies. In such environments, agents can be embedded in diverse contexts and interact with agents of various types and behaviors. In this chapter, Oliveira and Purvis explore the necessity of dealing with openness in multi-agent systems and its relation with the agent's autonomy. They stress the importance of building coordination mechanisms capable of managing complex agent societies composed by autonomous agents and introduce their institutional environment approach, which includes the use of commitments and normative spaces. It is based on a metaphor in which agents may join an open system at any time, but they must obey regulations in order to maintain a suitable reputation, that reflects its degree of cooperation with other agents in the group, and make them a more desired partner for others. Colored Petri Nets are used to formalize a workflow in the institutional environment defining a normative space that guides the agents during interactions in the conversation space.

Next, in the following chapter, K. Kolomvatsos and S. Hadjiefthymiades present techniques and challenges for trusting agents in multi-agent environments. In such systems, there must be a way for an agent to make sure that it can trust another entity, which is a potential partner. Without trust, agents cannot cooperate effectively and without cooperation they cannot fulfill their goals. Many times, trust is based on reputation. They discuss main issues concerning reputation and trust in MAS. They present research efforts and give formalizations useful for understanding the two concepts.

The third chapter, by M. Nowostawski, presents some novel concepts of autonomy management in distributed computation and multi-agent systems. He re-introduces two concepts: relative autonomy and absolute autonomy. He argues that even though the concept of absolute autonomy does not make sense in computational settings, it is useful if treated as an assumed property of computational units. For example, the concept of autonomous agents facilitates more flexible and robust architectures. He adopts and discusses a new formalism based on results from the study of massively parallel multi-agent systems in the context of evolvable virtual machines. He also presents the architecture for building such architectures based on his multi-agent system KEA, where he uses extended notion of dynamic and flexibly linking. This provides a useful notion and the necessary means to establish autonomy in open distributed systems.

In the fourth chapter, M. Purvis et al., give an analysis of agent-based library management system using RFID technology. One of the major issues in large libraries is to track misplaced items. By moving from conventional technologies such as barcode-based systems to RFID-based systems and using software agents that continuously monitor and track the items in the library, they believe an effective library system can be designed. Due to constant monitoring, the up-to-date location information of the library items can be easily obtained.

The authors of the fifth chapter, S. Savarimuthu et al., present several original mechanisms to restrict exploitation and improve societal performance in multi-agent environments. Societies are made of different kinds of agents, some cooperative and some uncooperative. Uncooperative agents tend to reduce the overall performance of the society, due to exploitation practices. In the real world, it is not possible to decimate all the uncooperative agents; thus, the objective of this research is to design and implement mechanisms that will improve the overall benefit of the society without excluding uncooperative agents. The mechanisms that they have designed include referrals and resource restrictions. A referral scheme is used to identify and distinguish noncooperators and cooperators. Resource restriction mechanisms are used to restrict noncooperators from selfish resource utilization. Experimental results are presented describing how these mechanisms operate.

The sixth chapter is by B. Tony et al., and gives proof that norms can be shared expectations of behaviours that exist in human societies and can help societies by increasing the predictability of individual behaviours and by improving cooperation and collaboration among members. Norms have been of interest to multi-agent system researchers as software agents intend to follow certain norms. But, owing to their autonomy, agents sometimes violate norms, which needs monitoring. In order to build robust MAS that are norm compliant and systems that evolve and adapt norms dynamically, the study of norms is crucial. Their objective is to propose a mechanism for norm emergence in artificial agent societies and provide experimental results. They also study the role of autonomy and visibility threshold of an agent in the context of norm emergence.

In the last chapter in this section, S. DeLoach and M. Kumar present an overview of the multi-agent systems engineering methodology for analyzing and designing multi-agent systems. This methodology has been used to design systems ranging from a heterogeneous database integration system to a biologically based, computer virus-immune system to cooperative robotics systems. The authors also provide a case study of an actual system developed using their methodology in an effort to help demonstrate the practical aspects of developing such systems.

The last section consists of six chapters which are related to *Fuzzy-based and other Methods for Integration*.

The first chapter, by F. Čapkovič, presents an approach based on Petri nets for modeling and analysing agent behaviour. The agents and agent systems are understood here as Discrete-Event Systems (DES). The approach is based on the place/transition Petri Nets (PN) that yield both the suitable graphical or mathematical description of DES and the applicable means for testing the DES properties, as well as for the synthesis of the agent's behaviour. The reachability graph of the PN-based model of the agent system and the space of feasible states are found. Control interferences are obtained on the base of the most suitable trajectory chosen from the set of feasible ones.

The second chapter, by M. Tabakow, includes a novel method of using fuzzy segmentation for color image enhancement to computed tomography perfusion images. The proposed image processing is used to improve the radiological analysis of the brain perfusion. Color image segmentation is a process of dividing the pixels of an image in several homogenously colored and topologically connected groups, called regions. As the concept of homogeneity in a color space is imprecise, a measure of dependency between the elements of such a space is introduced. The proposed measure is based on a pixel metric defined in the HSV color space. By this measure a fuzzy similarity relation is defined, which next is used to introduce a clustering method that generates a partition and so a segmentation. The achieved segmentation results are used to enhance the considered computed tomography perfusion images in purpose to improve the corresponding radiological recognition.

G. Vincenti's and G. Trajkovski's chapter presents a fuzzy mediation method for shared control and online learning. Fuzzy mediation differentiates itself from other algorithms, as this approach is dynamic in nature. The experiments reported in this work analyze the interaction of two distinct controllers as they try to maneuver an artificial agent through a path. Fuzzy mediation functions as a fusion engine to integrate the two inputs to produce a single output. Results show that fuzzy mediation is a valid method to mediate between two distinct controllers. The work lays the foundation for the creation of an effective tool that uses positive feedback systems instead of negative ones to train human and nonhuman agents in the performance of control tasks.

In the fourth chapter, A. Jatowt et al. present a method for analysing data stored in Web archives which preserve content of the Web, and investigating the methodology required for successful knowledge discovery from this data. The Web is a useful data source for knowledge extraction, as it provides diverse content virtually on any possible topic. They call the collection of such Web archives past Web, a temporal structure composed of the past copies of Web pages. First, they discuss the character of the data and explain some concepts related to utilizing the past Web, such as data collection, analysis and processing. Next, they introduce examples of two applications, temporal summarization and a browser for the past Web.

The next chapter is by D. Król and proposes a generic framework in C# to distribute and compute tasks defined by users. Unlike the more popular models, such as middleware technologies, his multi-node framework is task-oriented desktop grid. In contrast with earlier proposals, this work provides simple architecture to define, distribute and compute applications. The results confirm and quantify the usefulness of such ad-hoc grids. Although significant additional experiments are needed to fully characterize the framework, the simplicity of how they work in tandem with the user is the most important advantage of his current proposal.

And, last but not least, the chapter by X. Xie et al. includes an interesting survey on the application of economic and market theory for grid computing. In recent years, it has been observed that commercial companies are slowly shifting from owning their own IT assets in the form of computers, software and so forth, to purchasing services from utility providers. Technological advances, especially in the area of grid computing, have been the main catalyst for this trend. The utility model may not be the most effective model and the price still needs to be determined at the point of usage. In general, market-based approaches are more efficient in resource allocations, as it depends on price adjustment to accommodate fluctuations in the supply and demand. Therefore, determining the price is vital to the overall success of the market.

The material of each chapter of this volume is self-contained. The editors hope that the book with many papers provided by leading experts from all over the world can be useful for graduate and PhD students in computer science; participants of courses in Knowledge Management, Collective Intelligence, and Multi-agent Systems; and researchers and all readers working on knowledge management and intelligent systems.

The editors would like to thank the authors who present very interesting research results in their chapters. We are indebted to them for their reliability and hard work done in due time. We are looking forward to the same fruitful collaboration during the next edition, which is planned for the near future. We cordially thank the reviewers for their detail and useful reviews. Special thanks are also given to the IGI Global Team members for their friendly help and excellent editorial support in preparing the final version of this volume.

Dariusz Król and Ngoc Thanh Nguyen
Wrocław, February 2008

Section I
Advanced Methods for Integration

Chapter I
Logical Inference Based on Incomplete and/or Fuzzy Ontologies

Juliusz L. Kulikowski
Polish Academy of Sciences, Poland

ABSTRACT

In this chapter, a concept of using incomplete or fuzzy ontologies in decision making is presented. A definition of ontology and of ontological models is given, as well as their formal representation by taxonomic trees, bi-partite graphs, multigraphs, relations, super-relations and hyper-relations. The definitions of the corresponding mathematical notions are also given. Then, the concept of ontologies representing incomplete or uncertain domain knowledge is presented. This concept is illustrated by an example of decision making in medicine. The aim of this chapter is to give an outlook on the possibility of ontological models extension in order to use them as an effective and universal form of domain knowledge representation in computer systems supporting decision making in various application areas.

INTRODUCTION

The concept of *ontology* co-opted by computer specialists from ancient philosophy means organization of concepts in domains which might encompass selected application areas: management, law, engineering, medicine, and so forth (Chute, 2005; Pisanelli, 2004). As such, ontology of a domain is a form of computer-acceptable representation of knowledge about a part of an abstract or real world being an object of consideration or decision making. In general, an ontology Ω can be presented in the form of a set C of concepts and a finite family of *ontological models* M_k, $k = 1,2,\ldots,K$, defined as relationships described on selected subsets of C. The relationships may be of various kinds; however, taxonomies T_i, $i=1,2,\ldots,I$, of the concepts are mandatory elements of the ontology. The aim of this chapter is contributing to this concept in the particular cases when ontologies are being used in computer-based decision supporting systems have not been enough finely described. The chapter is organized as follows. In the beginning a concept of ontological models and their application to decision making are presented. Here, the models based on taxonomic trees, graphs, multigraphs, relations and hyperrelations are shortly described. Nondeterministic ontological models, including fuzzy models and models based on a concept of semi-ordering of syndromes of relations, are described next. Short conclusions are collected in the last section of the chapter. Our aim in this chapter is the presentation of intuitive aspects of the proposed approach to decision making, rather than revealing its strong theoretical backgrounds.

ONTOLOGIES AND ONTOLOGICAL MODELS

Taxonomies

In the simplest cases, the idea of ontology can be reduced to a *taxonomy of concepts* assigned to objects, phenomena or processes appearing in an examined part of abstract or of real world and being analyzed from some fixed points of view. For instance, in sociological investigations a concept of *People living in the town* can be specified by a structure called a *rooted tree,* as shown in Figure 1.

Figure 1. Example of a taxonomic tree based on the attribute "Status"

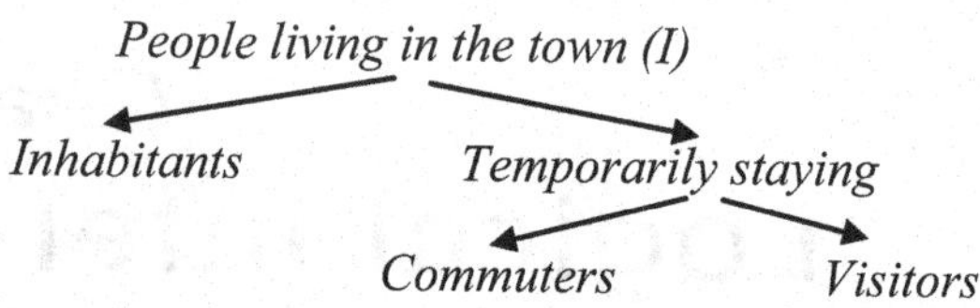

Figure 2. A taxonomic tree based on the attribute "Gender"

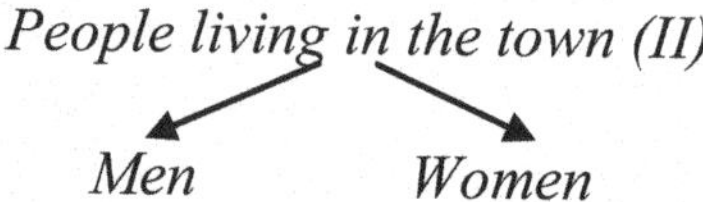

However, the same concept may be presented in several other ways (Figures 2 and 3) and so forth. The roots of the trees have been assigned above to the basic concept *People living in the town,* while the subjected nodes correspond to some subordered concepts. It is also assumed that on each level of any tree the subordered concepts totally cover the corresponding higher-level concept. So-interpreted rooted trees are called *taxonomic trees.* The fact that even in this simple case the part of real world under examination is represented by an ontology consisting not of a single but of several semantically linked taxonomic trees is worthy of being remarked. In general, formal structures constituting ontologies (in the above-defined, narrow sense) will be called *ontological models.* This given ontology

Figure 3. A taxonomic tree based on the attribute "Age"

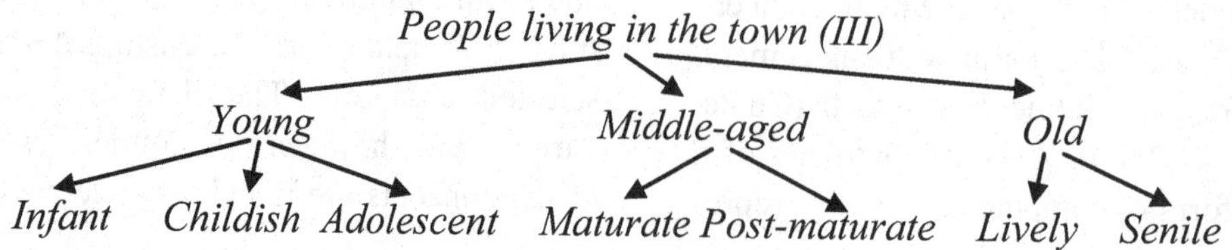

thus consists of three ontological models having the form of taxonomic trees, linked semantically because their roots have been assigned to the same top-level concept.

And still, the class of problems whose solution might be supported by this ontology is rather poor. It might contain, for example, designing a database of inhabitants of the town, planning some social activities or investments in the town, or it might be used in any deliberations concerning the population of the town. However, more advanced applications of this ontology are limited by its evident deficiencies:

1. The taxonomic trees contain no information about the statistical structure of the world as a composition of designates (real entities) represented by a given tree;
2. No relationships between the concepts belonging to different taxonomic trees have been described by the ontology; and
3. Taxonomic tree do not define concepts, but only characterize hierarchical relationships between higher level and lower level concepts.

Graphs

Ontologies reduced to taxonomic trees only are thus rather ineffective in real world description and as tools supporting decision making. Let us also remark that trees in their graphical form are suitable to be analyzed by a man in the case of low numbers of nodes, and for computer-aided analysis they should be represented in the form of digital data structures.

However, trees are a sort of *graph*, the last being formally described by a triple (Tutte, 1984):

$$G = [C, \Lambda, \varphi] \qquad (1)$$

where C denotes a set of *nodes*, Λ stands for a set of *edges* and φ is a function (*incidence function*) assigning edges to some ordered pairs of nodes so that any edge can be assigned to at most one pair of nodes. An edge l_{ij} assigned to the pair $[c_i, c_j]$ of nodes is called *outgoing from* c_i and *incoming to* c_j.

There are several possibilities of defining a *tree* as a sort of graph. The simplest one is based on a statement that a graph becomes a tree if the number of its nodes is 1 larger than the number linking those edges. A tree is called a *rooted tree* if: 1) it contains exactly one node, called a *root*, to which no incoming edge is assigned and 2) to each other node exactly one in-coming edge is assigned. The nodes of a rooted tree to which no outgoing edges have been assigned are called *leafs* of the tree.

The taxonomies of an ontology are represented by rooted trees whose roots have been assigned to the top-level concepts, while other nodes correspond to the subordered concepts. Any concept in a taxonomic tree is characterized by its *level-number*, that is, the number of edges connecting the corresponding node with the root. For example, in the given taxonomy of *People living in the town (I)* the concept *Inhabitants* is a first-level, while *Visitors* is a second-level one. The top-level concepts are 0-level ones.

On the basis of graph algebra operations (Kulikowski, 1986) simple ontological models represented by graphs can be used to create more sophisticated ontological models. For instance, several taxonomic trees corresponding to the same top-level concept can be represented in the form of a unified taxonomic tree. This can be illustrated in the case of two taxonomic trees. For this purpose, a Cartesian product of the trees (in general, of the graphs) can be used. Let $G^{(1)} = [C^{(1)}, \Lambda^{(1)}, \varphi^{(1)}]$, $G^{(2)} = [C^{(2)}, \Lambda^{(2)}, \varphi^{(2)}]$ be two graphs. Their Cartesian product $G = G^{(1)} \times G^{(2)}$ is defined as a graph such that:

1. The set of its nodes $C = C^{(1)} \times C^{(2)}$, which means that each node of G is an ordered pair of some nodes of $G^{(1)}$ and $G^{(2)}$;

2. The set of its edges $\Lambda = \Lambda^{(1)} \times \Lambda^{(2)}$; and
3. Its incidence function φ assigns an edge $l_{prqs} = [l^{(1)}_{pr}, l^{(2)}_{qs}]$ to the pair of nodes $c_{pr} = [c^{(1)}_{p}, c^{(2)}_{r}]$, $c_{qs} = [c^{(1)}_{q}, c^{(2)}_{s}]$, if and only if $l^{(1)}_{pr}$ is assigned by $\varphi^{(1)}$ to the pair of nodes $[c^{(1)}_{p}, c^{(1)}_{r}]$ and $l^{(2)}_{qs}$ is assigned by $\varphi^{(2)}$ to the pair of nodes $[c^{(2)}_{q}, c^{(2)}_{s}]$.

For example, a Cartesian product of the first two taxonomic trees based on the attributes *"Status"* and *"Gender"* takes the form in Figure 4.

For the sake of formal accuracy, it has been assumed that the graphs $G^{(1)}$ and $G^{(2)}$ admit existence of edges of the form $l^{(1)}_{ii}$, $l^{(2)}_{jj}$ (*loops*) linking each node with itself.

Using graphs (instead of trees only) in ontologies provides some additional possibilities to describe relationships between concepts. For example, let us take once more into consideration the first two taxonomic trees canceled to their upper two levels. Let $A = \{a_1, a_2, \ldots, a_K\}$ be a set of persons living in the given town. They can be classified according to the above-given ontology, that is, assigned to the leaves of the taxonomic trees. However, we would like to represent the persons and the assigned to them attributes: *Gender* = {*M – Man, W – Woman*}, *Status* = {*I – Inhabitant, TS – Temporarily staying*} in the form of a more concise structure. For this purpose, a graph *G'* will be constructed whose set of nodes *C'* consists of three disjoint subsets: $C' = A \cup \{M, W\} \cup \{I, ST\}$ and the incidence function φ admits edges connecting only persons with their attributes so that each person is connected with exactly two attributes: first, belonging to the subset {*M, W*} and second belonging to {*I, ST*}, as illustrated in Figure 5.

This bipartite graph represents a distribution of the attributes *Gender* and *Status* in a subset *A* of *persons living in the town*. However, it is not a tree, as it can be proved by counting and comparing the numbers of its nodes and edges. Using graphs as ontological models makes it possible using typical algebraic operations on graphs to construct more sophisticated models as compositions of some simpler ones. This can be illustrated by the following example.

Let *G'* be the bipartite graph illustrated in Figure 5 and *G"* be a bipartite graph representing the distribution of the attribute *Age* = {*Y – Young, MA – Middle aged, O – Old*} in the defined set *A* of elements (persons), as shown in Figure 6.

The *sum of graphs* $G' \cup G''$ can be defined as a graph $G^* = [C^*, \Lambda^*, \varphi^*]$ such that $C^* = C' \cup C''$, $\Lambda^* = \Lambda' \cup \Lambda''$, and φ^* is an incidence function such that to a pair of nodes an edge is assigned if it is assigned by at least one of the incidence functions, φ' or φ'' (if two different edges to the given pair of nodes have been assigned by both incidence functions, then the problem, whose edges should be finally assigned to it, can be arbitrarily solved).

Using the definition of a sum of graphs to the graphs shown in Figure 5 and Figure 6, one

Figure 4. Cartesian product of two taxonomic trees

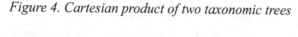

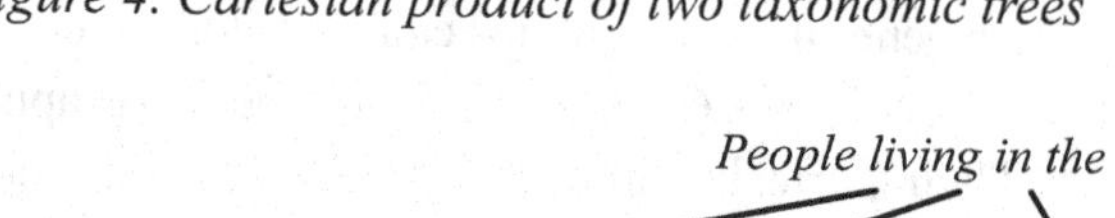

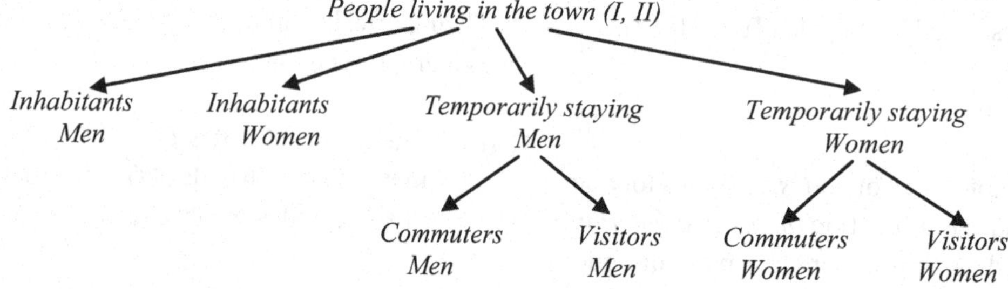

Figure 5. A bipartite graph representing distribution of two attributes

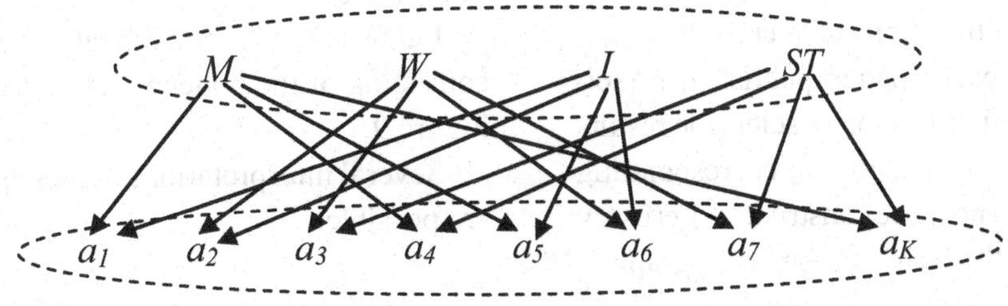

Figure 6. A bipartite graph representing distribution of the attribute Age

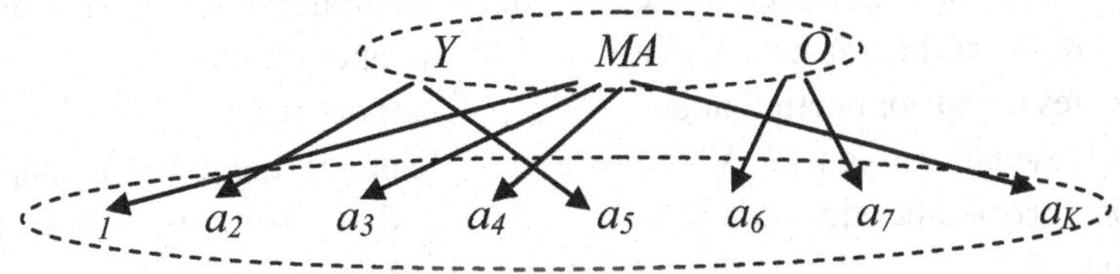

Figure 7. Sum of graphs: A bipartite graph representing a distribution of three attributes

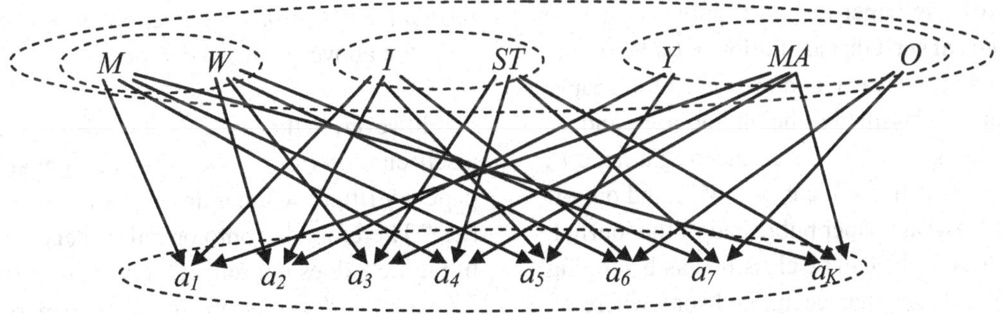

obtains a graph G^* illustrating the distribution of three attributes, shown in Figure 7.

In similar way, ontological models describing distribution of higher numbers of attributes over fixed sets of elements using the algebra of graphs operations can be constructed. Such models may have the form of bipartite graphs, the subset of nodes representing the attributes being subdivided into mutually disjoint lower-level subsets of values of the given attributes. For effective calculations, the graphs should be stored in computer memory in the form of the corresponding connection matrices or incidence matrices (Kulikowski, 1986). However, the graphs presented in Figures 5, 6 and 7 as ontological models are rather untypical, because the idea of ontological models consists in knowledge presentation in aggregated form rather than by individual listing of instances. The algebra of graphs provides us with more universal and flexible tools for ontological models construction than taxonomic trees. Alas, in certain cases this tool is not quite suitable to a presentation of knowledge about the real world, as it can be shown by the following example.

Let us assume that a problem consists in description of the impact of papers published in a scientific journal on distribution of scientific results in the world. For this purpose, a corresponding ontological model should be constructed. Let us try to construct it in the form of a bipartite graph:

$$G = [V \cup T, \Lambda' \cup \Lambda'', \varphi], \qquad (2)$$

where V is a subset of nodes assigned to affiliations of authors, T is a subset of nodes assigned to the topics covering the profile of the journal, Λ' is a subset of oriented edges (arcs) connecting nodes belonging to V with these belonging to T, Λ'' is a complementary subset of oriented edges connecting nodes belonging to T with these belonging to V and φ is an incidence function such that:

1. An edge (arc) l'_{ip} is connecting a node v_i, $v_i \in V$, with a node t_p, $t_p \in T$, if and only if at least one paper has been published in the journal such that affiliation of (at least one) its authors was v_i and the topic of the paper can be classified as belonging to t_p; and
2. An edge (arc) l''_{qj} is connecting a node t_q, $t_q \in T$, with a node v_j, $v_j \in V$, if and only if at least one paper published in the journal, whose topic can be classified as belonging to t_q, $t_q \in T$, has been cited somewhere by an author whose affiliation was v_j, $v_j \in V$.

A hypothetical part of such a graph is shown in Figure 8.

Let us take into consideration a partial graph consisting of the nodes v_2, v_5, v_8 and t_2 shown in Figure 9.

Several interpretations of these partial graphs are possible:

1. An author from v_2 has published in the journal a paper on t_2;
2. An author from v_5 has published in the journal a paper on t_2;
3. An author from v_8 has published in the journal a paper on t_2;
4. authors from v_2 and v_5 have commonly published in the journal a paper on t_2;
5. authors from v_2 and v_8 have commonly published in the journal a paper on t_2;
6. authors from v_5 and v_8 have commonly published in the journal a paper on t_2;
7. authors from v_2, v_5 and v_8 have commonly published in the journal a paper on t_2; or
8. an author from v_5 has cited at least one of the above-mentioned papers.

However, in the last case, it is not clear: was it a self-citation (four possibilities) or a citation of papers written by other authors (three possibilities)? Therefore, the ontological model presented in Figure 8 does not reflect all types of scientific information distribution caused by papers published in the given journal.

Figure 8. A bipartite graph representing affiliation of authorship and citations of papers

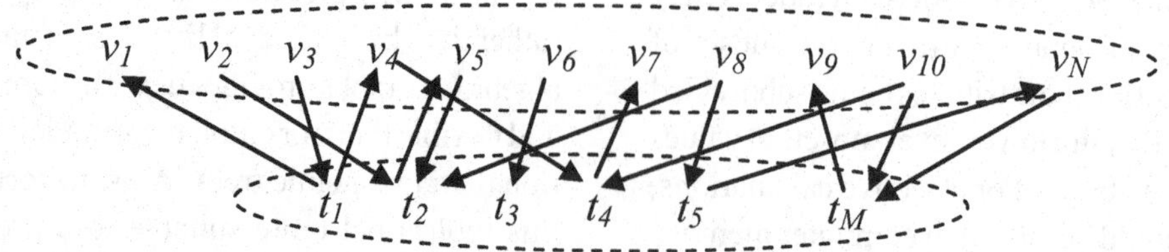

Figure 9. A partial graph of the graph shown in Figure 8

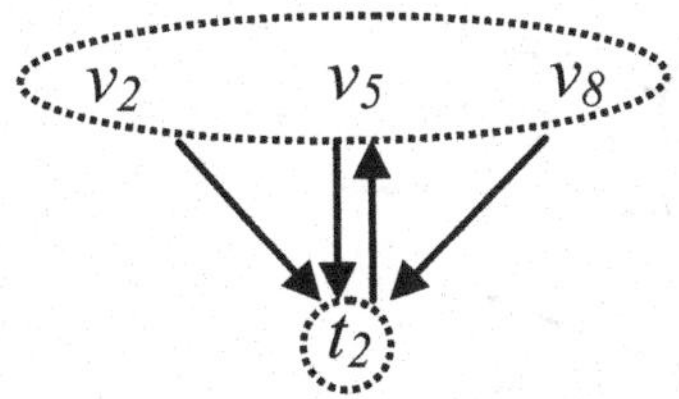

Multigraphs

In general, many relationships existing in the real world cannot be adequately presented by ontological models given in the form of graphs. Some larger possibilities are offered using *multigraphs*, that is, graphs whose incidence function admits more than one edge to any given pair of nodes. This can be illustrated by the following example.

Let us take into consideration a problem of young population flow and migration between the schools in a certain region. For analysis of the problem, an ontology consisting of several ontological models should be created, such as:

a. Taxonomic models of a regional population of pupils and students (sexuality, social background, etc.);
b. Taxonomic model of regional schools of any types and levels; or
c. Ontological model describing the flow of young population between the schools.

Our attention here will be focused on the last ontological model. For this purpose, it will be defined as a *weighted multigraph*:

$$M = [\Sigma, F, R^{+}, \varphi] \tag{3}$$

where:

- Σ is a set of nodes assigned to the regional schools (extended by adding the category *"Other"* for the schools outside the region);
- F is a set of oriented edges (arcs) assigned to the flows of pupils and students between the schools within the region, as well as coming from outside or going away from the region; the edges should also indicate a subclassification of flows based on the taxonomies following from the type a) models;
- R^{+} is a non-negative real half-axis used as a scale of flow intensity; and
- φ is a multigraph (vector) incidence function assigning to any pair of nodes $[S_i, S_j]$ an arc $f^{(r)}_{ij} \in F$ and a value $u^{(r)}_{ij} \in R^{+}$ if and only if between the corresponding pair of schools a flow of intensity $u^{(r)}_{ij}$ of the r-th category pupils (students) takes place.

A part of a multigraph of this type is illustrated in Figure 10. For the sake of simplicity multiple arcs have been replaced by the single ones and the denotations of arcs have been reduced to the weights of arcs (flow intensities in *persons/year*) presented in a concise symbolic form (in fact, they are numerical vectors whose components correspond to different sorts of pupils, for example, to *Boys* and *Girls*).

This model makes it possible to show, for example, which universities in the region are directly supplied with former pupils by given secondary schools or what is a social background of pupils or students entering the given schools. However, on the basis of this model it is not possible give a reply to a question such as, which elementary schools educate the highest percentage of pupils who graduate from the universities? The inadequacy of the above-described ontological model to answer these kinds of questions consists in the fact that graphs as well as multigraphs describe relationships between pairs of objects only, while our question concerns relationships among (in the simplest case) triples of elements:

Figure 10. A simplified partial multigraph representing the flow (migration) of pupils (students) between regional schools

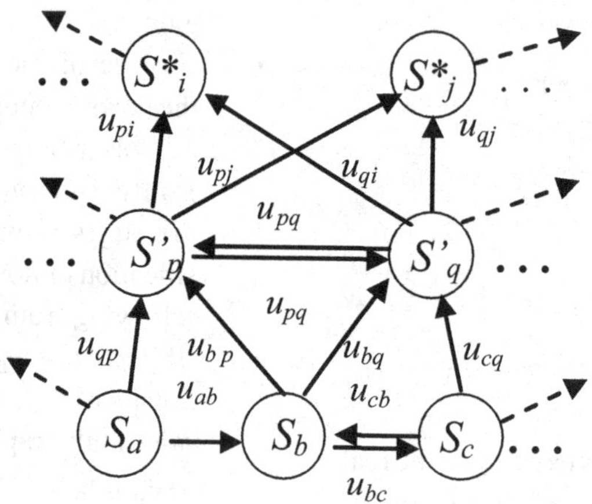

[*elementary school, secondary school, university*]. The information about a detailed structure of flows entering a given node is lost as a result of aggregation of flow components, and it is not possible to reconstruct it by any outgoing flows' components analysis.

Relations

If $Q_1, Q_2,\ldots, Q_n$ are some nonempty sets taken in the given linear order and $C = Q_1 \times Q_2 \times\ldots\times Q_n$ is their Cartesian product, then any subset:

$$r \subseteq C \qquad (4)$$

is called a relation described on the (linearly ordered) family of sets $[Q_1, Q_2,\ldots, Q_n]$. According to the definition, r is a set of n-tuples of the form $[a, b,\ldots,h]$ such that $a\in Q_1$, $b\in Q_2,\ldots$, and $h\in Q_n$, called sometimes *syndromes* of the relation.

For a fixed linearly ordered family of sets and the corresponding Cartesian product C it is possible to take into consideration a family Φ of all possible subsets of C including C itself and an empty subset $\varnothing$. Φ is thus a family of all possible relations that can be defined on the given family of sets. On the other hand, it is possible to apply to it the general set-algebraic rules (Rasiowa & Sikorski, 1968) which in this case becomes an algebra of relations described on the family of sets $[Q_1, Q_2,\ldots, Q_n]$. Moreover, this algebra can also be extended on all relations described on any subsets of this family assuming that the original linear order has been preserved (Kulikowski, 1972). The *extended algebra of relations*, being in fact a sort of Boolean algebra, becomes a flexible tool not only for description of relations between any final number of arguments, but also for the creation of more sophisticated relations as algebraic compositions of some simpler ones, as well as for the creation of higher-order relations (*superrelations*) defined as relations between relations (Kulikowski, 1992).

Multi-argument relations cannot be easily presented in graphical form. However, there are several methods of description of a new relation:

- By listing the syndromes of the relation;
- Be presenting it as an algebraic composition of some other, known relations; or
- By presenting a testing function making it possible to decide whether the relation is satisfied by any given syndrome.

The first method can be illustrated by the following example. The problem of young population flow and migration between the schools will be considered again. We would like to create an ontological model making possible the investigation of contribution of elementary schools in the region to the educational productivity of universities, taking into account the sex of the graduate students. For this purpose five sets will be taken into consideration:

- $Q_1 = \{B, G\}$ describing sex (*Boys, Girls*);
- $Q_2 = \{S_a, S_b,\ldots,S_h\}$ describing elementary schools in the given region;
- $Q_3 = \{S'_p, S'_q,\ldots, S'_t\}$ describing secondary schools;
- $Q_4 = \{S^*_i, S^*_j,\ldots,S^*_k\}$ describing universities; and
- $Q_5 \equiv R^+$ a non-negative real half-axis representing flow intensities.

On the basis of the Cartesian product $C = Q_1 \times Q_2 \times Q_3 \times Q_4 \times Q_5$ it can be defined a relation r given in the form of a list of syndromes of the form

$$v = [x, S_\alpha, S'_\beta, S^*_\gamma, w], \tag{5}$$

where $x \in Q_1$, $S_\alpha \in Q_2$, $S'_\beta \in Q_3$, $S^*_\gamma \in Q_4$, $w \in Q_5$. Each syndrome represents a component of the flow with the additional characterizing it parameters. The relation can easily be represented in computer, however, it cannot be so easily plotted on a plane. Answering the former question: what is the contribution of a given elementary school, let it be S_α, to supplying a given university, let it be S^*_γ, with, say, girl students (G) is then reduced to selection from r, a subrelation $r' \subseteq r$ consisting of all syndromes of the form

$$v' = [G, S_\alpha, \mathrm{F}, S^*_\gamma, w], \tag{6}$$

where F denotes an undefined data value (here denoting any secondary school). The final answer can be reached by summing over F all values w of the syndromes of r'.

As mentioned before, the algebra of relations makes possible the combining of ontological models in order to get more suitable forms of reality description. For example, if $r^{(\kappa)}$, $r^{(\lambda)}$ are two relations of similar structure described on the same family of sets $[Q_1, Q_2,\ldots, Q_n]$, then a sum of relations

$$r = r^{(\kappa)} \cup r^{(\lambda)} \tag{7}$$

is a relation consisting of all syndromes satisfying $r^{(\kappa)}$ or $r^{(\lambda)}$. In the above-described example, if $r^{(\kappa)}$ and $r^{(\lambda)}$ describe the flow of pupils (students) in two consecutive school years, then r describes it in the two school years taken together.

Another situation arises if the relations $r^{(\kappa)}$ and $r^{(\lambda)}$ are described on different families of sets, say, respectively, on $[Q^{(\kappa)}_1, Q^{(\kappa)}_2,\ldots, Q^{(\kappa)}_n]$ and $[Q^{(\lambda)}_1, Q^{(\lambda)}_2,\ldots, Q^{(\lambda)}_m]$. In this case, assuming that both families are conformably ordered, the algebraic operations can be defined according to the extended relations algebra rules (Kulikowski, 1992). In particular, a sum of relations can be defined as a relation described on the sum of families of sets $[Q^{(\kappa)}_1, Q^{(\kappa)}_2,\ldots, Q^{(\kappa)}_n] \cup [Q^{(\lambda)}_1, Q^{(\lambda)}_2,\ldots, Q^{(\lambda)}_m]$ consisting of syndromes such that each syndrome even 1) in its part belonging to $[Q^{(\kappa)}_1, Q^{(\kappa)}_2,\ldots, Q^{(\kappa)}_n]$ satisfies $r^{(\kappa)}$, or 2) in its part belonging to $[Q^{(\lambda)}_1, Q^{(\lambda)}_2,\ldots, Q^{(\lambda)}_m]$ it satisfies $r^{(\lambda)}$.

As an example, let a problem of air-passengers flow intensity in selected airports be considered. For this purpose, a set $A = \{a_1, a_2, \ldots, a_K\}$ of international airports will be considered. It will be multiplied in three versions: as departure airports A', as transit airports A^* and destination airports A''. In addition, a set V of flow intensity values, $V \equiv R^+$, where R^+ is a non-negative half-axis, will be taken into account. Then two Cartesian products will be constructed: $C = A' \times A'' \times R^+$, $C^* = A' \times A^* \times A'' \times R^+$. Let us also select a subset $D \subset A$ of particular interest, say, of international airports in a certain country. On the basis of C

two relations can be described: 1) r' describing direct flights starting from any airport of D, $D \subset A'$, and terminating in any airport of A', and 2) r'' describing direct flights starting from any airport of A' and terminating in any airport of D, $D \subset A.''$ In addition, on the basis of C^* a relation r^* describing transit flights from any airport of A' through any airport of D to any airport of A'' will be described. The syndromes of the above-mentioned relations thus indicate the names of starting, transit or terminating airports between which the flights took place within a certain time-period, as well as intensity of the corresponding flow of passengers. Let us assume that a total flow of passengers through the airports of D are of interest. Then, an extended algebraic sum of relations: $r = r' \cup r^* \cup r''$ should be taken into account and the corresponding arithmetic sum of intensities should be calculated. The syndromes of r are quadruples of a general form: *starting airport, transit airport, terminate airport, intensity of the flow of passengers*, such that exactly one starting, transit or terminate airport belongs to D, and the other airports within the set A are unlimited.

In a similar way, extended intersection of relations can be used in ontological models creation. For example, let us take into consideration a family $F = [Q_1, Q_2, Q_3, Q_4, Q_5, Q_6]$ of sets where Q_1 denotes a set of names of *teachers*, Q_2 a set of *subjects*, Q_3 a set of *scholar classes*, Q_4 a set of *classrooms*, Q_5 a set of *weekdays* and Q_6 a set of *scholar hours*. On the Cartesian product $C' = Q_1 \times Q_2 \times Q_3$ it may be defined as a relation r' between *teachers, subjects* and *scholar classes*. On the Cartesian product $C'' = Q_2 \times Q_3 \times Q_5$ a relation r'' between *subjects, scholar classes* and *weekdays* in a similar way can be defined. At last, a relation r''' between *classrooms, weekdays* and *scholar hours* on the Cartesian product $Q_4 \times Q_5 \times Q_6$ can be established. The relations r', r'' and r''' can be established independently of each on each other by taking into account some constraints imposed on the corresponding syndromes. Then, a problem arises of the construction of a relation r containing all syndromes consisting of *teachers, subjects, scholar classes, classrooms, weekdays* and *scholar hours* satisfying the constraints. This relation can be defined as an extended algebraic intersection of relations:

$$r = r' \cap r'' \cap r''' \quad (8)$$

whose syndromes, by definition, projected on C' satisfy the relation r', projected on C'' satisfy r'', and projected on C''' satisfy r'''. Then, finally, on the basis of the relation r, an optimized timetable can be constructed.

It might seem that the extended algebra of relations is a tool sufficient enough to construct a large class of ontological models. The following examples show that it is not quite so.

Hyper-Graphs

Let C be a set of scholar handbooks offered at a book market. A problem of recommending collections of handbooks for teaching given subjects during a multiyear education process will be considered. For this purpose, an ontology describing the regional educational subsystem should be constructed. However, our attention will be focused on ontological models describing the admissible collections of handbooks satisfying some educational requirements. Otherwise speaking, it is necessary to select according to some educational criteria a family of subsets of C assuming that the subsets are not obviously mutually disjoint. The first possibility is to construct a hyper-graph (Berge, 1973) whose nodes are assigned to the elements of C and any subset of nodes assigned to the handbooks satisfying the educational criteria constitutes a hyper-edge of the hyper-graph. Such hyper-graphs can be represented by a diagram, shown in Figure 11.

In this diagram vertical lines represent nodes, while dots lying on horizontal lines represent hyper-edges. A serious shortcoming of hyper-

Figure 11. Diagram of a hyper-graph

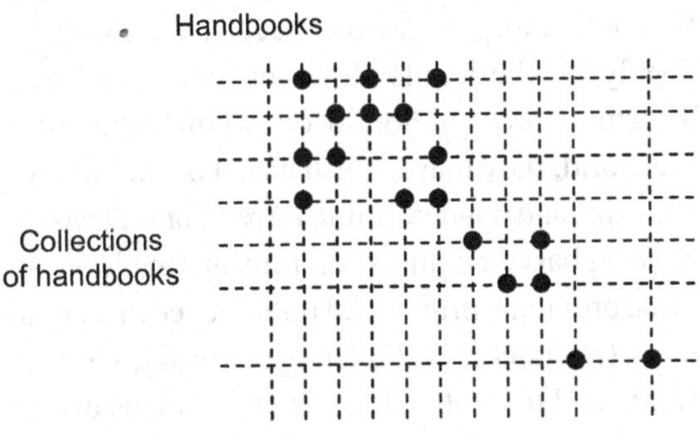

graphs used as ontological models exists in their inability to describe an order (if any exists) of the nodes belonging to the same hyper-edge, and of belonging to several hyper-edges where the orders are different. This will be illustrated by the next example.

It will be taken into consideration a medical clinic specialized in a certain sort of diagnostic and therapeutic services. It is desired to create an ontology describing admissible processes of individual medical treatment of patients. For example, each process of this type should start by registration of the patient, then a series of diagnostic procedures should be followed by the proper medical treatment, and at last the process is finished by discharging the patient from the clinic. It is possible to construct a directed graph representing admissible logical sequences of operations of which any instance of medical treatment process consists. An example of such a graph is given in Figure 12. The nodes represent here operations, while directed edges (arcs) link pairs of operations which can be performed just after the former one.

Figure 12. A directed graph representing admissible logical sequences of operations

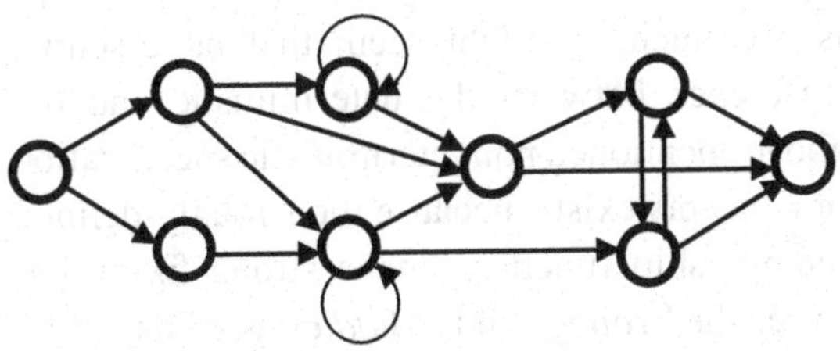

Any instance of the process can be embedded in the graph. Each process starts at the extreme-left node and is finished in the extreme-right node. The loops existing at two nodes show that the given operations can be repeated. However, it is not possible to separate from the graph the admissible instances of the process only. For example, it is not evident whether some operations can be repeated one, two or more times and whether or not they can be repeated independently on the preceding subsequences of operations. A more complete ontological model of medical treatment processes should thus represent all medically or organizationally admissible sequences of operations of various lengths, which can be embedded in the above-presented graph.

Hyper-Relations

Let $F = [Q_1, Q_2,\ldots, Q_n]$ be a finite family of sets. A family K_F of all subfamilies of F including F itself and an empty family $\varnothing$ will be considered. Then, each subfamily $H_g \subseteq F, H_g \in K_F$, g denoting the subfamilies, creates a family U_g of all linearly ordered subfamilies of sets obtained as a result of all possible permutations of H_g. Each element of U_g is thus a linearly ordered subfamily of F and, as such, a Cartesian product of its elements can be constructed. Next, on the basis of this Cartesian product some relations can be created. The syndromes of each such relation are thus some finite strings of elements belonging to and taken in the order of sets constituting the given Cartesian product. A *first-type hyper-relation* (a *h*-relation) is then defined as any sum (in the set algebra sense) of relations defined on any subfamilies H_g created in the above-defined way (Kulikowski, 2006). The following example should make it clearer. There will be taken into consideration:

- A family of sets $F = \{A, B, D\}$;
- A family of its subfamilies $K_F = \{\varnothing, \{A\}, \{B\}, \{D\}, \{A,B\}, \{A,D\}, \{B, D\}, \{A, B, D\}\}$;
- Selected subfamilies of sets $H_6 = \{A, D\}$, $H_7 = \{B, D\}$;
- Families of permutations of the subfamilies H_6 and H_7:
  $U_6 = \{[A, D], [D, A]\}$, $U_7 = \{[B, D], [D, B]\}$;
- Cartesian products based on U_6 and U_7:
  $C_{6,1} = A{\times}D$, $C_{6,2} = D{\times}A$, $C_{7,1} = B{\times}D$, $C_{7,2} = D{\times}B$;
- Selected relations described on the above-given Cartesian products:
  $r' \subseteq C_{6,1}$, $r'' \subseteq C_{6,2}$, $r''' \subseteq C_{7,2}$;
- *h*-relations:
  $H_1 = A \cup D \cup r' \cup r''$, $H_2 = r' \cup r'' \cup r'''$, and so forth.

The syndromes of H_1 are linearly ordered strings consisting of one or two element while all syndromes of H_2 are strings consisting of two elements.

On the basis of any given family F of sets, a universe U_F of all possible *h*-relations created on the basis of F can be considered. The elements of U_F (i.e., *h*-relations) being defined as some sets are subjected to the set algebra rules, which in this case can be interpreted as an algebra of *h*-relations. This makes it possible to create more sophisticated *h*-relations as algebraic compositions of some simpler ones. Hyper-relations, as well as the algebra of hyper-relations, are thus a flexible tool for the creation of ontological models, more powerful than graphs or relations.

NONDETERMINISTIC ONTOLOGIES

Until now, ontologies consisting of deterministic models were considered. We tried to show that decision making based on ontologies may be improved by using ontological models suitable to the description of the area of interest with a required accuracy. "Suitable" means here the covering of the area of interest without making it too large, based not on aggregated concepts, nor going too deeply into the details. However, ontologies being a form of presentation of our knowledge about the world, they may be also based on ambiguous concepts and nondeterministic relations. Decision making based on uncertain information is one of basic problems in artificial intelligence investigations (Bubnicki, 2002; Grzegorzewski, Hryniewicz, & Gil, 2002; Rutkowski, Tadeusiewicz, Zadeh, & Zurada, 2006). Only certain aspects of this problem, strongly connected with using nondeterministic ontological models in decision making, will be considered here.

Fuzzy Ontological Models

Let us go back to the taxonomic trees shown in Figure 2 and Figure 3. In the first case, the concepts *Men* and *Women* are strongly defined and the respective ontological model is no doubt deterministic. On the other hand, the concepts *Young, Middle-aged* and *Old* used in the second ontological model can be interpreted:

a. Deterministically, as:
   Young ≡ [aged not more than 18 years],
   Middle-aged ≡ [aged more than 18 and not more than 60 years],
   Old ≡ [aged more than 60 years]; or
b. Nondeterministically, say, using a fuzzy sets approach (Zadeh, 1975a, 1975b, 1975c) and the membership functions shown in Figure 13.

In the second case, a particular case of nondeterministic ontologies, a *fuzzy ontology,* is presented. It might seem that no essential difference between the deterministic and the above-mentioned nondeterministic specification of concepts exists, because the strongly-defined membership functions make strong fixing between the *"Young"* and *"Middle-aged"* as well as

Figure 13. Fuzzy specification of the subconcepts of Age

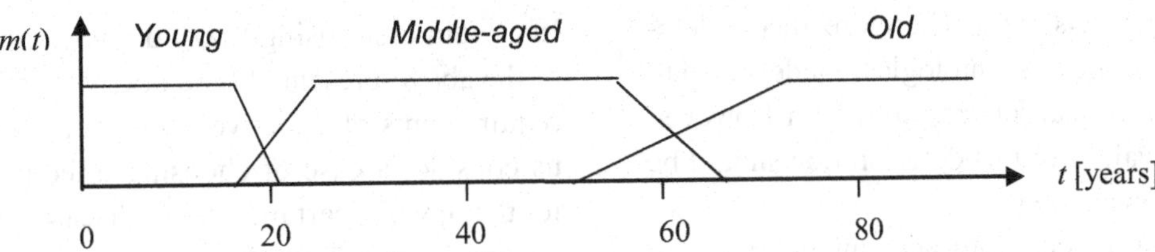

between the *"Middle-aged"* and *"Old"* concepts possibly due to a *defuzzyfication,* that is, to an operation consisting in fixing strong limits between the concepts. However, a difference between the deterministic and fuzzy ontology becomes evident if a practical decision based on fuzzy ontology is to be made. For example, if building of a network of sport fields for young people in the town is considered, then the fuzzy concept of *"young"* better suits to a characterization of the expected users of the fields than the deterministic one. The problem is that even if a border between *Young* and *Middle-aged* at the age of 18 years was fixed, not all people younger than 18 years would like to attend the sport fields and, on the other hand, some people older than 18 years would like to attend them. This example shows that the *fuzzy* or *nondeterministic* ontological model does not mean *worse* than a *deterministic* one in the case if it better describes the state of our knowledge about the area of interest and more exact knowledge is not available.

One should distinguish between decision making based on incomplete and on fuzzy ontology. Let us consider a taxonomic subtree of liver diseases (Coté, 1975).

Let it be known that a) a certain drug *D* is effective and recommended in *icterus hepatogenes* and rather ineffective in the case of *icterus hepatocellularis* therapy, and b) in a given population *p%* of patients diagnosed as affected with *liver jaundice* are in fact suffering from *icterus hepatogenes* and (100-*p*)% are suffering from *icterus hepatogenes.* Then, if a patient has been roughly diagnosed as affected with *liver jaundice* without indication of the type of jaundice and he is recommended to take the drug *D*, the decision is based on an incomplete model, canceled to its higher level ontological model. The expected effectiveness of the therapy in this case is at most *p*%. On the other hand, if diagnostic methods used to discriminate between the *hepatogenes* and *hepatocellularis icterus* work with *q% specifity* (i.e., the percentage of patients diagnosed as affected by a given disease really suffering from it), the given patient has been diagnosed as affected by *icterus hepatogenes* and, consequently, he has been recommended to take *D,* then the expected effectiveness of this therapy will be at most *q*%. The ontological model on which this decision is based is complete; however, if it is interpreted

Figure 14. A selected taxonomic subtree of liver diseases

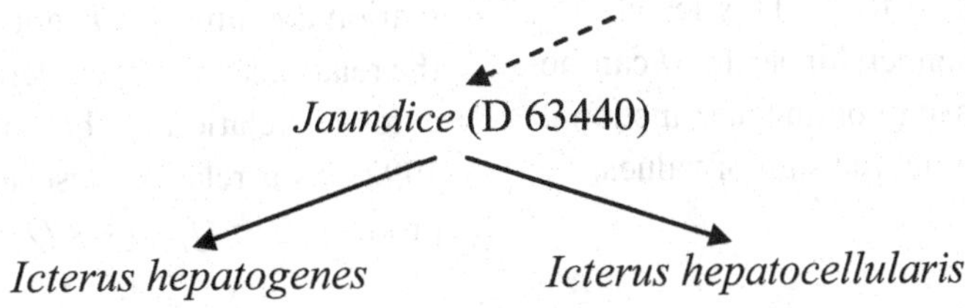

as a taxonomy of possible diseases in patients diagnosed as affected by one or another type of liver jaundice, it is fuzzy. Therefore, incompleteness and fuzziness of ontological models lead to deterioration of decisions based on them. However, numerical values of this deterioration should be differently evaluated.

It is easy to take into account fuzziness in ontological models based on relations or hyperrelations. For this purpose, a set M defining an linearly ordered numerical scale of *weights* of syndromes will be defined. If $C = Q_1 \times Q_2 \times ... \times Q_n$ is a Cartesian product of n given sets on which a relation r has been defined, then an extended Cartesian product $C^* = C \times M$ and a relation $r^* \subseteq C^*$ can be taken in consideration. The syndromes of r^* have the form:

$$\sigma^* = [\sigma, \mu], \quad (9)$$

where $\sigma \in C$, $\mu \in M$. The component μ in the simplest case may describe a *membership level* of σ as a syndrome of the fuzzy relation r^*. The membership level, in general, is not subjected to any additional constraints: it is used only to a relative assessment of the syndromes of r^*. If, for example $\sigma^*_1 = [\sigma_1, \mu_1]$ and $\sigma^*_2 = [\sigma_2, \mu_2]$ such that $\mu_1 < \mu_2$ are given, then this means that σ_1 in a certain sense is "less credible" than σ_2 as a syndrome of the relation. According to the context, "less credible" may mean: "less frequently," "with lower probability," "guaranteed by less known experts," and so forth. In many cases, such fuzzification of ontological models is sufficient as a basis of decision making. Let us remark that in the above-described example exact numerical membership levels are unimportant for decision making, because $\mu_1 < \mu_2$ holds for 0.1 < 0.15, 2 < 3, 46% < 58%, and so forth. This leads to a conclusion that the membership scale M can be defined up to any increasing continuous functional transformation preserving the sign of values.

Semi-Ordered Ontological Models

In certain cases using fuzzy ontological models of the above-presented type does not satisfy the requirements of effective decision making. Let us consider a case of choosing effective drugs for therapy of a certain class of diseases. For this purpose, two relation-based ontological models will be taken into account. First, there will be considered the following sets:

- Q_1 a set of available drugs;
- Q_2 a set of medical indications (diseases) for applying the drugs;
- Q_3 medical contraindications for applying the drugs; and
- Q_4 cost of the drug.

On the basis of these sets, a relation r' can be defined assuming that if there are more than one medical indication of contraindication for applying a given drug, then they should be presented by several relation syndromes. In addition, the following sets will be considered:

- Q_1 a set of drugs (as before);
- Q'_2 a set of diseases;
- Q'_3 a set of additional patients' health state characteristics; and
- M a scale of *credibility values.*

On the basis of these sets a fuzzy relation r'' can be defined, the component μ, $\mu \in M$, expresses the credibility (a positive real value between 0 and 1) that the drug is effective if the patient has been properly diagnosed and his additional health state characteristics have been correctly described. In order to combine information contained in r' and r'' an intersection of the relations $r = r' \cap r''$ defined, according to the extended relation algebra rule (Kulikowski, 1972, 1992), as a relation described on the Cartesian product: $C = Q_1 \times Q_2 \times Q_3 \times Q_4 \times Q'_2 \times Q'_3 \times M$, such that its syndromes projected on $Q_1 \times Q_2 \times Q_3 \times Q_4$ satisfy r' and projected on $Q_1 \times Q'_2 \times Q'_3 \times M$ satisfy r'' will be constructed. The relation r is fuzzy due to the credibility component $\mu \in M$ of its syndromes. However, it is not quite suitable to the requirements of making decisions about recommendation of a drug for the given patient. This is because it is not quite sure that: 1) the real patient's disease is identical to the result of diagnosis (the element of Q'_2) and, as a consequence, whether there is a full consistency in the syndromes between the elements of Q_2 and Q'_2, and 2) for similar reasons, whether there is a full consistency between the elements of Q_3 and Q'_3. Therefore, the relation r should be extended by adjoining to it two components: *a*) a measure ν of logical consistency between the syndrome's components of Q_2 and Q'_2, and *b*) a measure ρ of logical consistency between the syndrome's components of Q_3 and Q'_3. The way of defining the *logical consistency* is not substantial here. We would like only to show that the extended relation r^* contains three parameters, μ, ν and ρ, causing its fuzziness. At last, it becomes necessary to establish a method of relative assessment of the syndromes according to the values of the weight vectors $\mathbf{w} = [\mu, \nu, \rho]$. For this purpose, an additional ontological model can be created: a linear 3-dimensional semi-ordered real vector space $K^{(3)}$. One possibility of doing this exists in defining $K^{(3)}$ as a Kantorovich space (Kantorovich, Vulich, & Pinsker, 1950). As a component of ontology, $K^{(3)}$ represents the preferences established by the decision-makers (medical doctors, in the above-described example) for choosing the best decision(-s) from those, indicated by the fuzzy relational ontological model r^*. From a formal point of view, the principle of semi-ordering of vectors in a K-space consists in defining in a given linear vector space a *non-negative cone* K^+, its mirror-reflection being denoted by K^-, as illustrated in Figure 15.

Figure 15. Illustration of a 3-dimensional Kantorovich space

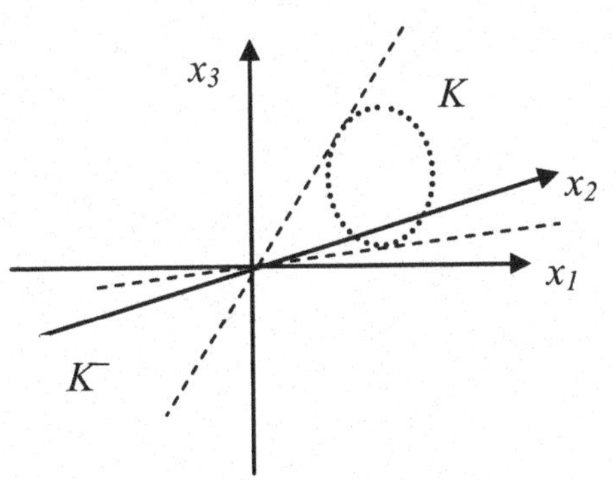

If $\mathbf{v}^{(1)}$ and $\mathbf{v}^{(2)}$ are two vectors belonging to the K-space and their difference satisfies the condition:

$$\mathbf{v}^{(1)} - \mathbf{v}^{(2)} \in K^+ \tag{10}$$

then it is said that $\mathbf{v}^{(1)}$ *is preceded by* $\mathbf{v}^{(2)}$ ($\mathbf{v}^{(1)}$ *is preferred with respect to* $\mathbf{v}^{(2)}$, what can be shortly denoted as $\mathbf{v}^{(1)} \prec \mathbf{v}^{(2)}$. If neither $\mathbf{v}^{(1)} \prec \mathbf{v}^{(2)}$ nor $\mathbf{v}^{(1)} \prec \mathbf{v}^{(2)}$, then it is said that $\mathbf{v}^{(1)}$ and $\mathbf{v}^{(2)}$ are *mutually incomparable*. In the last case, some additional criteria should be used in order to select the best decision from a subset of mutually incomparable ones.

CONCLUSION

It was shown in the chapter that ontologies used as a support in computer-aided decision making usually consist of several ontological models being a form of presentation of knowledge about a given area of interest. Ontological models can be constructed on the basis of various formal models: taxonomic trees, graphs, multigraphs, relations, hyper-relations, and so forth. However, deterministic models not always describe adequately the state of our knowledge about the area of interest. That is why in certain cases canceled or otherwise incomplete ontological models as well as nondeterministic models should be used. The nondeterministic ontological models, in the simplest case, may be presented as fuzzy models, that is, models based on fuzzy concepts in the Zadeh sense. In a more general case, nondeterministic models can be presented in the form of nondeterministic relations, that is, relations whose syndromes have been semi-ordered. In particular, the concept of a semi-ordered linear vector space to construction of nondeterministic ontological models can be used.

REFERENCES

Berge, C. (1973). *Graphs and hypergraphs.* Amsterdam: North-Holland.

Bubnicki, Z. (2002). *Uncertain logics, variables and systems.* Springer-Verlag. LNICS No. 276.

Chute, C.G. (2005). Medical concept representation. In H. Chen et al. (Eds.), *Medical informatics: Knowledge management and data mining in biomedicine* (pp. 163-182). Springer-Verlag.

Coté, R.A. (Ed.). (1975). SNOMED: Systematized nomenclature of medicine. *Diseases.* ACP.

Grzegorzewski, P., Hryniewicz, O., & Gil, M. A. (Eds.). (2002). *Soft methods in probability, statistics and data analysis.* Heidelberg, Germany: Physica Verlag.

Kantorovich, L.B., Vulich, B.Z., & Pinsker, A.G. (1950). *Functional analysis in semi-ordered spaces* (in Russian). Moscow: GITTL.

Kulikowski, J.L. (1972). *An algebraic approach to the recognition of patterns.* CISM Lecture Notes No. 85. Wien: Springer-Verlag.

Kulikowski, J.L. (1986). *Outline of the theory of graphs (*in Polish). Warsaw, Poland: PWN.

Kulikowski, J.L. (1992). Relational approach to structural analysis of images. *Machine Graphics and Vision, 1*(1/2), 299-309.

Kulikowski, J.L. (2006). Description of irregular composite objects by hyper-relations. In K. Wojciechowski et al. (Eds.), *Computer vision and graphics* (pp. 141-146). Springer-Verlag.

Pisanelli, D.M. (Ed.). (2004). *Ontologies in medicine.* Amsterdam: IOS Press.

Rasiowa, H., & Sikorski, R. (1968). *The mathematics of metamathematics.* Warsaw: PWN.

Rutkowski, L., Tadeusiewicz, R., Zadeh, L.A., & Zurada, J. (Eds.). (2006). *Artificial intelligence and soft computing–ICAISC 2006.* Berlin: Springer-Verlag.

Tutte, W.T. (1984). *Graph theory.* Menlo Park, CA: Addison-Wesley.

Zadeh, L.A. (1975a). The concept of a linguistic variable and its application to approximate reasoning. Part I. *Information Science, 8*, 199-249.

Zadeh, L.A. (1975b). The concept of a linguistic variable and its application to approximate reasoning. Part II. *Information Science, 8*, 301-357.

Zadeh, L.A. (1975c). The concept of a linguistic variable and its application to approximate reasoning. Part III. *Information Science, 9*, 43-80.

Chapter II
Using Logic Programming and XML Technologies for Data Extraction from Web Pages

Amelia Bădică
University of Craiova, Romania

Costin Bădică
University of Craiova, Romania

Elvira Popescu
University of Craiova, Romania

ABSTRACT

The Web is designed as a major information provider for the human consumer. However, information published on the Web is difficult to understand and reuse by a machine. In this chapter, we show how well established intelligent techniques based on logic programming and inductive learning combined with more recent XML technologies might help to improve the efficiency of the task of data extraction from Web pages. Our work can be seen as a necessary step of the more general problem of Web data management and integration.

INTRODUCTION

The Web is extensively used for information dissemination to humans and businesses. For this purpose, Web technologies are used to convert data from internal formats, usually specific to data base management systems, to suitable presentations for attracting human users. However, the interest has rapidly shifted to make that information available for machine consumption by realizing that Web data can be reused for various problem solving purposes, including common tasks like searching and filtering, and also more complex tasks like analysis, decision making, reasoning and integration.

For example, in the e-tourism domain one can note an increasing number of travel agencies offering online services through online transaction brokers (Laudon & Traver, 2004). They provide useful information to human users about hotels, flights, trains or restaurants, in order to help them plan their business or holiday trips. Travel information, like most of the information published on the Web, is heterogeneous and distributed, and there is a need to gather, search, integrate and filter it efficiently (Staab et al., 2002) and ultimately to enable its reuse for multiple purposes. In particular, for example, personal assistant agents can integrate travel and weather information to assist and advise humans in planning their weekends and holidays. Another interesting use of data harvested from the Web that has been recently proposed (Gottlob, 2005) is to feed business intelligence tasks, in areas like competitive analysis and intelligence.

Two emergent technologies that have been put forward to enable automated processing of information published on the Web are semantic markup (W3C Semantic Web Activity, 2007). and Web services (Web Services Activity, 2007). However, most of the current practices in Web publishing are still being based on the combination of traditional HTML-lingua franca for Web publishing (W3C HTML, 2007) with server-side dynamic content generation from databases. Moreover, many Web pages are using HTML elements that were originally intended for use in structure content (e.g., those elements related to tables), or for layout and presentation effects, even if this practice is not encouraged in theory. Therefore, techniques developed in areas like information extraction, machine learning and wrapper induction are still expected to play a significant role in tackling the problem of Web data extraction.

Data extraction is related to the more general problem of information extraction that is traditionally associated with artificial intelligence and natural language processing. Information extraction was originally concerned with locating specific pieces of information in text documents written in natural language (Lenhert & Sundheim, 1991) and then using them to populate a database or structured document. The field then expanded to cover extraction tasks from Web documents represented in HTML and attracted other communities including databases, electronic documents, digital libraries and Web technologies. Usually, the content of these data sources can be characterized as neither natural language, nor structured, and therefore usually the term semi-structured data is used. For these cases, we consider that the term data extraction is more appropriate than information extraction and consequently, we shall use it in the rest of this chapter.

A wrapper is a program that is used for performing the data extraction task. On one hand, manual creation of Web wrappers is a tedious, error-prone and difficult task because of Web heterogeneity in both structure and content. On the other hand, construction of Web wrappers is a necessary step to allow more complex tasks like decision making and integration. Therefore, a lot of techniques for (semi-)automatic wrapper construction have been proposed. One application area that can be described as a success story for machine learning technologies is wrapper induction for Web data extraction. For a recent overview of state-of-the-art approaches in the field see Chang, Kayed, Girgis, and Shaalan (2006).

In this chapter, we propose a novel class of wrappers, L-wrappers (i.e., logic wrappers), that fruitfully combine logic programming paradigm with efficient XML processing technologies (W3C Extensible Markup Language (XML), 2007). Our wrappers have certain advantages over existing proposals: i) they have a declarative semantics, and therefore their specification is decoupled from their implementation; ii) they can be generated using techniques and algorithms inspired by inductive logic programming (ILP hereafter); iii) they are implemented using XSLT – the "native" language for processing XML documents (W3C Extensible Stylesheet Language Family (XSL), 2007); and iv) they have also a visual notation making them easier to read and understand than their equivalent XSLT coding.

The chapter is structured as follows. We start with a brief review of logic programming, XML technologies and related approaches to Web data extraction. Then, we discuss flat relational and hierarchical approaches to Web pages conceptualization for data extraction. We follow with a concise definition of L-wrappers covering both their textual and visual representations. Both flat and hierarchical cases are considered. Next, we discuss efficient algorithms for semi-automatic construction of L-wrappers. Then, we present an approach for implementing L-wrappers using XSLT transformation language. The last two sections of this chapter contain some pointers to future works, as well as a list of concluding remarks.

BACKGROUND

The goal of this section is to briefly review the main ingredients of our approach to Web data extraction, that is, XML technologies and logic programming. Finally, as the application of logic programming and XML to information extraction is not entirely new, we briefly provide an literature overview of related proposals.

XML Technologies for Data Extraction

The Web is now a huge information repository that is characterized by i) high diversity, that is, the Web information covers almost any application area, ii) disparity, that is, the Web information comes in many formats ranging from plain and structured text to multimedia documents and iii) rapid growth, that is, old information is continuously being updated in form and content and new information is constantly being produced.

The HTML markup language is the lingua franca for publishing information on the Web, so our core data sources are in fact HTML documents. HTML was initially devised for modeling the structure and content of Web documents, rather than their presentation layout. However, with the advent of graphic Web browsers, software providers like Microsoft or Netscape added many features to HTML that were mainly addressing the visual representation and interactivity of Web documents, rather than their structure and content. The effects of this process were that initially HTML was developed (and consequently used) in a rather unsystematic way. However, starting with HTML 4.01, W3C consortium enforced a rigorous standardization process of HTML that ultimately resulted in a complete redefinition of HTML as an XML application, known as XHTML.

In our work we make the assumption that Web documents already are or can be converted through a preprocessing stage to well-formed XML before being actually processed for extraction of interesting data. While clearly, data extraction from HTML can benefit from existing approaches for information extraction from unstructured texts, we state that preprocessing and conversion of HTML to a structured (i.e., tree-like or well-formed XML) form has certain obvious advantages: i) an extracted item can depend on its structural context in a document, while this information is lost in the event the tree document is flattened as a string; ii) data extraction from XML documents can benefit from the plethora of XML query and transformation languages and tools.

A Web document is composed of a structural part and a content part. The structural part consists of the set of document nodes or elements. The document elements are nested into a tree-like structure. The content part of a document consists of the actual text in the text elements and the attribute-value pairs attached to the other document elements.

We model semistructured Web documents as labeled ordered trees. The node labels of a labeled ordered tree correspond to HTML tags. In particular, a text element will be considered to have a special tag text. Let Σ be the set of all node labels of a labeled ordered tree. For our purposes, it is convenient to abstract labeled ordered trees as sets of nodes on which certain relations and functions are defined. Note that in this chapter we are using some basic graph terminology as introduced in Cormen, Leiserson, and Rivest (1990).

Figure 1 shows a labeled ordered tree with 25 nodes and tags in the set Σ = {a, b, c}.

Intuitively, a wrapper takes a labeled ordered tree and returns a subset of extracted nodes. An extracted node can be viewed as representing the whole subtree rooted at that node. The structural context of an extracted node is a complex condition that specifies i) the tree delimiters of the extracted information, according to the parent-child and next-sibling relationships (e.g., is there a parent node ?, is there a left sibling ?) and ii) certain conditions on node labels and their position (e.g., is the tag label *td* ?, is it the first child ?). This conditions are nicely captured as conjunctive queries represented using logic programming (see next section).

Logic Programming for Representation and Querying of Web Documents

The rapid growth of the Web gave a boost to research on techniques to cope with the information flood. At the core of the various applications that include tasks like data retrieval, data extraction, and text categorization there are suitable representations of Web documents to allow their efficient structured querying and processing. In this subsection we show how logic programming can be used to achieve this desiderate.

Logic programming (Sterling & Shapiro, 1994) was originally developed within the artificial intelligence community to help with the implementation of natural language processing tools. However, its attractive features including declarative semantics, compact syntax, built-in reasoning capabilities, and so forth, together with efficient compilation techniques, made logic programming a suitable paradigm for the development of high-level general-purpose programming languages; see, for example, the Prolog language. Moreover, during the last decade applications of logic programming spread also to the areas of the Web and the Semantic Web (Alferes, Damasio, & Pereira, 2003).

A logic program is a set of logic statements that are classified as facts, rules and queries. Facts and rules are used to describe the problem

Definition 1 (Labeled ordered tree) A *labeled ordered tree* is a tuple $t = \langle T, E, r, l, c, n\rangle$ such that:

i. (T, E, r) is a rooted tree with root $r \in T$. Here, T is the set of tree nodes and E is the set of tree edges.
ii. $l : T \rightarrow \Sigma$ is a node labeling function.
iii. $c \subseteq T \times T$ is the parent-child relation between tree nodes, that is, $c = \{(v, u) \mid \text{node } u \text{ is the parent of node } v\}$.
iv. $n \subseteq T \times T$ is the next-sibling linear ordering relation defined on the set of children of a node. For each node $v \in T$, its k children are ordered from left to right, that is, $(v_i, v_{i+1}) \in n$ for all $1 \leq i < k$.

Figure 1. Document as labeled ordered tree

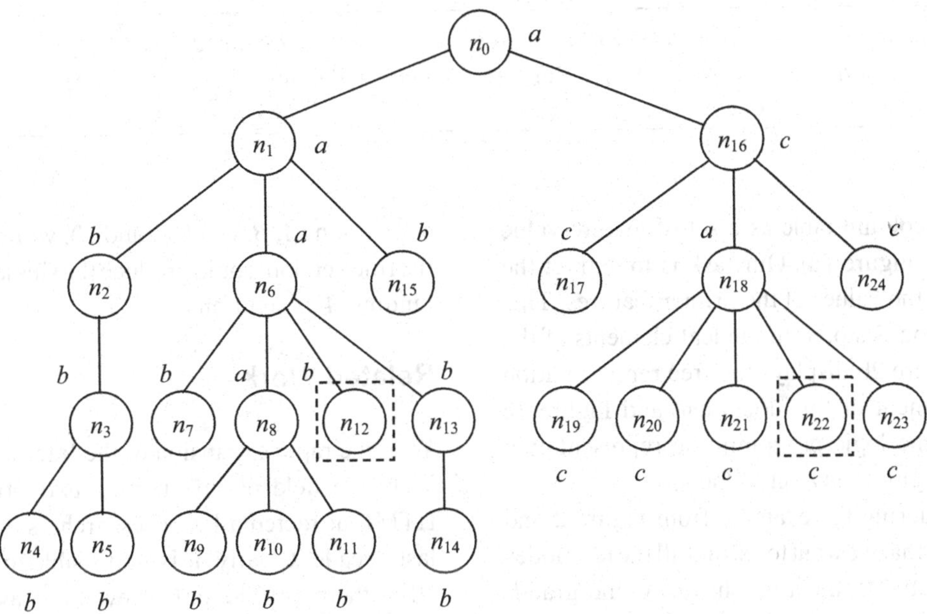

domain, while queries are used to pose specific problem instances and to retrieve the corresponding solutions as query answers. Intuitively, the computation associated to a logic program can be described as the reasoning process for determining suitable bindings for the query variables such that the resulting instance of the query is entailed by the facts and rules that comprise the logic program.

Consider, for example, the logic programming representation of a Web document tree. We assign a unique identifier (an integer value) to each node of the tree. Let N be the set of all node identifiers.

The structural component of a Web document can be represented as a set of facts that use the following relations Box 1.

In order to represent the content component of a Web document, we introduce two sets: the set S of content elements that denote strings attached to text nodes and values assigned to HTML attributes, and the set A of HTML attributes. With these notations, the content part of a Web document tree can be represented using two relations (Box 2).

Consider the Hewlett Packard's Web site of electronic products and the task of data extraction from a product information sheet for Hewlett Packard printers. The printer information is displayed

Box 1.

i. *child* $\subseteq \mathcal{N} \times \mathcal{N}$ defined as *child*(*P*,*C*) $\leftrightarrow$ *P* is the parent of *C*.
ii. *next* $\subseteq \mathcal{N} \times \mathcal{N}$ defined as *next*(*L*,*R*) $\leftrightarrow$ *L* is the left sibling of *R*.
iii. $tag_\sigma \subseteq \mathcal{N}$, $\sigma \in \Sigma$ defined as $tag_\sigma(N)$ $\leftrightarrow$ the tag of node *N* is σ.
iv. *first* $\subseteq \mathcal{N}$ defined as *first*(*X*) $\leftrightarrow$ *X* is the first child of its parent node.
v. *last* $\subseteq \mathcal{N}$ defined as *last*(*X*) $\leftrightarrow$ *X* is the last child of its parent node.

Box 2.

i. *content* $\subseteq \mathcal{N} \times \mathcal{S}$ defined as *content*$(N,S) \leftrightarrow S$ is the string contained by text node N.
ii. *attribute_value* $\subseteq \mathcal{N} \times \mathcal{A} \times \mathcal{S}$ defined as *attribute_value*$(N,A,S) \leftrightarrow S$ is the text value of attribute A of node N.

in a two-column table as a set of feature-value pairs (see Figure 2a). Our task is to extract the names or the values of the printer features. This information is stored in the leaf elements of the page. Figure 2b displays the tree representation of a fragment of this document and Figure 3c displays the logic programming representation of this fragment as a set of facts.

Considering the example from Figure 2 and assuming that we want to extract all the text nodes of this Web document that have a grand-grand-parent of type table that has a parent that has a right sibling, we can use the following query. Note that for expressing logic programs we are using the standard Prolog notation (Sterling, 1994):

```
? tag(A,text),child(B,A),child(C,B),
 child(D,C),tag(D,table),child(E,D).next(E,F).
```

The query can be more conveniently packed as a rule as follows:

```
extract(A) :-
  tag(A,text),child(B,A),child(C,B),
  child(D,C),tag(D,table),child(E,D),next(E,F).
```

The rule representation has at least three obvious advantages: i) modularity: the knowledge embodied in the query is encapsulated inside the body of the predicate extract; ii) reusability: the query can be more easily reused rather than having to fully copy the conjunction of conditions and iii) information hiding: the variables occurring in the right-hand side of the rule are hidden to the user, that is, running the initial version of the query would produce a tuple of variable bindings as solution (A, B, C, D, E, and F), while running the rule version would produce the single variable binding A as solution.

Related Works

With the rapid expansion of the Internet and the Web, the field of information extraction from HTML attracted a lot of researchers during the last decade. Clearly, it is impossible to mention all of their work here. However, at least we can try to classify these works along several axes and select some representatives for discussion.

First, we have focused our research on information extraction from HTML using logic representations of tree (rather than string) wrappers that are generated automatically using techniques inspired by ILP. Second, both theoretical and experimental works are considered.

Freitag (1998) is one of the first papers describing a "relational learning program" called SRV. It uses an ILP algorithm for learning first order information extraction rules from a text document represented as a sequence of lexical tokens. Rule bodies check various token features like length, position in the text fragment, if they are numeric or capitalized, and so forth. SRV has been adapted to learn information extraction rules from HTML. For this purpose, new token features have been added to check the HTML context in which a token occurs. The most important similarity between SRV and our approach is the use of relational learning and an ILP algorithm. The difference is that our approach has been explicitly devised to cope with tree structured documents, rather than string documents.

Chidlovskii (2003) describes a generalization of the notion of string delimiters developed for information extraction from string documents (Kushmerick, 2000) to subtree delimiters for information extraction from tree documents. The paper describes a special purpose learner that constructs a structure called candidate index based on tree data structures, which is very different from our approach. Note however, that the tree leaf delimiters described in that paper are very similar to our information extraction rules. Moreover, the representation of reverse paths using the symbols $\uparrow$, $\leftarrow$ and $\rightarrow$ can be easily simulated by our rules using the relations *child* and *next* in our approach.

Xiao, Wissmann, Brown, and Jablonski (2001) proposes a technique for generating XSLT-patterns from positive examples via a GUI tool and using an ILP-like algorithm. The result is a NE-agent (i.e., name extraction agent) that is capable of extracting individual items. A TE-agent (i.e., term extraction agent) then uses the items extracted by NE-agents and global constraints to fill-in template slots (tuple elements according to our terminology). The differences in our work are that XSLT wrappers are learned indirectly via L-wrappers, and our wrappers are capable of extracting tuples in a straightforward way, and therefore TE-agents are not needed. Additionally, our approach covers the hierarchical case, which is not addressed in Xiao et al. (2001).

Lixto (Baumgartner, Flesca, & Gottlob, 2001) is a visual wrapper generator that uses an internal logic programming-based extraction language called Elog. In Elog, a document is abstracted as a tree (similar to our work), rather than a string. Elog is very versatile by allowing the refinement of the extracted information with the help of regular expressions and the integration between wrapping and crawling via links in Web pages. The differences between Elog and L-wrappers are at least two fold: i) L-wrappers are only devised for the extraction task and they use a classic logic programming approach, for example, an L-wrapper can be executed without any modification by a standard Prolog engine. Elog was devised for both crawling and extraction and has a customized logic programming-like semantics, that is more difficult to understand; and ii) L-wrappers are efficiently implemented by translation to XSLT, a standard language for transforming XML documents, while for Elog the implementation approach is different (a custom interpreter has been devised from scratch).

Thomas (2000) introduces a special wrapper language for Web pages called token-templates. Token-templates are constructed from tokens and token-patterns. A Web document is represented as a list of tokens. A token is a feature structure with exactly one *type* feature. Feature values may be either constants or variables. Token-patterns use operators from the language of regular expressions. The operators are applied to tokens to extract relevant information. The only similarity between our approach and this approach is the use of logic programming to represent wrappers.

Laender, Ribeiro-Neto, and Silva (2002) describes the DEByE (i.e., Data Extraction By Example) environment for Web data management. DEByE contains a tool that is capable of extracting information from Web pages based on a set of examples provided by the user via a GUI. The novelty of DEByE is the possibility to structure the extracted data based on the user perception of the structure present in the Web pages. This structure is described at the example collection stage by means of a GUI metaphor called nested tables. DEByE also addresses other issues needed in Web data management, like automatic examples generation and wrapper management. Our L-wrappers are also capable of handling hierarchical information. However, in our approach, the hierarchical structure of information is lost by flattening during extraction (see the printer example where tuples representing features of the same class share the feature class attribute).

Sakamaoto (2002) introduces tree wrappers for tuples extraction. A tree wrapper is a sequence of tree extraction paths. There is an extraction path for each extracted attribute. A tree extraction path is a sequence of triples that contain a tag, a position and a set of tag attributes. A triple matches a node based on the node tag, its position among its siblings with a similar tag and its attributes. Extracted items are assembled into tuples by analyzing their relative document order. The algorithm for learning a tree extraction path is based on the composition operation of two tree extraction paths. Note also that L-wrappers use a different and richer representation of node proximity and therefore, we have reason to believe that they could be more accurate (this claim needs, of course, further support with experimental evidence). Finally, note that L-wrappers are fully declarative, while tree wrappers combine declarative extraction paths with a procedural algorithm for grouping extracted nodes into tuples.

A new wrapper induction algorithm inspired by ILP is introduced in Anton (2005). The algorithm exploits traversal graphs of documents trees that are mapped to XPath expressions for data extraction. However, that paper does not define a declarative semantics of the resulting wrappers. Moreover, the wrappers discussed in Anton (2005) aim to extract only single items, and there is no discussion of how to extend the work to tuples extraction.

Stalker (Muslea, Minton, & Knoblock, 2001) uses a hierarchical schema of the extracted data called *embedded catalog formalism* that is similar to our approach. However, the main difference is that Stalker abstracts the document as a string rather than a tree and therefore their approach is not able to benefit from existing XML processing technologies. Extraction rules of Stalker are based on a special type of finite automata called *landmark automata*, rather than logic programming, as our L-wrappers.

Concerning theoretical work, Gottlob and Koch (2004) is one of the first papers that analyzes seriously the expressivity required by tree languages for Web information extraction and its practical implications. Combined complexity and expressivity results of conjunctive queries over trees, that also apply to information extraction, are reported in Gottlob, Koch, and Schultz (2004).

CONCEPTUALIZING WEB PAGES FOR DATA EXTRACTION

Many Web pages are dynamically generated by filling in HTML templates with data obtained from relational data bases. We have noticed that most often such Web documents can be successfully abstracted as providing relational data as sets of tuples or records. Examples include search engines' answer pages, product catalogues, news sites, product information sheets, travel resources, multimedia repositories, Web directories, and so forth.

Sometimes, however, Web pages contain hierarchically structured presentations of data for usability and readability reasons. Moreover, it is generally appreciated that hierarchies are very helpful for focusing human attention and management of complexity. Therefore, as most Web pages are developed by knowledgeable specialists in human-computer interaction design, we expect to find this approach in many designs of Web interfaces to data-intensive applications.

Flat Relational Conceptualization

We adopt a standard relational model by associating to a Web data source a set of distinct attributes. Let $\mathcal{A}$ be the set of all attribute names and let $D \subset \mathcal{A}$ be the set of relational attributes associated to a given Web data source. An extracted tuple can be defined as a function $tuple : D \to \mathcal{N}$, such that for each attribute $a \in D$, $tuple(a)$ represents the document node extracted from the Web source as an instance of attribute a. Note that in practice, instead of an extracted node, a user is rather interested to get the HTML content of the node.

Let us consider, for example, the problem of extracting printer information from Hewlett Packard's Web site. The printer information is represented in multisection, two column HTML tables (as shown in Figure 2a). Each row contains a pair consisting of a feature name and a feature value. Consecutive rows represent related features that are grouped into feature classes. For example, there is a row with the feature name 'Processor speed' and the feature value '300 Mhz.' This row has the feature class 'Speed/monthly volume.' So, actually, this table contains triples consisting of a feature class, a feature name, and a feature value. The set of relational attributes is D = {*feature-class,feature-name,feature-value*}. The document fragment shown in Figure 2a contains three tuples:

(*feature-class*: 'Speed/monthly volume,' *feature-name*: 'Print speed, black (pages per minute),' *feature-value*: 'Up to 50 ppm')

(*feature-class*: 'Speed/monthly volume,' *feature-name*: 'First page out, black,' *feature-value*: '8 secs')

(*feature-class*: 'Speed/monthly volume,' *feature-name*: 'Processor speed,' *feature-value*: '300 Mhz')

Note that in this example some tuples may have identical feature classes. More generally, for some documents, distinct tuples might have identical attribute instances. Clearly, this happens when the document has a hierarchical structure. For such cases, a hierarchical conceptualization of the Web data source is more appropriate (see the next section).

Let us now show how logic programming can be employed to conveniently define wrappers for data extraction from Web pages that have been conceptualized as flat relational data sources. Anticipating (see next section on logic wrappers), we shall call such programs *logic wrappers* or *L-wrappers*.

A L-wrapper for extracting relational data operates on a target Web document represented as a labeled ordered tree and returns a set of relational tuples of nodes of this document.

A L-wrapper for the printers example shown in Figure 2b is (FN = feature name, FV = feature value) (Box 3).

Definition 2 (L-wrapper as set of logic rules) A *L-wrapper* can be defined formally as a set of patterns represented as logic rules. Assuming that $\mathcal{N}$ is the set of document tree nodes and Σ is the set of HTML tags, a *L-wrapper* is a logic program defining a relation $extract(N_1, \ldots, N_k) \subseteq \mathcal{N} \times \ldots \times \mathcal{N}$. For each clause, the head is $extract(N_1, \ldots, N_k)$ and the body is a conjunction of literals in the set {*child, next, first, last*, $(tag_\sigma)_{\sigma \in \Sigma}$}. Number k of extracted attributes is called *wrapper arity* and is equal to the number of elements of set D.

Box 3.

```
extract(FN,FV) :-
    tag(FN,text),text(FV),child(C,FN),child(D,FV),child(E,C),child(H,G),child(I,F),
    child(J,I),next(J,K),child(F,E),child(G,D),first(J),child(K,L),child(L,H).
```

Figure 2. Web document fragment

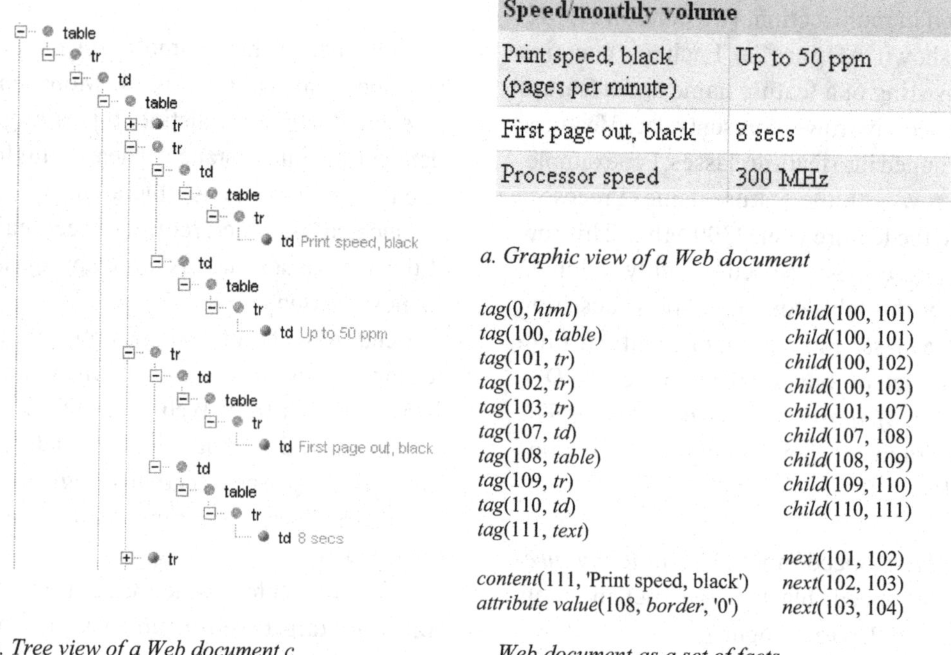

Speed/monthly volume	
Print speed, black (pages per minute)	Up to 50 ppm
First page out, black	8 secs
Processor speed	300 MHz

a. Graphic view of a Web document

b. Tree view of a Web document c

. Web document as a set of facts

Figure 3. A hierarchical HTML document and its schema

Red apple

weight	120
color	red
diameter	8
Limited stock, order today !	

Lemon

weight	70
color	yellow
height	7
width	4
Limited stock, order today !	

a. Sample HTML document containing hierarchically structured data

fruits

*

fruit

fruit-name

features

*

feature

feature-name

feature-value

b. Hierarchical schema

This rule extracts all the pairs of text nodes such that the grand-grand-grand-grandparent of the first node (*J*) is the first child of its parent node and also the left sibling of the grand-grand-grand-grandparent of the second node (*K*).

Hierarchical Conceptualization

In this subsection, we propose an approach for utilizing L-wrappers to extract hierarchical data. The advantage would be that extracted data will be suitably annotated to preserve its hierarchical structure, as found in the Web page. Further processing of this data would benefit from this additional metadata to allow for more complex tasks, rather than simple searching and filtering by populating a relational database. For example, one can imagine the application of this technique to the task of ontology extraction, as ontologies are assumed to be natively equipped with the facility of capturing taxonomically structured knowledge.

Let us consider a very simple HTML document that contains hierarchical data about fruits (see Figure 3a). A fruit has a name and a sequence of features. Additionally, a feature has a name and a value. This is captured by the schema shown in Figure 3b. Note that this representation allows features to be fruit-dependent; for example, while an apple has an average *diameter*, a lemon has both an average *width* and an average *height*.

Abstracting the hierarchical structure of data, we can assume that the document shown in Figure 3a contains triples consisting of a fruit name, a feature-name and a feature-value. However, this approach has at least two drawbacks: i) redundancy, because distinct tuples might contain identical attribute instances, and ii) the intrinsic hierarchical structure of the data is lost, while it might convey useful information.

Following the hierarchical structure of this data, the design of a L-wrapper of arity 3 for this example can be done in two stages: i) derive a wrapper W_1 for binary tuples (*fruit-name*, *list-of-features*); and ii) derive a wrapper W_2 for binary tuples (*feature-name*, *feature-value*). Note that wrapper W_1 is assumed to work on documents containing a list of tuples of the first type (i.e., the original target document), while wrapper W_2 is assumed to work on document fragments containing the list of features of a given fruit (i.e., a single table from the original target document). For example, wrappers W_1 and W_2 can be defined as logic programs as shown in Box 4.

Note that for the combination of W_1 and W_2 into a single L-wrapper of arity 3, we need to extend the definition of a L-wrapper by adding a new argument to relation *extract* for representing the root node of the document fragment to which the wrapper is applied, that is, instead of *extract*(N_1, …, N_k) we shall now have *extract*(R, N_1, …, N_k), R is the new argument. Moreover, it is required that for all $1 \leq i \leq k$, N_i is a descendant of R in the document tree. The resulted solution is shown in Box 5.

The final wrapper (assuming the index of document root node is 0) is shown in Box 6.

Box 4.

```
extr _ fruits(FrN,FrFs) :-
   tag(FrN,text),child(A,FrN),child(B,A),next(B,FrFs),child(C,FrFs),tag(C,p).
extr _ features(FN,FV) :-
   tag(FN,text),tag(FV,text),child(A,FN),child(B,FV),next(A,B),
   child(C,B),tag(C,tr).
```

Box. 5

```
% ancestor(Ancestor,Node).
ancestor(N,N).
ancestor(A,N) :-
   child(A,B),ancestor(B,N).
extr _ fruits(R,FrN,FrFs) :-
   ancestor(R,FrN),ancestor(R,FrFs),extr _ fruits(FrN,FrFs).
extr _ features(R,FN,FV) :-
   ancestor(R,FN),ancestor(R,FV),extr _ features(FN,FV).
```

Box 6.

```
extract(FrN,FN,FV) :-
   extr _ fruits(0,FrN,FrFs),extr _ features(FrFs,FN,FV).
```

Figure 4. Wrapping hierarchically structured data

```
extract_all(Res) :-
  extr_fruits_all(0,Res).
extr_fruits_all(Doc,fruits(Res)) :-
  findall(
    fruits(name(FrN),FrFs),
    (extr_fruits(Doc,NFrN,NFrFs),
     content(NFrN,FrN),
     extr_features_all(NFrFs,FrFs)),
    Res).
extr_features_all(Doc,features(Res)) :-
  findall(
    feature(name(FN),value(FV)),
    (extr_features(Doc,NFN,NFV),
     content(NFN,FN),text(NFV,FV)),
    Res).
```

a. Hierarchical wrapper

```
?-extract_all(Res).
Res = fruits(
  [fruit(name(`Red apple`),
    features(
      [feature(name(`weight`),value(`120`)),
       feature(name(`color`),value(`red`)),
       feature(name(`diameter`),value(`8`))
      ])),
   fruit(name(`Lemon`),
    features(
      [feature(name(`weight`),value(`70`)),
       feature(name(`color`),value(`yellow`)),
       feature(name(`height`),value(`7`)),
       feature(name(`width`),value(`4`))
      ])])
```

b. Hierarchical extracted data

While simple, this solution has the drawback that, even if it was devised with the idea of hierarchy in mind, it is easy to observe that the hierarchical nature of the extracted data is lost.

Assuming a Prolog execution engine of L-wrappers, we can solve the drawback using the *findall* predicate. *findall(X, G, Xs)* returns the list *Xs* of all terms *X* such that goal *G* is true (it is assumed that *X* occurs in *G*). The solution and the result are shown in Figure 4. Note that we assume that i) the root node of the document has index 0, and ii) predicate *content*(*TextNode*, *Content*) is used to determine the content of a text node.

Definition 3 (Pattern graph) Let $\mathcal{W}$ be a set denoting all vertices. A pattern graph G is a quadruple $\langle A, V, L, \lambda_a \rangle$ such that $V \subseteq \mathcal{W}$, $A \subseteq V \times V$, $L \subseteq V$ and $\lambda_a : A \rightarrow \{\text{'c'}, \text{'n'}\}$. The set $\mathcal{G}$ of pattern graphs is defined inductively as follows:

i. If $v \in \mathcal{W}$ then $\langle \emptyset, \{v\}, \{v\}, \emptyset \rangle \in \mathcal{G}$

ii. If $G = \langle A, V, L, \lambda_a \rangle \in \Gamma$, $v \in L$, and $w, u_i \in \mathcal{W} \setminus V$, $1 \le i \le n$ then a) $G_1 = \langle A \cup \{(w, v)\}, V \cup \{w\}, (L \setminus \{v\}) \cup \{w\}, \lambda_a \cup \{((w, v),\text{'n'})\} \rangle \in \mathcal{G}$; b) $G_2 = \langle A \cup \{(u_1, v), \ldots, (u_n, v)\}, V \cup \{u_1, \ldots, u_n\}, (L \setminus \{v\}) \cup \{u_1, \ldots, u_n\}, \lambda_a \cup \{((u_1, v),\text{'c'}), \ldots, ((u_n, v),\text{'c'})\} \rangle \in \mathcal{G}$; c) $G_3 = \langle A \cup \{(w, v), (u_1, v), \ldots, (u_n, v)\}, V \cup \{w, u_1, \ldots, u_n\}, (L \setminus \{v\}) \cup \{w, u_1, \ldots, u_n\}, \lambda_a \cup \{((w, v),\text{'n'}), ((u_1, v),\text{'c'}), \ldots, ((u_n, v),\text{'c'})\} \rangle \in \mathcal{G}$.

LOGIC WRAPPERS AS DIRECTED GRAPHS

In this section, we take a graph-based perspective in defining L-wrappers as sets of patterns. Within this framework, a pattern is a directed graph with labeled arcs and vertices that corresponds to a rule in the logic representation. Arc labels denote conditions that specify the tree delimiters of the extracted data, according to the parent-child and next-sibling relationships (e.g., is there a parent node?, is there a left sibling?, etc.). Vertex labels specify conditions on nodes (e.g., is the tag label *td*?, is it the first child?, etc.). A subset of graph vertices is used for selecting the items for extraction

Intuitively, an arc labeled '*n*' denotes the „next-sibling" relation, while an arc labeled '*c*' denotes the "parent-child" relation. As concerning vertex labels, label '*f*' denotes "first child" condition, label '*l*' denotes "last child" condition and label $\sigma \in \Sigma$ denotes "equality with tag σ" condition.

Patterns are matched against parts of a target document modeled as a labeled ordered tree. A successful matching asks for the labels of pattern vertices and arcs to be consistent with the corresponding relations and functions over tree nodes. The result of applying a pattern to a labeled ordered tree is a set of tuples of extracted nodes.

Patterns can be concisely defined in two steps: i) define the pattern graph together with arc labels that model parent-child and next-sibling relations, and ii) extend this definition with vertex labels that model conditions on vertices, extraction vertices and assignment of extraction vertices to attributes.

Intuitively, if $\langle A, V, L, \lambda_a \rangle$ is a pattern graph, then V denotes its set of vertices, A denotes its set of arcs, $L \subseteq V$ are its leaves (vertices with in-degree 0) and λ_a distinguishes between parent-child (labeled with 'c') and next-sibling (labeled with 'n') arcs. Note also that a pattern graph is tree shaped with arcs pointing up.

Note that according to definition 4, we assume that extraction vertices are among the leaves of the pattern graph, that is, an extraction pattern does not state any condition about the descendants of an extracted node. This is not restrictive in the context of patterns for information extraction from Web documents.

Definition 4 (L-wrapper pattern) Let $\mathcal{A}$ be the set of attribute names. An *L-wrapper pattern* is a tuple $p = \langle V, A, U, D, \mu, \lambda_a, \lambda_c \rangle$ such that $\langle A, V, L, \lambda_a \rangle$ is a pattern graph, $U = \{u_1, u_2, \ldots, u_k\} \subseteq L$ is the set of pattern extraction vertices, $D \subseteq \mathcal{A}$ is the set of attribute names, $\mu : D \rightarrow U$ is a one-to-one function that assigns a pattern extraction vertex to each attribute name, and $\lambda_c : V \rightarrow C$ is the labeling function for vertices. $C = \{\emptyset, \{\text{'f'}\}, \{\text{'l'}\}, \{\sigma\}, \{\text{'f'},\text{'l'}\}, \{\text{'f'},\sigma\}, \{\text{'l'},\sigma\}, \{\text{'f'},\text{'l'}, \sigma\}\}$ is the set of conditions, where σ is a label in the set Σ of tag symbols.

Figure 5. L-wrapper both as directed graph and as logic program

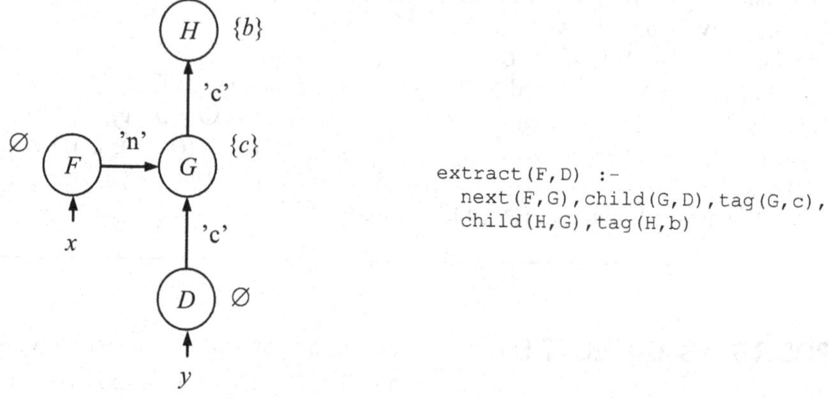

a.L-wrapper as directed graph *b. Same L-wrapper as logic program*

Definition 5 (L-wrapper) An *L-wrapper* of arity k is a set of $n \geq 1$ patterns $W = \{p_i | p_i = \langle V_i, A_i, U_i, D, \mu_i, \lambda_a^i, \lambda_c^i \rangle$, such that each p_i has arity k, for all $1 \leq i \leq n$. The set of tuples extracted by W from a labeled ordered tree t is the union of the sets of tuples extracted by each pattern p_i, $1 \leq i \leq n$, i.e. $Ans(W, t) = \cup_{1 \leq i \leq n} Ans(p_i, t)$.

Definition 6 (Schema tree) Let $\mathcal{W}$ be a set denoting all vertices. A schema tree $\mathcal{S}$ is a directed graph defined as a quadruple $\langle A, V, L, \lambda_a \rangle$ s.t. $V \subseteq \mathcal{W}$, $A \subseteq V \times V$, $L \subseteq V$ and $\lambda_a : A \rightarrow \{\text{'*'}, \text{'1'}\}$. The set of schema trees is defined inductively as follows:

i. For all $n \geq 1$, if $u, v, w_i \in \mathcal{W}$ for all $1 \leq i \leq n$ then $S = \langle A, V, L, \lambda_a \rangle$ such that $V = \{u, v, w_1, \ldots, w_n\}$, $A = \{(u.v), (v,w_1), \ldots, (v,w_n)\}$, $L = \{w_1, \ldots, w_n\}$, $\lambda_a((u,v)) = \text{'*'}$, and $\lambda_a((v,w_1)) = \ldots = \lambda_a((v,w_n)) = \text{'1'}$ is a schema tree.
ii. If $\mathcal{S} = \langle A, V, L, \lambda_a \rangle$ is a schema tree, $n \geq 1$, $u \in L$ and $v, w_i \in \mathcal{W} \setminus V$ for all $1 \leq i \leq n$ then $\mathcal{S}' = \langle A', V', L', \lambda_a' \rangle$ defined as $V' = V \cup \{v, w_1, \ldots, w_n\}$, $A' = A \cup \{(u, v), (v,w_1), \ldots, (v,w_n)\}$, $L' = (L \setminus \{u\}) \cup \{w_1, \ldots, w_n\}$, $\lambda_a'((u, v)) = \text{'*'}$, $\lambda_a'((v,w_1)) = \ldots = \lambda_a'((v,w_n)) = \text{'1'}$ and $\lambda_a'(a) = \lambda_a(a)$ for all $a \in A \setminus A'$ then $\mathcal{S}'$ is also a schema tree.

Figure 5 shows a single-pattern L-wrapper of arity 2 represented both as a directed graph and a logic program. The extraction vertices are marked with small arrows (vertices F and D on that figure). Note also that the figure shows the attributes extracted by the extraction vertices (attribute x extracted by vertex F and attribute y extracted by vertex D).

If p is a pattern and t is a Web document represented as a labeled ordered tree, we denote with $Ans(p,t)$ the set of tuples extracted by p from t. For a formal definition of function Ans see Bădică (2006).

A *L-wrapper* can comprise more patterns so it can be defined formally as a set of extraction patterns that share the set of attribute names. This idea is captured by definition 5.

Let us now formally introduce the concept of *hierarchical logic wrapper* or *HL-wrapper*. We generalize the data source schema from flat relational to hierarchical and we attach to this schema a set of L-wrappers.

If Σ is a set of tag symbols denoting schema concepts and S is a schema tree then a pair consisting of a schema tree and a mapping of schema tree vertices to Σ is called a *schema*. For example, for the schema shown in Figure 3b, Σ = {*fruits, fruit, features, feature, feature-name, feature-value*} (note that on that figure labels '1' are not explicitly shown). For a L-wrapper corresponding to the relational case if D is the set of attribute names, then $\Sigma = D \cup \{result, tuple\}$. Here, *result* denotes the tag of the root element of the output XML document containing the extracted tuples and *tuple* is a tag used to demarcate each extracted tuple; see example in Bădică (2006). Also, it is not difficult to see that in an XML setting a schema nicely corresponds to the document type definition of the output document that contains the extracted data.

An HL-wrapper for the example document considered in this paper consists of: i) schema shown in Figure 3b, ii) L-wrapper W_1 assigned to the vertex labeled with symbol *fruit,* and iii) L-wrapper W_2 assigned to vertex labeled with symbol *feature*.

EFFICIENT ALGORITHMS FOR AUTOMATED CONSTRUCTION OF LOGIC WRAPPERS

Inductive logic programming is one of the success stories in the application area of wrapper induction for information extraction. However, this approach suffers from two problems: high computational complexity with respect to the

Definition 7 (HL-wrapper) A HL-wrapper consists of a schema and an assignment of L-wrappers to split vertices of the schema tree. A vertex v of the schema tree is called *split vertex* if it has exactly one incoming arc labeled '*' and $n \geq 1$ outgoing arcs labeled '1.' An L-wrapper assigned to v must have arity n to be able to extract tuples with n attributes corresponding to outgoing neighbors of vertex v.

Definition 8 (Extraction path) An *extraction path* is a labeled directed graph that is described as a list $[t_0, t_1, \ldots, t_k]$, $k \geq 0$ with the following properties:

i. Each element t_i, $0 \leq i \leq k$ is a list $[v_{-l}, \ldots, v_{-1}, v_0, v_1, \ldots, v_r]$, $l \geq 0$, $r \geq 0$ such that: i) v_i, $-l \leq i \leq r$ are vertices; ii) (v_i, v_{i+1}), $-l \leq i < r$ are arcs labeled with 'n', and iii) for each pair of adjacent lists t_i and t_{i+1}, $1 \leq i < k$ in the extraction path, (v_0^{i+1}, v_0^i) is an arc labeled with 'c'.
ii. Vertex labels are defined as: i) if $l > 0$ then v_{-l} is labeled with a subset of {'f', σ}, $\sigma \in \Sigma$; ii) v_i, $-l < i < 0$ is labeled with a subset of {σ}, $\sigma \in \Sigma$; iii) if $r > 0$ then v_r is labeled with a subset of {'l', σ}, $\sigma \in \Sigma$; iv) v_i, $1 < i < r$ is labeled with a subset of {σ}, $\sigma \in \Sigma$; v) if l, $r > 0$ then v_0 is labeled with a subset of {σ}, $\sigma \in \Sigma$; if $l = 0$; $r > 0$ then v_0 is labeled with a subset of {'f',σ}, $\sigma \in \Sigma$; if $l > 0$; $r = 0$ then v_0 is labeled with a subset of {'l',σ}, $\sigma \in \Sigma$; if $l = r = 0$ then v_0 is labeled with a subset of {'f', 'l',σ}, $\sigma \in \Sigma$.

Vertex v_0^k (i.e. v_0 in list t_k) is matched against the extraction node and consequently is called *extraction vertex*.

Figure 6. Extraction paths for example nodes in Figure 1

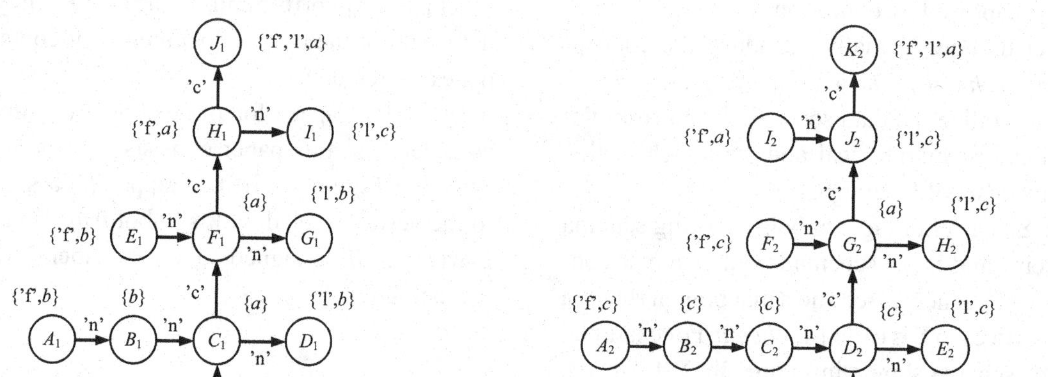

a. Extraction path for example node n_{12}

b. Extraction path for example node n_{22}

Definition 9 (Height and widths of an extraction path) Let $p = [t_0, t_1, \ldots, t_k]$ be an extraction path.

i. The value *height(p)* = *k* is called the *height* of *p*.
ii. The value $left(p) = \max_{0 \le i \le k} left(t_i)$ is called the *left width* of *p*. The value $right(p) = \max_{0 \le i \le k} right(t_i)$ is called the *right width* of *p*.

Definition 10 (Bounded extraction path) Let *H*, *L*, *R* be three positive integers. An extraction path *p* is called *(H, L, R)-bounded* if $height(p) \le H$, $left(p) \le L$ and $right(p) \le R$.

number of nodes of the target document and to the arity of the extracted tuples. In this chapter, we address the first problem by proposing a path generalization algorithm for learning rules to extract single information items. The algorithm produces a pattern (called extraction path) from positive examples and is proven to have good computational properties. The idea of this algorithm can also be used to devise an algorithm for learning tuples extraction paths from positive examples. Finally, note this algorithm can also be adapted to generate multiple pattern L-wrappers from both positive and negative examples.

Extraction Paths

Basically, an extraction path is a L-wrapper pattern for extracting single items, that is, a pattern of arity equal to 1. See examples in Figure 6.

Consider an extraction path $p = [t_0, t_1, \ldots, t_k]$. For a list $t_i = [v_{-l}, \ldots, v_{-1}, v_0, v_1, \ldots, v_r]$ in *p* let $left(t_i) = l$ and $right(t_i) = r$. The following definition introduces height, together with left and right widths of an extraction path.

In practice it is useful to limit the height and the widths of an extraction path, yielding a bounded extraction path.

Note that the extraction path shown in Figure 6a is a (3, 2, 1)-bounded extraction path. Moreover, if we restrict $H = 2$ and $L = 1$, then nodes J_1 and

A_1 will be pruned, resulting a (2, 1, 1)-bounded extraction path, that obviously is less constrained than the initial path.

Learning Extraction Paths

The practice of Web publishing assumes dynamically filling in HTML templates with structured data taken from relational databases. Thus, we can safely assume that a lot of Web data is contained in sets of documents that share similar structures. Examples of such documents are: search engines' result pages, product catalogues, news sites, product information sheets, travel resources, and so forth.

We consider a Web data extraction scenario which assumes the manual execution of a few extraction tasks by the human user. An inductive learning engine could then use the extracted examples to learn a general extraction rule that can be further applied to the current or other similar Web pages.

Usually the extraction task is focused on extracting similar items (like book titles in a library catalogue or product features in a product information sheet). One approach to generate an extraction rule from a set of examples is to discover a common pattern of their neighboring nodes in the tree of the target document.

In what follows, we discuss an algorithm that takes: i) an XML document (possibly assembled from many Web pages, previously preprocessed and converted to well-formed XML) that is modeled as a labeled ordered tree t; ii) a set of example nodes $\{e_1, e_2, \ldots, e_n\}$; and iii) three positive integers H, L, R, and that produces an (H, L, R)-bounded extraction path p that generalizes the set of input examples. Intuitively, this technique is guaranteed to work if we assume that semantically similar items will exhibit structural similarities in the target Web document. This is a feasible assumption for the case of Web documents that are generated on-the-fly by filling in HTML templates with data taken from databases. Moreover, based on experimental results recorded in our work (Bădică, 2006), we have noticed that in practice an extraction rule only needs to check the proximity of nodes. This explains why we focused on the task of learning bounded extraction paths.

The basic operation of the learning algorithm is the generalization operator of two extraction paths. This operator takes two extraction paths p_1 and p_2 and produces an appropriate extraction path p that generalizes p_1 and p_2.

Figure 7. Extraction path generalization algorithm

```
LEARN(p_1, ..., p_n, n)
p ← p_1
for i = 2, n do
    p ← GEN-PATH(p, p_i)
return p

GEN-PATH(p_1, p_2)
let p_1 = [t^1_0, ..., t^1_{k_1}]
let p_2 = [t^2_0, ..., t^2_{k_2}]
k ← min(k_1, k_2)
i ← k, i_1 ← k_1, i_2 ← k_2
while i ≥ 0 do
    t_i ← GEN-LEVEL(t^1_{i_1}, t^2_{i_2})
    i ← i − 1, i_1 ← i_1 − 1, i_2 ← i_2 − 1
return p = [t_0, ..., t_k]

GEN-LEVEL(t_1, t_2)
let t_1 = [v^1_{-l_1}, ..., v^1_{-1}, v^1_0, ..., v^1_{r_1}]
let t_2 = [v^2_{-l_2}, ..., v^2_{-1}, v^2_0, ..., v^2_{r_2}]
l ← min(l_1, l_2)
r ← min(r_1, r_2)
for i = 0, r do
    v_i ← GEN-VERTEX(v^1_i, v^2_i)
for i = 1, l do
    v_{-i} ← GEN-VERTEX(v^1_{-i}, v^2_{-i})
return t = [v_{-l}, ..., v_{-1}, v_0, ..., v_r]

GEN-VERTEX(v_1, v_2)
let λ_1 be the label of v_1
let λ_2 be the label of v_2
λ ← λ_1 ∩ λ_2
return node v with label λ
```

Figure 8. Generalized extraction path for example from Figure 1

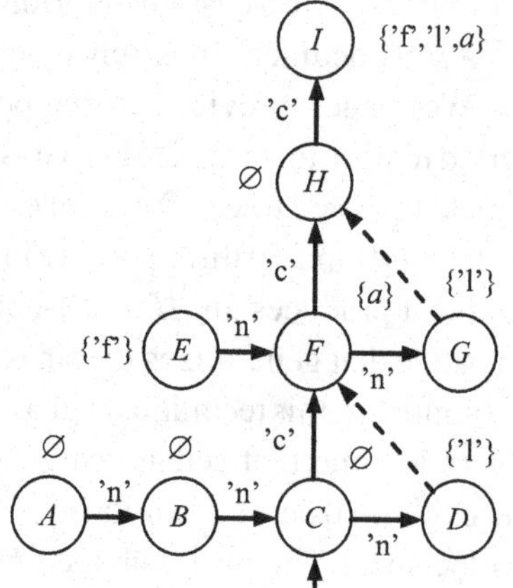

a. Generalized extraction path

```
//*[local-name() = 'a']
[not (preceding-sibling::*)]
[not (following-sibling::*)]/*/*
[not (following-sibling::*)]/
preceding-sibling::*[1]
[local-name() = 'a']
[preceding-sibling::*[1]
[not (preceding-sibling::*)]]/*
[not (following-sibling::*)]/
preceding-sibling::*[1]
[preceding-sibling::*[1]/
preceding-sibling::*[1]]
```

b. XPath code

The idea of the learning algorithm is as follows. For each example node we generate a bounded extraction path (of given input parameters *H*, *L*, *R*) by following sibling and parent links in the document tree. We initialize the output path with the first extraction path and then we proceed by iterative application of the generalization operator to the current output path and the next example extraction path, yielding a new output path. The result is a bounded extraction path that represents an appropriate generalization of the input examples.

The generalization of two paths assumes the generalization of their elements, starting with the elements containing the extraction vertices and moving upper level by level. The generalization of two levels assumes the generalization of each pair of corresponding vertices, starting with vertices with index 0 and moving to the left and respectively to the right in the lists of vertices. Generalization of two vertices is as simple as taking the intersection of their labels. The algorithm is shown in Figure 7.

Function LEARN generalizes the extraction paths of the example nodes. We assume that paths $p_1, \ldots, p_n$ are generated as bounded extraction paths before function LEARN is called. Function GEN-PATH takes two extraction paths p_1 and p_2 and computes their generalization *p*. Function GEN-LEVEL takes two lists of vertices t_1 and t_2 that are members of the extractions paths and computes a generalized list *t* that is member of the generalization path. Function GEN-VERTEX takes two vertices v_1, v_2 and computes a generalized vertex *v*.

It is not difficult to see that the execution of algorithm LEARN takes time $O(n \times H \times (L + R))$ because GEN-VERTEX takes time $O(1)$, GEN-LEVEL takes time $O(L+R)$ and GEN-PATH takes time $O(H \times (L + R))$. Note also that if we set $H = L = R = 1$, then the complexity of the algorithm is $O(n \times H^* \times W^*)$ where H^* and W^* are the height and the width of the target document tree.

Consider again the labeled ordered tree shown in Figure 1 and the example nodes marked with dashed rectangles (n_{12} and n_{22}). The extraction paths corresponding to these nodes are shown in Figure 6. The result of applying the generalization algorithm on those paths is shown in Figure 8a.

Figure 9. Algorithm for translating an extraction path into XPath

```
PATH-TO-XPATH(p)
let p = [t_0, ..., t_k]
xp ← "/"
for i = 0, k do
    let t_i = [v_-l, ..., v_-1, v_0, ..., v_r]
    t ← t_i
    xp ← xp+ "/*" + COND(v_r)
    for j = r − 1,0 do
        xp ← xp+ "/preceding-sibling::*[1]" +
            COND(v_j)
    if l > 0 then
      xp ← xp+ "[preceding-sibling::*[1]" +
          COND(v_-1)
      for j = 2,l do
          xp ← xp+ "/preceding-sibling::*[1]" +
              COND(v_-j)
      xp ← xp+ "]"
return xp
```

```
COND(v)
let λ be the label of v
xc ← ""
if there is σ ∈ λ ∩ Σ then
    xc ← xc+
        "[local-name()=σ]"
if there is 'f' ∈ λ ∩ Σ then
    xc ← xc+
        "[not (preceding-sibling::*)]"
if there is 'l' ∈ λ ∩ Σ then
    xc ← xc+
        "[not (following-sibling::*)]"
return xc
```

XML TECHNOLOGIES FOR DATA EXTRACTION

In this section we describe an approach for the efficient implementation of L-wrappers using XSLT transformation language—a standard language for transforming XML documents. We start with introducing $XSLT_0$ – an expressive subset of XSLT that has a formal operational semantics. Then we describe the algorithm for translating L-wrappers into $XSLT_0$ programs.

Translating Extraction Paths to XPath

An extraction path can be translated to an XPath query. The XPath query can be embedded into an XSLT stylesheet to finally extract the information and store it into a database or another structured document (W3C Extensible Stylesheet Language Family (XSL), 2007).

Figure 9 shows an algorithm for translating an extraction path into an XPath query. The translation algorithm takes an extraction path $p = [t_0, \ldots, t_k]$ and explores it starting with t_0 and moving downwards to t_k. For each element t_i, $0 \le i \le k$, the algorithm maps t_i to a piece of the output XPath query. Actually the algorithm takes the following route of vertices: $v_{r_0}^0 \rightarrow \ldots \rightarrow v_0^0 \rightarrow v_{r_1}^1 \rightarrow \ldots \rightarrow v_0^1 \rightarrow \ldots v_{r_k}^k \rightarrow \ldots \rightarrow v_0^k$. Note that when moving from element i to element i+1, $0 \le i < k$, the algorithm takes the route $v_0^i \rightarrow v_{r_{i+1}}^{i+1}$ (opposite direction of dotted arrows in figure 8a) rather than the route $v_0^i \rightarrow v_0^{i+1}$. For each vertex v_0^i, $0 \le i \le k$, the algorithm also generates a condition that accounts for their left siblings by taking the route $v_0^i \rightarrow v_{-1}^i \rightarrow \ldots \rightarrow v_{-l_i}^i$. It is easy to see that if p is an (H, L, R)-bounded extraction path then the time complexity of the translation algorithm PATH-TO-XPATH is $O(H \times (L + R))$.

Figure 8b shows the result of applying this algorithm to the extraction path from figure 8a. The algorithm will explore the following route of vertices: $I \rightarrow H \rightarrow G \rightarrow F \rightarrow D \rightarrow C$. For each vertex the algorithm generates a location step comprising an axis specifier, a node test and a sequence of predicates written between `'['` and `']'`. The node test is always `*`. The axis specifier is determined by the relation of the current vertex with its preceding vertex on the route explored by the translation algorithm. For example, the axis specifier that is generated for vertex F is `preceding-sibling::`. In this later case, an additional predicate `[1]` that constraints the selection of exactly the preceding node, is added. The algorithm generates also a predicate for each element of the label of a vertex. For example, predicate `[local-name() = 'a']` is generated for vertex *F*, that checks the node tag, and predicate `[not (following-sibling::*)]` is generated for vertex *G*, that checks if the matched node is the last child of its parent node. Moreover, for vertices *F* and *C* the algorithm generates an additional predicate that accounts for their left siblings *E* (of *F*) and respectively $B \rightarrow A$ (of *C*). For example, additional predicate `[preceding-sibling::*[1][not (preceding-sibling::*)]]` is generated for vertex *F*. This predicate checks if the document node matched by vertex *F* has a predecessor and if the predecessor is the first child of its parent node.

Note that running the XPath query from figure 4b on the document represented by the labeled ordered tree from figure 1 produces the following two answers `/a[1]/a[1]/a[1]/b[3]` and `/a[1]/c[1]/a[1]/c[4]` that correspond to nodes n_{12} and n_{22}.

XSLT$_0$ Transformation Language

XSLT$_0$ is a subset of XSLT that retains most of its features and additionally has a formal operational semantics, and a cleaner and more readable syntax. In what follows, we just briefly review XSLT$_0$ and its pseudocode notation. For more details on its abstract model and formal semantics, the reader is invited to consult Bex, Maneth, and Neven (2002).

An XSLT$_0$ program is a set of transformation rules. A rule can be either selecting or constructing. In what follows, we focus only on constructing rules, as we are only using constructing rules in translating L-wrappers into XSLT$_0$.

A (q,σ) constructing rule has the following general form:

template $q(\sigma, x_1, \ldots, x_n)$
vardef
  $y_1 := r_1; \ldots ; y_m := r_m$
return
  if c_1 **then** $z_1; \ldots$; **if** c_k **then** z_k
end

Here:

i. q is an XSLT mode (actually a constructing mode) and σ is a symbol in $\Sigma \cup \{\text{'*'}\}$;
ii. $x_1, \ldots, x_n, y_1, \ldots, y_m$ are variables;
iii. Each r_i is an expression (possibly involving variables $x_1, \ldots, x_n, y_1, \ldots, y_{i-1}$) that evaluates to a data value (i.e., the value of an attribute or the content of an element);
iv. Each c_i is a test (possibly involving variables $x_1, \ldots, x_n, y_1, \ldots, y_m$) and thus it evaluates either to true or false; and
v. Each z_i is a forest (i.e., a (possibly empty) sequence of tree fragments) that is created by the constructing rule. The leaves of this forest are expressions of the form $q'(p, z)$ such that q' is a constructing mode, p is an XPath pattern (possibly with variables), and z is a sequence (possibly empty) of variables from the set $\{x_1, \ldots, x_n, y_1, \ldots, y_m\}$.

Additionally, we require the existence of a constructing mode *start* such that each z_i of a (*start*, σ) constructing rule is a tree (rather than a forest). This constraint ensures that the output of an XSLT$_0$ program is always a tree. Also, for each rule, if none of tests c_i succeeds, the empty forest will be output. Finally, to ensure that the model is deterministic, we require that for any two (q_1, σ_1) and (q_2, σ_2) rules either $q_1 \neq q_2$ or $q_1 = q_2$ and $\sigma_1 \neq \sigma_2$ and $\sigma_1 \neq$ '*' and $\sigma_2 \neq$ '*'.

An XSLT$_0$ program defines a computation as a sequence of steps, that transforms an input labeled ordered tree t with root r into an output tree. At each step, the computation state is recorded as a tree such that some of its leaves are configura-

Figure 10. Algorithm for translating an L-wrapper into XSLT$_0$

```
GEN-WRAPPER(W)
let w ∈ L_W
L ← L_W \ {w}
GEN-FIRST-TEMPLATE(w)
while L ≠ ∅ do
    w_0 ← w
    V ← ∅
    let w' be another vertex in L \ {w_0}
    let w be the first common ancestor of w_0 and w'
    if w_0 ∈ U_W then
        var ← VAR-GEN(μ_W(w_0))
        GEN-TEMPLATE-WITH-VAR(w_0 ⇝ w, w' ⇝ w, var, V)
        V ← V ∪ {var}
    else
        GEN-TEMPLATE-NO-VAR(w_0 ⇝ w, w' ⇝ w, V)
    w ← w'
    L ← L \ {w}
GEN-LAST-TEMPLATE(w,V)
```

tions (the rest of its nodes are tag symbols). A configuration is a triple (u, q, d) such that u is a node of t, q is a mode and d is a sequence of data values. The initial state is a tree consisting of a single node $(r, start, \varepsilon)$, where ε denotes the empty sequence.

Let us assume that current state contains a configuration (u, q, d) as one of its leaf nodes. A step consists in transforming the current tree into a new tree by applying a (q, σ) rule such that either σ equals the label of u or σ = '*'. The intuition behind the application of a constructing rule is as follows. First, variables y_i and conditions c_j are evaluated. Let us assume that, as result of evaluating tests c_j-s, forest z' with leaves of the form $q'(p, z)$, is returned. Second, pattern p is applied to node u of t yielding a sequence of nodes $u_1, \ldots, u_l$ in document order. Third, a new forest f is computed by substituting the variables in z with their values and leaves $q'(p,z)$ of z with sequences of configurations $(u_1, q', d'), \ldots, (u_l, q', d')$. Here, d' are the new values assigned to variables x_i and y_j. Fourth, next state is computed by substituting configuration (u, q, d) of the current state with f. The computation stops when the current state is a labeled ordered tree (i.e. it does not contain any configurations as leaves). The result of a computation is the tree representing its final state.

Mapping L-Wrappers to XSLT$_0$

Let $W = \langle V, A, U, D, \mu, \lambda_a, \lambda_c \rangle$ be a single-pattern L-wrapper and let $L \subseteq V$ be the set of leaves of its pattern graph. Recall that we assumed that all extraction vertices are in L, that is $U \subseteq L$. Note that if u and v are vertices of the pattern graph then $u \rightsquigarrow v$ denotes the path from vertex u to vertex v in this graph. For example, referring to Figure 5a, $F \rightsquigarrow H = F, G, H$.

Let $L = \{w_1, \ldots, w_n\}$ be the leaves and let w be the root of the pattern graph. The idea of the translation algorithm is as follows. We start from root w and move down in the graph to w_1, that is, $w_1 \rightsquigarrow w$. Then, we move from w_1 to w_2 via their closest common ancestor w_1' that is, $w_1 \rightsquigarrow w_1'$ and $w_2 \rightsquigarrow w_1'$, …, and we move from w_{n-1} to w_n via their closest common ancestor w_{n-1}' that is, $w_{n-1} \rightsquigarrow w_{n-1}'$ and $w_n \rightsquigarrow w_{n-1}'$.

For example, referring to Figure 5a, we start from the root H and move downward to leaf F, that is, $F \rightsquigarrow H$. Then, we move from F to D via their closest common ancestor G, that is, $F \rightsquigarrow G$ and $D \rightsquigarrow G$.

Template rules are generated according to this traversal of the pattern graph. The first rule is generated according to path $w_1 \rightsquigarrow w$. The next $n-1$ rules are generated according to paths $w_i \rightsquigarrow w_i'$ and $w_{i+1} \rightsquigarrow w_i'$, $1 \leq i \leq n-1$. Finally, the last rule is generated for vertex w_n. Thus, a total of n1 rules are generated.

The resulting GEN-WRAPPER translation algorithm is shown in Figure 10. Note that function VAR-GEN generates a new variable name based on the attribute associated to an extraction vertex.

GEN-FIRST-TEMPLATE algorithm generates the first template rule. Let p_1 be an XPath pattern that accounts for the conditions on the path $w_1 \rightsquigarrow w$. Then, the template rule that is firstly generated has the following form:

template *start(/)*
return
  `result` ($sel_{w_1}(p_1)$)
end

GEN-TEMPLATE-WITH-VAR and GEN-TEMPLATE-NO-VAR algorithms generate a template rule for paths $w_i \rightsquigarrow w_i'$ and $w_{i+1} \rightsquigarrow w_i'$, $1 \leq i \leq n-1$, depending on whether vertex w_i is an extraction vertex or not. Note that a new variable is generated only if w_i is an extraction vertex, to store extracted data. Let p_{i+1} be the XPath pattern that accounts for the conditions on the paths $w_i \rightsquigarrow w_i'$ and $w_{i+1} \rightsquigarrow w_i'$.

The structure of a template rule for the case when a new variable *var* is generated is shown below:

template sel_{w_i}('*', V)
vardef
  var := *content*(.)
return
  $sel_{w_{i+1}}(p_{i+1}, V \cup \{var\})$
end

The structure of a template rule for the case when no new variable is generated is shown below:

template sel_{w_i}('*', V)
return
  $sel_{w_{i+1}}(p_{i+1}, V)$
end

GEN-LAST-TEMPLATE algorithm generates the last template rule. The constructing part of this rule fully instantiates the returned tree fragment, thus stopping the transformation process of the input document tree. Depending on whether vertex w_n is an extraction vertex or not, this template rule generates or does not generate a new variable. We assume that the set D of attribute names is $\{d_1, \ldots, d_n\}$. Note that because a new ariable is generated for each extraction vertex, it follows that the number of generated variables is n.

If a new variable *var* is generated, then the last generated rule has the following form:

template sel_{w_n}('*', V)
vardef
  var := *content*(.)
return
  `tuple` (($d_1(var_1), \ldots d_n(var_n)$))
end

Here $V \cup \{var\} = \{var_1, \ldots, var_n\}$.

If no new variable is generated, then the last generated rule has the following form:

```
template sel_wn('*', V)
return
    tuple((d1(var1), . . . dn(varn))
end
```

Here $V = \{var_1, \ldots, var_n\}$.

Example L-Wrappers in XSLT

Let us first consider an example for the flat relational case. Applying algorithm GEN-WRAPPER to the L-wrapper shown in Figure 5a, we obtain an $XSLT_0$ program comprising three rules:

```
template start(/)
return
    result((selx(p1))
end
template selx('*')
vardef
    vx := content(.)
return
    sely(p2, vx)
end
template sely('*', vx)
vardef
    vy := content(.)
return
    tuple(x(vx),y(vy))
end
```

XPath pattern p_1 = `//b/c/preceding-sibling::*[1]` is determined by tracing the path $H \rightsquigarrow G \rightsquigarrow F$ in the pattern graph (see Figure 5a). XPath pattern p_2 = `following-sibling::*[1]/*` is determined by tracing the path $F \rightsquigarrow G \rightsquigarrow D$ in the pattern graph (see Figure 5a).

The XSLT code for this wrapper is shown in Box 7.

Applying this XSLT transformation to the document shown Box 8.

Let us now consider the hierarchical example shown in Figure 3. The HTML code corresponding to the view shown Figure 3a is shown in Box 9.

An XSLT implementation for an HL-wrapper can be obtained by combining the idea of the hierarchical Prolog implementation with the translation of L-wrappers to XSLT outlined in the previous section.

A single-pattern L-wrapper for which the pattern graph has n leaves, can be mapped to an $XSLT_0$ stylesheet consisting on n1 constructing rules. In our example, applying this technique to each of the wrappers W_1 and W_2 (devised for the hierarchical source from Figure 3) we get three rules for W_1 (start rule, rule for selecting *fruit name* and rule for selecting *features*) and three rules for W_2 (start rule, rule for selecting *feature name* and rule for selecting *feature value*). Note that in addition to this separate translation of W_1 and W_2, we need to assure that W_2 selects feature names and feature values from the document fragment corresponding to a given fruit, that is, the document fragment corresponding to the features attribute of wrapper W_1. This effect can be achieved by plugging in the body of the start rule corresponding to wrapper W_2 into the body of the rule for selecting features, in-between tags <features> and </features> (see example below). Actually, this operation corresponds to realizing a join of the wrappers W_1 and W_2 on the attribute *features* (assuming L-wrappers are extended with an argument representing the root of the document fragment to which they are applied.

The resulting HL-wrapper expressed in XSLT is shown in Box 10.

Note that this output faithfully corresponds to the hierarchical schema shown in Figure 3b.

For wrapper execution, we can use any of the available XSLT transformation engines. In our experiments we have used Oxygen XML editor (Oxygen XML Editor, 2007), a tool that incorporates some of these engines (see Figure 11).

Box 7.

```
<?xml version="1.0" encoding="UTF-8"?>
<xsl:stylesheet xmlns:xsl="http://www.w3.org/1999/XSL/Transform" version="1.0">
  <xsl:template match="/">
    <result>
      <xsl:apply-templates mode="selx"
      select="//b/c/preceding-sibling::*[1][local-name()='t']"/>
    </result>
  </xsl:template>
  <xsl:template match="*" mode="selx">
    <xsl:variable name="var _ x">
      <xsl:value-of select="normalize-space(.)"/>
    </xsl:variable>
    <xsl:apply-templates mode="display" select="following-sibling::*[1]/*">
      <xsl:with-param name="var _ x" select="$var _ x"/>
    </xsl:apply-templates>
  </xsl:template>
  <xsl:template match="*" mode="display">
    <xsl:param name="var _ x"/>
    <xsl:variable name="var _ y">
      <xsl:value-of select="normalize-space(.)"/>
    </xsl:variable>
    <tuple>
      <x>
        <xsl:value-of select="$var _ x"/>
      </x>
      <y>
        <xsl:value-of select="$var _ y"/>
      </y>
    </tuple>
  </xsl:template>
</xsl:stylesheet>
```

FUTURE TRENDS AND CONCLUSION

In this chapter, we discussed a new class of wrappers for data extraction from semistructured sources inspired by logic programming, or L-wrappers. We described an inductive learning algorithm to generate extraction paths and also showed how to map the resulting extraction rules to XSLT stylesheets for efficient data extraction from Web sources. Our discussion covered two cases: i) extraction of relational tuples from flat relational Web data sources, and ii) extraction of hierarchical data from hierarchical Web data

Box 8.

```
<?xml version="1.0" encoding="UTF-8"?>
<a>
  <b>
    <t>x1</t>
    <c>
      <t>y1</t>
    </c>
  </b>
  <b>
    <t>x2</t>
    <d>
      <t>y2</t>
    </d>
  </b>
  <b>
    <t>x3</t>
    <c>
      <t>y3</t>
    </c>
  </b>
  <b>
    <t>x4</t>
    <c>
      <t>y4</t>
    </c>
  </b>
</a>
```

produces the following extracted data as output.

```
<?xml version="1.0" encoding="utf-8"?>
<result>
 <tuple>
   <x>x1</x>
   <y>y1</y>
 </tuple>
 <tuple>
   <x>x3</x>
   <y>y3</y>
 </tuple>
 <tuple>
   <x>x4</x>
   <y>y4</y>
 </tuple>
</result>
```

sources. These ideas are currently being implemented in a tool for information extraction. As future work, we plan to finalize and evaluate the implementation and also to give a formal proof of the correctness of the mapping of L-wrappers to XSLT.

Currently, our approach is semi-automated rather than fully automated. There are two tasks that must be performed manually by the user: i) definition of the schema of the Web data source, either flat relational or hierarchical; and ii) extraction of a few examples. We plan to address these issues in our future research work.

REFERENCES

Alferes, J. J., Damásio, C. V., & Pereira, L. M. (2003). Semantic Web logic programming tools. In F. Bry, N. Henze, & J. Maluszynski (Eds.), *Principles and practice of Semantic Web reasoning* (pp. 16-32). LNCS 2901. Springer-Verlag.

Anton, T. (2005). XPath-Wrapper Induction by generalizing tree traversal patterns. In M. Bauer, B. Brandherm, J. Fürnkranz, G. Grieser, A. Hotho, A. Jedlitschka, A. Kröner (Eds.), L-wrappers: Concepts, properties and construction. A declarative approach to data extraction from Web sources. *Soft Computing—A Fusion of Foundations, Methodologies and Applications*, *11*(8), 753-772.

Baumgartner, R., Flesca, S., & Gottlob, G. (2001). The Elog Web extraction language. In R. Nieuwenhuis, & A. Voronkov (Eds.), *Logic for programming, artificial intelligence, and reasoning* (pp.

Box 9.

```
<?xml version="1.0" encoding="UTF-8"?>
<html>
 <head>
   <title>Fruits.</title>
 </head>
 <body>
   <p>
    <h2>
      <th>Red apple</th>
    </h2>
    <table border="2">
      <tbody>
       <tr>
         <td>weight</td>
         <td>120</td>
       </tr>
       <tr>
         <td>color</td>
         <td>red</td>
       </tr>
       <tr>
         <td>diameter</td>
         <td>8</td>
       </tr>
      </tbody>
      <tbody>
       <tr>
         <td colspan="2">Limited stock, order today !</td>
       </tr>
      </tbody>
    </table>
   </p>
   <p>
    <h2>
      <th>Lemon</th>
    </h2>
    <table border="2">
      <tbody>
       <tr>
         <td>weight</td>
         <td>70</td>
```

continued on following page

Box 9. continued

```
        </tr>
        <tr>
          <td>color</td>
          <td>yellow</td>
        </tr>
        <tr>
          <td>height</td>
          <td>7</td>
        </tr>
        <tr>
          <td>width</td>
          <td>4</td>
        </tr>
      </tbody>
      <tbody>
        <tr>
          <td colspan="2">Limited stock, order today !</td>
        </tr>
      </tbody>
    </table>
   </p>
 </body>
</html>.
```

Figure 11. Wrapper execution inside Oxygen XML editor

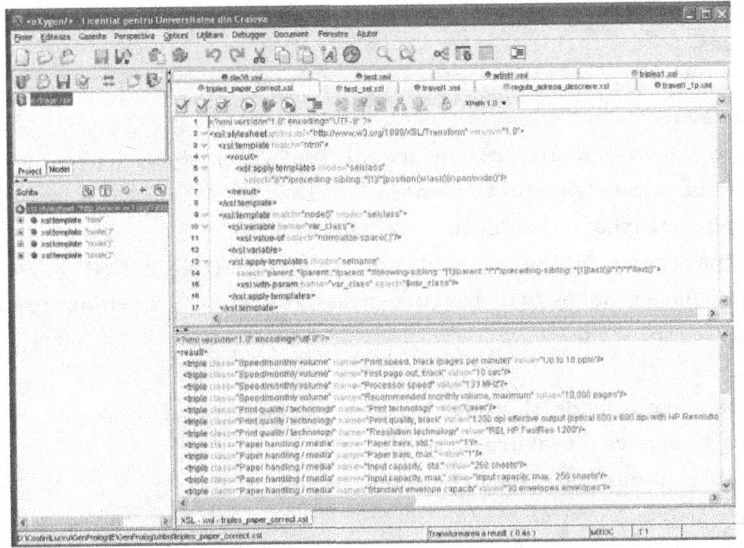

Box 10.

```
<?xml version="1.0" encoding="UTF-8"?>
<xsl:stylesheet xmlns:xsl="http://www.w3.org/1999/XSL/Transform" version="1.0">
  <xsl:template match="html">
    <fruits>
      <xsl:apply-templates select="//p/*/preceding-sibling::*[1]/*/text()"
        mode="select-fruit-name"/>
    </fruits>
  </xsl:template>
  <xsl:template match="node()" mode="select-fruit-name">
    <xsl:variable name="var-fruit-name" select="."/>
    <xsl:apply-templates mode="select-features"
      select="parent::*/parent::*/following-sibling::*[position()=1]">
      <xsl:with-param name="var-fruit-name" select="$var-fruit-name"/>
    </xsl:apply-templates>
  </xsl:template>
  <xsl:template match="node()" mode="select-features">
    <xsl:param name="var-fruit-name"/>
    <xsl:variable name="var-features" select="."/>
    <fruit>
      <name>
        <xsl:value-of select="normalize-space($var-fruit-name)"/>
      </name>
      <features>
        <xsl:apply-templates
          select="$var-features//tr/*/preceding-sibling::*[1]/text()"
          mode="select-feature-name">
        </xsl:apply-templates>
      </features>
    </fruit>
  </xsl:template>
  <xsl:template match="node()" mode="select-feature-name">
    <xsl:variable name="var-feature-name" select="."/>
    <xsl:apply-templates mode="select-feature-value"
      select="parent::*/following-sibling::*[position()=1]/text()">
      <xsl:with-param name="var-feature-name" select="$var-feature-name"/>
    </xsl:apply-templates>
  </xsl:template>
  <xsl:template match="node()" mode="select-feature-value">
    <xsl:param name="var-feature-name"/>
    <xsl:variable name="var-feature-value" select="."/>
    <feature>
```

continued on following page

Box 10. continued

```
      <name>
        <xsl:value-of select="normalize-space($var-feature-name)"/>
      </name>
      <value>
        <xsl:value-of select="normalize-space($var-feature-value)"/>
      </value>
    </feature>
  </xsl:template>
</xsl:stylesheet>
```

and it produces the following output when applied to the fruits example.

```
<?xml version="1.0" encoding="utf-8"?>
<fruits>
 <fruit>
   <name>Red apple</name>
   <features>
    <feature>
      <name>weight</name>
      <value>120</value>
    </feature>
    <feature>
      <name>color</name>
      <value>red</value>
    </feature>
    <feature>
      <name>diameter</name>
      <value>8</value>
    </feature>
   </features>
 </fruit>
 <fruit>
   <name>Lemon</name>
   <features>
    <feature>
      <name>weight</name>
      <value>70</value>
    </feature>
    <feature>
      <name>color</name>
      <value>yellow</value>
    </feature>
```

continued on following page

Box 10. continued

```
    <feature>
      <name>height</name>
      <value>7</value>
    </feature>
    <feature>
      <name>width</name>
      <value>4</value>
    </feature>
   </features>
 </fruit>
</fruits>
```

548-560). LNAI 2250. Springer-Verlag.

Bădică, C., Bădică, A., Popescu, E., & Abraham, A. (2007). L-wrappers: Concepts, properties and construction. A declarative approach to data extraction from web sources. *Soft Computing—A Fusion of Foundations, Methodologies and Applications, 11*(8), 753-772.

Bex, G.J., Maneth, S., & Neven, F. (2002). A formal model for an expressive fragment of XSLT. *Information Systems, 27*(1), 21-39.

Chang, C.-H., Kayed, M., Girgis, M.R., & Shaalan, K. (2006). A survey of Web information extraction systems. *IEEE Transactions on Knowledge and Data Engineering, 18*(10), 1411-1428.

Chidlovskii, B. (2003). Information extraction from tree documents by learning subtree delimiters. In *Proceedings of IJCAI-03: Workshop on Information Integration on the Web (IIWeb-03),* (pp. 3-8).

Cormen, T.H., Leiserson, C.E., & Rivest, R.R. (1990). *Introduction to algorithms.* MIT Press.

Freitag, D. (1998). Information extraction from HTML: Application of a general machine learning approach. In *Proceedings of AAAI'98,* (pp. 517-523).

Gottlob, G. (2005). Web data extraction for business intelligence: The Lixto approach. In G. Vossen, F. Leymann, P.C. Lockemann, & W. Stucky (Eds.), *Datenbanksysteme in business, technologie und Web, 11. Fachtagung des GI-Fachbereichs "Datenbanken und Informations-systeme" (DBIS)* (pp. 30-47). Lecture Notes in Informatics 65, GI.

Gottlob, G., & Koch, C. (2004). Monadic datalog and the expressive power of languages for Web information extraction. *Journal of the ACM, 51*(1), 74-113.

Gottlob, G., Koch, C., & Schulz, K.U. (2004). Conjunctive queries over trees. In *Proceedings of PODS' 2004,* (pp. 189-200). ACM Press.

Kushmerick, N. (2000). Wrapper induction: Efficiency and expressiveness. *Artificial Intelligence, 118*(1-2), 15-68.

Laender, A.H.F., Ribeiro-Neto, B., & Silva, A.S. (2002). DEByE—data extraction by example. *Data & Knowledge Engineering, 40*(2), 121-154.

Laudon, K.C., & Traver, C.G. (2004). *E-commerce, business, technology, society* (2nd ed.). Pearson Addison-Wesley.

Lenhert, W., & Sundheim, B. (1991). A performance evaluation of text-analysis technologies. *AI Magazine*, *12*(3), 81-94.

Muslea, I., Minton, S., & Knoblock, C. (2001). Hierarchical wrapper induction for semistructured information sources. *Journal of Autonomous Agents and Multi-Agent Systems*, *4*(1-2), 93-114.

Oxygen XML Editor. (2007). Retrieved April 2, 2008, from http://www.oxygenxml.com/

Staab, S., Werthner, H., Ricci, F., Zipf, A., Gretzel, U., Fesenmaier, D.R., et al. (2002). Intelligent systems for tourism. *IEEE Intelligent Systems*, *6*(17), 53-64.

Sterling, L., & Shapiro, E. (1994). *The art of Prolog* (2nd ed.). The MIT Press.

Thomas, B. (2000). Token-templates and logic programs for intelligent Web search. *Intelligent Information Systems, Special Issue: Methodologies for Intelligent Information Systems*, *14*(2/3), 241-261.

Web Services Activity. (2007). Retrieved April 2, 2008, from http://www.w3.org/2002/ws/

W3C Extensible Markup Language (XML). (2007). Retrieved April 2, 2008, from http://www.w3.org/XML/

W3C Extensible Stylesheet Language Family (XSL). (2007). Retrieved April 2, 2008, from http://www.w3.org/Style/XSL/

W3C Semantic Web Activity. (2007). Retrieved April 2, 2008, from http://www.w3.org/2001/sw/

W3C HTML. (2007). Retrieved April 2, 2008, from http://www.w3.org/html/

Xiao, L., Wissmann, D., Brown, M., & Jablonski, S. (2001). Information extraction from HTML: Combining XML and standard techniques for IE from the Web. In L. Monostori, J. Vancza, & M. Ali (Eds.), *Engineering of intelligent systems: Proceedings of the 14th International Conference on Industrial and Engineering Applications of Artificial Intelligence and Expert Systems, IEA/AIE 2001*, (pp. 165-174). Lecture Notes in Artificial Intelligence 2070, Springer-Verlag.

Chapter III
A Formal Analysis of Virtual Enterprise Creation and Operation

Andreas Jacobsson
Blekinge Institute of Technology, Sweden

Paul Davidsson
Blekinge Institute of Technology, Sweden

ABSTRACT

This chapter introduces a formal model of virtual enterprises, as well as an analysis of their creation and operation. It is argued that virtual enterprises offer a promising approach to promote both innovations and collaboration between companies. A framework of integrated ICT-tools, called Plug and Play Business, which support innovators in turning their ideas into businesses by dynamically forming virtual enterprises, is also formally specified. Furthermore, issues regarding the implementation of this framework are discussed and some useful technologies are identified.

INTRODUCTION

Innovations are important to create both private and social values, including economic growth and employment opportunities. From an innovator's perspective, there are some common obstacles for realizing the potential of innovations such as shortage of time to spend on commercialization activities, lack of business knowledge, underdeveloped business network and limited financial resources (Tidd, Bessant, & Pavitt, 2005). Thus, the innovator requires support to develop the innovation into business, something often seen as the specific role of the entrepreneur, which is to search, discover, evaluate opportunities and marshal the financial resources necessary, among other things (Leibenstein, 1968). Playing the role of an innovator or entrepreneur in the networked economy requires a global outlook. New trade and production patterns, as well as the emergence of new markets point toward a more efficient use of global resources. Information and communications technology (ICT) already plays an important role as a facilitator in this development. We believe that better economic growth can be achieved when the innovator and the entrepreneur can compete and collaborate in order to solve problems on a global market place.

In realizing innovations, small and medium sized enterprises (SME) are particularly important. In response to fast changing market conditions, most enterprises and especially the SMEs need ICT-infrastructures that consider their size as well as high specialization and flexibility. While allowing them to maintain their business independence, ICT-based innovation support should help SMEs reach new markets and expand their businesses (Cardoso & Oliveira, 2005). On this topic, one promising approach for SMEs is to participate in computer-supported collaborative networks that will act as breeding environments for the formation of dynamic virtual enterprises (cf. Ecolead, 2008).

The vision of Virtual Enterprises, or more generally Collaborative Networks, is constituted by a variety of entities (e.g., organizations and people) that are largely autonomous, geographically distributed and heterogeneous in terms of their operating environment, culture, social capital and goals (Camarinha-Matos & Afsarmanesh, 2006). The idea of participating in highly dynamic coalitions of enterprises that are formed according to the needs and opportunities of the market, as well as remaining operational as long as these opportunities persist, put forward a number of potential benefits, some of which are related to agility, innovation management, resource optimization and the adoption of complementary roles based on core competencies (Camarinha-Matos & Afsarmanesh, 2003).

The lack of appropriate theoretic definitions and formal models has been argued to be one of the main weaknesses in the area of collaborative networks and virtual enterprises (Camarinha-Matos & Afsarmanesh, 2003, 2005, 2006). In fact, D'Atri and Motro (2007, p. 21) point out that "while the essential principles of virtual enterprises are mostly agreed upon, a formal model of virtual enterprises has been curiously missing." In this article, we provide a formal model of virtual enterprises and their most crucial tasks. We also propose a formal framework of integrated ICT-support intended to enable secure and agile virtual enterprise creation and operation. We call this framework Plug and Play Business (Davidsson et al., 2006).

In the next section, we provide a formal description of the concept of virtual enterprise followed by a formal analysis of Plug and Play Business. Next, we discuss the usefulness of emerging technology trends relevant for the implementation of Plug and Play Business software. In the end, we present some conclusions and ideas for future work.

PROBLEM DESCRIPTION

The concept of virtual enterprise (see Figure 1) has been applied to many forms of collaborative business relations, like outsourcing, supply chains or temporary consortiums. As is emphasized by Camarinha-Matos and Afsarmanesh (2005), it is important for the companies in a virtual enterprise to share data and information, and to communicate with each other efficiently and securely. A virtual enterprise is typically defined as "a temporary alliance of enterprises that come together to share skills or core competencies and resources in order to better respond to business opportunities and whose cooperation is supported by computer networks" (Camarinha-Matos & Afsarmanesh, 2005, p. 440). We can describe a virtual enterprise a as a tuple:

$ve = \langle A, R, AR, CI, S, G \rangle$

where

- $A = \{a_1,...,a_n\}$ is the set of actors (typically enterprises) in *ve*. An actor can be described as a tuple:

  $a_i = \langle I_i, T_i, C_i, G_i \rangle$

  Where I_i are the relevant information systems needed in *vei*, T_i is the set of resources of the actor, C_i is the set of core competencies of the actor and G_i is the set of individual goals of the actor.
- $R_i = \{r_1,...,r_m\}$ is the set of roles that the actors can play in the *vei*. Each actor in the virtual enterprise can play one or more roles, for example, innovator or supplier/provider of for example, goods, services, expertise, and so forth. The choice of role depends on the virtual enterprise goal(s), the actor's core competencies, resources and individual business goals.
- AR is a set of triples $\langle a_k, r_j, O_j^k \rangle$ where $a_k \in A$ and $r_j \in R$ that is, the actors and their roles in the virtual enterprise and the set of obligations, O_j^k, that is associated with the actor's role in the virtual enterprise.
- *CI* is a set of communication infrastructures needed for operating the virtual enterprise.
- *S* is a set of states of affairs that hold at each time in *ve*.
- *G* is a set of goals of the virtual enterprise that is derived from the business opportunities that motivate the initiation of the virtual enterprise.

According to literature (cf. Camarinha-Matos & Afsarmanesh, 2006), there are two critical stages in the lifecycle of collaborative networks and virtual enterprises when transforming the enterprise from a business opportunity to a successful business collaboration. These stages are the creation and the operation phases. For reasons of completeness, we add the phase of virtual enterprise definition, in which the preconditions for the creation phase are specified.

Figure 1. An example of a virtual enterprise including a communication infrastructure connecting three actors (where each actor is a company that consists of the information resource (i) and the spe-cial skill or competence required in the cooperation) collaborating in the roles of transporter, retailer, and producer

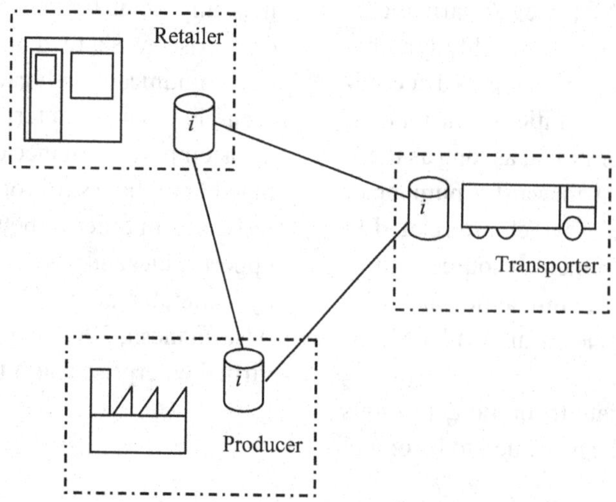

Virtual Enterprise Definition

In the definition process, the business opportunity is described in terms of roles and goals of the virtual enterprise. This process emanates from the detection of a business opportunity and results in a set of goals, G and roles, R, which are necessary for the fulfillment of the virtual enterprise.

Virtual Enterprise Creation

During the creation process, the virtual enterprise is formed. Given the set of goals and roles specified in the definition phase, the virtual enterprise initiator determines the set of actors, A, maps the actors to the roles, AR, and selects the communication infrastructures, CI, to be used in the virtual enterprise. The creation process is thus initiated when a set of goals and roles for the virtual enterprise has been specified and it is terminated when an agreement concerning the actors, their roles and the communication infrastructures in the virtual enterprise has been reached.

Virtual Enterprise Operation

When an agreement concerning the roles and obligations in the virtual enterprise has been reached, the operation phase can be initiated. We regard operation as a process that, given a set of actors, their roles and a set of communication infrastructures, fulfills the goals, G, of the virtual enterprise. Operation is initiated when the communication infrastructures are in place to support the actors in their roles to reach the agreed goals and it is terminated when the goals of the virtual enterprise are fulfilled. Note that virtual enterprise operation may include both multilateral and resource-sharing collaboration.

On Requirements of ICT-Tools that Support Virtual Enterprises

It is clear that the vision of virtual enterprises can be realized with the help of ICT. We will therefore specify a set of quality attributes (i.e., nonfunctional requirements) for such a framework of ICT-tools. Based on interviews with SMEs and on previous work by, for example, Camarinha-Matos and Afsarmanesh (2003, 2005), we believe that the following quality attributes are important:

- **Scalability.** Some virtual enterprises may be large in that the number of involved companies can be large, whereas some virtual enterprises may be small in that the number of involved companies can be small. Hence, ICT-solutions must be scalable to the shifting number of enterprises within the virtual enterprises.
- **Flexibility.** Being adaptable or variable is important due to the heterogeneity of companies (especially given that SMEs belong to the target group of virtual enterprises), relationships and actors in a virtual enterprise. Hence, software must be flexible to the varying needs of the intended virtual enterprise organizations.
- **Performance.** Although there may not be many hard real-time requirements for such a software to meet, response times and other delays must be kept on reasonable levels.
- **Cost.** High costs associated with joining and participating in a collaboration alliance is considered an obstacle for any growing network (Shapiro & Varian, 1999). Some of the envisioned benefits of virtual enterprises are low preparation and transactions costs as well as decreased time to market.
- **Usability.** A user-friendly interface is crucial in order to get interaction from the humans involved in the chain of collaboration.
- **Security.** The prevention and detection of unauthorized actions is a key feature if trust

is to be established among the parties in a virtual enterprise.

PLUG AND PLAY BUSINESS

The concept of Plug and Play Business (Davidsson et al., 2006; Jacobsson & Davidsson, 2006) relies on an integrated set of ICT-tools that support innovators in turning their ideas into businesses by forming virtual enterprises for interorganizational and interoperable collaboration. We envision Plug and Play Business as a software framework that helps companies, SMEs in particular, in realizing innovations and thus developing their business potential.

After having deployed the Plug and Play Business software, companies are connected to a networked community where all participants share one common goal; namely to increase business. In that way, the purpose of Plug and Play Business is to stimulate the realization of innovations without interfering with the individual goals of the Plug and Play Business companies. Together with the autonomy, heterogeneity and possibly conflicting goals of the involved parties of a Plug and Play Business community, this requires ICT-solutions that are able to handle dynamically evolving and distributed business partnerships and processes that cross the borders of various enterprises. Thus, the interoperability between the information systems of the involved enterprises belongs to the technological core of the concept of Plug and Play Business.

In addition to the concept of virtual enterprises, another important concept for implementing Plug and Play Business is Internet communities. Enterprises dynamically join a Plug and Play Business community by installing and running the Plug and Play Business software and by describing and validating the resources of the enterprise, for example, production capacity, distribution network, intellectual capital, and so forth. The community is dynamic in the sense that enterprises may (in principle) join and leave the community at any time. To enhance security, a gate-keeper facility that regulates the entering and leaving of the community is included in the community. Formally, a Plug and Play Business community, p, can be described as a tuple:

$$p = \langle A, R, VE, S, l, CI, gk \rangle$$

where

- $A = \{a_1, ..., a_n\}$ is the set of actors (typically enterprises) in the community. An actor in the Plug and Play Business community can be described as a tuple:

  $$a_i = \langle I_i, T_i, C_i, G_i, h_i, b_i \rangle$$

  Compared to the definition of actors in virtual enterprises, we add h_i, which is the person representing the actor/enterprise and b_i, which is the Plug and Play Business client software (an intelligent agent supporting the (agent) communication language, l_i) acting on behalf of the actor/enterprise.
- $R = \{r_1, ..., r_m\}$ is the set of roles that the actors can play,
- $VE = \{ve, ..., vel\}$ is the set of virtual enterprises currently active in the community,
- S is a set of states of affairs that hold at any time in p,
- l is the agent communication language used by the agents B. We will assume that l includes a set of relevant interaction protocols, a set of relevant ontologies and possibly other things necessary to perform useful communication,
- CI is a set of communication infrastructures needed for operating the community, and
- gk is the gate-keeper facility that regulates the entering (and leaving) of actors to (and from) the community. In order to become a member of p there is a set of criteria that must be fulfilled, for example, corporate identi-

fication number must be declared, the roles the actor is willing to play should be stated and information systems must be specified. Thus, some of the aims of the gate-keeper are to ensure that this type of information is available to the Plug and Play Business community and to verify the identity of the actors. Possibly, the gate-keeper may also be equipped with capabilities of handling different levels of memberships with different sets of norms in order to cope with the varying needs of potential participants and members. The gate-keeper could also inform the potential member about general rules that hold in the community and require the potential member to comply with them.

Note that all these entities change dynamically, but with different frequency. New virtual enterprises may be formed (and dissolved) relatively frequently, actors enter and leave the community every now and then, and new roles may be added although this is not expected to happen often. In Figure 2, we illustrate an example of a Plug and Play Business community.

In Plug and Play Business, the interactions between participants in the community as well as in the virtual enterprises are role-based. Each actor plays one or more roles, for example, innovator, raw material producer, transporter, product designer, logistics provider, marketer, financier, retailer, and so forth. The choice of role depends on the company's core competencies and business intentions. An important role in the life cycle of businesses is the entrepreneur and we make a distinction between this role and that of the innovator. One of the main purposes of Plug and Play Business is to automate as much of the entrepreneurial role as possible, thus increasing the probability of turning an innovative idea into a business.

Figure 2. An example of a Plug and Play Business community (p) where i is the relevant information systems needed in the virtual enterprise (ve1), T is the resources of the actor/enterprise, C is the core competencies of the actor/enterprise, G is the goals of the actor/enterprise, gk is the gate-keeper, h is the person representing the actor/enterprise, and b is the Plug and Play Business client. The figure also contains three examples of roles played by the actors (a)

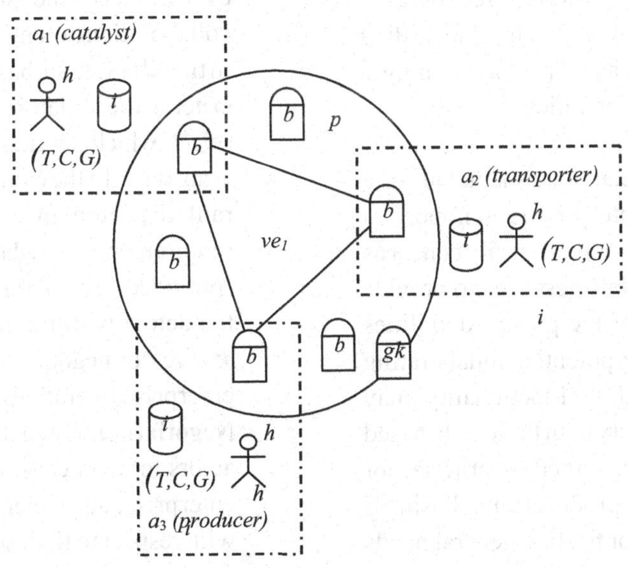

Virtual Enterprise Definition

In the virtual enterprise definition phase, a member of the Plug and Play Business community, typically an innovator, may at any time initiate an attempt to form a collaborative coalition between the members. This process may be viewed analogous to crystallization, where a catalyst (innovator) initiates a process resulting in a precise form of collaboration, that is, the formation of a virtual enterprise. The main role of the entrepreneur, which to a large extent is automated by the Plug and Play Business software, is to drive this process. It may be a more or less elaborate process starting with just a seed of an innovative idea without any predefined business structure, or it may be a full-fledged business idea with well-defined needs to be met by potential collaborators. In this phase, the catalyst, Ω, where $\Omega \in A$, describes the business opportunity in terms of goals, G and roles, R, of the virtual enterprise. Because this is a highly complex task, h_Ω will be the main contributor, whereas b_Ω primarily will provide structural support.

Virtual Enterprise Creation

The virtual enterprise creation phase (where crystallization takes place) consists of three subtasks and is initiated by a catalyst. We have identified the following functions as helpful in forming a successful collaborative coalition.

- **Finding.** To find candidates suitable for a potential virtual enterprise is an important function. This function primarily concerns the catalyst of the business idea to provide the requirements of the preferred abilities of the roles for the potential collaborating partners. The finding functionality may include the possibility both for search, based on specific needs specified by criteria, for example, type of products and business model, as well as for posting general needs or ideas that other members may suggest solutions or resources for. Further, Plug and Play Business software should provide the feature of suggesting actors for collaboration based on, for example, content-based recommendation and collaborative recommendations. The function of finding requires that Ω has a list of the roles that must be filled in order to get an operating virtual enterprise. This list is provided by h_Ω, that is, the person representing Ω in the definition phase. Then, for each of the roles, the task for b_Ω is to find the set of candidate actors K where $K \subset A$ that are able to play the role.
- **Selection/Evaluation.** When a set of potential collaborators has been found they need to be evaluated. This requires support for using track records and potentially support for certification schemes of, for instance, the trustworthiness of the actors. Further, decision support for evaluating trade-offs between a number of characteristics are needed, for example, trade-offs between cost of product/service, cost of transportation and time to delivery of product/service. Which actors to choose for the creation of a virtual enterprise should be based on the evaluation and the estimated future value of collaboration potential with the other actors in the alliance. So, based on some evaluation criteria, the initiator of the virtual enterprise selects which candidate actors to start negotiating with. In the evaluation task, Ω should rank the actors in K according to a set of requirements Q_r where $Q_r = \{q_1, q_2, \ldots q_k\}$ (provided by h_Ω). Based on this, Ω selects the actors with the highest rank k where $k \in K$ for negotiating on terms for virtual enterprise operation.
- **Negotiation.** When the catalyst has selected actors for the necessary roles of the virtual enterprise, agreements between the actors with respect to their roles, their obligations,

the communication infrastructure and the goals of the virtual enterprise need to be settled. The Plug and Play Business software should provide support for different types of contracts of agreements including support for intellectual property rights. The goal of negotiation is to establish an agreement between Ω and k concerning k's set of obligations, O_k. These obligations should of course be consistent with the set of goals, G of the *ve* and the set of goals of G_k.

Virtual Enterprise Operation

When the creation phase is finished and a virtual enterprise is formed, the Plug and Play Business software should provide support also for the operation phase, that is, the collaboration between the parties of the virtual enterprise. This support may be on a quite shallow level, for example, transactions of information between actors. On a deeper level, the Plug and Play Business software should support and facilitate complex coordination and synchronization of activities. A wide range of information types needs to be transferred in an efficient way in order to reduce the administrational costs of the actors as well as reducing the risk of inaccuracy in information. The management of the virtual enterprise requires support for controlling the flow of activities between the involved actors. It concerns activities with potential long-term consequences (e.g., initiating product development) as well as regular business activities (e.g., decisions of production and distribution). With respect to the operation phase, Plug and Play Business software must support:

- **Information resource-sharing.** This is related to the content and purpose of the exchanged information with tasks ranging from administrative information exchange to complex operations planning. An example of a simple administrative task is ordering and invoicing, whereas a more complex task may concern making critical information available to the collaborating partners in order to improve operations by better and more efficient planning and scheduling, that is, resource optimization.
- **Multilateral collaboration.** The more parties involved in the collaboration, the more complex the solutions may be. The simplest case concerns cooperation between only two enterprises, whereas the general case involves a large number of enterprises collaborating with each other in different ways (many-to-many collaboration).

We separate between two levels of collaboration: administrational and operational. They are defined by the type of interaction protocols they support. Administrational collaboration includes only protocols using the "weaker" *performatives*, such as, *ask*, *tell*, *reply*, and so forth. Let us call this set of interaction protocols *IPW*. Operational collaboration supports protocols also using the *performatives* that actually manipulate the receiver's knowledge, such as, *insert*, where the sender requests the receiver to add the content of the message to its knowledge base, and *delete*, where the sender requests the receiver to delete the content of the message from its knowledge base. Let us call this set of interaction protocols *IPS*. Moreover, we make a distinction between bilateral and multilateral collaboration. Thus, we have four types of collaboration within the operation phase:

- Bilateral administrational collaboration between two actors a_i and a_j (where $a_i \in A$ and $a_j \in A$) in a virtual enterprise *ve* should support the use of a set of interaction protocols, IPW_{ij} where $IPW_{ij} \subset IPW$, between the two actors' information systems (i_i and i_j) and mediated by the actors' Plug and Play Business client software (b_i and b_j).
- Multilateral administrational collaboration between a set of actors A_u (where $A_u \subset A$

in a virtual enterprise *ve* should support the use of a set of interaction protocols, IPW_u where $IPW_u \subset IPW$, between all the actors' information systems and mediated by the actors' Plug and Play Business client software.

- Bilateral operational collaboration between two actors a_i and a_j (where $a_i \in A$ and $a_j \in A$) in a virtual enterprise *ve* should support the use of a set of interaction protocols, IPS_{ij} where $IPS_{ij} \subset IPS$, between the two actors' information systems (i_i and i_j) and mediated by the actors' Plug and Play Business client software (b_i and b_j).
- Multilateral operational collaboration between a set of actors A_u (where $A_u \subset A$ in a virtual enterprise *ve* should support the use of a set of interaction protocols, IPS_u where $IPS_u \subset IPS$ between all the actors' information systems and mediated by the actors' Plug and Play Business client software.

Implementation Issues

Enhancing security and trust has been identified as a key issue in making virtual enterprises reach their potential (Camarinha-Matos & Afsarmanesh, 2005, 2006). One example of improving security is the gate-keeper facility that uses identification and authentication mechanisms regulates the entering (and leaving) of enterprises and registers them as members of the community. The gate-keeper can also inform the potential members about general rules that hold in the community and require the potential member to comply with them before being allowed to enter the Plug and Play Business community. Moreover, there may be a need for a surveillance mechanism that monitors the behavior of members in the community. The purpose of such a mechanism is to block unauthorized users and to detect and cope with malicious behavior, thereby incorporating security management into the Plug and Play Business software. A more detailed study on security analysis is provided by Jacobsson and Davidsson (2007).

One of the intended key advantages with Plug and Play Business software is to lower the costs of collaboration (including, e.g., preparation, transaction and search costs), which is particularly important in order to be accepted by SMEs. Another intended key advantage is the "plug'n'play" aspect, that is, the user-friendly interface of the software, which must be carefully considered in the software design. Response times and other delays must be kept on a reasonable level, thereby addressing performance requirements. The choice of system architecture is closely related to the system's performance in terms of a number of the previously mentioned attributes. Compared to a centralized architecture, a distributed architecture supports many of the quality attributes, for example, flexibility, scalability and dynamicity. Also, the risk of single point of failure and traffic bottlenecks may be avoided increasing the robustness of the system.

A decentralized paradigm such as peer-to-peer (P2P) may be preferable for the Plug and Play Business software because no central authority determines how the participants interact or coordinates them in order to accomplish some task. A P2P infrastructure self-configures and nodes can coordinate autonomously in order to search for resources, find actors and interact together. The heterogeneity of enterprises and relationships between the enterprises thereby have the potential to be maintained. P2P is a paradigm that allows the building of dynamic overlay networks and it can be used in order to realize an environment that manages a dynamic network of business relations. Dealing with business sensitive assets (e.g., innovators' knowledge), searching and retrieval of contents, as well as discovery, composition and invocation of new services, should be made secure and trustable. The P2P infrastructure realizes an environment in which every organization can make its knowledge and services available to other organizations keeping control over them.

In a P2P infrastructure, each organization can autonomously manage this task without having to delegate it to an external central authority that could be perceived as less trusted than the organization itself and should be the object of an external (to the collaborating network) agreement between all the involved organizations.

Related Work

Camarinha-Matos and Afsarmanesh (2003, p. 2) state that "there is a need for flexible and generic infrastructures to support the full life cycle of virtual enterprises, namely the phases of creation, operation and dissolution." We believe that the Plug and Play Business has the potential of constituting such an infrastructure. Moreover, they provide further motivation to our work by emphasizing the need for research on generic, interoperable, pervasive, free (low cost) and invisible (user-friendly) infrastructures that include methods for the creation of business (e.g., negotiation, methodologies for transforming existing organizations into a virtual enterprise-ready format) and business collaboration (e.g., coordinated and dynamic resource sharing, administration and management of distributed activities and risk management).

In recent years, a rich literature on the topic of collaborative networks and virtual enterprises has emerged. The concept of Plug and Play Business is similar to the concept of Virtual Breeding Environments as described by Camarinha-Matos and Afsarmanesh (2003, 2005). A virtual breeding environment represents an association or a cluster of organizations and their related supporting institutions that have both the potential and the will to cooperate with each other through the establishment of a "base" long-term cooperation agreement and interoperable infrastructure. When a business opportunity is identified by one member (acting as a broker similar to the catalyst within Plug and Play Business communities), a subset of these organizations can be selected, thus forming a virtual enterprise. Plug and Play Business is different from virtual breeding environments in that dynamic and temporary alliances can be formed within the community whenever a business opportunity is detected. Thus, Plug and Play Business also supports short-term collaboration. Another distinction is that Plug and Play Business emphasizes the importance of promoting innovations by automating as much as possible of the entrepreneurial role in the virtual enterprise, thereby promoting economic growth and employment.

Plug and Play Business has some resemblance to the work described by Chituc and Azevedo (2005) in that dynamic collaboration processes for agile virtual enterprises are emphasized. However, their work excludes crucial aspects such as the dynamic creation of virtual enterprises and security management.

USEFUL TECHNOLOGIES

Technological support for the creation and operation phases of virtual enterprises is arising in many forms. Cardoso and Oliveira (2005, p. 1) state that "the most ambitious technologies intend to automate (part of) the process of creation and operation of virtual enterprises, mainly through multi-agent technology approaches, where each agent can represent each of the different enterprises." This is also the overall intention with Plug and Play Business software. Because agents are autonomous, can interact with other agents, and enable approaching distributed problems by means of negotiation and coordination capabilities, they are fit for the tasks within Plug and Play Business.

Based on the requirements and attributes mentioned previously, we hereby make a brief review of some relevant technologies (including multi-agent technologies) that are useful when developing Plug and Play Business software.

Finding, Evaluating and Selecting Potential Partners

The tasks of finding and evaluating (e.g., business partners) have been the object of a lot of research within the area of recommendation systems (cf. Adomavicius & Tuzhilin (2005)). Here, the main idea is to automate the process of "word-of-mouth" by which people recommend products or services to one another. Recommendation systems are usually classified based on how they are constructed into three categories:

- Content-based recommendation, which is based on previous interests of actors;
- Collaborative recommendation, which is based on preference of similar actors; and
- Hybrid recommendations, which is a combination of the two previous ones.

So far, recommendation systems have successfully been deployed primarily in consumer markets (see, for instance the collaborative filtering system at book dealer Amazon.com). As most existing recommendation systems are not developed for business-to-business applications, they generally exclude the negotiation process. Because recommendation systems are already deployed in large-scale consumer systems it can be assumed that they enable scalability, flexibility, usability and cost-efficiency. Thus, they may be a beneficial alternative to use when meeting the requirements of virtual enterprise creation; more specifically the finding and evaluation stages. Also, because they can take the history of a potential collaborator into account, they may also contribute to the enhancement of security and trust.

In the area of intelligent agents, middle agents or brokering agents have been used to locate other agents in an open environment like the Internet (Wiederhold, 1992; Wong & Sycara, 2000). Here, each agent in the community typically advertises its capabilities to some broker. These brokering agents may simply be match-makers or yellow page agents who match advertisements to requests for advertised capabilities. Brokering agent systems are able to cope quickly and robustly with a rapidly fluctuating agent population (Wooldridge, 2004), which indicates both a high level of flexibility, scalability, robustness and performance. This makes them appropriate to use in Plug and Play Business software.

Establishing an Agreement

There is a long tradition in the area of agent-based systems of studying how to reach agreements, for instance, using the Contract Net protocol (Smith, 1980) and computational auctions (Rosenschein & Zlotkin, 1994). Auctions are generally considered to be a useful technique for allocating resources to agents (Wooldridge, 2004), however, they are too simple for many settings as they are mainly concerned with the allocation of goods or resources. For more general settings, where agents must reach agreements on matters of mutual interest and including complex constraints, richer techniques for reaching agreements are required. Here, negotiation may be a promising alternative. Four different components are relevant for the Plug and Play Business setting:

- A negotiation set, which represents the space of possible obligations that agents can make;
- A protocol, which defines the legal obligations that the agents can make;
- A collection of strategies, one for each agent, which determines what obligations the enterprises will make; and
- A rule that determines when the negotiation is over and the deal has been closed.

Here, the concept of obligations is an important component in that it specifies the commitments that the members (or the agents acting on behalf of their owners) have against each other. Substantial work on obligations in normative multiagent systems has been done (cf. the work by Boella and van der Torre (2004) and López y López, Luck and d'Inverno (2006)). In the area of electronic contracts, which are to be regarded as virtual representations of traditional contracts, that is, "formalizations of the behavior of a group of agents that jointly agree on a specific business activity" (Cardoso & Oliveira, 2005, p. 6). Electronic contracts usually have a set of identified roles to be fulfilled by the parties involved in the relation. Three types of norms are specified within a contract structure, namely obligation, permission or prohibition. Plug and Play Business software primarily adopts the concept of obligations, that is, that an agent (the Plug and Play Business software) has an obligation toward another agent to bring about a certain state of affairs (the goal(s) of the virtual enterprise) before a certain deadline. However, what more types of norms (e.g., permissions and prohibitions) and norm-enhancing mechanisms (e.g., promoter and defender functionality) that should be included in the definition of Plug and Play Business remains to be determined.

Agent-based auctions, negotiation protocols and electronic contracts may be sound technologies to enable the establishment of agreements within Plug and Play Business since intelligent agents can be designed to cope with individual goals and conflicting behavior (which certainly may occur in the Plug and Play Business community).

Operation

Several examinations on current state of the art technologies useful for building ICT-infrastructures with the purpose of business collaboration within virtual enterprises have been undertaken (cf. Camarinha-Matos & Afsarmanesh (1999, 2003, 2005)). Some common conclusions are that multi-agent technology constitutes a promising contributor to the development of support infrastructures and services. Also, Internet and Web technologies, such as Web services, represent a fast growing sector with large potential in interenterprise collaboration support. However, further support for multilateral collaboration is necessary. A number of other emerging technologies, for example, service-oriented architectures, the semantic Web and countless collections of software standards (cf. the ebXML framework) are likely to provide important contributions.

It seems that Microsoft's BizTalk Server is the most sophisticated solution for interenterprise collaboration widely available. BizTalk is based upon a central server through which all exchanged information passes, it uses XML and supports the main protocols for e-mail and http. However, BizTalk supports multilateral collaboration only to some extent and it is not fit for interoperable information resource sharing. Being a centralized proprietary client-server solution, it has several disadvantages, such as making the actors dependent of third party, being expensive and having possible risks for communication bottlenecks, thereby failing to meet requirements such as scalability, flexibility, robustness, cost and security.

Another possibility is to use computational auctions (Rosenschein & Zlotkin, 1994; Yamamoto, 2004). They can be used within the collaboration task as a method for dynamically solving resource allocation within the virtual enterprise. Possibly, auctions can also be deployed within multilateral administrational collaboration when allocating work tasks between partners of a virtual enterprise. However, as stated by Camarinha-Matos and Afsarmanesh (2005, p. 447) "publicly funded research should avoid approaches that are too biased by existing technologies."

We believe that there are some technologies that may be useful for the collaboration task within Plug and Play Business software. One promising alternative for multilateral collaboration is the use of decentralized intelligent agents. In previous work (Davidsson, Ramstedt, & Tornquist, 2005), we have described a general wrapper agent solution based on open source freeware that makes it possible (in principle) for any business system to exchange (administrational) information with any other business system. In Carlsson, Davidsson, Jacobson, Johansson, and Persson (2005), we suggest further improvements to the wrapper agent technology by addressing security issues as well as an extended, possibly dynamic, set of involved companies and higher levels of cooperation (i.e., operational resource sharing).

CONCLUSION AND FUTURE WORK

One of the weaknesses in the area of virtual enterprises and collaborative networks is the lack of appropriate theoretic definitions, formal models and consistent modeling paradigms. The main contribution of this article is a formal model of virtual enterprise definition, creation and collaboration as well as their associated tasks. We have also formally described Plug and Play Business, which is a set of integrated ICT-tools that support innovators in turning their ideas into businesses by forming virtual enterprises for interorganizational and interoperable collaboration.

In approaching a technology platform for Plug and Play Business software, we have made an assessment of useful technologies and related work. Based on this review, we can conclude that some of the evaluated technologies may be used for the tasks of Plug and Play Business software. With respect to finding and evaluating partners for a virtual enterprise, recommendation systems show numerous fruitful examples that can be applied. For the process of establishing an agreement between the catalyst and the highest ranked actor in the evaluation process, the Contract Net protocol and broker agents may be promising alternatives. Relevant approaches for supporting virtual enterprise operation include Microsoft's BizTalk solution, wrapper agents and computational auctions.

The next step will mainly focus on further analyzing the components of the Plug and Play Business software and on refining the requirements that were only briefly discussed in this article. We will also continue to perfect the formal framework presented above. In particular, we will further develop the roles and different types of obligations in the Plug and Play Business community as well as in the virtual enterprises. We also intend to implement a proof of concept of the Plug and Play Business concept and to evaluate its viability.

ACKNOWLEDGMENT

This work has been partially funded by the project "Integration of business information systems," financially supported by "Sparbanksstiftelsen Kronan." The authors would also like to thank all the members in the project.

REFERENCES

Adomavicius, G., & Tuzhilin, A. (2005). Toward the next generation of recommender systems: A survey of the state-of-the-art and possible extensions. *IEEE Transactions on Knowledge and Data Engineering, 17*(6), 734-749.

Boella, G., & van der Torre, L. (2004). Virtual permission and authorization in policies for virtual communities of agents. In G. Moro, S. Bergamaschi, & K. Aberer (Eds.), *Proceedings of the Agents and P2P Computing Workshop at the 3rd International Joint Conference on Autonomous Agents and Agent Systems,* (pp. 86-97). New York.

Camarinha-Matos, L.M., & Afsarmanesh, H. (1999). The virtual enterprise concept. In *Proceedings of the IFIP Working Conference on Infrastructures for Virtual Enterprises: Networking Industrial Enterprises,* (pp. 3-14). Deventer, The Netherlands: Kluwer Academic.

Camarinha-Matos, L.M., & Afsarmanesh, H. (2003). Elements of a base VE infrastructure. *Journal of Computers in Industry*, *51*(2), 139-163.

Camarinha-Matos, L.M., & Afsarmanesh, H. (2005). Collaborative networks: A new scientific discipline. *Journal of Intelligent Manufacturing*, *16*, 439-452.

Camarinha-Matos, L.M., & Afsarmanesh, H. (2006). A modeling framework for collaborative networked organizations. In L.M. Camarinha-Matos, H. Afsarmanesh, & M. Ollus (Eds.), *Network-centric collaboration and supporting frameworks* (pp. 3-14). Boston: Springer Science Business Media.

Cardoso, L.C., & Oliveira, E. (2005). Virtual enterprise normative framework within electronic institutions. In M.-P. Gleizes, A. Omicini, & F. Zambronelli (Eds.), *Engineering societies in the agent world, Lecture notes in artificial intelligence* (Vol. 3451, pp. 14-32). Berlin: Springer-Verlag.

Carlsson, B., Davidsson, P., Jacobsson, A., Johansson, S.J., & Persson, J.A. (2005). Security aspects on inter-organizational cooperation using wrapper agents. In K. Fischer, A. Berre, K. Elms, & J.P. Müller (Eds.), *Proceedings of the Workshop on Agent-based Technologies and Applications for Enterprise Interoperability at the 4th International Joint Conference on Autonomous Agents and Agent Systems,* (pp. 13-25). Utrecht, The Netherlands: University of Utrecht.

Chituc, C.-M., & Azevedo, A.L. (2005). Enablers and technologies supporting self-forming networked organizations. In H. Panetto (Ed.), *Interoperability of enterprise software and applications* (pp. 77-89). London: Hermes Science.

D'Atri, A., & Motro, A. (2007). VirtuE: A formal model of virtual enterprises for information markets. *Journal of Intelligent Information Systems.*

Davidsson, P., Hederstierna, A., Jacobsson, A., Persson, J.A., et al. (2006). The concept and technology of plug and play business. In Y. Manolopoulos, J. Filipe, P. Constantopoulos, & J. Cordeiro (Eds.), *Proceedings of the 8th International Conference on Enterprise Information Systems Databases and Information Systems Integration,* (pp. 213-217).

Davidsson, P., Ramstedt, L., & Törnquist, J. (2005). Inter-organization interoperability in transport chains using adapters based on open source freeware. In D. Konstantas, J.-P. Bourrières, M. Léonard, & N. Boudjlida (Eds.), *Interoperability of enterprise software and applications* (pp. 35-43). Berlin: Springer-Verlag.

Electronic Business using eXtensible Markup Language (ebXML). Retrieved April 2, 2008, from http://www.ebxml.org/

European collaborative networked organizations leadership initiative (ECOLEAD). Retrieved April 2, 2008, from http://www.ecolead.org

Jacobsson, A., & Davidsson, P. (2006). An analysis of plug and play business software. In R. Suomi, R. Cabral, J.F. Hampe, A. Heikkilä, J. Järveläinen, & E. Koskivaára (Eds.), *Project e-society: Building bricks* (pp. 31-44). New York: Springer Science Business Media.

Jacobsson, A., & Davidsson, P. (2007). Security issues in the formation and operation of virtual enterprises. In L. Kutvonen, P. Linnington, J.-H. Morin, & S. Ruohomaa (Eds.), *Proceedings of the Second International Workshop on Interoperability Solutions to Trust, Security, Policies and QoS for Enhanced Enterprise Systems at the Third International Conference on Interoperability for Enterprise Applications and Software,* (pp. 55-66).

Leibenstein, H. (1968). Entrepreneurship and development. *The American Economic Review*, 58, 72-83.

López y López, F., Luck, M., & d'Inverno, M. (2006). A normative framework for agent-based systems. *Computational and Mathematical Organization Theory, 12*, 227-250.

Microsoft BizTalk Server (BizTalk). Retrieved April 2, 2008, from http://www.microsoft.com/biztalk/

Rosenschein, J.S., & Zlotkin, G. (1994). *Rules of encounter: Designing conventions for automated negotiation among computers.* Cambridge, MA: MIT Press.

Shapiro, C., & Varian, H.R. (1999). *Information rules: A strategic guide to the network economy.* Boston: HBS Press.

Smith, R.G. (1980). The contract net protocol: High-level communication and control in a distributed problem solver. *IEEE Transactions on Computers, C-29*(12), 1104-1113.

Tidd, J., Bessant, J., & Pavitt, K. (2005). *Managing innovation–integrating technological, market and organizational change.* Chichester, West Sussex: John Wiley & Sons.

Wiederhold, G. (1992). Mediators in the architecture of future information systems. *IEEE Transactions on Computers, 25*(3), 38-49.

Wong, H.-C., & Sycara, K.P. (2000). A taxonomy of middle-agents for the Internet. In *Proceedings of the 4th International Conference on Multi-Agent Systems,* (pp. 465-466).

Wooldridge, M. (2004). *An introduction to multiagent systems.* Chichester, West Sussex: John Wiley & Sons.

Yamamoto, L. (2004). Automated negotiation for on-demand inter-domain performance monitoring. In *Proceedings of the 2nd International Workshop on Inter-domain Performance and Simulation,* (pp. 159-169).

Chapter IV
Application of Uncertain Variables to Knowledge-Based Resource Distribution

Donat Orski
Wroclaw University of Technology, Poland

ABSTRACT

The chapter concerns a class of systems composed of operations performed with the use of resources allocated to them. In such operation systems, each operation is characterized by its execution time depending on the amount of a resource allocated to the operation. The decision problem consists in distributing a limited amount of a resource among operations in an optimal way, that is, in finding an optimal resource allocation. Classical mathematical models of operation systems are widely used in computer supported projects or production management, allowing optimal decision making in deterministic, well-investigated environments. In the knowledge-based approach considered in this chapter, the execution time of each operation is described in a nondeterministic way, by an inequality containing an unknown parameter, and all the unknown parameters are assumed to be values of uncertain variables characterized by experts. Mathematical models comprising such two-level uncertainty are useful in designing knowledge-based decision support systems for uncertain environments. The purpose of this chapter is to present a review of problems and algorithms developed in recent years, and to show new results, possible extensions and challenges, thus providing a description of a state-of-the-art in the field of resource distribution based on the uncertain variables.

INTRODUCTION

Among many theories of uncertainty (Klir, 2006) developed for different applications the uncertain variables introduced by Bubnicki (2001a, 2001b) may be considered as a useful tool for modeling expert's knowledge in knowledge-based decision systems. In the definition of the uncertain variable $\bar{x}$ we consider two soft properties: "$\bar{x} \cong x$" which means "$\bar{x}$ is approximately equal to x" or "x is the approximate value of $\bar{x}$," and "$\bar{x} \,\tilde{\in}\, D_x$" which means "$\bar{x}$ approximately belongs to the set D_x" or "the approximate value of $\bar{x}$ belongs to D_x." The *uncertain variable* $\bar{x}$ is defined by a set of values X (real number vector space), the function $h(x) = v(\bar{x} \cong x)$ (i.e., the *certainty index* that $\bar{x} \cong x$, given by an expert) and the following definitions for $D_x, D_1, D_2 \subseteq X$:

$$v(\bar{x} \,\tilde{\in}\, D_x) = \max_{x \in D_x} h(x)$$

$$v(\bar{x} \,\tilde{\notin}\, D_x) = 1 - v(\bar{x} \,\tilde{\in}\, D_x),$$

$$v(\bar{x} \,\tilde{\in}\, D_1 \vee \bar{x} \,\tilde{\in}\, D_2) = \max\{v(\bar{x} \,\tilde{\in}\, D_1)\ v(\bar{x} \,\tilde{\in}\, D_2)\},$$

$$v(\bar{x} \,\tilde{\in}\, D_1 \wedge \bar{x} \,\tilde{\in}\, D_2)$$

$$= \begin{cases} \min\{v(\bar{x} \,\tilde{\in}\, D_1), v(\bar{x} \,\tilde{\in}\, D_2)\} & \text{for} \quad D_1 \cap D_2 \neq \varnothing \\ 0 & \text{for} \quad D_1 \cap D_2 = \varnothing. \end{cases}$$

The function $h(x)$ is called a *certainty distribution.* Let us consider a plant with the input vector $u \in U$ and the output vector $y \in Y$, described by a relation $R(u, y; x) \subset U \times Y$ (*relational knowledge representation*) where the vector of unknown parameters $x \in X$ is assumed to be a value of an uncertain variable described by the certainty distribution $h(x)$ given by an expert. If the relation R is not a function, then the value u determines a set of possible outputs $D_y(u; x) = \{y \in Y : (u, y) \in R(u, y; x)\}$. For the requirement $y \in D_y \subset Y$ given by a user, we can formulate the following **decision problem**: For the given $R(u, y; x)$, $h(x)$ and D_y one should find the decision u^* maximizing the certainty index that the set of possible outputs approximately belongs to D_y (i.e., belongs to D_y for an approximate value of $\bar{x}$). Then

$$u^* = \arg\max_{u \in U} v[D_y(u; \bar{x}) \,\tilde{\subseteq}\, D_y] = \arg\max_{u \in U} \max_{x \in D_x(u)} h(x)$$

where $D_x(u) = \{x \in X : D_y(u; x) \subseteq D_y\}$. It is easy to see that u^* maximizes $v[u \,\tilde{\in}\, D_u(\bar{x})]$ where $D_u(x)$ is a set of all u such that the implication $u \in D_u(x) \to y \in D_y$ is satisfied. The uncertain variables are dedicated to analysis and decision problems (Bubnicki, 2002, 2004a) in a class of systems containing a decision plant described by a relational knowledge representation with unknown parameter characterized by an expert.

An important example for such a class of decision plants may be a *complex of operations.* It consists of operations characterized by their execution times, and the execution time of a particular operation depends on the amount of a resource allocated to the operation. All the operations use the same kind of a resource which is continuous and may be distributed among operations in any way. In the knowledge-based approach under consideration, this relationship has a form of a relation and an unknown parameter in this relation is assumed to be a value of an uncertain variable characterized by an expert. The decision problem consists then in finding a resource allocation to the operations optimizing a given *performance index* and satisfying the user's requirement typically concerning the execution time of the whole set of operations. Because the resource distribution is based on uncertain knowledge, certainty indexes should be used in decision problem formulations.

Complexes of operations with operations characterized by their execution times are decision plants different than *activity networks* widely used in production or project management (e.g., Banaszak & Jozefowska, 2003). In these networks, the set of activities (production operations or project tasks) is depicted by a graph and the activities are represented by arcs assigned probability distributions describing random durations (execution times) of the activities. The PERT method developed in the 1950s for the analysis of activity networks is based on such a probabilistic description, which seems inadequate in most real world situations. Its extension known as PERT/cost (e.g., Berman, 1964) may be applied also in decision problems, but the allocation is determined in two steps and in the first step typical PERT network analysis and determination of a critical path should be performed. Thus, PERT/cost inherits drawbacks of the PERT method. Models and methods developed for complexes of operations in the 1960s provide analytical formulas and decision algorithms solving resource distribution problems in a unified way on the basis of analytical relationships between operations' execution times and resources allocated to them. If, in the case of an activity network, the execution times are described by experts, the formalism of fuzzy numbers and fuzzy CPM or fuzzy PERT/cost methods may be used (e.g., Mon, Cheng, & Lu, 1995; Fargier, Galvagnon, & Dubois, 2000). If in the case of a complex of operations the parameters in analytical formulas for execution times are described by experts, the formalism of uncertain variables should be used (e.g., Bubnicki, 2003; Orski 2005a, 2005b, 2006a).

In the latter case, for a parallel and for a cascade structure of a complex of operations resource distribution algorithms have been developed (Orski, 2006a), examined (Bubnicki, 2004b; Orski, 2005a), and a method for improving their quality has been proposed (Orski, Sugisaka, & Graczyk, 2006b). The purpose of this chapter is to present a review of problems and algorithms developed in recent years, and to show new results and extensions related to the following directions of current research:

i. Designing knowledge-based resource distribution algorithms for mixed structures of a complex of operations,
ii. Exploiting a concept of three-level uncertainty, which may be considered a knowledge integration method in case of multiple experts, and developing resource distribution algorithms taking into account not only uncertain but also random parameters,
iii. Applying *C*-uncertain variables, and
iv. Applying a learning system to improve the knowledge obtained from the experts.

We present the description of a complex of operations and three formulations of the resource distribution problem. Then, solution algorithms and simple numerical examples are shown for the complex of parallel operations and for the complex of cascade operations. Next, descriptions of typical mixed structures of operations are given, we show how resource distribution algorithms may be derived from formulas determined for parallel and for cascade operations, and present simple numerical examples. We then devote time to the problem of evaluating and improving quality of resource allocation based on experts' knowledge, and include results of simulations. Finally, we address issues related to topics (ii), (iii) and (iv) in the list of current research directions.

KNOWLEDGE REPRESENTATION AND RESOURCE DISTRIBUTION PROBLEMS

Let us consider a complex of k operations described by a set of inequalities

$$T_i \leq \varphi_i(u_i, x_i),\ i \in \overline{1,k} \qquad (1)$$

where T_i is the execution time of the i-th operation, φ_i is a decreasing function of u_i, u_i is the amount of a resource assigned to the i-th operation, the parameter $x_i \in R^+$ is unknown and assumed to be a value of an *uncertain variable* $\overline{x}_i$ described by a certainty distribution $h_i(x_i)$ given by an expert, and $(\overline{x}_1, ..., \overline{x}_k) \triangleq \overline{x}$ are independent variables.

The total amount of a resource to be distributed among the operations is limited to U, hence every resource allocation $(u_1, ..., u_k) \stackrel{\Delta}{=} u$ must satisfy the constraints:

$$u_i \geq 0 \text{ for each } i \quad \text{and} \quad u_1 + u_2 + ... + u_k = U. \tag{2}$$

Let us denote by T the execution time of the whole complex of operations. It is given by the function

$$T = f(T_1, T_2, ..., T_k) \tag{3}$$

depending on the structure of the complex of operations. In typical situations, functions φ_i in (1) are the following:

$$\varphi_i(u_i, x_i) = \frac{x_i}{u_i}, \quad i \in \overline{1,k}, \tag{4}$$

and certainty distributions h_i are assumed to be triangular, as shown in Figure 1.

For the given values $U > 0$ and $\alpha > 0$ (required project completion time), one may determine the certainty index

$$v[T(u;\overline{x}) \tilde{\leq} \alpha] \stackrel{\Delta}{=} v(u;\alpha,U) \tag{5}$$

of the property "the execution time T is approximately less or equal to α for the allocation u and the uncertain variable $\overline{x}$" where $T(u;x) = f[\varphi_1(u_1,x_1), \varphi_2(u_2,x_2), ..., \varphi_k(u_k,x_k)]$ is the upper bound function for T. The description of a complex of operations directly corresponds to the description of a decision plant presented in the previous section, that is, allocation u corresponds to plant input, execution time T corresponds to plant output y, the set $(0, \alpha]$ corresponds to the set D_y required by a user, and the inequality $T \leq T(u;x)$ defines the relation $R(u, y; x)$. The following three versions of the **resource distribution problem** may be formulated, depending on which variable in formula (5) is chosen for optimization:

Version I. Given U and α, find u maximizing certainty index v (5), subject to the constraints (2).
Version II. Given U and $\overline{v}$ (*certainty threshold*, that is, $v \geq \overline{v}$ is required), find u minimizing α in (5), subject to the constraints (2).
Version III. Given α and $\overline{v}$, find u minimizing U in (5), subject to the constraints (2).

Performance indexes to be maximized in the three problems stated above are v, α^{-1} and U^{-1}, respectively. Solution algorithms are based on

Figure 1. Triangular certainty distribution for the uncertain variable $\overline{x}_i$

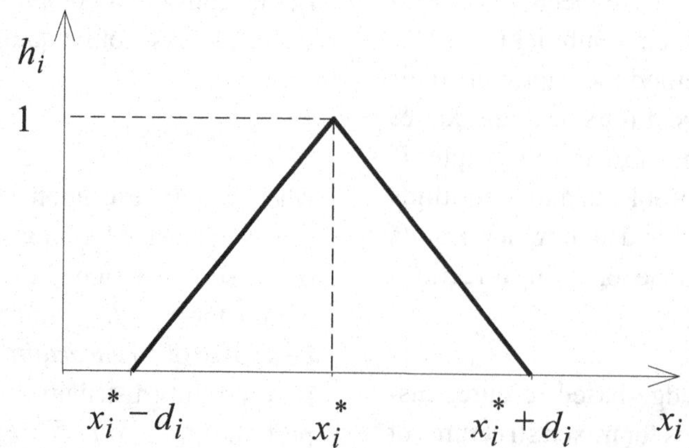

certainty indexes defined individually for particular operations. They are determined (Bubnicki, 2003) by the formulas

$$v_i(u_i;\alpha_i) = v[\varphi_i(u_i,\bar{x}_i) \tilde{\leq} \alpha_i] = \tag{6}$$

$$\max_{x_i \in D_i(u_i;\alpha_i)} h_i(x_i) = \begin{cases} 1 & \text{for } x_i^* \leq \alpha_i u_i \\ \frac{1}{d_i}\alpha_i u_i - \frac{x_i^*}{d_i} + 1 & \text{for } x_i^* - d_i \leq \alpha_i u_i \leq x_i^* \\ 0 & \text{otherwise} \end{cases}$$

where

$D_i(u_i;\alpha_i) = \{x_i \in R^1 : \varphi_i(u_i,x_i) \leq \alpha_i\} = (0,\ \alpha_i u_i]$, $i \in \overline{1,k}$,

and α_i are required completion times for particular operations given directly by a user or needed to be calculated based on α and a structure of the complex of operations.

RESOURCE DISTRIBUTION FOR PARALLEL AND CASCADE OPERATIONS

In this section, main results obtained for a complex of parallel operations and for a complex of cascade operations are shown. The more detailed considerations are presented in Orski (2006). Two typical structures reflecting constraints imposed on a sequence of operations' executions are illustrated in Figure 2. In the case a) operations are executed independently, whereas in the case b) there is a cascade of executions and the execution of any operation may begin only if preceding operations have been completed.

The execution time T of the whole set of operations is given by the following functions (3)

$$T = \max\{T_1, T_2,..., T_k\} \text{ or } T = T_1 + T_2 + ... + T_K$$

in the case a) or b), respectively.

Algorithms for Parallel Operations

Version I. Given U and α, find u maximizing v (5), subject to the constraints (2). Because the required execution times α_i are equal to α, $v(u;\alpha,U) = v\{[\varphi_1(u_1,\bar{x}_1) \tilde{\leq} \alpha)] \wedge [\varphi_2(u_2,\bar{x}_2) \tilde{\leq} \alpha)] \wedge ... \wedge [\varphi_k(u_k,\bar{x}_k) \tilde{\leq} \alpha)]\} = \min_i v_i(u_i;\alpha)$ and finding the solution

$$u^* = \arg\max_u v(u;\alpha,U)$$

is based on certainty indexes (6). It may be proved (Bubnicki, 2004a) that the optimal distribution should satisfy the set of equations

$$v_1(u_1;\alpha) = v_2(u_2;\alpha) = ... = v_k(u_k;\alpha).$$

Hence, it is easy to obtain the following analytical result in the case when $0 < v < 1$:

Figure 2. Operations of a) parallel and b) cascade structure

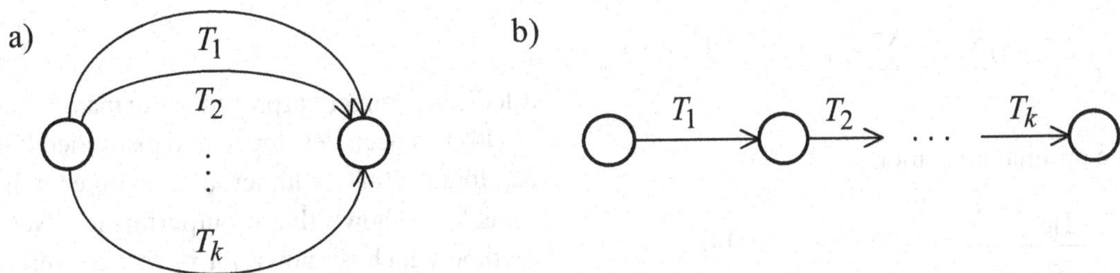

$$u_i^* = \frac{1}{\alpha}[d_i(\alpha U - \sum_{j=1}^{k} x_j^*)(\sum_{j=1}^{k} d_j)^{-1} + x_i^*] \triangleq \Psi_i(U;\alpha,x^*,d) \quad i \in \overline{1,k} \tag{7}$$

$$v(u^*) = [\alpha U - \sum_{i=1}^{k}(x_i^* - d_i)](\sum_{i=1}^{k} d_i)^{-1} \tag{8}$$

The resource distribution algorithm $\Psi_i(U;\alpha,x^*,d)$ is a linear function of U with parameters depending both on the value α given by a user and on values $(x_1^*,...,x_k^*) \triangleq x^*$, $(d_1,...,d_k) \triangleq d$ given by experts.

Version II. Given U and $v = \overline{v}$ (the required *certainty threshold*), find u minimizing α in (5), subject to the constraints (2). Using (8) with $\overline{v}$ instead of $v(u^*)$ one determines the shortest possible execution time

$$\alpha_{\min} = \frac{1}{U}[(\overline{v}-1)\sum_{i=1}^{k} d_i + \sum_{i=1}^{k} x_i^*] \tag{9}$$

and the optimal allocation

$$u_i^* = \frac{(\overline{v}-1)d_i + x_i^*}{(\overline{v}-1)\sum_{j=1}^{k} d_j + \sum_{j=1}^{k} x_j^*} U, \quad i \in \overline{1,k}. \tag{10}$$

It may be noted that, as in version I, the resource distribution algorithm depends on numerical values given both by a user and by experts.

Version III. Given α and $\overline{v}$, find u minimizing U in (5), subject to the constraints (2). Using (8) with $\overline{v}$ instead of $v(u^*)$ one determines the smallest amount of a resource

$$U_{\min} = \frac{1}{\alpha}[(\overline{v}-1)\sum_{i=1}^{k} d_i + \sum_{i=1}^{k} x_i^*] \tag{11}$$

and the optimal allocation

$$u_i^* = \frac{(\overline{v}-1)d_i + x_i^*}{\alpha}, \quad i \in \overline{1,k}. \tag{12}$$

In version III, the resource distribution algorithm depends on numerical values given both by a user and by experts, and is independent of U assumed to be unknown.

Algorithms for Cascade Operations

Version I. One should determine $u^* = \arg\max_u v(u;\alpha,U)$ using certainty indexes (6) and values

$$\alpha_i = \frac{d_i}{u_i}(\alpha - \sum_{j=1}^{k}\frac{x_j^*}{u_j})(\sum_{j=1}^{k}\frac{d_j}{u_j})^{-1} + \frac{x_i^*}{u_i}$$

determined for particular operations by using an optimal decomposition of α (Orski, 2006), that is, such that

$$v_1(u_1,\alpha_1) = v_2(u_2,\alpha_2) = ... = v_k(u_k,\alpha_k) \tag{13}$$

and taking into account that $\alpha_1 + \alpha_2 + ... + \alpha_k = \alpha$. If $0 < v < 1$, it is given by the formula

$$v(u;\alpha,U) = 1 - (\sum_{i=1}^{k}\frac{x_i^*}{u_i} - \alpha)(\sum_{i=1}^{k}\frac{d_i}{u_i})^{-1} \tag{14}$$

which cannot be maximized analytically and a numerical procedure should be used. Based on an analytical solution

$$u_i^* = \frac{\sqrt{x_i^*}}{\sum_{j=1}^{k}\sqrt{x_j^*}} U, \quad i \in \overline{1,k}$$

obtained for a special case of triangular certainty distributions, that is,

$$\frac{x_1^*}{d_1} = \frac{x_2^*}{d_2} = ... = \frac{x_k^*}{d_k},$$

a dedicated numerical procedure for maximization of (14) has been developed and examined (Orski & Hojda, 2007). Numerous computer simulations have shown that it outperforms a Newton method which is known for its fast convergence but requires application of a penalty function for the constraints (2). Our numerical procedure does not need a penalty function and is of much less computational complexity even when compared to a Newton method without a penalty function.

Version II. In a way analogous to that for parallel operations, using (14) with $\overline{v}$ instead of $v(u)$, one determines the shortest possible execution time as the following function of u:

$$\alpha_{\min}(u) = \sum_{i=1}^{k} \frac{(\overline{v}-1)d_i + x_i^*}{u_i} \qquad (15)$$

Minimization of this function with respect to u, subject to the constraints (2), yields

$$\alpha_{\min} = \frac{[\sum_{i=1}^{k}\sqrt{(\overline{v}-1)d_i + x_i^*}\,]^2}{U},$$

$$u_i^* = \frac{\sqrt{(\overline{v}-1)d_i + x_i^*}}{\sum_{j=1}^{k}\sqrt{(\overline{v}-1)d_j + x_j^*}}\,U, \; i \in \overline{1,k}. \qquad (16)$$

Based on the above formulas a numerical-analytical algorithm for maximization of (14) in version I has been developed and examined (Orski & Hojda, 2007). In a numerical part of the algorithm, the first equation in (16) is solved with respect to $\overline{v} = v^*$ for $\alpha_{\min} = \alpha$, and in an analytical part, the optimal allocation is determined by using the second formula in (16) with v^* in the place of $\overline{v}$. Simulation experiments have shown that its performance is comparable to that of the dedicated numerical procedure.

Version III. Using the results (16) obtained in version II, with α instead of $\alpha_{\min}$, one determines the smallest amount of a resource

$$U_{\min} = \frac{1}{\alpha}[\sum_{i=1}^{k}\sqrt{(\overline{v}-1)d_i + x_i^*}\,]^2 \qquad (17)$$

and

$$u_i^* = \frac{1}{\alpha}\sqrt{(\overline{v}-1)d_i + x_i^*}\sum_{j=1}^{k}\sqrt{(\overline{v}-1)d_j + x_j^*}, \; i \in \overline{1,k} \qquad (18)$$

It is worth noting that optimal distribution problems in versions II and III are solvable analytically both for parallel and cascade operation, whereas the problem in version I is more complicated from a computational point of view and a numerical or numerical-analytical procedure should be applied for cascade operations.

RESOURCE DISTRIBUTION FOR MIXED STRUCTURES OF OPERATIONS

In this section, we will show how the formulas and algorithms presented previously may be applied in knowledge-based resource distribution in a complex of operations of neither parallel, nor cascade structure. We will consider simple examples of such a mixed structure and a more general case of a cascade-parallel complex of operations. Generally, for the determination of solution algorithms in mixed structures, for cascade operations one should use formulas presented in the last section, and for parallel ones, the formulas presented in the section preceding it.

Algorithms for a Simple Mixed Structure ($k = 3$)

Let us take into account a complex of $k = 3$ operations of structure presented in Figure 3. In this case, the formulas corresponding to (1), (2) and (3) are as follows:

$$T_1 \le \frac{x_1}{u_1}, \quad T_2 \le \frac{x_2}{u_2}, \quad T_3 \le \frac{x_3}{u_3},$$

$u_1, u_2, u_3 \ge 0$, $u_1 + u_2 + u_3 = U$ and $T = \max\{T_1, T_2\} + T_3$.

Figure 3. Complex of $k = 3$ operations of a mixed structure

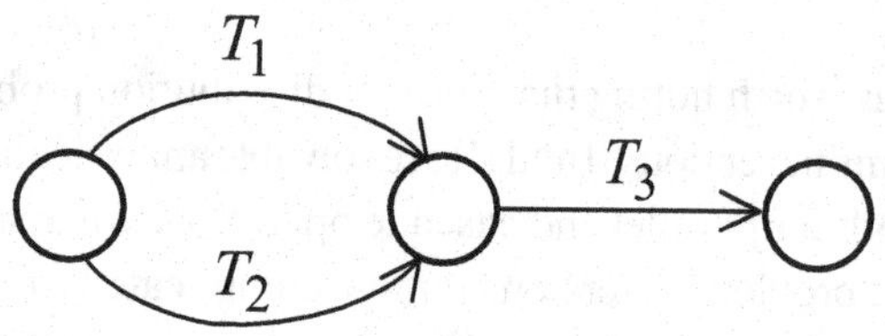

Now, the certainty index (5)

$$v(u;\alpha,U) = v\{[\varphi_1(u_1,\bar{x}_1) \tilde{\leq} \alpha_1] \wedge [\varphi_2(u_2,\bar{x}_2) \tilde{\leq} \alpha_2] \wedge [\varphi_3(u_3,\bar{x}_3) \tilde{\leq} \alpha_3]\} = \min_{i\in\{1,2,3\}} v_i(u_i,\alpha_i)$$

where, according to the structure of the complex in Figure 3, $\alpha_1 = \alpha_2$ and $\alpha_1 + \alpha_3 = \alpha$.

Version I. We may apply (8) for the two parallel operations and introduce $v_{12}(u_1,u_2;\alpha_1) =$

$$1 - \frac{x_1^* + x_2^* - \alpha_1(u_1 + u_2)}{d_1 + d_2} = 1 - (\frac{x_{12}^*}{u_{12}} - \alpha_1)(\frac{d_{12}}{u_{12}})^{-1}$$

as the aggregated certainty index, where $x_{12}^* = x_1^* + x_2^*$, $d_{12} = d_1 + d_2$ and $u_{12} = u_1 + u_2$. Consequently,

$$v(u;\alpha,U) = \min\{v_{12}(u_{12};\alpha_1),\ v_3(u_3;\ \alpha_3)\}$$

where

$$v_3(u_3;\alpha_3) = 1 - (\frac{x_3^*}{u_3} - \alpha_3)(\frac{d_3}{u_3})^{-1}$$

denotes the certainty index for the one cascade operation obtained from (14). Now, using (14) for the cascade connection of aggregated parallel operations and a single cascade operation one obtains

$$v(u;\alpha,U) =$$

$$= 1 - (\frac{x_{12}^*}{u_{12}} + \frac{x_3^*}{u_3} - \alpha)(\frac{d_{12}}{u_{12}} + \frac{d_3}{u_3})^{-1}$$

$$= 1 - (\frac{x_1^* + x_2^*}{u_1 + u_2} + \frac{x_3^*}{u_3} - \alpha)(\frac{d_1 + d_2}{u_1 + u_2} + \frac{d_3}{u_3})^{-1}$$

Using (2) we substitute $u_{12} = U - u_3$ and maximize $v(u_3;\alpha,U)$ with respect to $u_3 \in (0,U)$. The maximization is reduced to solving a quadratic equation with respect to u_3 and an analytical result may be obtained. However, it is pointless to present it, since we expect that for more complicated mixed structures we will not be able to use it for obtaining analytical solutions, and we will have to use numerical procedures anyway.

Version II. Using (9) for the two parallel operations and using (16) for the one cascade operation yields

$$\alpha_{1,\min} = \frac{(\bar{v} - 1)(d_1 + d_2) + x_1^* + x_2^*}{u_1 + u_2}$$

and

$$\alpha_{3,\min} = \frac{(\bar{v} - 1)d_3 + x_3^*}{u_3},$$

respectively. Again, we may treat the two aggregated parallel operations as a single cascade operation with $x_{12}^* = x_1^* + x_2^*$, $d_{12} = d_1 + d_2$ and $u_{12} = u_1 + u_2$. Let us denote $c_i \stackrel{\Delta}{=} (\bar{v} - 1)d_i + x_i^*$, $i \in \{1,2,3\}$ Because $\alpha_{\min} = \alpha_{1,\min} + \alpha_{3,\min}$ then, according to (15),

$$\alpha_{\min}(u) = \frac{c_1 + c_2}{u_{12}} + \frac{c_3}{u_3}$$

and using the first formula in (16) yields

$$\alpha_{\min} = \frac{(\sqrt{c_1 + c_2} + \sqrt{c_3})^2}{U}$$

Figure 4. Complex of $k = 4$ operations of a mixed structure

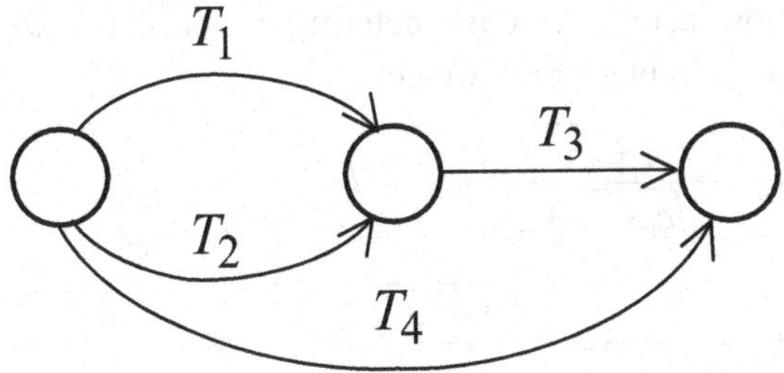

From the second formula in (16) we obtain

$$u_3^* = \frac{\sqrt{c_3}}{\sqrt{c_1+c_2}+\sqrt{c_3}}U\ ,\ u_{12} = \frac{\sqrt{c_1+c_2}}{\sqrt{c_1+c_2}+\sqrt{c_3}}U$$

and application of (10) results in

$$u_1^* = \frac{c_1}{c_1+c_2+\sqrt{c_1+c_2}\sqrt{c_3}}U,$$

$$u_2^* = \frac{c_2}{c_1+c_2+\sqrt{c_1+c_2}\sqrt{c_3}}U.$$

Version III. The solution algorithm is based on aggregation analogous to that in version I or in version II. For two aggregated parallel operations from (11) we get $u_{12,\min} = \frac{c_1+c_2}{\alpha_1}$, and for the cascade operation from (17) we have $u_{3,\min} = \frac{c_3}{\alpha_3}$. Then, using formulas (17) and (18) for a complex of cascade operations gives the following results:

$$U_{\min} = \frac{(\sqrt{c_1+c_2}+\sqrt{c_3})^2}{\alpha},\ u_3^* = \frac{c_3+\sqrt{c_1+c_2}\sqrt{c_3}}{\alpha}$$

$$u_{12,\min} = \frac{c_1+c_2+\sqrt{c_1+c_2}\sqrt{c_3}}{\alpha}$$

Because $u_{12,\min} = \frac{c_1+c_2}{\alpha_1}$, then $\alpha_1 = \frac{\sqrt{c_1+c_2}}{\sqrt{c_1+c_2}+\sqrt{c_3}}\alpha$ and from (12) for two parallel operations we get

$$u_1^* = \frac{c_1(\sqrt{c_1+c_2}+\sqrt{c_3})}{\alpha\sqrt{c_1+c_2}},\ u_2^* = \frac{c_2(\sqrt{c_1+c_2}+\sqrt{c_3})}{\alpha\sqrt{c_1+c_2}}$$

Example 1. Let $\alpha = 50$, $\overline{v} = 0.8$ and h_1, h_2, h_3 be triangular with $x_1^* = 10$, $d_1 = 5$, $x_2^* = 51$, $d_2 = 10$, $x_3^* = 54$, $d_3 = 20$. Then, in version III one obtains $u_1^* = 0.35$, $u_2^* = 1.89$, $u_3^* = 2.07$ and $U_{\min} = 4.31$.

Algorithms for a Simple Mixed Structure ($k = 4$)

Let us take into account a complex of $k = 4$ operations of structure presented in Figure 4. In this case, the formulas corresponding to (1), (2) and (3) are as follows:

$$T_1 \le \frac{x_1}{u_1},\ T_2 \le \frac{x_2}{u_2},\ T_3 \le \frac{x_3}{u_3},\ T_4 \le \frac{x_4}{u_4}$$

$$u_1, u_2, u_3, u_4 \ge 0,\ \ u_1+u_2+u_3+u_4 = U$$

and $T = \max\{\max\{T_1, T_2\} + T_3,\ T_4\}$.

Now, the certainty index (5) $v(u;\alpha,U) = \min_{i\in\{1,2,3,4\}} v_i(u_i,\alpha_i)$ where, according to the structure of the complex in Figure 4, $\alpha_1 = \alpha_2$, $\alpha_1 + \alpha_3 = \alpha$, and $\alpha_4 = \alpha$. Because three of four operations are connected in the same way as in Figure 3, we may use the results obtained for these three operations and apply formulas presented for parallel operations to the fourth operation connected in parallel.

Version I. The certainty index is given by the formula

$$v(u;\alpha,U) = \min\{v_{123}(u_{12},u_3;\alpha),\ v_4(u_4;\ \alpha)\}$$

where $v_{123}(u_{12},u_3;\alpha)$ denotes the certainty index for the complex of three operations, that is,

$$v_{123}(u_{12},u_3;\alpha) = 1-(\frac{x_{12}^*}{u_{12}}+\frac{x_3^*}{u_3}-\alpha)(\frac{d_{12}}{u_{12}}+\frac{d_3}{u_3})^{-1}$$

and from (8)

$$v_4(u_4;\alpha) = \frac{\alpha u_4 - (x_4^* - d_4)}{d_4} = 1 - \frac{x_4^* - \alpha u_4}{d_4}$$

We cannot use formula (8) further for a parallel connection of the fourth operation and the structure composed of three other operations, because $v_{123}(u_{12}, u_3; \alpha)$ does not represent a certainty index for a single operation being a result of aggregation. We may use (13), which holds also for parallel operations, and maximize $v_{123}(u_{12}, u_3; \alpha)$ subject to the constraints (2) and the additional constraint $v_{123}(u_{12}, u_3; \alpha) = v_4(u_4; \alpha)$. The maximization may be reduced to solving two dependent quadratic equations in such a way that the solution of the first equation is a parameter in the second equation. In general, we will have to use a numerical optimization.

Version II. Using results presented previously for the structure composed of three operations, we have

$$\alpha_{13,\min} = \frac{c_{123}}{u_{123}}, \ \alpha_{4,\min} = \frac{c_4}{u_4}$$

where $u_{123} = u_1 + u_2 + u_3$ is the amount of a resource used in a subcomplex of three operations aggregated in such a way that $c_{123} = (\sqrt{c_1 + c_2} + \sqrt{c_3})^2$ and $\alpha_{4,\min}$ stands for the execution time of the fourth operation parallel to the subcomplex of aggregated operations. Then, using the formula (9) for $\alpha_{\min}$ in case of parallel operations we obtain

$$\alpha_{\min} = \frac{c_{123} + c_4}{U} = \frac{(\sqrt{c_1 + c_2} + \sqrt{c_3})^2 + c_4}{U}$$

and from (10) we obtain

$$u_4^* = \frac{c_4}{(\sqrt{c_1 + c_2} + \sqrt{c_3})^2 + c_4} U,$$

$$u_{123} = \frac{(\sqrt{c_1 + c_2} + \sqrt{c_3})^2}{(\sqrt{c_1 + c_2} + \sqrt{c_3})^2 + c_4} U$$

Now, using previously determined solutions and u_{123} in place of U we get

$$u_3^* = \frac{c_3 + \sqrt{c_1 + c_2}\sqrt{c_3}}{(\sqrt{c_1 + c_2} + \sqrt{c_3})^2 + c_4} U,$$

$$u_{12} = \frac{c_1 + c_2 + \sqrt{c_1 + c_2}\sqrt{c_3}}{(\sqrt{c_1 + c_2} + \sqrt{c_3})^2 + c_4} U$$

$$u_1^* = \frac{c_1(\sqrt{c_1 + c_2} + \sqrt{c_3})}{\sqrt{c_1 + c_2}[(\sqrt{c_1 + c_2} + \sqrt{c_3})^2 + c_4]} U,$$

$$u_2^* = \frac{c_2(\sqrt{c_1 + c_2} + \sqrt{c_3})}{\sqrt{c_1 + c_2}[(\sqrt{c_1 + c_2} + \sqrt{c_3})^2 + c_4]} U.$$

Version III. Using the results for the subcomplex composed of three operations we have

$$u_{123,\min} = \frac{(\sqrt{c_1 + c_2} + \sqrt{c_3})^2}{\alpha}, \ u_{4,\min} = u_4^* = \frac{c_4}{\alpha},$$

and application of (11) as for parallel operations leads to

$$U_{\min} = \frac{(\sqrt{c_1 + c_2} + \sqrt{c_3})^2 + c_4}{\alpha}.$$

The distribution of $u_{123,\min}$ among operations from the subcomplex is, obviously, the same. This results from the fact that now we have one additional parallel operation which needs additional resources to achieve the goal defined in the same way. This would not be the case if additional operation was cascade and not parallel.

Example 2. Let the numerical data be the same as in example 1, and for the additional operation we have a triangular h_4 with $x_4^* = 30$ and $d_4 = 20$. Then, in version III one obtains $u_1^* = 0.35, u_2^* = 1.89$, $u_3^* = 2.07$, $u_4^* = 0.5$ and $U_{\min} = 4.81$.

Algorithms for a Cascade-Parallel Structure

Let us consider a complex of $k= ml$ operations of structure presented in Figure 5. This structure may represent, for example, a supply chain with m production stages. In this multistage production system, a production task at each stage is performed by a cluster (subcomplex) of l production units (operations). Now, we will use double index notation where T_{ij} denotes the execution time of the j-th operation in the i-th subcomplex, u_{ij} is the amount of a resource allocated to this operation, and the unknown parameter x_{ij} is a value of an uncertain variable $\overline{x}_{ij}$ with a given triangular certainty distribution defined by x^*_{ij} and d_{ij}. All uncertain variables are assumed to be independent. The formulas corresponding to (1), (2) and (3) are as follows:

$$T_{ij} \le \frac{x_{ij}}{u_{ij}}, \; i \in \overline{1,m}, \quad j \in \overline{1,l},$$

$$u_{ij} \ge 0, \sum_{i=1}^{m}\sum_{j=1}^{l} u_{ij} = U,$$

and

$$T = \sum_{i=1}^{m} \max_{j \in \overline{1,l}} T_{ij}.$$

The solution of a resource distribution problem in all three versions will now require application of a two-level approach. This approach may be briefly described as finding a solution to the upper level optimization problem (for the whole complex of operations) by solving lower level optimization problems (for particular subcomplexes) and by using these solutions at the upper level. Solving optimization problems at both levels is then coordinated. For particular subcomplexes, formulas presented for parallel operations should be used at the lower level, whereas for the upper level optimization problem we will use formulas presented for cascade operations.

Version I. Using (8) we may express the certainty index for the i-th subcomplex by using U_i and α_i in place of U and α, respectively. Then

$$v_i(u^*_{i1},...,u^*_{il};\alpha_i) = [\alpha_i U_i - \sum_{j=1}^{l}(x^*_{ij} - d_{ij})](\sum_{j=1}^{l} d_{ij})^{-1}$$

where, as in (2), $U_i = u_{i1} + u_{i2} + ... + u_{il}$. The parallel operations may be aggregated by introducing $x^*_i = \sum_{j=1}^{l} x^*_{ij}$ and $d_i = \sum_{j=1}^{l} d_{ij}$. Aggregation leads to a definition of the certainty index for the i-th subcomplex the same as for a single operation in (6),

$$v_i(U_i;\alpha_i) = \frac{1}{d_i}\alpha_i U_i - \frac{x^*_i}{d_i} + 1 \cdot$$

A lower level distribution algorithm is analogous to (7), that is,

Figure 5. Complex of $k = m \cdot l$ operations of a cascade-parallel structure

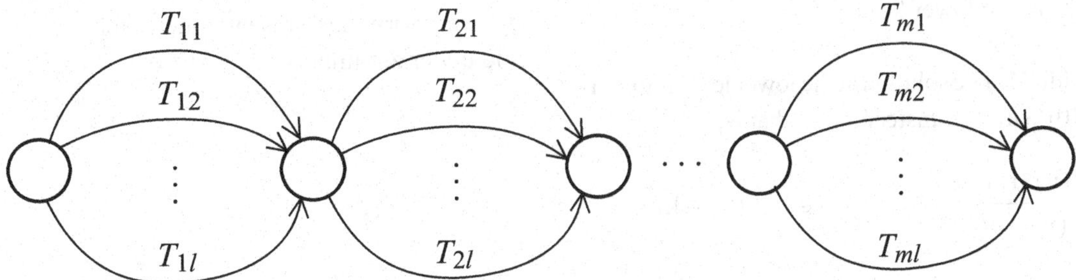

$$u_{ij}^* = \frac{1}{\alpha_i}(d_{ij}\frac{\alpha_i U_i - x_i^*}{d_i} + x_{ij}^*)$$

where α_i is a result of optimal decomposition defined in the same way as for cascade operations with U_i instead of u_i. The amount of a resource in the *i*-th subcomplex U_i is an unknown variable coordinating solution algorithms at the both optimization levels. Its optimal value U_i^* should be determined at the upper level. Because particular subcomplexes are treated as single operations with individual certainty indexes $v_i(U_i;\alpha_i)$, we may apply formula (14) for cascade operations with $U_1, U_2, ..., U_m$ in place of $u_1, u_2, ..., u_k$ which gives

$$v(U_1,...,U_m;\alpha,U) = 1 - (\sum_{i=1}^{m}\frac{x_i^*}{U_i} - \alpha)(\sum_{i=1}^{m}\frac{d_i}{U_i})^{-1}.$$

Maximization of $v(U_1,...,U_m;\alpha,U)$, subject to the constraints $U_i > 0$ and $\sum_{i=1}^{m} U_i = U$, may be performed analytically only under simplifying assumptions, for example, $\frac{x_1^*}{d_1} = \frac{x_2^*}{d_2} = ... = \frac{x_m^*}{d_m} = \gamma$. Then

$$U_i^* = \frac{\sqrt{x_i^*}}{\sum_{p=1}^{m}\sqrt{x_p^*}}U, \; \alpha_i = \alpha\frac{x_i^*}{U_i^*}(\sum_{p=1}^{m}\frac{x_p^*}{U_p^*})^{-1},$$

$$u_{ij}^* = \frac{d_{ij}}{d_i}U_i^* + \frac{x_{ij}^* - \gamma d_{ij}}{\alpha_i}.$$

Otherwise, numerical optimization at the upper level should be performed and its results should be used in analytical resource distribution algorithms at the lower level.

Version II. The solution at the lower level is given by (10) with U_i instead of *U*, that is,

$$u_{ij}^* = \frac{(\overline{v}-1)d_{ij} + x_{ij}^*}{(\overline{v}-1)d_i + x_i^*}U_i, \qquad i \in \overline{1,m}, \quad j \in \overline{1,l}.$$

Optimal values U_i^* are determined at the upper level, by using the following formula, analogous to the second one in (16):

$$U_i^* = \frac{\sqrt{(\overline{v}-1)d_i + x_i^*}}{\sum_{p=1}^{m}\sqrt{(\overline{v}-1)d_p + x_p^*}}U, \; i \in \overline{1,m}.$$

The shortest execution time is described directly by the first formula in (16).

Version III. The solution at the lower level is given by a formula analogous to (12),

$$u_{ij}^* = \frac{(\overline{v}-1)d_{ij} + x_{ij}^*}{\alpha_i}, \; i \in \overline{1,m}, \quad j \in \overline{1,l},$$

and the smallest amount of a resource in the *i*-th aggregated subcomplex is given by the formula

$$U_{i,\min} = \frac{(\overline{v}-1)d_i + x_i^*}{\alpha_i}.$$

Optimal values of coordinating variables are calculated at the upper level by using (18) for the cascade operations, that is,

$$U_{i,\min}^* = \frac{1}{\alpha}\sqrt{(\overline{v}-1)d_i + x_i^*}\sum_{p=1}^{m}\sqrt{(\overline{v}-1)d_p + x_p^*}, \; i \in \overline{1,m}.$$

The smallest total amount of a resource $U_{\min}$ satisfying a user's requirement is described directly by (17).

Let us, finally, note that the presented simple mixed structure of a complex of $k = 3$ operations may be considered as a special case of the cascade-parallel structure. Therefore, the results presented for that structure in versions I, II and III of a resource distribution problem may be obtained by using formulas presented here, for $m = l = 2$ and under assumption $x_{22}^* = d_{22} = 0$.

KNOWLEDGE QUALITY AND SYSTEM'S PERFORMANCE

If a deterministic model of the complex of operations existed and was known, we might use it for determination of accurate resource distribution algorithms. In the knowledge-based approach presented in this chapter it is, however, assumed that only an uncertain nondeterministic description obtained from human experts is available. Then, it may be expected that quality of a resource allocation determined depends directly on quality of knowledge acquired from the experts. The following sections present a method for evaluation quality of knowledge, show effects of possible inaccuracy in experts' opinions and present a concept of the adaptive resource distribution system in which some parameters may be adjusted so as to improve system's performance.

Quality of Knowledge-Based Resource Distribution

Without a loss of generality, we will discuss and illustrate this important issue for a complex of cascade operations and version I of the resource distribution problem. Let us assume that the exact deterministic descriptions of the operations have a form of the equations

$$T_i = c_i u_i^{-\lambda}, \quad i = 1, 2, ..., k$$

where u_i is the amount of a resource allocated to the i-th operation, $0 < \lambda < 1$. If the parameters c_i are known, the optimal allocation $\overline{u}$ minimizing execution time (3) and satisfying the constraints (2) may be determined in an analytical form

$$\overline{u}_i = \sqrt[\lambda+1]{c_i}\,(\sum_{j=1}^{k} \sqrt[\lambda+1]{c_j})^{-\lambda} U, \quad i = 1, 2, ..., k\,,$$

and the minimal execution time (3) is given by the formula

$$\overline{T} = (\sum_{i=1}^{k} \sqrt[\lambda+1]{c_i})^{\lambda+1} U^{-\lambda}.$$

If the values c_i are unknown, we use the description given by experts and apply the allocation u^*. The execution time is then the following

$$T^* = c_1(u_1^*)^{-\lambda} + c_2(u_2^*)^{-\lambda} + ... + c_k(u_k^*)^{-\lambda}.$$

For the evaluation of the result of allocations based on experts' knowledge a quality index

$$\frac{T^*}{\overline{T}} \triangleq Q$$

Figure 6. Relationship between Q and x_2^* for different α

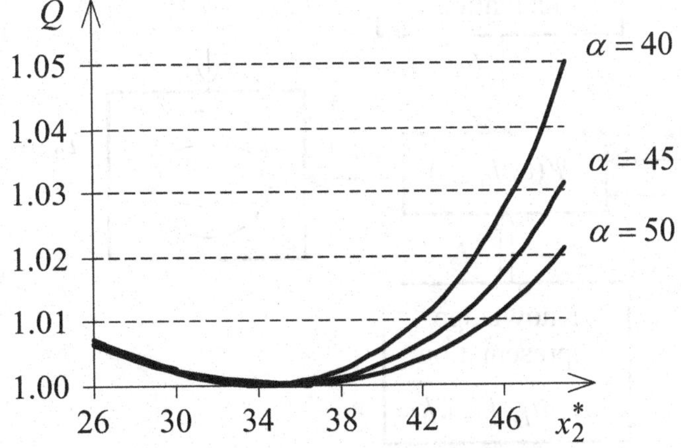

may be proposed. The value Q evaluates the allocation based on the knowledge in the form of the inequalities (1) and the certainty distributions h_i (i.e., the knowledge given by experts) and consequently, evaluates the quality of experts. Quality evaluation based on the index Q may be performed for the given values c_i and λ. Hence, it can be used to investigate the influence of the parameters in h_i on the quality of allocation and to compare execution times obtained with different experts. Figure 6 presents the influence of x_2^* on Q for the following data (Orski, 2005b): $U = 1$, $\alpha = 50$, $x_1^* = 4$, $d_1 = 3$, $d_2 = 21$, and for $c_1 = 4$, $c_2 = 36$, $\lambda = 0.95$. It may be observed that for a wide interval of x_2^* the quality of the knowledge-based resource allocation is high enough to be accepted (i.e., $Q \leq 1.02$). However, one can note that within this interval parameters quite strongly influence the quality index, and that there exist optimal values of the parameters, for which $Q \approx 1$. Therefore, further improvement is still possible.

Another observation is that less realistic requirement results in longer execution time. However, unrealistic values of α do not influence quality of allocation based on knowledge of good experts, that is, experts giving near-optimal values of the parameters.

Adaptive Resource Distribution System

Based on expert's knowledge a resource distribution algorithm $\Psi(U;b)$, $b \in B$ is a vector of parameters, should be determined. We have presented resource distribution algorithms determined for different versions of the resource distribution problem formulation and for basic structures of a complex of operations. For example, the resource distribution algorithm $\Psi(U;b)$ determined for the complex of k parallel operations in version I of the problem was given by the following formulas:

$$\Psi_i(U;\alpha,x^*,d) = \frac{d_i}{d_1 + \ldots + d_k}(U - \frac{x_1^* + \ldots + x_k^*}{\alpha}) + \frac{x_i^*}{\alpha}, \; i \in \overline{1,k}\,.$$

Performance of a decision support system using imprecise knowledge is, of course, worse than would be performance of a decision support system using an exact description. Two ways of improving

Figure 7. Adaptive knowledge-based resource distribution system based on uncertain variables

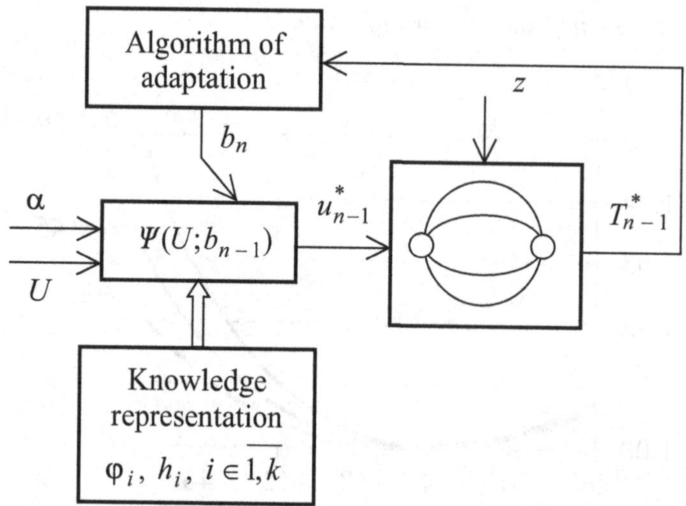

it may be indicated: (i) taking advantage of better experts or (ii) applying adaptation of a knowledge-based resource distribution algorithm Ψ. In case (ii) adaptation requires availability of a real-life complex of operations or its simulator and may consist in *step by step* changing of the parameters b, using performance evaluation and starting from the values of b resulting from the experts' description. It was suggested that a performance index $Q = T^*/\overline{T}$ be used to evaluate performance of a knowledge-based resource distribution algorithm for a complex of parallel operations. Index Q may be useful for comparing different experts, under assumption that a deterministic model of the plant is known. In this chapter, we assume that a deterministic model is unknown, so $\overline{T}$ cannot be calculated. However, under assumption that a real-life complex of operations is available, we suggest application of an adaptation process using a performance index $Q = T^*$, which leads to an adaptive system, presented in Figure 7.

The following algorithm of adaptation based on stochastic approximation method is proposed

$$b_{n+1} = b_n + \gamma_n \beta_n, \quad n = 0,1,...$$

where

$$\beta_{n,i} = \frac{Q(b_n + \overline{\delta}_i) - Q(b_n)}{\delta}, \quad i = 1,2,...,r$$

is the estimation of $\left.\frac{\partial Q}{\partial b_i}\right|_{b_i = b_{n,i}}$; $\overline{\delta}_i$ is a vector of all components equal to 0 except the i-th one equal to δ (testing step), and r is a number of parameters adjusted in the resource distribution algorithm. The coefficient $\gamma_n > 0$ should satisfy the following relationships to ensure the convergence of an adaptation process

$$\lim_{n\to\infty} \gamma_n = 0, \; \sum_{n=0}^{\infty} \gamma_n = \infty, \; \sum_{n=0}^{\infty} \gamma_n^2 < \infty.$$

The application of the adaptation process is possible if distribution of a resource and execution of the operations are activities performed repeatedly. In the presented above algorithm of adaptation, it is necessary to take into account specific features of the problem under consideration:

1. The algorithm Ψ may be given by three different formulas corresponding to three parts of (6).

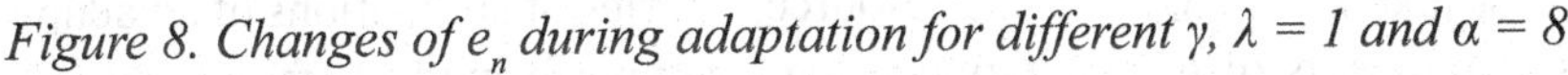

Figure 8. Changes of e_n during adaptation for different γ, λ = 1 and α = 8

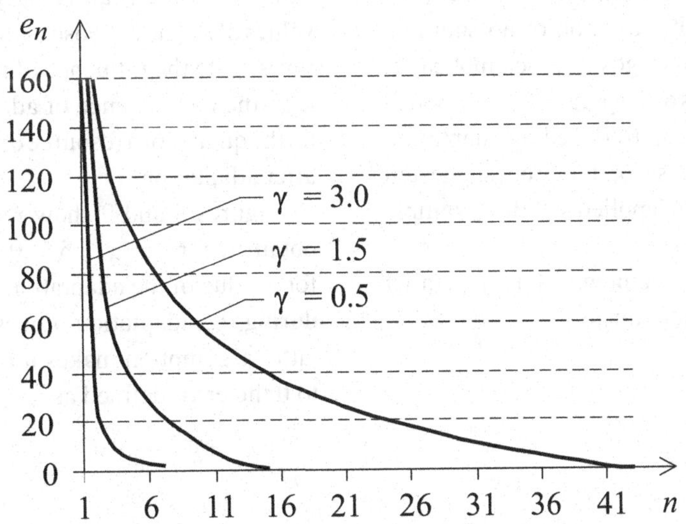

Figure 9. Changes of e_n during adaptation for different λ, γ = 3 and α = 8

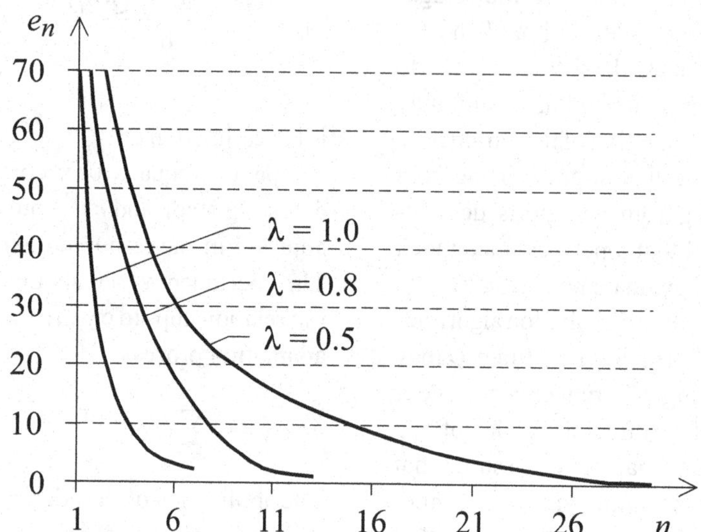

2. The constraints (2) should be satisfied in every step of the adaptation process when using a real-life complex of parallel operations.

As components of b, we may take $k-1$ of k parameters $\underline{d}_i \triangleq \frac{d_i}{d_1+...+d_k}$, $i \in \overline{1,k}$, and change values of parameters d_i accordingly at each step of the adaptation process, or take $k-1$ of k parameters x_i^*, $i \in \overline{1,k}$, as components of b, and change values of parameters d_i when it is necessary to avoid the situation of no solution. In both approaches, changes to d_i cannot cause any changes to values of the parameters $\underline{d}_i$ which are subject to changes introduced by adaptation algorithm only. The second of the suggested approaches has been applied in the example presented below.

Let us consider a complex of two parallel operations of execution times

$$T_1 = \frac{8}{u_1^\lambda}, T_2 = \frac{2}{u_2^\lambda},$$

described by experts in the form of inequalities

$$T_1 \le \frac{x_1}{u_1}, \quad T_2 \le \frac{x_2}{u_2}$$

where x_1 and x_2 are values of uncertain variables $\overline{x}_1$ and $\overline{x}_2$ characterized by triangular certainty distributions (Figure 1) with $x_1^* = 8$, d_1=6, $x_2^* = 8$ and $d_2 = 2$. For $U = 1$ (normalized total amount of a resource) a series of simulations of an adaptation process was performed. Different real-life situations were examined by choosing different values of λ. The purpose of the simulations was to investigate the influence of the coefficient γ_n and λ on the convergence of adaptation process and on the quality of a resource distribution algorithm after adaptation.

Figures 8 and 9 show results of simulations obtained for $b = x_1^*$, δ = 0.1 (testing step) and for a value of the adaptation coefficient constant during the adaptation process, that is, $\gamma_n \triangleq \gamma$. The latter assumption makes it impossible to reduce to 0 the error defined as

$$e_n = \frac{T_n^* - \overline{T}}{\overline{T}} \cdot 100\% \ ,$$

but is sufficient to satisfy the following stop condition

$$\beta_{n+1} \cdot \beta_n \leq 0.$$

According to this condition, the adaptation process is terminated after N steps if the correction to b_N is of a different sign than was the correction to b_{N-1}. The stop condition was suggested based on the observation that with constant γ a satisfactorily small error value was achieved after N adaptation steps during which the adaptation process was monotonic.

We may see that e_n decreases in a monotonic way, which is an advantage of the method proposed. The error decreases faster for larger values of γ and for values of λ being closer to that suggested by an expert.

In the framework of a research on adaptive resource distribution systems a promising approach using artificial neural networks has been developed and reported (Orski et al., 2006). In one method, a specifically designed artificial neural network plays a role of an analytical resource distribution algorithm with weighting parameters adjusted during the learning process, using external trainer (expert) or a real-life complex of operations. In the other method, artificial neural network plays a role of an analytical algorithm of adaptation adjusting parameters of an analytical resource distribution algorithm. In the latter method, a general-purpose multilayer network may be used together with its learning algorithm. Both methods have been verified through a number of simulations which have proven that these methods may be successfully applied also in case where external disturbances influence a real-life complex of operations.

OTHER PROBLEMS AND EXTENSIONS

This section contains a presentation of three concepts which, when desired or necessary, may be applied to a knowledge-based resource distribution. The first one introduces an extension of the knowledge representation by assuming that the parameters in triangular certainty distributions given by experts are values of *random variables.* The second concept consists in employing a *complex definition* of an uncertain variable in place of a *basic definition* of an uncertain variable used to formulate and solve the resource distribution problems. The third concept is similar to that of the adaptive system, but instead of adjusting parameters in the analytical resource distribution algorithm, the parameters of triangular certainty distributions are modified based on an observation of current resource consumption and time elapsed, that is, *learning* a knowledge representation is performed.

Three-Level Uncertainty

Let us consider a set of k operations described by inequalities (1) where T_i is the execution time of the *i*-th operation, u_i is the amount of a resource allocated to the *i*-th operation, an unknown parameter $x_i \in R^1$ is a value of an *uncertain variable* $\bar{x}_i$ described by a certainty distribution $h_i(x_i; w_i)$ given by an expert, and $w_i \in W_i$ is a *random vector variable* $\tilde{w}_i$ described by probability density $f_i(w_i)$. Both $(\tilde{w}_1, ..., \tilde{w}_k) = \tilde{w}$ and $(\bar{x}_1, ..., \bar{x}_k) = \bar{x}$ are vectors of independent variables. The execution time T is described by a relation defined by (1) and (3) with the vector of uncertain parameters $\bar{x}$ described by $h(x; w)$ where w is a value of a random variable $\tilde{w}$ described by $f(w)$. Assumption about randomness of w may be justified when we have a representative group of experts randomly chosen from a whole "population" of experts, each of them suggesting values w. Based on their opinions estimates of probability densities $f_i(w_i)$ may be calculated, which results in knowledge integration. We can distinguish two levels of uncertainty concerning the unknown parameters (Bubnicki, 2004a). In fact, we have three levels of uncertainty concerning the complex of operations:

1. Relational level described by (1) and (3).
2. Uncertain level described by $h(x;w)$.
3. Random level described by $f(w)$.

For the given values U and α one may determine the certainty index

$$v[T(u;\bar{x}) \tilde{\leq} \alpha] \triangleq v(u;w,\alpha,U)$$

analogous to (5). Then, for the given certainty distributions h_i and probability densities f_i, $i \in \overline{1,k}$, the following three versions of the resource distribution problem may be formulated (Orski, 2007):

Version I. Given U and α, find u maximizing the expected value of $v(u;w,\alpha,U)$, subject to the constraints (2).

Version II. Given U and $v(u;w,\alpha,U)=\bar{v}$ (the required *certainty threshold*), find u minimizing the expected value of α, subject to the constraints (2).

Version III. Given α and $\bar{v}$, find u minimizing the expected value of U, subject to the constraints (2).

Let us present now a solution algorithm to a resource distribution problem formulated in version I for a complex of parallel operations described by (1) and (4), for triangular certainty distributions $h_i(x_i;w_i)$ with random parameters $w_i=(w_{i1},w_{i2})=(x_i^*,d_i)$ and $f_i(w_i)=f_{i1}(x_i^*)\cdot f_{i2}(d_i)$, $i \in \overline{1,k}$. One should determine

$$u^* = \arg\max_u \mathrm{E}[v(u;\tilde{w},\alpha,U)],$$

that is, the allocation maximizing an expected value of the certainty index $v(u;w,\alpha,U)$, subject to the constraints(2). Because $v(u;w,\alpha,U)=\min_i v_i(u_i;w_i)$ then $\mathrm{E}[v(u;\tilde{w},\alpha,U)] = \mathrm{E}[\min_i v_i(u_i;\tilde{w}_i)]$, where $v_i(u_i;w_i)=v[\varphi_i(u_i,\bar{x}_i) \tilde{\leq} \alpha] = 1-(w_{i1}-\alpha u_i)w_{i2}^{-1}$.

The solution may be based on expected values $\mathrm{E}[v_i(u_i;\tilde{w}_i)] \triangleq e_i(u_i)=a_i u_i + b_i$, where

$$a_i = \alpha \int_{R^1} w_{i2}^{-1} f_{i2}(w_{i2}) dw_{i2},$$

$$b_i = 1 - \int_{R^1}\int_{R^1} w_{i1} w_{i2}^{-1} f_{i1}(w_{i1}) f_{i2}(w_{i2})\, dw_{i1} dw_{i2}$$

may be obtained in an analytical or numerical way. It is easy to note that the optimal allocation should satisfy the set of equations $e_1(u_1)=e_2(u_2)=\ldots=e_k(u_k)$. Then, in case $0<v<1$:

$$u_i^* = \left(U + \sum_{j=1}^{k} \frac{b_j - b_i}{a_j}\right)\left(\sum_{j=1}^{k} \frac{a_i}{a_j}\right)^{-1},\ i \in \overline{1,k},$$

$$e_i(u_i^*) = \left(U + \sum_{j=1}^{k} \frac{b_j}{a_j}\right)\left(\sum_{j=1}^{k} \frac{1}{a_j}\right)^{-1}.$$

Application of *C*-Uncertain Variables

Apart from a basic definition of an uncertain variable, which is widely used in all applications of the uncertain variables, a complex definition has been introduced (Bubnicki, 2004a) as the so-called *C*-uncertain variable. According to this definition, the *C*-certainty index of the same property as in (5) would be defined in the following way:

$$v_c[T(u;\bar{x}) \tilde{\leq} \alpha] = \frac{1}{2}\{ v[T(u;\bar{x}) \tilde{\leq} \alpha] + 1 - v[T(u;\bar{x}) \tilde{>} \alpha]\} \triangleq v_c(u;\alpha,U).$$

This means that in the calculation of the certainty index (6) for the i-th operation, both

$$D_i(u_i;\alpha_i) = \{x_i \in R^1 : \varphi_i(u_i,x_i) \leq \alpha_i\} = (0, \alpha_i u_i]$$

and its complement

$$\bar{D}_i(u_i;\alpha_i) = \{x_i \in R^1 : \varphi_i(u_i,x_i) > \alpha_i\} = (\alpha_i u_i, \infty)$$

should be taken into account. Consequently, both parts of $h_i(x_i)$, that is, for $x_i \leq x_i^*$ and for $x_i > x_i^*$, will be taken into account, which results in making better use of expert's knowledge. In case of a basic definition of an uncertain variable only one part of the certainty distribution ("half" the expert's knowledge) is used in the determination of the resource allocation. In the determination of $v_c(u;\alpha,U)$ and in the solution of resource distribution problem the following C-certainty indexes for particular operations will be used:

$$v_{ci}(u_i;\alpha_i)=\frac{1}{2}\{v[\varphi_i(u_i,\bar{x}_i)\tilde{\leq}\alpha_i]+1-v[\varphi_i(u_i,\bar{x}_i)\tilde{>}\alpha_i]\}=$$

$$\frac{1}{2}\{\max_{x_i\in D_i(u_i;\alpha_i)} h_i(x_i)+1-\max_{x_i\in \bar{D}_i(u_i;\alpha_i)} h_i(x_i)\}=$$

$$\begin{cases} 1 & \text{for } x_i^*+d_i\leq\alpha_i u_i \\ \frac{1}{2}(\frac{1}{d_i}\alpha_i u_i-\frac{x_i^*}{d_i}+1) & \text{for } x_i^*-d_i\leq\alpha_i u_i\leq x_i^*+d_i \\ 0 & \text{otherwise} \end{cases}$$

A very important and useful theorem may be proved, defining a set of equations which should be satisfied by the solution of a resource distribution problem in version I:

$$v_{c1}(u_1;\alpha_1)=v_{c2}(u_2;\alpha_2)=...=v_{ck}(u_k;\alpha_k).$$

For example, for the complex of parallel operations, applying the above set of equations and definitions of $v_{ci}(u_i;\alpha_i)$ one obtains the same resource distribution algorithm as given by (7) for the basic definition of the uncertain variable. This time, however, (8) is replaced with

$$v_c(u^*)=\frac{1}{2}[\alpha U-\sum_{i=1}^{k}(x_i^*-d_i)](\sum_{i=1}^{k}d_i)^{-1}$$

for $\sum_{i=1}^{k}(x_i^*-d_i)\leq\alpha U\leq\sum_{i=1}^{k}(x_i^*+d_i)$. The difference between applications of basic and C-uncertain variables is more evident in version II of the resource distribution problem. Based on the above formula for $v_c(u^*)$, one obtains

$$\alpha_{c\min}=\frac{1}{U}[(2\bar{v}_c-1)\sum_{i=1}^{k}d_i+\sum_{i=1}^{k}x_i^*]$$

and the optimal allocation

$$u_{ci}^*=\frac{(2\bar{v}_c-1)d_i+x_i^*}{(2\bar{v}_c-1)\sum_{j=1}^{k}d_j+\sum_{j=1}^{k}x_j^*}U, \qquad i\in\overline{1,k}.$$

For example, for $\bar{v}_c=1$ we get $\alpha_{c\min}=\frac{1}{U}(\sum_{i=1}^{k}x_i^*+\sum_{i=1}^{k}d_i)$ and

$$u_{ci}^*=\frac{x_i^*+d_i}{\sum_{j=1}^{k}x_j^*+\sum_{j=1}^{k}d_j}U,$$

whereas for $\bar{v}=1$ the results are as follows: $\alpha_{\min}=\frac{1}{U}\sum_{i=1}^{k}x_i^*$ and

$$u_i^*=\frac{x_i^*}{\sum_{j=1}^{k}x_j^*}U.$$

This simple example shows that if maximum certainty threshold is required, then in case of C-uncertain variables both x_i^* and d_i are taken into account, whereas in case of a basic definition of the uncertain variable only the most certain values x_i^* are used.

Learning System for Resource Distribution

The purpose of this ongoing research is to explore possibilities of using actual information on the execution of all operations, obtained at a current moment of time, to update the initial knowledge obtained from experts. The updated knowledge then should be a basis for the determination of the allocation of unused resources, that is, for the redistribution of resources. The problems addressed in the framework of this research are the following:

i. Evaluation of execution of operations completed and of operations being executed;

ii. Knowledge validation;
iii. Knowledge updating by proper adjustment of the certainty distributions, based on the current result of evaluation; and
iv. Resource redistribution based on updated knowledge.

The redistribution of resources referred to in (iv) should be done using resource distribution algorithms presented previously, but issues indicated in (i)-(iii) are new and will be briefly explained below for version III of the resource distribution problem.

(i) Execution Evaluation for a Single Operation

Let us assume that solving the distribution problem (version III) for the whole set of operations with given $\overline{v}$ and α resulted in values u_i^* and α_i, $i \in \overline{1,k}$. Then, *after* execution of the *i*-th operation the following effects may be observed in terms of execution time and resources used:

a. $T_i = \alpha_i$ and $u_i < u_i^*$ (amount of resources smaller than allocated was sufficient); and
b. $T_i \geq \alpha_i$ and $u_i = u_i^*$ (all allocated resources have been consumed).

In version II we would have $T_i \leq \alpha_i$ and $u_i \geq u_i^*$. Both for the case a) and b), based on (6), the performance index

$$\delta_i = \frac{\alpha_i u_i^*}{\overline{v} T_i u_i}$$

may be suggested. It takes a value greater than 1 when the execution time T_i is more optimistic than the required α_i or when the amount of resources actually used u_i is smaller than the amount of resources allocated u_i^*. The above definition accommodates the intuition that the requirement $T_i \leq \alpha_i$ is, in fact, weakened if the certainty threshold $\overline{v} < 1$.

(ii) Knowledge Validation for a Single Operation

On the basis of (1) and (4) one can get the inequality

$$x_i \geq T_i u_i$$

describing possible values of the unknown parameter x_i. This new knowledge resulting from observed values T_i, u_i may be compared to the knowledge given by an expert in the form of $h_i(x_i)$. Let us note (see Figure 1) that only if $T_i u_i \leq x_i^* - d_i$ we may say that the expert's knowledge is valid (thoroughly), and if $T_i u_i \geq x_i^* + d_i$ – that it is invalid (thoroughly), whereas in other cases it may be considered valid to some extent. Then, it seems reasonable to list and analyze all other cases of $T_i u_i$, so as to refer to them when introducing and verifying rules for knowledge updating.

(iii) Knowledge Updating for a Single Operation

We may verify whether the knowledge about x_i used so far is consistent with the current observation and, if not, try to correct it. The similar idea described in Bubnicki (2005) consists in passing to the expert results of current observation and obtaining from her/him a new certainty distribution for the uncertain variable $\overline{x}_i$. In the approach presented here, we assume that the certainty distribution obtained initially from an expert is then corrected automatically, without any further consultation usually unavailable in real-world situations. The learning (knowledge updating) procedure may be based on the performance index δ_i, which is a kind of a measure of discrepancy between the expected effects of the operation and the actual effects observed after its execution. From the economical point of view, we do not want to have $\delta_i > 1$, and would rather use only as much resources as necessary to achieve T_i assumed to be satisfactory to a user. Because $\delta_i < 1$ means that the observed execution time T_i was not satisfactorily short, we may use $\delta_i^* = 1$ as a desired value of the performance index and suggest that a general rule for knowledge updating should be based on the actual error value $\varepsilon_i = (\delta_i - \delta_i^*)$. Then, certainty distribution parameters would be modified so as to reduce ε_i. Particular modification algorithms are under design.

CONCLUSION

The purpose of the research presented in this chapter is to develop methods and algorithms useful for the implementation in a computer system supporting resource distribution in uncertain environments. This computer system based on experts' knowledge, that is, the expert system, could assist managers in an initial phase of planning resource acquisition and distribution on a customer order with possibly imprecisely defined requirements (α, or α and $\bar{v}$). It appears that it would be reasonable and desired to use this expert system in business applications to support managers in the contract negotiation phase.

However, the final step still has to be done. The methodology for determining resource distribution algorithms, presented for simple cases of a mixed structure and for a case of cascade-parallel structure, should be a basis for developing a general resource distribution algorithm solving distribution problems in a uniform way. With this general algorithm, the expert system's designer would not have to go into a detailed analysis of a structure of the complex of operations so as to appropriately perform decomposition or aggregation and then apply algorithms for parallel or cascade subcomplexes.

It is expected that resource distribution problems in versions II and III should be solvable analytically for any mixed structure of a complex of operations, whereas finding resource allocation in version I will always involve numerical or numerical-analytical procedures. Then, the expert system should integrate:

i. Knowledge on the structure of a complex of operations, for example, expressed by using a typical graph theory model like arcs-to-nodes adjacency matrix;
ii. Knowledge on particular operations obtained from experts; and
iii. A general solution algorithm, that is, a knowledge processing algorithm using (i) and (ii).

The general solution algorithm (iii) would further integrate analytical methods for subcomplexes of parallel (versions I, II and III) or cascade (versions II and III) operations, and numerical or numerical-analytical procedures dedicated for the subcomplex of cascade operations (version I).

When implemented, the expert system may be further extended by adding adaptation or learning functionality. Particular methods and algorithms have been already developed and examined by computer simulations (adaptation) or are being developed and will be available in the nearest future (learning). These extensions would allow improvement of system's performance by reducing the initial uncertainty, based on online observation of executions of real-life operations. Therefore, such an extended expert system could assist not only the contract negotiation phase, but also the phase of carrying out the tasks required to fulfill the contract.

Further enhancements of the expert system could be related to the concepts of applying random variables and of applying the formalism of *C*-uncertain variables. These would allow integration of knowledge obtained from multiple experts and better exploration of this knowledge, which could result in better quality of the resource distribution determined and executed even before application of adaptation or learning.

The presented approach may be extended to new and practically important problems, not addressed in this chapter, of (i) resource distribution in complexes of operations with execution times described by two inequalities (instead of a single one) or (ii) resource distribution with a combined performance index, for example, taking into account both execution time and an amount of a resource.

REFERENCES

Banaszak, Z., & Jozefowska, J. (Eds). (2003). *Project driven manufacturing*. Warsaw, Poland: WNT.

Berman, E. B. (1964). Resource allocation in a PERT network under continuous activity time-cost functions. *Management Sciences, 10*(4).

Bubnicki, Z. (2001a). Uncertain variables and their applications for a class of uncertain systems. *International Journal of Systems Science, 32*(5), 651-659.

Bubnicki, Z. (2001b). Uncertain variables and their application to decision making. *IEEE Transactions on SMC, Part A: Systems and Humans, 31*(6), 587-596.

Bubnicki, Z. (2002). Uncertain variables and their applications for control systems. *Kybernetes, 31*(9/10), 1260-1273.

Bubnicki, Z. (2003). Application of uncertain variables to a project management under uncertainty. *Systems Science, 29*(2), 65-79.

Bubnicki, Z. (2004a). *Analysis and decision making in uncertain systems*. Berlin, London, New York: Springer-Verlag.

Bubnicki, Z. (2004b). Quality of an operation system control based on uncertain variables. In M. H. Hamza (Ed.), *Modelling, identification and control* (pp. 148-153). Zurich: Acta Press.

Bubnicki, Z. (2005). Application of uncertain variables in learning algorithms for uncertain systems. In G. Lasker (Ed.), *Advances in Computer Cybernetics: Proceedings of InterSymp 2005*, (pp. 25-29), Tecumseh: IIAS.

Fargier, H., Galvagnon, V., & Dubois, D. (2000). Fuzzy PERT in series-parallel graphs. In *Proceedings of the IEEE International Conference on Fuzzy Systems*, (pp. 717-722). San Antonio, TX: IEEE Press.

Klir, G. (2006). *Uncertainty and information: Foundations of generalized information theory*. Hoboken, NJ: Wiley Interscience.

Mon, D.-L., Cheng, C.-H., & Lu, H.-C. (1995). Application of fuzzy distributions on project management. *Fuzzy Sets and Systems, 73*, 227-234.

Orski, D. (2005a). Quality of cascade operations control based on uncertain variables. *Artificial Life and Robotics, 9*(1), 32-35.

Orski, D. (2005b). Application of uncertain variables to planning resource allocation in a class of research projects. In *Proceedings of the 18th International Conference on Systems Engineering*, (pp. 238-243). Los Alamitos: IEEE CS Press.

Orski, D. (2006). Application of uncertain variables in decision problems for a complex of operations. *IIAS Transactions on Systems Research and Cybernetics, VI*(1), 19-24.

Orski, D. (2007). Knowledge-based decision making in a class of operation systems with three-level uncertainty. *Acta Systemica, VII*(1), 19-24.

Orski, D., & Hojda, M. (2007). Application of uncertain variables to decision making in a class of series-parallel production systems. In A. Grzech (Ed.), *Proceedings of the 16th International Conference on Systems Science*, (Vol. II, pp. 131-142). Oficyna Wydawnicza PWr: Wroclaw.

Orski, D., Sugisaka, M., & Graczyk, T. (2006). Neural networks in adaptive control for a complex of operations with uncertain parameters. *Systems Science, 32*(2), 19-35.

Chapter V
A Methodology of Design for Virtual Environments

Clive Fencott
University of Teesside, UK

ABSTRACT

This chapter undertakes a methodological study of virtual environments (VEs), a specific subset of interactive systems. It takes as a central theme the tension between the engineering and aesthetic notions of VE design. First of all method is defined in terms of underlying model, language, process model, and heuristics. The underlying model is characterized as an integration of Interaction Machines and Semiotics with the intention to make the design tension work to the designer's benefit rather than trying to eliminate it. The language is then developed as a juxtaposition of UML and the integration of a range of semiotics-based theories. This leads to a discussion of a process model and the activities that comprise it. The intention throughout is not to build a particular VE design method, but to investigate the methodological concerns and constraints such a method should address.

INTRODUCTION AND PROBLEM STATEMENT

Interactive systems (ISs) are becoming ubiquitous to the extent that there is the very real possibility of their disappearing altogether, at least in the sense of users' perceptions of them as entities worthy of conscious identification. This very ubiquity will largely be the result of effective design, which results in ISs becoming so embedded in our everyday lives that we use them without conscious thought. We can draw an analogy here with the electric motor, which pervades almost all everyday technologies and yet is hardly ever noticed. In the early twentieth century, it was possible to buy electric motors for the home along with a variety of attachments for food preparation, hair drying, vacuum cleaning, and so on. Today we buy specialized gadgets, many of which contain electric motors that go largely unnoticed by us. Even the mobile phone contains an electric motor that is weighted to spin off-centre in order to create the vibrations that can silently signify an incoming call.

Will this ever be the case with ISs? Will they ever be so effectively designed that they cease to attract conscious attention in their final ubiquity? Certainly, the theory of design for ISs is still in its infancy; hence the need for the present volume.

Before considering their design, we first need to make clear what we mean by ISs. Many systems are interactive but outside the remit of this book. Motor cars, power drills, electric kettles, and so on are all interactive systems that will not be the subject of this chapter. By ISs we surely mean interactive digital systems (IDSs) that make use of digital representations and operations on these in order to effectively perform their allotted tasks. IDSs will therefore identify everything from ATMs and remote controlled TV teletext systems to PC and game console applications to onboard computers in cars and fly-by-wire aircraft.

An interesting subset of IDSs are interactive digital environments (IDEs) by which we mean an IDS that creates a large-scale digital environment that takes time and effort to explore and otherwise interact with. Examples of IDEs are videogames and virtual environments (VEs) in general, computer-based learning applications, and large-scale sites on the World Wide Web. These are interesting because the scale and complexity of their content demands that their effective design transcend established user interface techniques. Indeed, for VEs the very term *design* is a problem because it has to be interpreted in two quite distinct ways. First of all there is the notion of designing something to create the desired perceptual and aesthetic responses: essential for computer games. Secondly, there is the engineering notion of design as the creation of plans and models from which to test and build the desired artefact and ensure its correct functioning. Both forms of design are of equal importance to the design of effective VEs. It is the tension between these two notions of design and the resolution of this 'design tension' that is the central problem addressed in this chapter.

The need to resolve or at least alleviate this tension leads to a consideration of methods for VE design. It is assumed by some that the design of effective VEs will necessitate a development methodology akin to those used (or not) by software engineers. This is not necessarily the case. A craft-based approach based on the application of good practice—perhaps acquired through some form of apprenticeship—might do equally well. The computer games industry seems to prosper on just such an approach. The approach taken in this chapter is that an appropriate form of development methodology for VEs is viable, but that that methodology needs to accommodate—and certainly not stifle—the creative flair that is at the heart of aesthetic design of such large and complex systems.

This chapter therefore concerns itself with the investigation of what form an appropriate design methodology for VEs would take and the obstacles to establishing such a methodology. It is thus primarily concerned with a methodology of design—in other words, the meta-study of VE design methods rather than the outline of a particular method, although this is an obvious objective.

This chapter first undertakes an overview of the meaning of the various terms involved in the discussion: method, methodology, model, and language, among others. It then goes on to discuss the particular form an 'underlying model' for a VE method would have to take. Following this the issue of the form a language for expressing VE design decisions might take with regard to the underlying model put forward in the third section is addressed. The chapter then goes on to establish a process model for VE design and the 'practice of methodology' it to a large extent determines. It finally attempts to address future trends in the field and is followed by a short conclusion to the issues raised.

TERMINOLOGY

A methodology of design for VEs concerns itself with the study of methods for the design of VEs; in other words, the nature, definition, and application of such methods. This notion of methodology, while being quite correct, is at variance with a related but somewhat different notion that commonly views a methodology as a configurable method. In this chapter we use the approach of the former in order reach some conclusions with respect to achieving the latter.

If we are considering the study of methods for VE design, what do we mean by method in the first place? In software engineering the concepts of method and model are commonly understood, although the formality with which they are defined and applied varies considerably.

With respect to the question posed above, we will adopt the definition of Kronlof (1993) who defined a method as consisting of the following:

- An underlying model
- A language
- A process model
- Heuristics

Fencott et al. (1994) discuss these terms in the context of investigating the integration of structured and formal methods for software engineering. Methods integration will also be at the heart of the investigations of this chapter. Before using this characterization of method to address VE design, we will discuss the concept of model in some detail as it appears twice above in seemingly different contexts.

Models have been at the heart of much of human understanding and enquiry from very ancient times. Cultures very often attempt to explain the world and human beings' place in it by means of complex mythologies. Such mythologies are essentially abstractions—etiological fables (Carruthers, 1998)—that allow complex and inexplicable phenomena to be understood in terms of a more accessible set of characters and stories set around them. Very often the underlying explanation of phenomena will map onto supernatural beings and phenomena which thus replace unfathomable cause with commonly held narrative.

With time, more rigorous forms of modelling were invented. The ancient Mesopotamians developed sophisticated mathematics as a technique for modelling trade involving large numbers of items and customers (Davis & Hersh, 1983). This early theory of mathematics was thus being used to build abstract models of trade and stock control. The ancient Greeks and following them the Arabic world continued to develop models—mathematical and otherwise—for a variety of phenomena ranging from cosmology to music and poetry. Meter and rhyming schemes for poetry, for example, are models that facilitate the construction of new poems within established forms. This leads us naturally to ask what we mean by the term model, and how and why models are so generally useful?

The Concise Oxford dictionary variously describes a model as "a representation of structure"; "a summary, epitome, or abstract"; and "something that accurately resembles something else." Formal logic uses the term *model* to mean the system of rules by which meaning is mapped onto the syntactic constructions expressed within a particular logic. It is thus possible for a model to be highly formal—that is, expressed in mathematics—or highly informal, but not presumably both. Scientific models may be more pragmatic in that they are related to some aspect of reality by means of observational data, which in turn causes the hypothesis upon which the model is constructed to be reformulated and so on. In other words they are empirical rather than strictly formal and thus sit somewhere between the extremes of the formal-informal axis.

As already mooted with respect to etiological fables, models may be quite instrumental in the sense that the application of the model as an analysis technique—and the results obtained therein—may be more important than the degree to which the model accurately reflects reality; psychoanalysis is an obvious example. Semiotics (Chandler, 2002) is perhaps another case in point because it has never been ascertained whether or not signs as defined by semioticians actually represent structures or functions within the human brain. There is some evidence to support this (e.g., Damasio, 1994). Nonetheless, semiotic analysis of communications artefacts—texts to semioticians—is a very valuable and general technique for gaining insights into the way in which humans communicate and make meaning using a whole range of media. Semiotics is very important to this chapter.

With respect to Kronlof's characterization of method, we can see that the term model is used in two rather different ways:

1. An 'underlying model' is a semantic structure to which terms of the language of the method are mapped in order to assign meaning to them.
2. A 'process model' is an abstract representation of the activities undertaken as part of the model along an expression of their ordering.

The first use of the term model given above is a formal notion, while the second is the more intuitive notion of an abstraction of some more complex system, both discussed in our aside above. If we were to take the language and its underlying model together, we would arrive at the second form of model which is essentially a notation for simplifying and elucidating a more complex system. But what language and underlying model are we to use for VE design? The role of the former is to facilitate the creation and expression of design decisions. The role of the latter is less obvious, but its nature has a direct bearing on the applicability of the method in general. The two parts of this question are addressed in the succeeding sections of this chapter.

The process model of a method is most often expressed as a simple diagram, a graph where the nodes name particular activities and the arcs indicate the relative ordering over time of these activities. The graph is thus a focused simplification of a complex set of activities and the relationships between them and their products. What process model might be suitable for VEs? Kaur (1998) put forward a tentative process model for VEs as an ordered list of activities. These activities and their ordering were deduced from questionnaire data drawn from a limited number of VE developers. Fencott (1999b) put forward a process model that was more representative of the design tensions inherent to VEs. We will return to the process model after the sections devoted to language and underlying model.

Heuristics are essentially advice and guidelines on the successful application of the model to real problems. In terms of VE design, we can observe that there are a lot of such heuristics around in terms of standalone advice that is almost invariably devoid of a methodological context with respect to VEs. There are exceptions to this, the 'SENDA' method of Sanchez-Segura et al. (2003; also, see Chapter 4) for example.

The 'design tension' identified above as the driving force in the methodology of VE design has its antecedents. In the early 1990s there was a debate as to whether formal methods or structured methods for software design were most appropriate. The former use logic and set theory to build mathematical models of software systems, while the latter use diagrams, pseudo code, and other 'non-formal' notations to the similar ends. Integrated methods research attempted to combine these approaches to maximize the strengths and minimize the weaknesses of both (Fencott et al., 1992, 1994). In this chapter we draw on the experiences gained in the earlier research in order to address the design tension directly.

In this section we have posed a number of questions with respect to a possible VE design method:

1. What language and underlying model are we to use for VE design?
2. What process model is appropriate for VE design?
3. What sort of heuristics do we need and are any of those extant adaptable to the model we hypothesize in 1 and 2 above?

In this chapter we specifically deal with Questions 1 and 2. Question 3 will be for future consideration, as it depends on the answers to Questions 1 and 2.

THE UNDERLYING MODEL

The question of what an underlying model might be for a VE design methodology might seem of purely theoretical interest, but attempting to answer it necessitates a consideration of the design tension highlighted in the previous two sections. We have to find an underlying model that expresses the meaning of a VE design in terms of both:

- **Engineering:** As a computer system composed of program and hardware, understood largely by those trained in computer science and related disciplines;
- **Aesthetics:** As an interactive communications medium, understood by those trained in the creative arts.

We appear to have confounded the issue, as we now seem to need an underlying model that not only addresses two different design issues, but that is understood differently by two quite different groups of professionals. Is one underlying model possible, and who on earth is going to understand it? In fact there have been various attempts to reconcile the two with varying degrees of success, but it's useful for our purposes to consider them separately for the time being.

We can begin to suggest possible underlying models, bearing in the mind the tension already identified. VEs and IDSs in general have interaction machines (IMs) as their underlying model (Goldin et al., 2001) in terms of computational functionality, but we also need a model that operates at the perceptual, meaning-making level. Semiotics (Chandler, 2002) is highly appropriate for the latter. Interaction Machines encompass a set of possible computational systems—more expressive than Turing Machines—that allow for the persistence of state and unlimited user inputs that characterize interactive media, IDSs in general and VEs in particular. Semiotics is the study of sign systems and the way humans find meaning in them. The two might not be so incompatible as a cursory glance might seem to suggest. We will briefly consider each separately and then consider their integration.

For much of the latter half of the twentieth century, it was the received wisdom that Turing Machines captured the notion and limits of what is computable. In the 1990s a number of researchers began to develop models which showed that Turing Machines were not expressive enough to model interactive computer systems. In fact it was shown that the simplest interactive program:

```
P  :=  input(x:Boolean);  output(x);
P
```

which recursively inputs a Boolean value for x and simply outputs that same value, cannot be programmed using any Turing Machine. That this is so even for a very simple datatype such as Boolean might be somewhat surprising. The reason is that although each input and output is finite—a requirement for conventional Turing Machine input—there might be an infinite number of them, and it is impossible to represent such an infinite set of choices on a sequential, yet infinite tape.

Goldin et al. (2001) have shown that Turing Machines can be extended to model interaction by defining Persistent Turing Machines (PTMs), which employ dynamic streams to model inputs and outputs, and a tape to remember the current state ready for the commencement of a new computation. PTMs are an example of the general class of IMs.

PTMs are certainly not the only possible characterization of IMs. We could, for instance, have used an approach based on concurrent systems in the manner of Milner (1989). In many respects this would be better as it not only captures the notion of VEs as IMs, but also allows us to consider them as being the composition of a number of embedded systems—autonomous agents and non-playable characters, for instance. PTMs are, however, better suited as a brief illustration of the concept for our present purposes.

Human beings ceaselessly work to find meaning in any situation they might find themselves, in any communications media they might find themselves using, and in even mundane situations such as walking down the street or sitting on a train or bus. Semiotics is the study of this meaning-making process, and signs are the basic unit of the theory (e.g., Eco, 1977; Barthes, 1987). The most common characterization of signs consists of two components, a:

- **Signifier:** That which we can perceive in the world around us using any of our senses;
- **Signified:** The meaning(s) we form in our minds as a response to perceiving the signified.

Communications artefacts, texts to semioticians, are made up of signs and can be anything we humans find meaningful, for instance: novels, films, body language and facial expressions, and VEs.

Semiotics provides us with a means of understanding the output of a VE, the digital displays, and the signs of intervention, as we shall call them, that the user generates by means of the input technology. VEs are a particular form of IM that attempt to restrict its users' environments to the digital displays it generates in response to user input. We thus have a partially closed system. Semiotics can provide a means of analysing how a user might make meaning out of such a system and thus make meaningful choices about how to interact with it. We can thus refer to our underlying model as a Semiotically Closed Interaction Machine (SCIM).

Figure 1 shows the relationship between semiotics and IMs. The two downward pointing arrows represent inputs by the user, in1 and in2. The horizontal, black arrows represent computation steps that result in the generation of new outputs, out2 and out3. The fuzzy, curved arrow represents the semiotic closure between out2 and in2; in other words the cognitive process of finding meanings in out2 and formulating a response to them as in2. On the one hand we have the human, meaning-making process and on the other the non-semiotic act of using the signs of intervention to create a new input to the IM and thus instigate a further macro-computation step. Note that the diagram is a simplification, as in VEs in general outputs may also be produced without direct input from the user.

In SCIMs such as VEs, the semiotic link is very strong, whereas in IDSs in general, the link may be far weaker and intermittent. There is no semiotic link between individual customer transactions at an ATM, for instance. There is also no recognizable semiotic link between a customer inserting his or her debit card, the PIN input, and the amount of money requested; ATMs are not SCIMs.

Both IMs and semiotics are appropriate as a choice of an integrated, underlying model because they do not constrain us to particular programming languages or computational platforms on the one hand, nor particular modes of communication on the other. That will be the business of the next section when we consider the nature of a language suitable for expressing VE design decisions.

An integrated underlying model is not the only approach. There is a field of enquiry called computational semiotics that has as one of its

Figure 1. Semiotically closed interaction machines

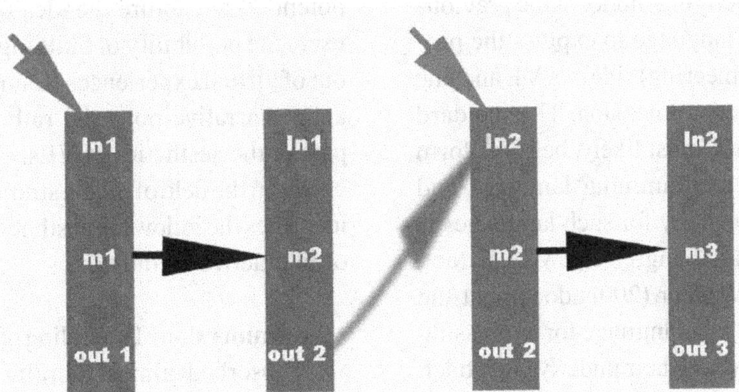

concerns the integration of semiotics and computer science; this can operate at the level of the underlying model or at the level of language within a methodological context while sometimes at both. For instance, Goguen (1999) defines 'algebraic semiotics' as semiotics formalized using the algebraic specification language OBJ. He has outlined the application of this formalism to user interface design and VE design. As another example, Doben-Henisch (1999) has attempted to integrate semiotics with Turing Machines. The problem with the latter is that Turing Machines are not expressive enough to model VEs. The problem with the former as an underlying model for VEs is that the formalism makes use of difficult mathematical concepts, such as category theory, which obscure the insights into the nature of VE that our integrated approach highlights—difficult, that is, for those VE designers without a strong mathematical background.

The integrated underlying model we have adopted is a very practical one, as it preserves the 'design tension' rather than allowing the engineering or the aesthetic dimension to dominate.

THE LANGUAGE

We move now to the nature of languages for expressing VE design decisions. A review of existing work on VE design (Fencott, 2003b) reveals that while there is a quantity of research and commentary on the human factors affecting design, for instance, there is very little that is directly relevant to VE content modelling, which is at the heart of this chapter. There are examples of the construction and application of methods or guidelines for realizing certain aspects of VE design; some of these are:

1. Various work on usability for VEs (e.g., Workshop on Usability Evaluation of Virtual Environments, 1998)
2. Structured methods for VEs (e.g., Workshop on Structured Design of Virtual Environments, 2001)
3. Various commentaries from the computer games world (e.g., Gammasutra, Rollings, & Adams, 2003)
4. Semiotics of games and new media (e.g., Lindley et al., 2001)

In light of the discussion in the previous section, we can make the following observations: 1 and 2 are insufficient to express VE design decisions because they do not address aesthetics adequately; 3 provides some very useful insights; 4 gives us a way to alleviate the inadequacies of 1 and 2.

If we continue with the integrated approach adopted for the underlying model in the previous section, we need a language to express the programming (the engineering) side of a VE and one to express its aesthetic dimension. The standard for the former should, most likely, be some form of object-oriented programming language and the standard methodology for such languages is the Unified Modelling Language (UML). In fact, Goldin, Keil, and Wegner (2001) document the suitability of UML as a language for expressing designs that have IMs as their underlying model. UML would seem a good choice of language for this aspect of VE design.

Aesthetics of VEs has been a constant theme of this chapter, and we now discuss them in some detail. Church calls for a set of "formal, abstract, design tools" (FADTs) that will not only guide the design of successful games, but which will also enable designers to compare and contrast computer games from diverse genres (Church, 1999). Church's FADTs are perhaps better understood as an aesthetic characterization of computer games and are:

- **Intention:** Being able to establish goals and plan their achievement;
- **Perceivable consequence:** A clear reaction from the game world to the action of the player;
- **Story:** The narrative thread, both designer-driven and user-driven, that binds events together.

Other computer games designers talk in a similar vein: of players needing to feel in control, of maintaining the emotional feel of a game and/or level, of providing suitable and timely rewards for effort, and of a perceivable gross structure that allows players to identify what is required of them at the beginning of a level, plan to achieve this, and understand the significance of their achievement (Saltzman, 1999). Intentions and perceivable consequences are the building blocks for this.

Brenda Laurel introduced the term 'narrative potential' to capture the idea that VEs can offer users the possibility of building their own stories out of virtual experiences (Laurel, 1992). We will adopt narrative potential rather than 'story' as part of the aesthetics of VEs.

From the field of media studies, Murray (1996) identifies the following aesthetic characterization of interactive media as:

- **Immersion:** The feeling of being completely absorbed (almost literally immersed) in the content (we will use the term presence for reasons detailed below);
- **Agency:** Being able to affect change in the VE;
- **Transformation:** Being able to become someone or something else.

Lombard and Ditton (1997) define presence as *the perceptual illusion of non-mediation*. This characterizes presence as the state of mind of a visitor to a VE as not noticing or choosing not to notice that that which they are experiencing and interacting with is artificially generated. They document the evaluation of the embodying interface of a VE in terms of presence seen largely as the degree of fidelity of sensory immersion. Much of the research to date into presence is particularly concerned with the embodying interface as well as researches into the mental state of people who are present in VEs. Immersion is thus the degree to which the technology of the embodying interface mediates the stimuli to the senses. Slater has shown that high degrees of sensory immersion heighten the emotional involvement with a VE (Slater et al., 1999).

However, as presence is a mental state, it is therefore a direct result of perception rather than sensation. In other words, the mental constructions that people build from stimuli are more important than the stimuli themselves. It is the patterns that we, as VE constructors, build into the various cues that make up the available sensory bandwidth for a given VE that help or hinder perception and thus presence. These patterns are the result of what is built into the VE and the way the user behaves in response to them. The fidelity of the sensory input is obviously a contributing factor, but by no means the most important. In the context of the working VE builder, being able to identify and make effective use of the causes of presence is more important than the nature of presence itself. This means that it is the effective consideration of the perceptual consequences of what we build into VEs that will give rise to the sense of presence that we are looking for. In this sense it is the content of VEs that has the greatest effect on the generation of presence. Thus, for our purposes, content is the object of perception.

Agency is the fundamental aesthetic pleasure of VEs and IDSs in general and the one from which all the others derive. Agency actually equates quite nicely to Church's intention and perceivable consequence; agency is in part the interplay between intention and perceivable consequence.

Transformation is important to many communications media. One of the great pleasures of novels is seeing the world through someone else's eyes, to view the world through the eyes of another creature, machine, or alien being. VEs in particular are ideally suited to this, and much of the success of 3D computer games is due to the player being able to be the hero or villain in some great and dangerous adventure. In such games the player cannot only play an alien, but through the real-time graphics actually see the world as the alien would see it. It seems certain, for instance, that one of the reasons for the success of the classic Hubble Space Telescope Virtual Training Environment (Loftin et al., 1994) was that members of the ground-based flight team could actually become astronauts for a while, and experience some of the drama and spectacle of a space walk. To the author's knowledge and despite the insightful research into the effect and effectiveness of the Hubble, the question "Did you enjoy being an astronaut for a change?" was never asked. Yet it seems highly likely that this was a major experience for the subjects.

Finally, in this brief review of aesthetics for IDSs, we must include Turkle's (1995) observation that being present with others—sentient beings, robots, creatures, and autonomous agents in general—is something that has drawn users to IDSs since the earliest days of Eliza and MUDs.

Bringing these various aesthetic viewpoints together, we can characterize the aesthetics of VEs as:

- **Agency:** Which itself consists of:
    - **Intention:** Being able to set goals and work towards their attainment.
    - **Perceivable consequence:** Being rewarded for one's mental and virtual activity by sensing the VE change appropriately as a result of the actions taken.
- **Narrative potential:** The sense that the VE is rich enough and consistent enough to facilitate purposive experience that will allow the user to construct her own narrative accounts of it.
- **Co-presence:** Being present with others.
- **Transformation:** Temporarily becoming someone or something else as a result of interacting with the VE.
- **Presence:** The perceptual illusion of non-mediation (Lombard & Ditton, 1997).

In terms of underlying theory, aesthetics are signifieds of a particular type; they are connotations that arise from interacting with VEs. Connotations, in semiotic theory, are deeper levels of meaning that humans build up from the level of denotation: the commonplace or everyday meanings of things.

On a more concrete level, Murray (1996) equates the structure of interactive media with the notion of the labyrinth and asserts that this structure works best when its complexity is somewhere between the 'single path maze' and the 'rhizome' or entangled Web. Aarseth (1999) has proposed the notion of cybertext to capture the class of texts, not just digital, which require the visitor to work to establish their own path(s) through the possibilities offered. He calls this class of text ergodic from the Greek words meaning work and path. So we have a notion of a labyrinth that requires effort to explore. Equating the structure of VEs in general with the notion of a labyrinth of effort would seem useful, but poses several questions. First of all, what are the actual components with which VE designers build such experiential labyrinthine structure? Second, how do VE designers structure a VE so that the visitor follows an appropriate path and, moreover, accumulates an appropriate set of experiences so as to discover and remember the intended purpose of the VE?

Fencott (1999a, 2003a, 2003b) draws on these various aesthetic views to define a model of VE content, Perceptual Opportunities (POs), which focuses on the aesthetic design of the perceptual experiences over time which users are intended to accumulate.

Figure 2 characterizes the breakdown of POs in terms of:

- **Sureties***:* Designed to deliver belief in a VE, equated with unconscious experience (e.g., Spinney, 1998; Blackmore, 1999)
- **Surprises:** Designed to deliver the essential purpose of the VE
- **Shocks:** Perceptual bugs that undermine the first two

Surprises are further broken down into:

- **Attractors:** Literally content that attracts attention
- **Connectors:** Content that supports the achievement of goals
- **Rewards:** Content that literally rewards users for effort

Attractors can be characterized in two ways: By the way they attract attention—they might be mysterious, awesome, active, alien, complex—collections of attractors—and so on. They can also be characterized by the basic emotions they stimulate, typically fear and desire. Rewards can be information, access to new areas of the VE, new activities enabled, and so on. Connectors can be as simple as railings, footpaths, and street signs, but can also be dynamic maps, indicators of

Figure 2. Perceptual opportunities

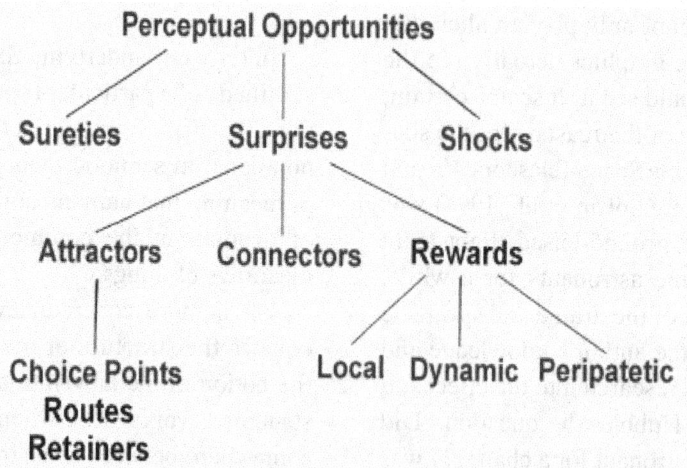

health, wealth, and so on. Attractors are the means by which users are led to form intentions. The perceivable consequences of a player attempting to realize an intention leads to the identification of rewards which leads to the identification of new attractors and so on. Thus agency and POs are very strongly associated.

POs can be organized into higher level structures, perceptual maps, which characterize patterns of behaviour that users exhibit when interacting with a VE. A perceptual map can be made up of:

- **Choice points:** Basically the choice between intentions stimulated by one or more attractors.
- **Challenge points:** Intentions that have to be satisfied.
- **Routes:** Linear sequences of attractors.
- **Retainers:** Mini-missions or mini-games, tightly grouped attractor-reward pairs, puzzles, and so forth.

The arrangement of such structures are thus a realization of Murray's rhizome and lead to the other aesthetic pleasures of narrative potential, co-presence, transformation, and presence.

In a later publication, the same author asserts that POs, as well as the aesthetics identified above, have semiotics as their underlying model (Fencott, 2003b). Essentially, POs and in particular surprises are connotations that humans derive through interacting with VEs. POs interface very closely with the aesthetic pleasure of agency, but at a more abstract level of VE content. Figure 3 illustrates the relationships between POs, aesthetics, and semiotics at the level of the language of a VE design method:

- The two arrows linking attractor and intention and perceivable consequence and reward are semiotic acts, meaning making, on the part of people interacting with a VE. Attractors and perceivable consequences are signifiers, while intentions and rewards are signifieds.

Figure 3. The code of interaction

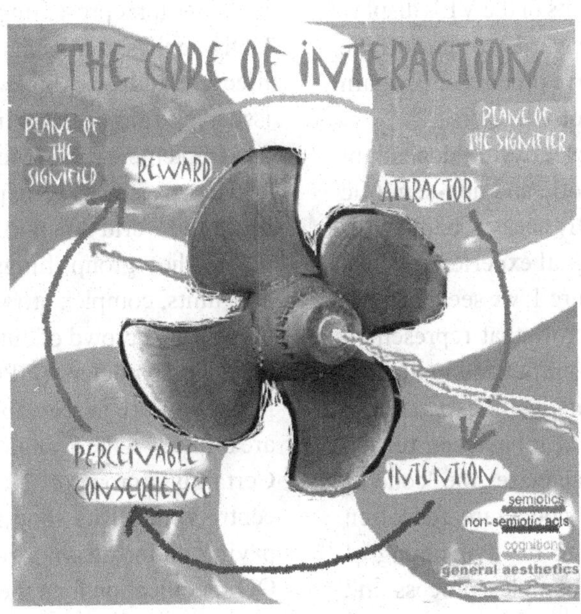

- The arrow linking reward and attractor indicates cognition, though of course cognition is a continuous process and not a segment of a cycle as this diagram would seem to suggest.
- The arrow linking intention and perceivable consequence represents what Tronstad (2001) calls non-semiotic acts that are essentially the site of the IM, the computer-based system in the wider IDE. The term non-semiotic is used because, while the user might draw some significations from pressing interface buttons and so on, the computer responds algorithmically.
- The arrow that runs through the cyclic plane of the above relationships from right to left represents the development over time of the other aesthetic properties of narrative potential, co-presence, transformation, and presence.

On the level of aesthetics and POs, we see the following. Having formed an intention, a user will provide input to the VE, which will trigger the execution of one or more calculations, non-semiotic acts on the part of the computer. This will result in a change (perceivable consequence) in the various digital elements of the VE's display which provide signifiers (rewards) to start off the whole semiotic and cognitive process once again through the identification of attractors.

Figure 3 shows quite clearly the dependant relationship between semiotic and non-semiotic acts, which Tronstad (2001) sees as being fundamental to interactive digital experience. If we compare Figure 3 with Figure 1, we see that what has changed is that the arrow that represented the semiotic closure of the output and input step in the latter has been dramatically expanded in the former; it is almost as if it has been turned 'inside out'. Figure 3 characterizes the 'code of interaction'. In semiotics, codes are the often innate rules that allow us to make meaning of signifiers. Interaction is a complex process and the diagram reflects this. So much so, in fact, that Fencott (2004) devotes a whole chapter to the 'code of interaction. In the context of our present discussions, the various components and the relationships between them that make up the code constitute the general aesthetic side of the language of our method.

Semiotics not only provide an underlying model for POs and aesthetics, they also operate at the level of the language of a method as well. In interacting with VEs we not only recognize the code of interaction—connotations specific to VEs and IDSs in general—but we also find meanings that correspond to world of the 'real' outside the world of the VE. We recognize shops and cars and people and furniture and so on and so on. It is semiotics itself that is used as 'language' in this type of meaning-making.

Therefore, in addressing the second question concerning the nature of the language of a VE design method, we now need to consider how POs, aesthetics, and semiotics on the one hand and UML on the other might work together. In this respect, the central issue that needs to be addressed concerns what we might call the 'object problem'. Objects or rather object-oriented design (OOD) might seem a very promising candidate for our language for representing VEs at the design phase. OOD applies at all stages in the VE production lifecycle, addresses both coding and user-cantered issues, and has been applied directly to VE design and implementation (McIntosh).

However, in the act of perception, people do not break the world down into nicely programmable units. They group things together into perceivable units, complex attractors, which focus their attention. A crowd of autonomous agents—non-playable characters (NPCs) in computer games parlance—are perceivable as a single entity, but are unlikely to be a single object in an OO model. Certainly a crowd of NPCs in a busy shopping centre with all its shop fronts, street furniture, paving, and so on is not going to be an object in an OO specification for a shopping centre. However, each of the entities that makes up the perceivable unit that is the crowded shopping centre will have to be identified in terms of its capacity (or not) for interaction as a basis for its incorporation into a functioning scene graph.

On the one hand, we have the Unified Modelling Language (UML), which models structural, engineering aspects of a scene graph, and on the other hand, we have POs and so on which model content at the level of perception, of aesthetics (Fencott, 1999b, 2003b). There is in fact a bridge, a semiotic bridge, which links the two, and this is Andersen's Computer Based Signs (CBSs) (Andersen, 1997), which model interactive aspects of individual signs (objects) in IDSs in general and thus VEs. Fencott (2003a) discusses this relationship and its relevance to VE design.

Andersen arrives at the following classification of signs in IDSs:

- **Interactive:** Signs that can be controlled by the user and can affect other signs; such signs are subject to the signs of intervention.
- **Actor:** Signs that to a limited extent are autonomous and can affect other signs.
- **Controller:** Signs that constrain other signs but do not themselves change nor can they be affected by other signs.
- **Object:** Signs that can be affected but cannot affect others.
- **Ghost:** A sign that affects others, but only becomes apparent by its effects on others; a sign particular to IDSs. Essentially a controller that signifies its presence solely through effect.
- **Layout:** Non-interactive signs.

CBSs essentially constitute six distinct classes that will be used to instance all objects in a VE implementation. The integration of the three elements of our language of VE design can now be summarized thus:

- Each content item in the perceptual model is assigned to a CBS class.
- Other aesthetic attributes of content items carry over directly to UML, that is, colour, form, and so on.
- General information in the perceptual model carries over to UML to become the game engine, the visualiser.
- Other such information carries over directly to UML in terms of the semiotic realization of the VE: mood, myths, and hyperrealities.

So the language of our VE design method is an amalgam of OO and POs and so on—with some bridging by CBSs. It is, in fact, an integrated method, a process, rather than a statically characterisable relationship. In this way we have carried the design tension identified early in this chapter, and clarified through to the language stage. It seems that we might be able to make this tension work for us rather than it being a hindrance to try to do away with.

THE PROCESS MODEL

The process model captures the relationship over time between the constituent activities of a method. In a sense it captures the essence of the 'practice of methodology', the choosing of how to apply a method. As part of a study of VE design practice, Kaur (1998) constructs the following outline VE design methodology:

1. Requirements specification;
2. Gathering of reference material from real-world objects;
3. Structuring the graphical model and, sometimes, dividing it between designers;
4. Building objects and positioning them in the VE;
5. Enhancing the environment with texture, lighting, sound, and interaction, and optimising the environment.

She also notes that there might be a narrative design component missing here, but this is probably because of the small scale of the VEs in the study. Certainly the narrative aspects of 3D game design are considered as soon as the principle subject and genre are established. Computer games are almost certainly the major examples of VEs large enough to benefit from software engineering practice.

With these arguments in mind, Fencott (1999b) offered a prototype design methodology for VEs which attempts to resolve the two-sided design problem for IDEs by juxtaposing structural and perceptual modelling, and attempting to empathize with current practice. The methodology is also based on practical experience gained in building a variety of desktop VEs, and in particular a virtual tourism project, as well as teaching VE design to several hundred undergraduate and master's students over a number of years. Figure 4 characterizes this suggested process model, and we now go on to revisit the original, tentative discussions that were offered in the 1999 paper, in the light of the discussions concerning the possible underlying model and language required for a design method for VEs laid out above.

In terms of the design tension, the route down the left-hand side of the diagram represents engineering design and the route down the right-hand side represents aesthetic design. The horizontal arrows represent interactions that seek to resolve the tension.

- **Requirements modelling** equates to Point 1 in Kaur's methodology above and parallels very closely the software engineering concept. One of the chief requirements is that purpose should be clearly established here. In terms of our integrated underlying model, we might here conduct as 'use case analysis' in UML and commence the analysis of our intended VE in terms of Barthe's notion of *myth*—connotations so seemingly natural as to be unquestioned (Barthe, 1987)—and perhaps Baudrillard's notion of hyperreality (Baudrillard, 1995). Both are concerned with the cultural basis upon which a VE's belief system will be grounded. At this stage we

Figure 4. A process model for VE design

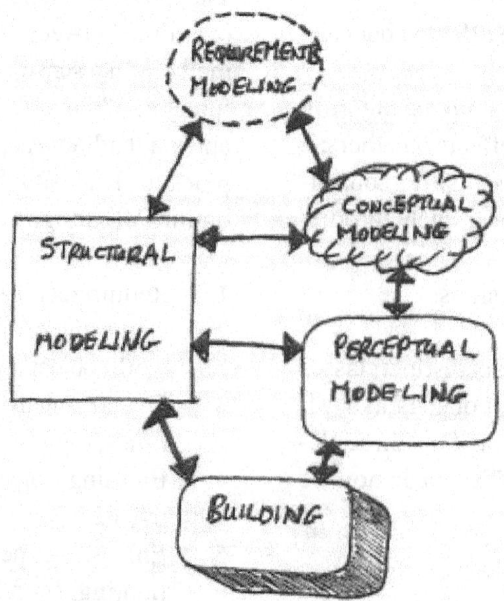

are thus making direct use of the underlying model and the techniques associated with it. The semiotic and software engineering viewpoints are left unresolved until the latter stages of structural and perceptual modelling.

- **Conceptual modelling** equates to Point 2 in Kaur's methodology and is effectively the background research activity common to many design projects, but in particular those with an aesthetic component. It is the gathering of materials, taking of photographs, sketches, sound and video recordings, and so forth. It might also include the construction of mood boards as well as potential storyboards. This is where the VE builder or builders get to know the world they have to build. Note that the world to be built might have no real-world counterpart, which will of course impact on the kinds of activities that might be undertaken here. The artists' accounts and the techniques employed by animators are sources of applicable techniques (e.g., Moser, 1996). An important outcome of this stage will be a choice of genre, to best achieve the purpose established at the requirements stage, with which to inform the nature of the meta-narrative structure to be developed in the perceptual modelling phase.

  The end of this phase is effectively concerned with the semiotic activity of translating the decisions concerning myth and/or hyperreality—from the requirements phase—into connotation, metaphor, and metonymy.
- **Perceptual modelling** is the act of building up a model of the nature of the perceptual opportunities and their inter-relationships. It equates very roughly to Point 5 in Kaur's methodology. It is of course modelling the intended users' experience of the VE. In Fencott (2003a) perceptual maps, for instance attractor graphs, are used to build up a meta-narrative structure of POs, analogous to the comprehensible labyrinth of Murray (1997), which are categorized according to the role they play in the planned scheme of possible user activity. Perceptual opportunities deal not only with conscious experience—derived from *the specifically designed infidelities* of Whitlock et al. (1996)—but also with unconscious experience, sureties, which deliver belief in the VE—perceptual realism in Lombard and Ditton (1997)—irrespective of any real-world counterpart. The existence and importance of unconscious experience is identified and modelled by considering sureties.
- **Structural modelling**, Point 3 in Kaur's methodology, covers a variety of activities that relate to the underlying realization of the VE that the delivery platform uses to construct the run-time sensory stimuli. Structural modelling would seem to commence alongside conceptual modelling and to run on alongside perceptual modelling. It starts with decisions on scale, the construction of plans, and diagrams. It draws on Andersen's CBSs to further decompose the perceptual map constructed in the perceptual modelling phase in terms of the way in which particular objects implement gross structure of attractors and rewards identified in perceptual modelling.

  The conclusion of the structural modelling phase will result in a scene graph diagram that lays out the code structure of the VE and its programmed behavioural components. In terms of software engineering practice, UML has already been identified as a candidate language here. In later stages, object models would lay out the actual structure of nodes in the scene graph as well as class diagrams for programmed components.
- **Building** here relates more closely to the software engineering coding phase that should occur after all requirements, specification, and design activities have been

completed. Building refers to authoring using a WIMP-based tool, direct coding of scene graph and program code itself, in VRML and Java/Javascript for example, and using an API such as World Tool Kit.

We will now consider some of the flows (arrows) in this process model, first of all the structural-conceptual flow. The conceptual modelling stage can deliver important high-level plans for the layout of the VE as well as the principle entities that will need to be present to reinforce the results of *use-case-analysis*, for instance. The structural-perceptual flow delivers object denotations to do with such attributes as appearance and sound. It will deliver object connotations concerned with the way objects contribute to the overall purpose of the VE. Importantly it will also—via CBSs—deliver attributes concerned with interactive capabilities of objects.

Finally, we note that we do not address the question of heuristics in this chapter. There are two reasons: first of all our method is too methodological at the moment to be able to be supported by practical advice; secondly there is a wealth of help and advice on VE design and games design in particular, and it will be necessary to investigate how it might integrate with the method under consideration.

FUTURE DESIGN METHODOLOGY

It is the author's suspicion that one of the foreseeable trends will be ever more sophisticated VE authoring tools, which will mean that the explicit use of OO techniques such as UML will be more and more hidden from the author. Much of the time the scene graph, whether at the level of an OO specification or at the level of OO coded implementation, will only be available to authors via specific views rather than as a coherent whole. More and more it will be the perceptual modelling and the interface between this and the structural views that will be made explicit and malleable. The process model discussed above shows the nature of this interface and provides clues as to how this might be supported by authoring technology.

However, in terms of authoring tools, there is a serious problem that we have not identified nor discussed so far. This is the problem of authoring agency, which lags far behind the authoring possibilities on offer for 3D modelling, texture mapping, shading, and rendering, to name but a few. Nothing approaching the sophisticated tools on offer for these exists for authoring agency. Typical examples of this are easy to find in a wide range of VE authoring tools for both games and VR. In the excellent Unreal Editor for example, the only agencies we can easily implement are such concerned with opening doors, travelling in lifts (elevators), and shooting guns. Unreal is a *first-person-shooter* and it has in-built agencies typical of its genre. If an author wants to implement additional agency, then she has to program it in Unrealscript, a Java variant.

Yet a theoretical analysis of games genres has shown that agency is exactly what characterizes games (Fencott, 2004). Any game design method should not only incorporate the analysis and design of appropriate agency in its process model, but should encourage authors to reconsider it throughout the lifecycle from early requirements analysis through to later modelling stages. By focusing on agency in terms of the aesthetic pleasures of intention and perceivable consequence, and in terms of the POs of attractors and rewards, the process model does indeed ask the designer to consider agency in a fundamental way that authoring tools do not at present support.

In terms of underlying theory, two significant trends can be identified. The growing interest in the investigation, formalization, and application of interaction machine theory (e.g., Goldin et al., 2001) and the emergence of semiotics and computational semiotics as a tool to analyse and design VEs and IDSs in general—for example the COSIGN series of conferences (COSIGN).

Of particular interest will be the further investigation of the possible integration of interaction machines and semiotics, which in effect amounts to the nature of the interplay between empirical computer science and interactive media aesthetics. As has already been pointed out, the tempting approach is to formalize semiotics as computation (e.g., Dogen-Henisch, 1999; Goguen, 1999), but this does not capture or investigate the playfully surprising relationship people have with IDSs and IDEs in particular.

CONCLUSION

A little tension can be a good thing; too much can be very destructive. We need to keep the VE design tension apparent throughout the analysis, conceptual, and perceptual design stages to be more or less resolved in the structural modelling stage. It must not be allowed to tear the process apart. On the other hand an imbalance biasing one pole of the tension or the other will result in an equally unbalanced VE—either well engineered and boring, or fascinating but badly made. The author believes that the design tension will manifest itself in a benign way in a well-designed VE and that users will recognize and appreciate that manifestation.

In effect VE design methodology is encouraging us to confront and meld a great rift in contemporary Western culture, namely that between the arts and the sciences. Of course, at present it is inviting us to do this in terms of two particular forms of abstraction which represent the two sides of the divide. That we should confront reality through virtual reality might come as a surprise, but the concept has been around since the early days of virtual environments and was clearly articulated by Lauria (1997) when she envisioned virtual reality as a 'metaphysical testbed'.

On a less grandiose scale, it may well be that no design method for VEs ever becomes a real practicality or if it does is ever widely adopted by the developer community. Surely, however, the investigation of the methodology of VE design will inform us far better than we are now as to the fundamental nature of VEs and thus be of benefit to us when we come to design future interactive systems.

REFERENCES

Aarseth, E.J. (1997). *Cybertext: Perspectives on Ergodic literature.* Baltimore, MD: John Hopkins University Press.

Andersen, P.B. (1997). *A theory of computer semiotics.* Cambridge University Press.

Barthes, R. (1987). *Mythologies.* New York: Hill and Wang.

Baudrillard, J. (1995). The Gulf War did not take place (trans Patton P.). Indiana University Press.

Blackmore, S. (1999). *The meme machine.* Oxford University Press.

Carruthers, M. (1998). *The craft of thought: Meditation, rhetoric, and the making of images, 400-1200.* Cambridge University Press.

Chandler, D. (2002). *Semiotics: The basics.* Routledge.

Church, D. (1999). Formal abstract design tools. *Games Developer Magazine,* (August).

COSIGN. Retrieved from www.cosignconference.org

Damasio, A.R. (1994). *Descarte's error: Emotion, reason and the human brain.* Papermac.

Davis, P.J., & Hersh, R. (1983). *The mathematical experience.* Pelican Books.

Doben-Henisch, G. (1999). Alan Mathew Turing, the Turing Machine, and the concept of sign. Retrieved from www.inm.de/kip/SEMIOTIC/

DRESDEN_ FEBR99/CS_Turing_and_Sign_febr99.html

Eco, U. (1977). *A theory of semiotics.* Macmillan Press.

Fencott, C. (1999a). Content and creativity in virtual environment design. *Proceedings of Virtual Systems and Multimedia '99,* University of Abertay Dundee, Scotland.

Fencott, C. (1999b). Towards a design methodology for virtual environments. *Proceedings of the International Workshop on User Friendly Design of Virtual Environments,* York, UK.

Fencott, C. (2003a). Virtual saltburn by the sea: Creative content design for virtual environments. *Creating and using virtual reality: A guide for the arts and humanities.* Oxbow Books, Arts and Humanities Data Service.

Fencott, C. (2003b). *Perceptual opportunities: A content model for the analysis and design of virtual environments.* PhD thesis, University of Teesside, UK.

Fencott, C. (2004). *Game invaders: Computer game theories.* In preparation.

Fencott, P.C., Fleming, C., & Gerrard, C. (1992). Practical formal methods for process control engineers. *Proceedings of SAFECOMP '92,* Zurich, Switzerland, October. City: Pergamon Press.

Fencott, P.C., Galloway, A.J., Lockyer, M.A., O'Brien, S.J., & Pearson, S. (1994). Formalizing the semantics of Ward/Mellor SA/RT essential model using a process algebra. Proceedings of Formal Methods Europe '94. *Lecture Notes in Computer Science, 873.* Berlin: Springer-Verlag.

Gammasutra. Retrieved from www.gamasutra.com

Goguen, J. (1999). An introduction to algebraic semiotics, with application to user interface design. Computation for metaphor, analogy and agents. *Springer Lecture Notes in Artificial Intelligence, 1562,* 242-291.

Goldin, D., Keil, D., & Wegner, P. (2001). An interactive viewpoint on the role of UML. *Unified Modelling Language: Systems analysis, design, and development issues.* Hershey, PA: Idea Group Publishing.

Goldin, D.Q., Smolka, S.A., Attie, P.C., & Wegner, P. (2001). Turing Machines, transition systems, and interaction. *Nordic Journal of Computing.*

Kaur, K. (1998). *Designing virtual environments for usability.* PhD Thesis, City University, London.

Kronlof, C. (1993). *Methods integration: Concepts and case studies.* New York: John Wiley & Sons.

Laurel, B. (1992). Placeholder. Retrieved from www.tauzero.com/Brenda_ Laurel/Placeholder/Placeholder.html

Lauria, R. (1997). Virtual reality as a metaphysical testbed. *Journal of Computer Mediated Communication, 3*(2). Retrieved from jcmc.huji.ac.il/vol3/issue2/

Lindley, C., Knack, F., Clark, A., Mitchel, G., & Fencott, C. (2001). New media semiotics—computation and aesthetic function. *Proceedings of COSIGN 2001,* Amsterdam. Retrieved from www.kinonet.com/conferences/cosign2001/

Loftin, R.B., & Kenney, P.J. (1994). *The use of virtual environments for training the Hubble Space Telescope flight team.* Retrieved from www.vetl.uk/edu/Hubble/virtel.html

Lombard, M., & Ditton, initial. (1997). At the heart of it all: The concept of telepresence. *Journal of Computer Mediated Communication, 3*(2). Retrieved September 1997 from jcmc.huji.ac.il/vol3/issue2/

McIntosh, P. Course notes on UML/VRML. Retrieved from www.public. asu.edu/~galatin/

Milner, R. (1989). *Communication and concurrency.* Englewood Cliffs, NJ: Prentice-Hall.

Moser, M.A. (1996). *Immersed in technology.* Boston, MA: MIT Press.

Murray, M. (1997). *Hamlet on the Holodeck: The future of narrative in cyberspace.* New York: The Free Press.

Rollings, A., & Adams E. (2003). *Andrew Rollings and Ernest Adams on games design.* New Riders.

Ryan, T. (1999). Beginning level design. Retrieved from www.gamasutra.com

Saltzman, M. (ed.). (1999). *Games design: Secrets of the sages.* Macmillan.

Sánchez-Segura, M.I., Cuadrado, J.J., de Antonio, A., de Amescua, A., & García L. (2003). Adapting traditional software processes to virtual environments development. *Software Practice and Experience, 33*(11). Retrieved from www3.interscience.wiley.com/cgi-bin/jhome/1752

Slater, M. (1999). Co-presence as an amplifier of emotion. *Proceedings of the Second International Workshop on Presence,* University of Essex, UK. Retrieved from www.essex.ac.uk/psychology/tapestries/

Spinney, L. (1998). I had a hunch.... *New Scientist,* (September 5).

Trondstad, R. (2001). Semiotic and nonsemiotic MUD performance. *Proceedings of COSIGN 2001,* CWI, Amsterdam.

Turkle, S. (1995). *Life on the screen: Identity in the age of the Internet.* Phoenix.

UML Version 1.1 Summary. Retrieved from www.rational.com/uml/resources/documentation/summary/

Whitelock, D., Brna, P., & Holland, S. (1996). *What is the value of virtual reality for conceptual learning? Towards a theoretical framework.* Retrieved from www.cbl.leeds.ac.uk/~paul/papers/vrpaper96/VRpaper.html

Workshop on Structured Design of Virtual Environments. (2001). *Proceedings of Web3D Conference,* Paderborn, Germany. Retrieved from www.c-lab.de/web3d/VE-Workshop/index.html

Workshop on Usability Evaluation for Virtual Environments. (1998). De Montforte University. Retrieved from www.crg.cs.nott.ac.uk/research/technologies/evaluation/workshop/workshop.html

This work was previously published in Developing Future Interactive Systems, edited by M.-I Sanchez-Segura, pp. 66-91, copyright 2005 by IGI Publishing, formerly known as Idea Group Publishing (an imprint of IGI Global).

Chapter VI
An Ontological Representation of Competencies as Codified Knowledge

Salvador Sanchez-Alonso
University of Alcalá, Spain

Dirk Frosch-Wilke
University of Applied Sciences, Germany

ABSTRACT

In current organizations, the models of knowledge creation include specific processes and elements that drive the production of knowledge aimed at satisfying organizational objectives. The knowledge life cycle (KLC) model of the Knowledge Management Consortium International (KMCI) provides a comprehensive framework for situating competencies as part of the organizational context. Recent work on the use of ontologies for the explicit description of competency-related terms and relations can be used as the basis for a study on the ontological representation of competencies as codified knowledge, situating those definitions in the KMCI lifecycle model. In this chapter, we discuss the similarities between the life cycle of knowledge management (KM) and the processes in which competencies are identified and assessed. The concept of competency, as well as the standard definitions for this term that coexist nowadays, will then be connected to existing KLC models in order to provide a more comprehensive framework for competency management in a wider KM framework. This paper also depicts the framework's integration into the KLC of the KMCI in the form of ontological definitions.

INTRODUCTION

Models of knowledge creation inside organizations are considered as dynamic processes of development that evolve over time (Cavaleri & Reed, 2000). These models provide a breakdown of the creation process in terms of concrete processes and elements that drive the overall production of knowledge as targeted to satisfy organizational expectations. For example, the knowledge life cycle (KLC) model of the Knowledge Management Consortium International (KMCI, http://www.kmci.org) distinguishes the knowledge processing environment (KPE) from the business processing environment (BPE), describing the latter as the context of actual usage and field assessment of the claims formulated, produced and evaluated in the former. As the KPE is divided into two subprocesses, namely knowledge production (KP) and knowledge integration, the existence of a BPE emphasizes the fact that knowledge codified in artefacts as part of KP processes and disseminated as part of KI processes will be subject to further validation in actual business experience.

Previous work has shown KLC models as a comprehensive framework for situating learning-oriented artefacts in an organizational context (Sanchez-Alonso & Frosch-Wilke, 2005; Sicilia, 2005). The work of Sicilia (2005) has demonstrated that the design and creation of learning resources as described by Downes (2004) is not essentially different from knowledge production. The integration processes, in particular, might be considered to subsume programmed organizational learning activities. Thinking about learning as an outcome of the need to acquire new competencies, learning activities inside the organization can be considered enablers of knowledge acquisition activities. In this context, the concept of competency becomes essential in the KLC model, both as a prerequisite to perform knowledge acquisition activities and as an outcome of these kinds of activities. Furthermore, meta-claims about the knowledge produced—in the case of competencies—may be interpreted as the recording of usage conditions, hypotheses, and assumptions on the acquisition of the competencies evaluated. In consequence, the concepts related with competency management can be put in connection with existing KLC models, in an attempt to provide a comprehensive framework for reuse-oriented competency management and KM. In this chapter, we approach the integration of concepts related to competencies into the framework of the KLC. This would clarify the relationships between knowledge management and competency definition standard efforts. The method to develop the conceptual integration is that of engineering an initial ontological description for the main concepts, connecting them to existing ontological databases. This continues existing work described by Sicilia, Lytras, Rodríguez, and García (2006) regarding the ontological description of learning activities as an extension of the ontology of KM described recently by Holsapple and Joshi (2004).

Formal ontologies (Baader, Calvanese, McGuinness, Nardi, & Patel-Schneider, 2003) are a vehicle for the representation of shared conceptualizations that is useful for technology-intensive organizations. Ontologies based on description logics (Gruber, 1995) or related formalisms provide the added benefit of enabling certain kinds of reasoning over the terms, relations, and axioms that describe the domain. A pragmatic benefit of the use of formal ontologies is that it is accompanied by a growing body of semantic Web (Berners-Lee, Lassila, & Hendler, 2001) tools, techniques, and knowledge. Previous work considered here as a point of departure (Sicilia, García, Sánchez-Alonso, & Rodríguez, 2004) has described the integration of e-learning technology concepts with the OpenCyc knowledge base, the open source version of the Cyc system (Lenat, 1995).

The rest of this chapter is structured as follows. The second section describes the knowledge life cycle of the KMCI, as this is the framework for the subsequent discussion. The third section includes a brief discussion on some current definitions of the term competency and details the most interesting efforts in the standardization of competency definitions. The fourth section shows how competencies can be integrated in the knowledge life cycle of the KMCI, while the fifth section provides a preliminary mapping of competency-related concepts to terms in upper ontologies. Finally, conclusions are provided in the last section.

THE KNOWLEDGE LIFE CYCLE OF THE KMCI

Knowledge management is an area built on the assumption that each and every organization has a certain amount of "valuable knowledge" that is worth being captured, catalogued, and preserved with the main aim of sharing it whenever it is necessary. However, first-generation KM, as it is referred by McElroy (1999) has not been considered fully satisfactory, which is probably due to an excessive emphasis on both knowledge integration and on the technology side as the answer to most questions. To many, this first generation of knowledge management has supposed little more than document management and imaging becoming the reason why some feel that KM is "an idea that amounts to little more than yesterday's information technologies trotted out in today's more fashionable clothes." Hopefully, a second generation of KM has emerged. This second generation of knowledge management is not so focused on the technology side but instead on the participants, the processes involved, and the social interactions and initiatives among them. The arrival of this second generation has introduced a number of new concepts and ideas, such as the knowledge life cycle, nested knowledge domains, containers of knowledge, organizational learning, the open enterprise, social innovation capital, and sustainable innovation, among others. While an in-depth discussion of this and other key ideas of this second generation of KM is out of the scope of this chapter, the interested readers are directed to the book by McElroy (2003).

The knowledge life cycle, one of the previously mentioned ideas introduced as part of the second generation of KM, is a new view of KM that emphasizes knowledge production in detriment of the knowledge integration. The following explanation by McElroy points out the differences between the first and the second generation of KM taking as criteria of comparison the KLC:

> *While practitioners of first-generation KM tend to begin with the rather convenient assumption that valuable knowledge already exists, practitioners of second generation KM do not. Instead, they–or we–take the position that knowledge is something that we produce in human social systems, and that we do so through individual and shared processes that have regularity to them. We can describe this process at an organizational level in the form of what is now being referred to as the knowledge life cycle, or KLC.* (McElroy, 2003)

From this perspective, the Knowledge Management Consortium International, a non-profit association of knowledge and innovation management professionals from around the world (www.kmci.org) based in the U.S., has developed a model of KLC that is shown in Figure 1 taken from McElroy (n.d.).

This model shows how the knowledge of an organization is held both subjectively in the minds of individuals and groups, and objectively in recorded or expressed form, shaping what is known as the distributed organizational knowledge base (DOKB) of the organization. The use of this knowledge in specific business environments can lead to outcomes that either satisfy expectations or fail to do so. The former outcomes, known as matches, reinforce knowledge previously used, thereby leading to its re-use, whereas the later ones, known as mismatches, lead to adjustments in a business processing environment. Adjustments

Figure 1. The knowledge life cycle (KLC) model of the KMCI

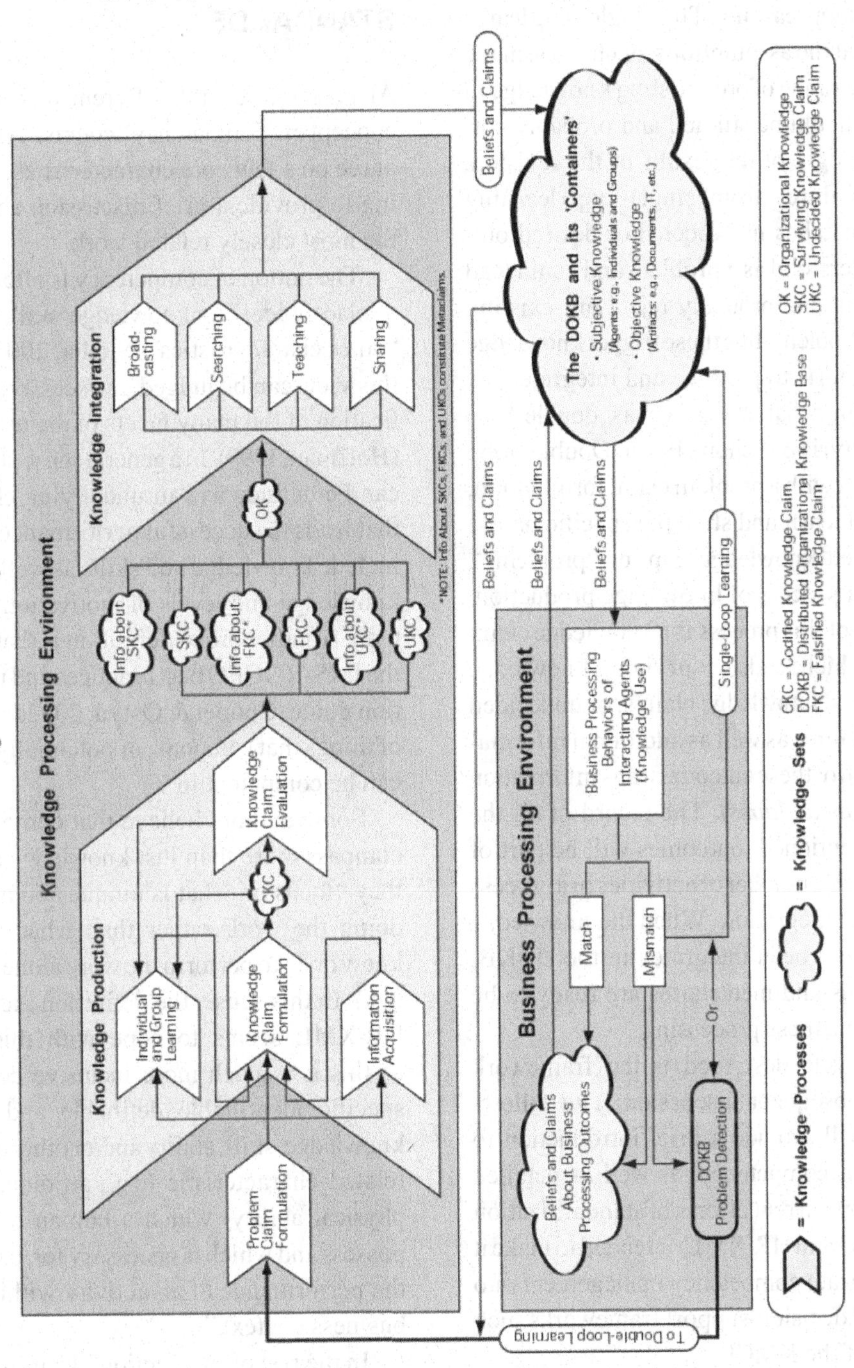

triggered by a mismatch introduce what is known as the single-loop learning. This single-loop learning means that the assumptions, or choices, made from within a range of pre-existing knowledge in the DOKB should be studied and probably corrected in the light of the results of the revision. Successive failures from single-loop learning to produce matches in expected or desired outcomes is understood as a problem and could lead to doubt about and probably reject pre-existing knowledge. Problems like these trigger knowledge processing efforts to produce and integrate new knowledge, in what is known as double-loop learning (Argyris & Schon, 1996). Double- loop learning starts with a problem claim formulation, an attempt to learn and state the specific nature of the detected knowledge gap (or "problem"), followed by a process of knowledge production. The outcome of this process is a knowledge claim evaluation, which leads to surviving knowledge claims, falsified knowledge claims, or undecided knowledge claims, as well as additional information about each of these outcomes (this information is known as *metaclaims*). The record of all the previously mentioned outcomes will be part of the DOKB after a number of activities in a process of knowledge integration. When the knowledge has successfully been integrated in the DOKB, the new claims and metaclaims are ready to be used in new business processing.

The life cycle described is the framework for all the subsequent discussion. The following section will provide a brief introduction to the concept of competency, as well as detailed information on current efforts of standardization (IMS-RDCEO and HR-XML) intended to make it easier to integrate competency management into workflow and decision-support frameworks such as the KLC of the KMCI.

COMPETENCY: DEFINITION AND STANDARDS

At present, several different definitions of the concept of "competency" coexist. Although most agree on a few core characteristics, it is interesting to provide a brief discussion about some of the most closely related work.

The notion of competency is often considered a "placeholder" for knowledge, skill, abilities, and "other characteristics" (Sicilia, 2005). However, this view can be judged an excessive oversimplification of the many facets of the use of the term (Hoffman, 1999). In a general sense, a competency can be defined as "an underlying characteristic that leads to successful performance, which may include knowledge and skills as well as bodies of knowledge and levels of motivation" (Rothwell, n.d.). Another broad definition is that included in the IMS-RDCEO best practices and implementation guide (Cooper & Ostyn, 2002c): "All classes of things that someone, or potentially something, can be competent in".

Some authors believe that competencies encompass more than just knowledge and skill, as they "focus on what is unique about individuals doing the work rather than what people must know or do to perform the work alone" (Rothwell, n.d.). In this sense, the definition included in the HR-XML seems to cope with this approach, as this is a much more inclusive definition: "A specific, identifiable, definable, and measurable knowledge, skill, ability and/or other deployment-related characteristic (e.g., attitude, behaviour, physical ability) which a human resource may possess and which is necessary for, or material to, the performance of an activity within a specific business context."

In the rest of this section, the most prominent approaches to competency standardization are studied. It should be remarked that, as it has been stated earlier, most agree on the core characteristics of competencies, even though all include their own definitions and consequently refer to the term competency from their own perspective.

IMS-RDCEO

The IMS consortium (http://www.imsglobal.org) provides a specification for competencies called "reusable definition of competency or educational objective (RDCEO)". IMS-RDCEO defines an information model for describing, referencing, and exchanging definitions of competencies, primarily in the context of online and distributed learning. This specification allows to formally represent the most important characteristics of a competency, and its main aim is to enable interoperability among learning systems that deal with competency information. The complete specification consists of three documents:

- IMS-RDCEO **information model** (Cooper & Ostyn, 2002a), including the complete description of the main elements of the specification: semantics, structure, data types, value spaces, multiplicity, and obligation. This information model is purposely extensible, minimalist, and model-neutral.
- IMS-RDCEO XML binding (Cooper & Ostyn, 2002b), constituting only one example of the possible bindings that might use the information model, is a binding of the Information Model to XML version 1.0.
- IMS-RDCEO best practices and implementation guide (Cooper & Ostyn, 2002c), a non-normative set of rules about the application of both the information model and the XML binding, as well as examples to, for example, illustrate how the conceptual framework maps to practical uses.

The information model defines a set of elements of information in five different categories that can be used to define a competency. Hence, competency data may include a definition of the competency, evidences of the competency, information about its context, and the scale (i.e., proficiency on a predetermined scale). Following this schema, a competency can be described by stating information in the following five main categories:

1. Identifier, subdivided into *catalog and entry*
2. Title
3. Definition
4. Description, subdivided into *model source* and *statement*
5. Metadata, subdivided into *RDCEO schema, RDCEO schema version,* and *additional metadata*

The definition of a competency, according to this schema, is shown in the following example, a simplification of a broader example taken from Cooper and Ostyn (2002c):

```
<?xml version="1.0" encoding="utf-8"?>
<rdceo xsi:schemaLocation="http://www.w3.org/XML/1998/namespace xml.xsd" xmlns="http://www.imsglobal.org/xsd/imsrdceo_rootv1p0" xmlns:xsi="http://www.w3.org/2001/XMLSchema-instance">
<identifier>http://www.imsglobal.org/fictional/rdceo_cat1.xml#pass_eg
</identifier>
  <title>
      <langstring xml:lang="en-US">Reading IMS specifications</langstring>
  </title>
  <description>
     <langstring xml:lang="en-US">
     Reads and understands IMS Global Learning specifications
   </langstring>
  </description>
  <definition>
    <model>IMS Competency WG</model>
    <statement statementname="Performance">
      <statementtext>
```

```
<langstring xml:lang="en-US">
Reads and understands IMS Global Learning
specifications
</langstring>
</statementtext>
</statement>
</definition>
<metadata>
<rdceoschema> IMS RDCEO </rdceoschema>
<rdceoschemaversion> 1.0 </rdceoschemaver-
sion>
</metadata>
</rdceo>
```

However, although IMS-RDCEO is explicitly intended to be integrated in the description of "learner profiles" and "learning objects" (Polsani, 2003), its underlying model provides similar capabilities to that of HR-XML, a general-purpose competency schema that will be detailed in the next section.

HR-XML

The HR-XML (http://www.hr-xml.org/) is an independent, non-profit consortium whose main aim is to enforce e-commerce and inter-company exchange of human resources data within a variety of business contexts. Represented by its membership in 22 countries, the main effort supported by this consortium is the development of standardized XML vocabularies for human resources, as well as standards for staffing and recruiting, compensation and benefits, and training and work force management. Major companies such as Addeco, Cisco Systems, PeopleSoft GmbH, IBM, Microsoft, and many others are currently members of the HR-XML Consortium.

Up to the present, the HR-XML Consortium has produced a library of more than 100 interdependent XML schemas that define the data elements for particular HR transactions, as well as options and constraints governing the use of those elements. It has also produced schemas covering major processes, as well as component schemas used across multiple business processes. For example, the assessments standard, facilitates employers to leverage the assessment tests, tools, and expertise offered by assessment service providers.

One of the schemas provided by the HR-XML Consortium is the competencies recommendation. This set of recommendations about competencies allows "the capture of information about evidence used to substantiate a competency and ratings and weights that can be used to rank, compare, and otherwise evaluate of the sufficiency or desirability of a competency" (Allen, 2006). The competencies schema is particularly relevant to processes involving the rating, measuring, comparing, or matching an asserted competency (for example, a skill claimed in a resume) against one that is demanded (for example, a skill required in a job description). This fact, added to the fact that this schema is intended as a module that can be incorporated within broader process-specific schemas, facilitates its use outside the HR domain as a general-purpose competency schema and makes it possible its integration in diverse frameworks. The only requirement for those frameworks is, of course, the use of some kind of competency management.

Figure 2, taken from Allen (2006), depicts the components of a competency after what is stated in the HR-XML recommendation. This standard defines a number of elements of information for each competency, as well as the structure and information of the competency evidences and weights, among other information.

The definition of a competency, according to this schema, is shown in the following example, again taken from Allen (2006):

```
<Competencyname = «Reading Comprehension»
description = «Understanding written sentences
and paragraphs
in work related documents»>
<CompetencyId id = «2.A.1.a»/>
```

Figure 2. Components of a competency after the HR-XML recommendation

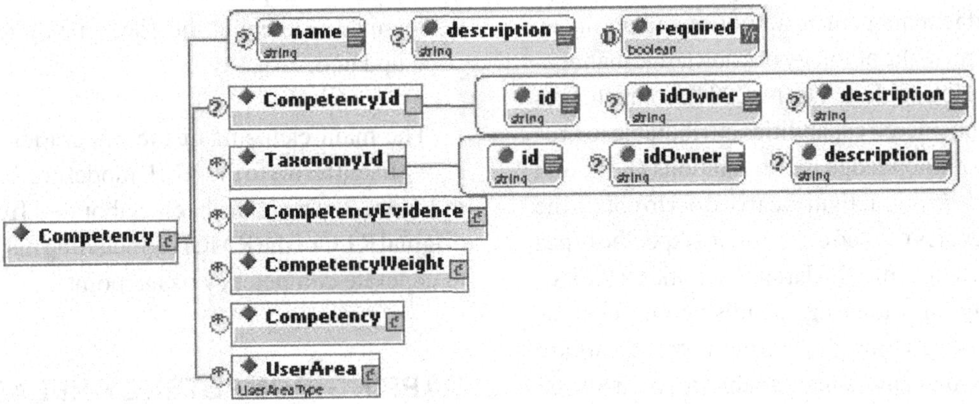

```
<TaxonomyId          id = «O*NET»
  idOwner = «National O*Net Consortium»
  description = «Occupational Information Network»/>
    <CompetencyWeight type = «x:Importance»>
  <NumericValue maxValue=«100» minValue=»1»>85</NumericValue>
    </CompetencyWeight>
    <CompetencyWeight type = «x:Level»>
  <NumericValue maxValue = «100» minValue=»1»>
57 </NumericValue>
    </CompetencyWeight>
  </Competency>
```

HR-XML can also be used as a wrapper of an RDCEO record by using a URN, as shown in the following example taken from (2006):

```
<Competency description="Can read and understand W3C Schema Language 1.0"
      name="Reads and Understands W3C Schema"
      xmlns:xsi="http://www.w3.org/2001/XMLSchema-instance"
      xsi:noNamespaceSchemaLocation="Competencies-1_0.xsd">
    <CompetencyId description="IMS Global Example Competency Catalogue"
        id="URN:X-IMS-PLIRID-V0::6ba7b8149dad11d180b400c04fd430c8"/>
    <!-- omitted evidence data etc. -->
  </Competency>
```

INTEGRATING COMPETENCIES IN THE KNOWLEDGE LIFE CYCLE (KLC)

In this section, the related concepts of competencies are described as the main elements to be integrated as resources in the KLC. Then, their integration inside the KLC model of the KMCI is described.

The process of acquisition of a competency (or knowledge in a broader sense) usually starts from a business need originated in the context of the organization. This need triggers a process of assessing whether the organization can deal with the given need or not, which is commonly referred to as knowledge gap analysis (Sunassee & Sewry, 2002). This assessment process essentially consists of matching the competencies required for the newly appointed needs with the available ones. When the result of this process is not satisfactory, a process of acquiring the competencies identified begins. After this process is considered finished, some kind of assessment would take place and, later on, an update of the registry of available competencies should be carried out. The newly acquired competencies might change the position of the organization to offer services or products, closing in this manner the so-called "knowledge acquisition loop."

As a knowledge acquisition endeavour, the just described cycle can be expressed in terms of knowledge management activities and products. According to the ontology of knowledge management by Holsapple and Joshi (2004), competences can be considered capabilities attributable to processors of knowledge representations (KR), and the final learning activities carried out to obtain the competencies needed can be seen as specific types of knowledge manipulation activities (KMA), consisting of knowledge acquisition or, eventually, transformation. Furthermore, processors are considered to have some capabilities as analysed by Sicilia (2005). This author identifies the terms (or, as it is called in the original work, abstract elements) related to competency management as a previous step to integrating them in the KLC:

- **Competency registry:** Not a term but instead a set of terms related to the description of competencies in detail, particularly those of the existing employees.
- **Needs:** an expression of the required competencies that can be represented in the form of triples (C: competency_description, L: level, I: intensity). According to Sicilia (2005), the level desired for the competency is expressed as an overall aggregate level that maps the levels of individuals inside the organization, whereas the intensity is an estimation of the part of the workforce that is needed to have the competency.
- **Available competencies:** A detailed record of employee's competencies.
- **Required competencies:** A subset of the needs after matching them with the competency registry. Aimed at describing needs not covered by the existing competencies.
- **Competency gap analysis:** A process used to obtain the required competencies. This process has a collection of needs and a competency registry as inputs and the required competencies as outputs.
- **Competencies update:** The process of creation or update of competency instances, aimed at keeping the competency registry updated.

The main elements of the integration of the above listed terms to the KML model are depicted in Figure 3, which has been elaborated from the original KLC of the KMCI by including mappings to concrete competency usage points.

MAPPING COMPETENCY-RELATED CONCEPTS TO TERMS IN UPPER ONTOLOGIES

Competency management can be integrated in the broader framework of a knowledge management lifecycle to provide guidance for Information System development and insights into notions of organizational value of competencies, among others. However, even though current standards for the description of competencies are intended to provide data aimed at being interchanged by machines, the information they contain is currently intended for human interpretation. Present practices result in data lacking machine-understandable characteristics, which seriously hampers their use in semantic Web environments. Ontologies can be used to improve the quality of competency descriptions, but "translating" current competency descriptions that conform to a given standard (such as HR-XML) to an ontology language is not enough by itself to provide computational semantics to those descriptions. The right step in this direction is the integration of competency terms with high-level terms and definitions in upper ontologies, as this constitutes an interesting direction for bringing explicit semantics to competency descriptions.

An upper ontology is a large general knowledge base that includes definitions of concepts, relations, properties, constraints, and instances, as

Figure 3. Mapping of the main terms of competency management to the KLC model

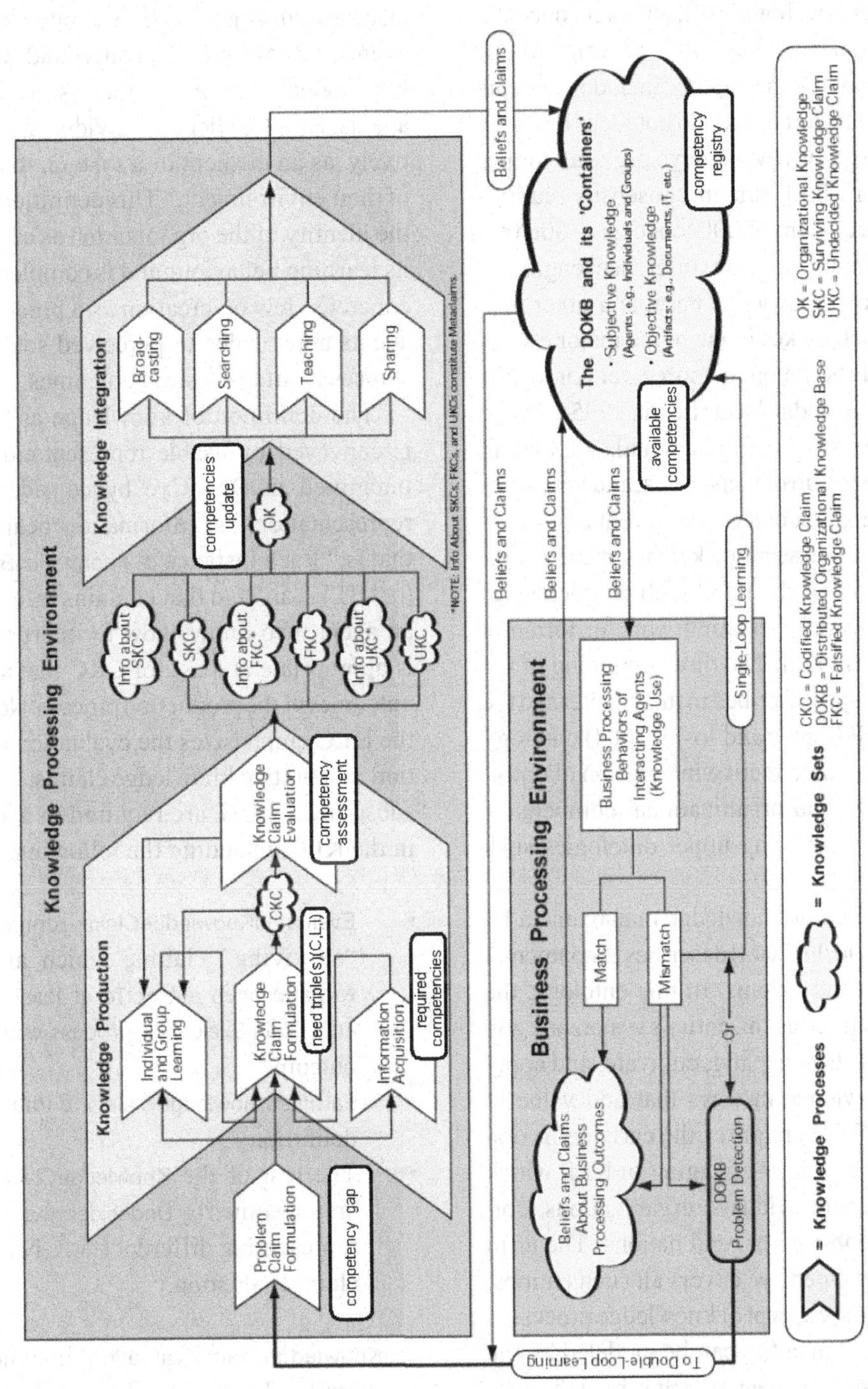

well as reasoning capabilities on these elements. Limited to generic, high-level, abstract concepts, general enough to address a broad range of domains, upper ontologies do not include concepts specific to given domains or do not focus on them. Opencyc (http://www.opencyc.org), an upper ontology "for all of human consensus reality," includes more than 47,000 concepts, 306,000 assertions about them, an inference engine, a browser for the knowledge base, and other useful tools, which makes it one of the major efforts in the field. It is the open source version of the larger Cyc knowledge base (Lenat, 1995), a huge representation of the fundamentals of human knowledge made up of facts, rules, and heuristics for reasoning about objects and events.

The rest of this section sketches the main integration points of the KLC with competencies in the framework of existing work in formally conceptualizing KM. The direct mapping of the essential concepts described in this chapter and the terms in the Holsapple and Joshi (2004) ontology of KM, enables an effective integration of ontology-based KM and organizational competency management in existing upper ontologies such as OpenCyc.

The ontology of knowledge management by Holsapple and Joshi (2004) describes fundamental KM concepts and axioms. In this ontology, the term KM is defined as "an entity's systematic and deliberate efforts to expand, cultivate, and apply available knowledge in ways that add value to the entity [..]." This requires the early definition of "entities" capable of engaging in KM, which include at least individuals, organizations, collaborating organizations, and nations. The term Organization in OpenCyc covers all such entities. Accordingly, the concept of knowledge processor as a member of an entity can be modeled by the concept of IntelligentAgent, which is by definition "capable of knowing and acting, and of employing its knowledge in its actions." Humans are by logical definition intelligent agents, and certain software pieces may also be, since they are not restricted to not being able to know. The subtype MultiIndividualAgent fits the definition of collective agents. According to Cavaleri and Reed (2000), knowledge creation is "the result of efforts by agents, acting either as individuals, or collaboratively, as an element of a system, to make sense of their environment." This definition focuses on the identity of the organization as a key driver of its learning behaviour and is complemented by a concrete view on creation as a process in which agents apply rules to perceived sets of circumstances to attain desired outcomes.

The definition of knowledge as "that which is conveyed by usable representations" can be integrated in OpenCyc by considering usable representations as information bearing things, that is, "Each instance of InformationBearingThing (or IBT) is an item that contains information (for an agent who knows how to interpret it)." This is appropriate at least for CKC that are tangible outcomes of the production process. Nevertheless, the KLC emphasizes the evaluation of information as tentative knowledge claims, so that terms subsumed by IBT are required to adequately fit in the KLC, including the following:

- EvaluatedKnowledgeClaim representing the "surviving" claims, which are required to have been subjectTo at least one KnowledgeClaimEvaluation process with a positive outcome.
- FalsifiedKnowledgeClaims, with the opposite definition.
- The rest of the KnowledgeClaim instances are subsumed by UndecidedKnowledgeClaim, representing different states before or after claim evaluation.

KnowledgeClaimEvaluation instances are a concrete kind of knowledge manipulation. The recognizable kinds of knowledge manipulation are referred to as KnowledgeManipulationActivities, and thus, CompetencyAssessment may be considered a subtype of KMA. In OpenCyc, activities are represented as Actions, collections of Events carried out (doneBy) a "doer." This generic concept of action can be specialized to represent KMA executions by restricting them to be carried out by intelligent agents. The predicate ibtUsed (subsuming the above-mentioned subjectTo) can be used to represent the knowledge representations manipulated by KMAs. In addition, since KM activities are deliberate, it is preferable to use the subclass PurposefulAction. Each of the processes in Figure 3 can be considered KMAs.

Competencies are represented in OpenCyc. However, the attribute Competence, subsumed by Quantity-ScriptPerformance (aimed at describing the manner in which an actor performs an action) and ScriptPerformanceAttributeType (aimed at describing the manner in which an action is performed), is defined as "a general attribute to define the level of skill with which an agent performs some task." For that reason, this notion of competency is considered too general and thus inadequate to define the concept competency as it has been used in this work.

The most accurate way to define competencies is that of defining OpenCyc Actions. Accordingly, predicates related to the definition, description, and use of competencies would be derived from the predicate SkillLevel. This OpenCyc predicate, as stated in the OpenCyc knowledge base, defines a relation between performers and types of actions in the following manner: some performer (probably, but not necessarily, an Agent) has the ability to play a given role in a specific type of Event with a certain level of PerformanceAttribute. For example:

(skillLevel MagicJohnson PlayingBasket performedBy Creativity #$High)

Meaning that, in general, Magic Johnson can play basket with great creativity. If this behaviour is translated to competency management, the knowledge about the fact that the employee Angela has a particular competence should be stated like this:

(skillLevel Angela SpeakingInPublic performedBy Competence #$VeryHigh)

In this example, the competency is represented by the action SpeakingInPublic, whereas the attribute Competence is just one qualifier to describe the manner in which the competency SpeakingInPublic is performed by the employee (others might be Charisma, Precision, Dexterity, or Gracefulness). This form of modeling competencies is similar to the manner in which competencies are defined in HR-XML and opens the door to a full description of other concepts related as the triples (competency, level, and intensity), easy to model in OpenCyc through a specifically-designed ternary predicate.

CONCLUSION

Competency management can be integrated in the broader framework of a knowledge management lifecycle to provide guidance for information system development and insights into notions of organizational value of competencies. Concretely, a feasible integration of such concepts into the KMCI KLC model has been described.

Current standards for the definition, sharing, and exchange of competencies, as well as the information about competencies that conform to this specification included in the DOKB of the organizations, are intended for interchange by machines, but instead they are currently intended for human interpretation only. Their main aim is to enable interoperability among systems that deal with competency information by providing a means for them to refer to common definitions with common meanings. However, these efforts insist in the construction of models of competencies but do not focus on semantic interoperability. The resulting ontological schemes shown in this chapter are intended as a foundation for further research and standardization activities.

The authors consider that an additional effort of integrating current standards in commonsense knowledge bases, such as OpenCyc, through formalizing concepts in ontology languages, can be particularly rewarding as it would provide competency management with the benefits of the Semantic Web vision.

ACKNOWLEDGMENT

This work is funded by the LUISA EU project (FP6-027149).

REFERENCES

Allen, C. (Ed.). (2003). HR-XML recommendation. Competencies (Measurable Characteristics). *Recommendation, 2006 Feb 28*. Retrieved March 1, 2006, from http://www.hr-xml.org/

Argyris, C., & Schon, D. A. (1996). *Organizational learning II: Theory, method, and practice*. Reading, MA: Addison Wesley.

Baader, F., Calvanese, D., McGuinness, D., Nardi, D., & Patel-Schneider, P. (Eds.). (2003). *The description logic handbook*. Theory, implementation and applications. Cambridge.

Berners-Lee, T., Hendler, J., & Lassila, O. (2001). The semantic Web. *Scientific American, 284*(5), 34-43.

Cavaleri, S., & Reed, F. (2000). Designing knowledge creating processes. *Knowledge and Innovation, 1*(1).

Cooper, A., & Ostyn, C. (Eds.). (2002a). *IMS reusable definition of competency or educational objective: Information model, (version 1.0), final specification*. Retrieved March 1, 2006, from http://www.imsglobal.org/competencies/rdceov1p0/imsrdceo_infov1p0.html

Cooper, A., & Ostyn, C. (Eds.). (2002b). *IMS reusable definition of competency or educational objective: XML binding, (version 1.0), final specification*. Retrieved March 1, 2006, from http://www.imsglobal.org/competencies/rdceov1p0/imsrdceo_bindv1p0.html

Cooper, A., & Ostyn, C. (Eds.). (2002c). *IMS reusable definition of competency or educational objective: Best practice and implementation guide, (version 1.0), final specification*. Retrieved March 1, 2006, from http://www.imsglobal.org/competencies/rdceov1p0/imsrdceo_bestv1p0.html

Gruber T. (1995). Towards principles for the design of ontologies used for knowledge sharing. *International Journal of Human-Computer Studies, 43*(5/6), 907-928.

Hoffmann, T. (1999). The meanings of competency. *Journal of European Industrial Training, 23*(6), 275-286.

Holsapple, C. W., & Joshi, K. D. (2004). A formal knowledge management ontology: Conduct, activities, resources, and influences. *Journal of the American Society for Information Science and Technology, 55*(7), 593-612.

Lenat, D. B. (1995). Cyc: A large-scale investment in knowledge infrastructure. *Communications of the ACM, 38*(11), 33-38.

McElroy, M. W. (1999, October). The second generation of knowledge management. *Knowledge Management Magazine*.

McElroy, M. W. (2003). *The new knowledge management—Complexity, learning, and sustainable innovation*. Boston: KMCI Press/Butterworth-Heinemann.

McElroy, M. W. (n.d.). *The knowledge life cycle (KLC)*. Retrieved March 1, 2006, from http://www.kmci.org/media/Knowledge_Life_Cycle.pdf

Polsani, P. R. (2003). Use and abuse of reusable learning objects. *Journal of Digital Information, 3*(4).

Rothwell, W. J. (n.d.). *A report on workplace learner competencies.* Retrieved March 1, 2006, from http://www.ilpi.wayne.edu/files/roth_present.pdf

Sánchez-Alonso, S., & Frosch-Wilke, D. (2005). An ontological representation of learning objects and learning designs as codified knowledge. *The Learning Organization, 12*(5), 471-479.

Sicilia, M. A. (2005). Ontology-based competency management: Infrastructures for the knowledge-intensive learning organization. In M. D. Lytras & A. Naeve (Eds.), *Intelligent learning infrastructures in knowledge intensive organizations: A semantic Web perspective* (pp. 302-324). Hershey, PA: Information Science Publishing.

Sicilia, M.A., García, E., Sánchez-Alonso, S., & Rodríguez, E. (2004). Describing learning object types in ontological structures: Towards specialized pedagogical selection. In *Proceedings of ED-MEDIA 2004: World Conference on Educational Multimedia, Hypermedia and Telecommunications*, 2093-2097.

Sicilia, M. A., Lytras, M., Rodríguez, E., & García, E. (2006). *Integrating descriptions of knowledge management learning activities into large ontological structures: A case study.* Data and Knowledge Engineering.

Sunassee, N., & Sewry, D. (2002). A theoretical framework for knowledge management implementation. In *Proceedings of the 2002 Annual Research Conference of the South African Institute of Enablement Through Technology* (pp. 235-245).

This work was previously published in Competencies in Organizational E-Learning: Concepts and Tools, edited by M. A. Sicilia, pp. 169-184, copyright 2007 by Information Science Publishing (an imprint of IGI Global).

Section II
Integration Aspects for Agent Systems

Chapter VII
Aspects of Openness in Multi-Agent Systems:
Coordinating the Autonomy in Agent Societies

Marcos De Oliveira
University of Otago, New Zealand

Martin Purvis
University of Otago, New Zealand

ABSTRACT

In the distributed multi-agent systems discussed in this chapter, heterogeneous autonomous agents interoperate in order to achieve their goals. In such environments, agents can be embedded in diverse contexts and interact with agents of various types and behaviours. Mechanisms are needed for coordinating these multi-agent interactions, and so far they have included tools for the support of conversation protocols and tools for the establishment and management of agent groups and electronic institutions. In this chapter, we explore the necessity of dealing with openness in multi-agent systems and its relation with the agent's autonomy. We stress the importance to build coordination mechanisms capable of managing complex agent societies composed by autonomous agents and introduce our institutional environment approach, which includes the use of commitments and normative spaces. It is based on a metaphor in which agents may join an open system at any time, but they must obey regulations in order to maintain a suitable reputation, that reflects its degree of cooperation with other agents in the group, and make them a more desired partner for others. Coloured Petri Nets are used to formalize a workflow in the institutional environment defining a normative space that guides the agents during interactions in the conversation space.

INTRODUCTION

The management of interoperation among agents is a complex task and requires robust techniques and methodologies to be applied in the development of reliable and open *Multi-agent Systems* (MAS). Generally, this category of computational systems is used to model distributed scenarios where heterogeneous software entities, *agents*, interact to pursue particular or common goals. Even more demanding is the modeling and implementation of features that give openness to those agents, allowing them to have the ultimate choice of obeying regulations or deal with possible sanctions imposed by the MAS norms. After all agents are *autonomous* entities and the biggest challenge is to have a *coordination system* where the agents can be free to decide what to do but at the same time be encouraged or seduced to obey the regulations of the *artificial society* where they are entering.

Agent frameworks, such as JADE (Bellifemine, Poggi, & Rimassa, 2000) and Opal (Nowostawski, Purvis, & Cranefield, 2002), are developed on top of abstract architectures such as the FIPA Abstract Architecture (Fipa) that aim to offer a standard mean for message exchanging between agents and with that ease the communication process in its lowest levels. *Agent communication languages,* such as FIPA-ACL (Fipa, 2002) and KQML (Finin, Labrou, & Mayfield, 1997), are introduced to define standard semantics for dialects used during agent communication, and *interaction protocols* attempt to balance the expressive power of such languages defining patterns of behaviour that agents must follow to engage in a communicative interaction with other agents as a way of building MAS that interoperate in "predictable" ways.

The individual, local, cultural and social aspects of the communicator are significant and can often outweigh conventional concerns of software developers, such as the development of appropriate syntax, efficient coding systems, and suitable terminologies. What is necessary is an infrastructure where the agents can rely on mechanisms that will implement or offer support for interoperation with efficiency.

Our approach aims to use institutional concepts for modeling open MAS as social groups formed by agents that establish a set of regulations to interact among them. As in ordinary institutional environments, those agents assume predefined roles in the MAS and will be supervised by some authorities, defined as *system agents*. The system agents are part of the infrastructure built to manage commitments among agents, or groups of agents, and agents' reputation in the institution.

Coloured Petri Nets (CPN) (Jensen, 1997) are used to model agent roles as well as all the interactive processes in the institution. CPN brings to our approach a well-defined semantics which builds upon true concurrency, hierarchical representations and an explicit description of both states and actions. Adding to that CPN has an elaborated set of computer tools supporting their drawing, simulation and formal analysis.

From the assumption that artificial agents can be part of a structured society that follows rules, we present an approach for the use of institutional environments, commitments and reputation models to organise MAS in social structures based on roles. One major concern in our approach is to grant an institutional environment where the degree of openness the member agents experience is part of the norms that regulates the MAS.

OPENNESS CONSIDERATIONS

How does one constrain an environment as heterogeneous as a human community? In different parts of the world different rules and norms are created so that the members of that group of people can feel a sense of security and order, so that they can carry on with their lives in a more predictable way. When a person needs some kind of service they know where to go, and if not they will ask another person or consult some kind of public catalogue. Once the person chooses to go to a place and make use of some sort of service he will make use of his past experiences and knowledge to carry on with actions, autonomously, and in a more or less standard way. Rules are there to be followed as well as to guide the members of a community so that they will have their rights observed in local and global aspects.

One of the main goals of our approach is to define and maintain an environment were agents will interact observing a set of rules or norms but not necessarily or compulsorily will have their actions restricted via some kind of interface agent, for example, the governors, in Electronic Institution Development Environment (EIDE) (Arcos, Esteva, Noriega, Rodriguez-Aguilar, & Sierra, 2005), or the controllers in Law-Governed Interaction (LGI) (Minsky & Ungureanu, 2000). Our approach is to offer mechanisms that will seduce the agents to play a cooperative role in the agent society, but ultimately the choice of cooperating or not cooperating will be its. That choice might generate the employment of sanctions by the system agents to noncooperative agents, and with that we will be able to use a model closer to the human way of organizing their societies.

Therefore, among the empirical concerns on electronic institutions (Noriega, 2006), we are more concerned with the choice the agents will have to make when joining a institution of agents in respect to the degree of freedom of speech that institution will allow, and by that we mean that as autonomous agents, ourselves humans, we have individual goals and beliefs and those are in the autonomous artificial agents that we build. They are there because the agent-oriented paradigm aims to model the human society model and for a completely constrained environment the object-oriented paradigm is already there to be used.

Such mechanisms as governors (Arcos et al., 2005) and controllers (Minsky & Ungureanu, 2000), to our view might compromise the agents' autonomy. We otherwise want to influence agents to behave according with the rules of the environment. That is the motivation of our work, to develop an environment were agents are influenced to cooperate and follow a predefined set of rules. That environment is organized based on institutional concepts with the definition of roles to joining agents. Those concepts create an artificial environment similar to real world institutions, where people can join to obtain or offer access to services.

Autonomous agents aim to copy human reasoning or strategies when interacting with others and deciding their course of action in the electronic environment. Interaction protocols are there to be used according with the speech acts, institutional actions or illocutions identified in the dialogs executed by agents. Those interaction protocols help to predict actions and model the conversation space before it actually happens, including the possible break of expected courses of action by agents in the electronic environment. Authorities in the institutions are available to audit interactions and observe rules in the society.

The rigid control of agents' interoperation may or may not grant rigid security for MAS, but it definitely compromises the degree of agent's autonomy. If agents are only allowed to follow the rules, part of the intrinsic characteristics of the multi-agent model might be overlooked and their capability to deal with real world situations diminished. We then advocate for a model where the environment allows for the definition of norms and sanctions applicable if they are violated. The performance of the system as a whole must be observed and autonomous agents are free to misbehave, but must recognise the possibility of loosing reputation points as well, which will make them less attractive as a partner for interoperation.

INSTITUTIONAL ABSTRACTION AND DESIGN

When modeling and implementing open agent systems that allow heterogeneous agents to join the system and perform tasks we use the abstraction of *institutional environments* and *normative spaces.* As the agents that join the institution are heterogeneous the necessity of the insertion of *social norms* in the system becomes evident. Norms are introduced to balance the functioning of the system and introduce a variable mechanism of control in the environment. With that mechanism the goals of the system as a whole are observed when autonomous agents want to achieve their own goals (Aldewereld et al., 2006).

The degree of openness observed in normative systems is variable. Norms define what is legal or illegal in the system and at the same time influence the agents to behave in a desired way, much like *legal frameworks* are developed to guide humans in the real world. It is important to observe the necessity of having this control over the autonomous agents to grant a sense of order to MAS, and a degree of openness as variable as the domain modeled needs or the system designer's desire.

Interaction and coordination are identified as major concerns when designing and deploying MAS, giving a distinct approach toward the modeling and design of distributed intelligent systems (DeLoach, 2002; F. Zambonelli, Jennings, & Wooldridge, 2003). Software engineering agent-oriented methodologies, such as Gaia (Wooldridge, Jennings, & Kinny, 2000), have been developed to observe the interaction between agents as a critical design aspect when building MAS.

The system organization as well influence the design of a MAS (DeLoach, 2002; Zambonelli, Jennings, & Wooldridge, 2001). A social setting is realized in the form of an environment where agents play roles and interact with each other pursuing individual or common goals. The organization has a defined structure, which define and enforce norms to manage the interoperations among agents. Norms are associated to roles agents assume in the system upon registration and will guide the agent behaviour in the system.

The operational use of norms in institutional environments is directly related to the context the agents interoperate is defined through ontologies (De Oliveira, Purvis, Cranefield, & Nowostawski, 2005); Grossi, Aldewereld, Vázquez-Salceda, & Dignum, 2006). Our approach is to use CPN (Jensen, 1997) to represent normative spaces that will guide agents throughout their useful existence in the institutional environment (De Oliveira, Purvis, Cranefield, & Nowostawski, 2004). Before defining the elements involved in the implementation of the institutional environment we introduce CPN in the next section.

COLOURED PETRI NETS AS A FORMALIZATION TOOL

CPN (Jensen, 1997) have some attractive properties for the representation of agents' conversations: they are expressive, have a history of successfully modeling dynamic processes, have a standard graphical presentation, as well as formal semantics. Compared to finite state machines broadly used for the representation of interaction protocols and behaviour patterns, for example, Dellarocas (2000), Artikis, Pitt, and Sergot (2002) and Arcos et al. (2005), they have the advantage of effectively representing concurrency and states together with actions that determine state change. In Cost, Chen, Finin, Labrou, and Peng (2000) a comprehensive analysis for the use of CPN to represent patterns of agent interaction is made and some agent development environments translates that assertion in reality as is the case of Duvigneau, Moldt, and Rolke (2002) and Cost et al. (2000). However, they have not applied the social model when designing MAS in their frameworks, and much less have they given the deserved consideration to the openness aspects we are interested in investigating.

Conversation policies (Bradshaw, Dutfield, Benoit, & Woolley, 1997); Elio & Haddadi, 1999) stress another existent layer of abstraction relevant to agent interoperation modeling and implementation, which suggests the relevance of cultural and social aspects of the communicator during the interaction. In Elio and Haddadi (1999) the separation of *task space* and *discourse space* goes a bit further in that direction. Our intent is to demonstrate the suitability of CPN for the modeling of institutional environments that use *social expectations* as a mechanism of control and management of open MAS.

With a static and dynamic specification of a process CPN unambiguously defines the behaviour of each net forming a foundation for formal analysis methods and allowing the development of CPN simulators. They describe explicitly both states and actions and offer a hierarchical description allowing the modeling of a large CPN by relating a number of small CPNs to each other, easing management and offering modulation. CPN simulators help to debug the net and facilitate the formal analysis of it by means of methods such as state spaces and place invariants, identifying undesired loops and deadlock conditions helping to eliminate humans errors introduced in the design process.

Figure 1 is a snapshot of a simple CPN designed with the open source CPN simulation framework JFern (Nowostawski, 2002). Figure 1 defines a very simple auction where commodities are selected as soon as they arrive in the *imput place*, the *transition* make bid do some processing with the token and the bid is put in the *outgoing arc*. For a brief introduction to the terminologies used in our CPN drawing, we define:

- **Tokens:** Represented by Java objects giving flexibility for the representation of simple structures as a *integer* or more complex ones as FIPA Messages (Fipa, 2002).
- **Places:** Net elements that represent achievable states. They are containers for tokens and are represented graphically by circles.
- **Transitions:** Net elements represented graphically by rectangles and that define actions triggered when the transitions are enabled.
- **Input Arcs:** Represented as arrows from input places to transitions. They are used to select a set of tokens from the input place, using guards and expressions.
    - **Input Arc Gards:** Boolean expression that must be evaluated to *true* for the transition to be triggered.
    - **Input Arcs Expressions:** Are used to select a set of tokens from the input place.

Figure 1. Simple auction

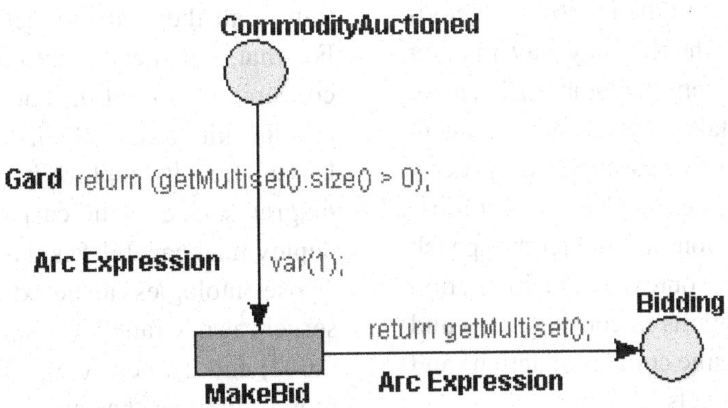

Figure 2. Player role and sub net

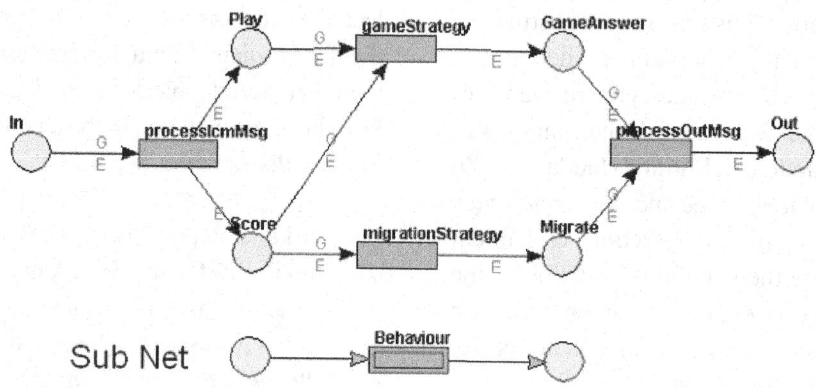

- **Output Arcs:** Represented as arrows from transitions to output places.
    - **Output Arcs Expressions:** These are used to generate output tokens which are being placed into output places.
- **Sub Nets:** Represented as double rectangles that specify a link from a higher hierarchy net to a lower hierarchy net; see Figure 2.

INSTITUTIONAL ENVIRONMENTS

In an institutional environment a group of persons agree to follow a set of regulations in order to develop a fruitful relationship with the others participating in the institution. That set of regulations is formed based on the actions each individual can perform in the institutional environment, according to the role they play. In other words, institutional actions (Colombetti, Fornara, & Verdicchio, 2002) taken by agents in an agent society must follow the rules imposed by a certain set of regulations defined by an institution. Those institutional actions are in fact the speech acts identified in the context of an institution that are used as illocutions in conversations and are used here to predefine courses of actions and build interaction protocols.

Institutional Acts

The institutional environment represents well defined groups of agents that together form organizations that follow a set of regulations, which specify how agents should undertake activities in a specific domain. Therefore, we use institutional actions to identify standard dialogs that take place in an institutional environment and define CPNs that will manage the interaction protocols necessary to achieve the wanted outcomes. Some examples of institutional actions would be to bid in an electronic auction institution, buy in an electronic commerce institution or to kick in an electronic soccer game institution.

The process of identification of the institutional acts in a domain requires analysis and evaluation of all the elements defined in that context. For that, we specify ontologies that define the concepts identified in that specific domain and relationships among them. We prefer to use ontological models graphically represented by UML diagrams, due to the easier reading and object deployment capabilities that those models offer. Those ontologies can be extracted from ontology servers as in Cranefield, Nowostawski, & Purvis (2002) and Cranefield & Purvis (2002), or from representations that are found in the semantic Web (Lee, Hendler, & Lassila, 2001).

Systems Elements

A significant characteristic of our approach is its open and distributed nature. Elements are organised in such a way that they are not compelled to report their actions to any other participating agent in the structure. We define our Institutional Environment (IE) as:

$IE = (O_s, N_s, I_a)$

Where:

- O_s stands for system ontology;
- N_s stands for normative space; and
- I_a stands for institutional actions.

The N_s is defined as:

$N_s = (R_s, R_e, O_c, W_f, L)$

Where:

- R_s stands for system roles;
- R_e stands for external roles and correspond to roles available to be played by agents that register to the institutional environment;
- O_c stands for context ontology;
- W_f stands for workflow; and
- L stands for content language and represents the language expressed in the content attribute of the FIPA messages exchanged among agents in the N_s.

The system level infrastructure identifies system agents that assume roles from R_s and are deployed to manage the interactions in the normative space; see Figure 3. Following is the description of such agents.

Institution Agent: The first step for an agent to get into the institution is to register with the Institution Agent and assume a role in the artificial society. Upon registering, the agent will gain

Figure 3. Normative space and conversation space

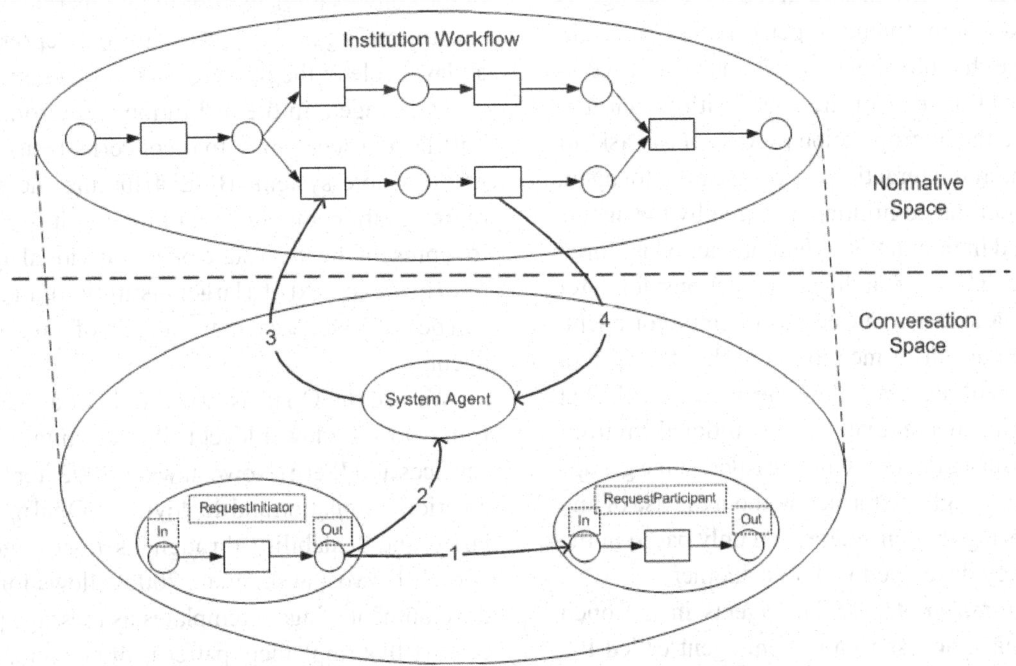

access to a representation of W_f in the form of a CPN that represents a suggested path or course of actions the recently registered agent should follow in order to obey the norms that regiment the institutional environment. The W_f can be as constrained as the MAS developer wants, it depends on the degree of restrictions, security or access he wants to deploy in the system, for example, demanding the use of monitored interactions for certain activities. It is important to mention here that the member agents will have their own goals and strategies to obtain them, and even though they have knowledge about the W_f, they still can participate freely in conversations without compulsorily following it, but everybody knows the rules of the game, though.

Monitor Agent: This system agent represents a monitoring authority in the institutional environment. It monitors certain activities, defined as institutional actions, according with the norms defined in the normative space by the W_f. The monitoring is done through the use of commitments and observation by the monitor agent of the commitments' life cycle. Once in the artificial society, the agent can commit itself with other agents to perform tasks and request other agents to make commitments to perform tasks for it. In the case that the agent does not have an acceptable level of trust of the agent with which it is starting the interoperation process, it can ask for a monitored interaction, where the monitor agent will audit the commitment shared by the agents engaged in interaction. That interaction is stored in a database of audited interactions for later examination, if requested to the monitor agent. It is important to mention here that the system agents will always follow the W_f and with that the norms that regulate the institutional environment. But again, knowing the rules, some groups of agents might be formed without the use of any monitoring in their interaction only based in the trust they developed with one another.

Reputation Agent: The agents in the open MAS have access to a system agent called the Reputation Agent. This agent is responsible for giving information about other agents that have been present at some time in the society and have developed a reputation. The agents are not obliged to report nor to consult the Reputation Agent prior to every interoperation they take part in. This offers flexibility and the possibility of open MAS implementations with verifying degrees of agent autonomy.

Agent Roles and the Conversation Space

In our approach, CPN are used to represent agent roles and agents conversations. Those two concepts together define what we call the Conversation Space of an institution; see Figure 3. In that way, we visualize an agent's role as part of a CPN that describes the messages exchanged by a specific group of agents in the institution. The overall institution conversations are represented as a CPN that has specific access points, where smaller CPNs, representing roles of individual participating agents, can be plugged in. With that approach, an agent will be involved in the part of the conversation appropriate for its role in the institution. Figure 2 depicts a CPN that represents a player role in the prisoners' dilemma game.

Every agent in the institutional environment will have a local state that concerns to its role in the whole system. By distributing the state representation among a set of tokens, it is easier to represent the local state of an individual agent role (in the context of a larger institution), and this can be useful for local management of individual agents.

We use the Opal (Nowostawski et al., 2002) framework for lower level FIPA communication services and JFern (Nowostawski, 2002) for CPN creation, simulation and deployment. Opal modularity and scalability through its micro-agents kernel (Nowostawski et al., 2002) allows for the development of agent templates as message processors that implement patterns of behaviour.

COMMITMENTS

In our approach, a commitment is an object created at the agent level and handled by the Auditor agent involved in the interaction. The Debtor is the agent that makes the commitment, and the Creditor is the agent relative to which the commitment is made.

We have adapted a model defined in Colombetti et al. (2002) to our needs. The commitment states used are: unset, pending, cancelled, active, violated, fulfilled and not fulfilled. The not fulfilled commitment state is introduced to differentiate a state identified when the commitment's task does not have a deadline associated with it and the commitment is never violated nor fulfilled, because commitment violation is directly related to a deadline.

The Monitor Agent is the system agent that manages the commitments in the MAS. We have defined a CPN that represents the auditor role and manages the commitment object from its instantiation to its storage in the commitment log, a data base of commitments that have reached one of its final states (cancelled, violated, fulfilled or not fulfilled).

The Monitor Agent will observe the normative space when monitoring interactions. Operations on commitments reflect in its state and together with the norms defined in the W_f a monitored agent can have from a change in its system reputation for better or worse or in an extreme case could lead to the banishment of the agent from the institutional environment.

The audit process performed by the Monitor Agent can be visualised as the proof given by the debtor agent that he performed some task. That would have the form of a commitment with condition equal to "true," acknowledging some information about a task the debtor should perform. Basically, the debtor would be committing itself formally to an authority in the MAS (represented by the Monitor Agent) that it did something. If an agent commits itself too often to false statements, the Monitor Agent should receive many complaints about that agent, which would lead the Monitor Agent to use its power to update the reputation of agents in the MAS.

The normative space guides the Monitor Agent during the monitoring process; its role is a commitment management protocol that manages commitment objects according to their life cycle. Participating agents can decide to break the protocol, but the normative space defines the level of tolerance to those actions, and depending on the application domain that the institutional environment is implementing, agent acts can decrease its reputation to such a level that other agents which consult the Reputation Agent will cease to communicate with the untrustworthy agent. The time factor is an important element in the representation of commitments. To express a commitment formally, it is necessary to find a representation mechanism able to handle the constraints of time found in the definition of commitments. We represent the idea of commitment's *timeout* and *deadline*. The timeout is related to the period of time available for the commitment to become active, and the deadline is related to the amount of time available for the commitment to be fulfilled.

Reputation Influence in the Institutional Environment

All the requisite institutional information is given to the agents upon their registration in the institutional environment, and will be available from the Institution Agent for consultation at any time. However, the Institution Agent does not maintain any control over the registered agents. In fact, it is the reputation that the agent has in the MAS that will determine its useful existence in the institution. Based on the trust the agents have for each other, they will interact or not. With time and the development of a number of interactions, agents can build up trust networks and establish trust relationships with each other.

The reputation the agent develops in the institutional environment, together with restrictions of the normative space, defined in the W_f, may restrict levels of access to certain resources. Such concepts as level of access to resources and services can be modeled in the normative space through the context ontology O_c and implemented in the W_f. Therefore, as all the agents are aware of the rules of the institutional environment they know that their actions might not only affect their reputation, but diminish the level of access to resources in the system.

We define a trust relationship ontology that defines the concept of trust and reputation and how they influence the system agents and can be used by external agents to compose their own definition of trust. That ontology is available for the agents that are joining the institutional environment to understand the information managed by the Reputation Agent.

The reputation update model sometimes needs to express characteristics of the context in which the institution was developed, and different reputation update models can be attached to the Reputation Agent CPN to express that. Another important aspect of that approach is that the external agents can have their own definition of trust and use the one defined by the institution to add information to it, or simply ignore it. The agents are not compelled to use the Reputation Agent before every transaction. They can have, for example, a history of their conversations internally and their own information about other agents and choose to refuse certain kinds of interactions from some agents, that they might not trust, but in case they are willing to use the information agent, it is available. Being aware of the W_f, they can calculate the risk of losing privileges in the institutional environment and act as they will.

DISCUSSION

Multi-agent system coordination methods are a research topic that has gained quite a portion of the research community's attention in the past years due to its suitability to design and build complex systems (Zambonelli et al., 2003). Usually, the MAS methodologies are based on agent-oriented foundations and are agent-centred instead of community- or socially-centred. Those methodologies focus more on the individual aspects of the agents, which bring very good contributions for the designing of agents, but lack in feasibility for the implementation of agent interactions in an organized way.

Agent infrastructures such as DARPA COABS (Kahn & Cicalese, 2003) and FIPA compliant platforms such as JADE (Bellifemine et al., 2000) and OPAL (Nowostawski et al., 2002), deal with many issues that are essential for open agent interactions. They use standards to manage communication, identification, synchronization and matchmaking. That infrastructure is necessary, but lies at a lower layer of abstraction than the one our model here presents.

An important development in the area of electronic institutions has been done in Arcos et al. (2005) Minsky and Ungureanu (2000) and Ricci and Omicini (2003). Even though, along with Arcos et al. (2005), we have the goal of implementing social centred frameworks to develop open MAS, our approach differs in many aspects from theirs. We adapt the work done in Colombetti et al. (2002) and Singh (1999) in the definition of our commitment-based infrastructure and use it to define objects whose active state changes according to the agent interactions with which it is associated. Again, in our architecture the three kinds of operational agent present use the concept of trust for better accommodating the openness that we claim to have in our model because we do not rely on interface agents to implement norms in our institutional environment. Instead of having a special operational agent, such as the governor (Arcos et al., 2005) and controller (Minsky & Ungureanu, 2000), which forces the agents to comply with the interactions in the institution, we use the concept of trust among agents and normative spaces so that different levels of security and other important constraints can be implemented in a way that the agent itself will decide which strategy to use in order to comply with the societal norms and avoid being penalized by losing reputation points, in a first instance, going to losing access to system resources until the banishment of the agent.

Another important aspect of our approach is that we do not use finite state machines to represent an electronic institution and the conversations in the institution. Our approach is the use of CPNs to represent the institution's normative space and conversation space, which includes the roles played by agents in the institution. By that, we seek the use of a formalism defined over concurrency concepts and powerful semantics relating states and actions.

CONCLUSION

Agent autonomy is an important aspect of the multi-agent paradigm. When coordinating the autonomy in MAS norms in institutional environments, bring a flexible mechanism for the definition of legality and illegality in an artificial social environment, as its use eases the coordination of open MAS composed by heterogeneous agents, allowing for the adaptation of degrees of openness according to the context being modeled.

REFERENCES

Aldewereld, H., Dignum, F., García-Camino, A., Noriega, P., Rodríguez-Aguilar, J., & Sierra, C. (2006). Operationalisation of norms for usage in electronic institutions. In *Proceedings of the Fifth International Joint Conference on Autonomous Agents and Multi-agent Systems*, (pp. 223-225).

Arcos, J. L., Esteva, M., Noriega, P., Rodriguez-Aguilar, J. A., & Sierra, C. (2005). Engineering open environments with electronic institutions. *Engineering Applications of Artificial Intelligence, 18*(2), 191-204.

Artikis, A., Pitt, J., & Sergot, M. (2002). Animated specifications of computational societies. In *Proceedings of the First International Joint Conference on Autonomous Agents and Multi-agent Systems* (Part 3, pp. 1053-1061).

Bellifemine, F., Poggi, A., & Rimassa, G. (2000). Developing multi-agent systems with JADE. In *Proceedings of the 7th International Workshop on Intelligent Agents VII: Agent Theories Architectures and Languages*, (pp. 89-103).

Bradshaw, J. M., Dutfield, S., Benoit, P., & Woolley, J. D. (1997). KAoS: Toward an industrial-strength open agent architecture. *Software Agents*, 375-418.

Colombetti, M., Fornara, N., & Verdicchio, M. (2002). The role of institutions in multiagent systems. In *Proceedings of the Workshop on Knowledge-based and Reasoning Agents, VIII AIIA*, (pp. 67-75).

Cost, R., Chen, Y., Finin, T., Labrou, Y., & Peng, Y. (2000). Using colored petri nets for conversation modeling. *Lecture Notes in Computer Science*, 178-192.

Cranefield, S., Nowostawski, M., & Purvis, M. (2002). Implementing agent communication languages directly from UML specifications. In *Proceedings of the First International Joint Conference on Autonomous Agents and Multi-agent Systems* (Part 2, pp. 553-554).

Cranefield, S., & Purvis, M. (2002). A UML profile and mapping for the generation of ontology-specific content languages. *Knowledge Engineering Review, 17*(1), 21-39.

De Oliveira, M., Purvis, M., Cranefield, S., & Nowostawski, M. (2004). Institutions and commitments in open multi-agent systems. In *Proceedings of IAT 2004: IEEE/WIC/ACM International Conference on Intelligent Agent Technology*, (pp. 500-503).

De Oliveira, M., Purvis, M., Cranefield, S., & Nowostawski, M. (2005). The role of ontologies when modelling open multi-agent systems as institutions. In R. P. Katarzyniak (Ed.), *Ontologies and soft methods in knowledge management* (Vol. 4, pp. 181-199). Adelide: Advanced Knowledge International.

Dellarocas, C. (2000, June). Contractual agent societies: Negotiated shared context and social control in open multi-agent systems. In *Proceedings of the Workshop on Norms and Institutions in Multi-agent Systems, 4th International Conference on Multi-agent Systems (Agents-2000)*, Barcelona, Spain.

DeLoach, S. (2002). Modeling organizational rules in the multi-agent systems engineering methodology. In *Proceedings of the 15th Canadian Conference on Artificial Intelligence,* Calgary, Canada.

Duvigneau, M., Moldt, D., & Rolke, H. (2002). Concurrent architecture for a multi-agent platform. In *Proceedings of the 2002 Workshop on Agent-oriented Software Engineering (AOSEÕ02), 2585.*

Elio, R., & Haddadi, A. (1999). On abstract task models and conversation policies. *Working Notes of the Workshop on Specifying and Implementing Conversation Policies* (pp. 89-98).

Finin, T., Labrou, Y., & Mayfield, J. (1997). KQML as an agent communication language. *Software Agents*, 291-316.

Fipa. (2002). *Fipa ACL Message Structure Specification.*

Grossi, D., Aldewereld, H., Vázquez-Salceda, J., & Dignum, F. (2006). Ontological aspects of the implementation of norms in agent-based electronic institutions. *Computational & Mathematical Organization Theory, 12*(2), 251-275.

Jensen, K. (1997). Coloured petri nets—Basic concepts, analysis methods and practical use: Basic concepts. *Monographs in theoretical computer science* (Vol. 1).

Kahn, M. L., & Cicalese, C. D. T. (2003). CoABS grid scalability experiments. *Autonomous Agents and Multi-agent Systems, 7*(1), 171-178.

Lee, T. B., Hendler, J., & Lassila, O. (2001). The Semantic Web. *Scientific American, 5*, 28-37.

Minsky, N., & Ungureanu, V. (2000). Law-governed interaction: A coordination and control mechanism for heterogeneous distributed systems. *ACM Transactions on Software Engineering and Methodology (TOSEM), 9*(3), 273-305.

Noriega, P. (2006). Fencing the open fields: Empirical concerns on electronic institutions. *LNAI: Coordination, organizations, institutions, and norms in multi-agent systems* (Vol. 3913, pp. 81-98).

Nowostawski, M. (2002). *JFern manual.*

Nowostawski, M., Purvis, M., & Cranefield, S. (2002). OPAL: A multi-level infrastructure for agent-oriented development. In *Proceedings of the First International Joint Conference on Autonomous Agents and Multi-agent Systems (AAMAS 2002),* (pp. 88-89).

Ricci, A., & Omicini, A. (2003). Supporting coordination in open computational systems with Tucson. In *Proceedings of WET ICE 2003*, (pp. 365-370).

Singh, M. P. (1999). An ontology for commitments in multiagent systems. *Artificial Intelligence and Law, 7*(1), 97-113.

Wooldridge, M., Jennings, N., & Kinny, D. (2000). The Gaia methodology for agent-oriented analysis and design. *Autonomous Agents and Multi-agent Systems, 3*(3), 285-312.

Zambonelli, F., Jennings, N., & Wooldridge, M. (2001). Organisational abstractions for the analysis and design of multi-agent systems. *Agent-Oriented Software Engineering, LNCS*, 98-114.

Zambonelli, F., Jennings, N., & Wooldridge, M. (2003). Developing multiagent systems: The Gaia methodology. *ACM Transactions on Software Engineering and Methodology (TOSEM), 12*(3), 317-370.

Chapter VIII
How Can We Trust Agents in Multi-Agent Environments? Techniques and Challenges

Kostas Kolomvatsos
National and Kapodistrian University of Athens, Greece

Stathes Hadjiefthymiades
National and Kapodistrian University of Athens, Greece

ABSTRACT

The field of Multi-agent systems (MAS) has been an active area for many years due to the importance that agents have to many disciplines of research in computer science. MAS are open and dynamic systems where a number of autonomous software components, called agents, communicate and cooperate in order to achieve their goals. In such systems, trust plays an important role. There must be a way for an agent to make sure that it can trust another entity, which is a potential partner. Without trust, agents cannot cooperate effectively and without cooperation they cannot fulfill their goals. Many times, trust is based on reputation. It is an indication that we may trust someone. This important research area is investigated in this book chapter. We discuss main issues concerning reputation and trust in MAS. We present research efforts and give formalizations useful for understanding the two concepts.

INTRODUCTION

The technology of Multi-agent systems (MAS) offers a lot of advantages in computer science and more specifically in the domain of cooperative problem solving. **MAS** are systems that host a number of autonomous software programs that are called agents. **Agents** act on behalf of their owners giving them access to information resources easily and efficiently. Users state their requirements and agents are responsible to fulfill them. Hence, MAS include many entities trying to solve their problems that are beyond of their capabilities. For this reason, in many cases, agents must cooperate with others in order to find the appropriate information and services to achieve their goals.

It is obvious that **MAS** are dynamic and distributed environments where agents may co-operate and communicate with others in order to complete their tasks. A key challenge arises from this nature of MAS. In such open systems, entities change their behavior dynamically. Thus, there is a requirement for trust between agents when they must exchange information Therefore, the basic question in such cases is: How and when can we trust an agent? Agents, in the majority of cases are selfish and their intentions and beliefs change continually.

We try to address this dilemma throughout this chapter. Specifically, we cover the fields of reputation and trust in MAS. This is an active research area, which is very important due to the fact that these two concepts are used in commercial applications. However, open issues exist in many cases, as it is difficult to characterize an agent as reliable or not.

In our work, we try to provide a detailed overview of reputation and trust models highlighting their importance to open environments. Due to the abundance of the relevant models, only the basic characteristics of models are discussed. We discuss basic concepts concerning MAS, reputation and trust. Accordingly, we present efforts, formalizations, and models related to the mentioned concepts. Finally, we discuss about trust engineering issues and we present future challenges and our conclusions.

BACKGROUND

Multi-Agent Systems (MAS)

Software agents and agency have been active research areas for many years due to their importance in various domains. The Web and the recently emerged Semantic Web are the most appropriate examples of such systems. In this section, basic characteristics of MAS are described. Our goal is to provide necessary knowledge about these systems and their requirements for security.

With the rapid evolution of the Internet, Software agents are a very important research area in Computer Science. **Software agents** are components of software or hardware which are capable of acting on behalf of a user in order to accomplish tasks (Nwana, 1996). The owner of an agent may be a human or another computational entity. Tasks are requested by the owners of agents in order to fulfill their needs. There are different kinds of agents. One can meet information agents that search for information sources, mobile agents that move from an environment to another, intelligent agents that can learn from their owners and the environment and so forth. For an extensive discussion of the different types of agents one can refer to Nwana (1996).

In the most cases, agents must deal with complicated tasks that demand cooperation with others. A **Multi-agent system (MAS)** can be defined as a loosely coupled network of problem solvers that interact to solve problems that are beyond the individual capabilities or knowledge of each problem solver (Durfee & Lesser, 1989). In such systems agents can cooperate or compete with others to complete their tasks. We must note that such systems are open. An open system is one in which the structure of the system is capable of dynamically changing (Sycara, 1998). In open MAS, the basic components may change over time such as information sources or agents' behaviors. From this point of view, it can be assumed that in open MAS (Huynh, Jennings, & Shadbolt, 2006):

- Agents have different owners and for this reason they are selfish and may be unreliable;

- There is no knowledge about the environment in which agents must interact with each other; and
- There is no central authority that controls the agents.

The last point is important for the cooperation among agents. Cooperation is often presented as one of the key concepts which differentiates MAS from other systems (Doran, Franklin, Jennings, & Norman, 1997). Through cooperation agents are able to obtain the necessary information needed for their tasks. Of course, interactions are the key issue for the cooperation.

It is obvious that there is an increasing need for the definition of trustworthy entities. In an open environment like MAS, agents change their intentions, goals and behaviors continually thus rendering imperative the need to define methods based on which each agent can be enabled to recognize nontrustworthy entities. The most important thing in such cases is to find ways to acquire information related to others' behavior. For example, an agent must communicate with the candidate partner or with others, in order to infer its trustworthiness. We describe methods to achieve this goal, and we give their basic characteristics.

Reputation in MAS

Reputation is an important factor in many research fields. Especially in computer science, reputation mechanisms are used either in research efforts or in commercial applications. In MAS, agents have to interact with others in order to fulfil their owners' needs for information. In such cases, reputation plays an important role.

According to a dictionary "**reputation is the state for a person of being held in high esteem and honour.**" From a social point of view "**reputation is the general estimation that the public has for a person**" (Wordnet, http://wordnet.princeton.edu).

In MAS, reputation refers to a perception that an agent has of the intentions and norms of another (Mui, Halberstadt, & Mohtashemi, 2002). This is critical for the cooperation among autonomous components in open environments, where the knowledge about the plans of others is limited.

One can find a categorisation of reputation in Wang and Vassileva (2003). Authors distinguish reputation models as centralised or decentralised, according to who has the responsibility to derive a reputation value. It should be noted that authors consider that trust is elicited through reputation. Therefore, their categorisation concerns both reputation and trust models.

- **Centralised.** In centralised reputation and trust models, the system is responsible to collect ratings for agents and publish them. Through this procedure, all ratings are evident to all members of the community and there is little need for communication among agents. Also, an aggregation procedure is performed by the system. The aggregation procedure aims to combine the different opinions in a final reputation level. Centralised models are characterised by simplicity and are mainly encountered in the area of e-commerce, where the main transactions are between sellers and buyers.
- **Decentralized or Distributed.** In decentralised systems there is not a central responsible authority and for this reason each agent develops its own reputation level for other community members. This means that there is an increased need for interactions between agents. Through them, agents form a subjective trust in their potential partners.

Mui et al. (2002) discern reputation based on which experiences and ratings are taken into consideration, and through what procedure information for the opponents is extracted. According to authors, reputation models can be divided into the following:

- **Individual Reputation.** Individual reputation is the description of the reputation level of a simple entity by another. This level is computed based on actions and information related to an agent and not a group of agents.
- **Group Reputation.** Group reputation depicts the social dimension of reputation. In these models, reputation is a function of the aggregated ratings taken from a group of entities. Entities rate others having their own experiences. These ratings may be utilised to provide information to an agent, when it needs to cooperate with others.
- **Direct Reputation.** Direct reputation is based on straight experiences with an entity. Usually, these observations are taken by interactions held between two entities. Direct reputation may be *observed* or *encounter-derived.* We have observed reputation when feedbacks through direct experiences of others consists a reference of the reputation of an agent. On the other hand, entities' ratings, after an interaction with others, may affect the reputation level of an agent. In this case, we have encounter-derived reputation.
- **Indirect Reputation.** With the lack of direct experiences for an entity, reputation can be derived from information gathered indirectly. There are three basic models for indirect reputation. The first model uses prior beliefs that agents carry about their interactions with strangers while the second model takes into consideration the group that an agent belongs to. Finally, the third model uses the information taken about an agent from the entities in the environment.

Trust in MAS

Trust is a common theme in computer science research, and refers to a range of different issues. It has important impact on domains such as security, e-commerce and Semantic Web. Trust is also an important concept for MAS. While in general trust refers to an aspect of the relationship of individuals, the concept has a completely different meaning depending on the context is used (Deriaz, 2006). Hence, trust has different meaning when we use it to characterize that humans' actions are trusted or when an agent decides to rely on another in order to obtain some resources.

Trust can be seen as the extent to which one entity intends to depend on somebody in a given situation (McKnight & Chervany, 1996). Trust can be defined as the belief that one can rely on someone else to accomplish a task. There is, however, a possibility that unfavourable issues can arise from interactions with a trusted person.

In this point, we describe a list of trust categories found in the literature.

According to Ramchourn, Huynh, and Jennings (2004) trust may be categorised, based on the part which decides the grade of trust, into the following:

- **Individual Level Trust:** Each agent decides which entity can be trusted based on its beliefs. These beliefs derive from interactions held between agents. Individual level trust can be further divided into:
  a. **Learning based.** Agents may interact with each other many times before deciding to trust someone. From this procedure, useful conclusions can be derived for the potential partners. Through repeated games, agents are able to analyze their opponents' moves in order to reach a conclusion. There are different kinds of metrics used in such models, as bi-stable values (good or bad) or fuzzy mechanisms with which one can decide to trust someone else for various acts.
  b. **Reputation based.** Reputation-based models use ratings from the members

of a community in order to derive a trust level. Important issues concerning this type of trust are the collection of ratings, their aggregation and the diffusion of members' opinions. First of all, an agent must collect ratings from the members of the community through the use of *referrals*. Referrals are the opinions of community members about a certain entity. After that, he must use an aggregation method in order to extract a result. In this point, there are issues that can complicate the whole procedure, as the lying witnesses or the absence of ratings. Finally, an important issue is the propagation of reputation in a community based on reputation scores that agents have for a set of other entities.

c. **Socio-cognitive based.** Contrary to previous types, where trust is computed taking into consideration the results and the components of interactions between agents, socio-cognitive-based trust is computed based on beliefs that an agent has for their opponents. These beliefs are: competence, willingness, persistence, and motivation belief.

- **System Level Trust:** Agents are selfish components which want to obtain as much profit as possible. For this reason, it is imperative to force them to follow some rules when interacting in the context of a system. This is the system level trust. In consequence, they will be trustworthy, thus minimizing the danger to interact with liars. System level trust can be further divided into:

a. **Truth-eliciting protocols.** These protocols may be used to elicit trustworthy behaviour of an agent. Agents must conform to certain protocols' steps in order to complete transactions in the system.

b. **Reputation mechanisms development.** Reputation may be used in system level trust. There are rules posed by the system concerning the three key elements of reputation models (collection of ratings, aggregation and propagation). In such systems, the entities responsible to store ratings may be centralised or decentralised. All agents working in the system have access to these entities either to read ratings or to publish their own.

c. **Security mechanisms development.** In these model types, a number of features are taken into consideration in order to provide a reliable security mechanism that ensures trust in system entities. The essential elements that make an agent trustworthy can be identity proof, access permissions, content integrity and content privacy (Poslad, Calisti, & Charlton, 2002). Additionally, certificates may be used to provide a higher level of security. The system forces members to give the necessary information as for the aforementioned elements in order to have an acceptable degree of safety.

Artz and Gil (2006) note that there are two methods to derive trust:

- **Based on credentials.** Credentials are elements that can be used to elicit information for an entity. A credential may be simple as a signature or complex relationships between elements in an open environment as the Semantic Web. For example, an agent may have an identifier with which may interact with others. This identifier may be used in a system to provide to an agent permissions or rights to work with specific information sources. An extensive review of systems that use credentials-based trust models is presented in Artz and Gil (2006).

- **Based on Reputation.** This model uses reputation to assign trust to members of a community. An agent utilizes personal experiences taken from interactions held between potential partners, and ratings from other members of the system. There are two ways to extract the trust level for an entity. One can rely on a central authority to have access in reputation ratings or on himself. Few efforts in literature use the first method. The second one describes the decentralised model where each entity must develop methods for the aggregation of ratings taken from the community.

In Osman (2006), authors specify the difference between trusting an agent, and trusting an interaction. They provide a categorization and a specific connection with the categories presented in Ramchourn et al. (2004). Two new categories are presented:

- **Local Deontic Level.** It concerns the constraints and permissions that an agent must follow when interacting in a multi-agent system. Agents are dynamic components and their goals, intentions and plans change continually. This means that when they work in an open environment, they have obligations, permissions and prohibitions, posed by the system. Through this, the system defines a security level concerning the transactions held among the potential partners.
- **Global Interaction Level.** Apart from the internal deontic model of each agent, there is another interaction model that specifies the rules based on which interactions are held. Every agent must conform to these rules in order to gain access to interactions with others, useful for the completion of its goals. Specifically, the interaction model is a protocol that determines steps to carry out interactions.

Grandison and Sloman (2000) presented a set of trust classes which are:

- **Provision Trust.** It describes the trust that an entity may have to a service provider.
- **Access Trust.** It describes the trust that an entity may have for the purposes of accessing resources.
- **Delegation Trust.** It describes the trust that an entity may have to an agent that works on its behalf.
- **Identity Trust.** It describes the belief that an identity is as claimed.
- **Context Trust.** It describes that the relying entity has confidence in a system in which transactions are held. Moreover, each entity can rely on system, when problems may arise in transactions.

Finally, another categorisation found in Wang and Vassileva (2003) has already been presented, where trust quantification may be held either by a central authority or by each agent individually.

A significant amount of work on trust has been performed in the area of Normative MAS (NMAS). NMAS is an extension of classical MAS which combines traditional MAS with normative systems where concepts such as obligations, commitments, permissions and rights are used to describe the behaviour of an entity (Boella, Van Der Torre, & Verhagen, 2006). Thus, every agent acting in a community has some obligations and commitments that should be fulfilled. In such systems, norms are defined to describe when the behaviour of an agent is acceptable. While the agent follows these norms its trust level increases in the community. In reverse, the agent's trust level decreases when its actions do not conform to the specified rules of normal behaviour. These rules aim to force agents to do the right thing cooperating with others in the broader environment. However, norms can be violated for various reasons and thus there is a dynamic trust valuation (Boella & Van Der Torre, 2005).

ISSUES CONCERNING REPUTATION

Reputation level can be derived based on three elements: the experiences of the evaluator, the referrals of others and the combination of the experiences and the referrals (Josang et al., 2006). There are methods that deal with all these three issues. Generally speaking, in MAS a set of agents $A=\{a_1, a_2, \ldots, a_n\}$ want to interact with others in order to complete their goals. For each potential partner, every agent must calculate its reputation degree which is extracted through a reputation function:

$$Reputation\_value=f(R, E, S) \quad (1)$$

where R represents ratings from other members of the community, E are the individual experiences taken from direct interactions with the target entities, and S represents ratings retrieved from the system.

The factor R is computed based on aggregation of ratings. Namely, there is an aggregation function which derives a final value from a set of witness agents $WA=\{wa_1, wa_2, \ldots, wa_m\}$.

$$R=g(r_1, r_2, \ldots, r_m) \quad (2)$$

where r_i denotes the referrals of the i_{th} witness agent. In literature, there are models that use only one of the above mentioned elements or a combination of them. As discussed below, every reputation model uses a function in order to calculate the final result, which follows the general form depicted in (1). For example, an agent may be based only on direct experiences with the target agent, without paying attention to the ratings of others. In order to define efficiently final reputation value, a model must be based on a combination of the above mentioned features (e.g., referrals, the system's ratings and direct interactions). The majority of systems in the literature follow this direction. Their difference is located to the form of the referred functions and the type of the computed values. Hence, in all models we can find a reputation function that produces a value that can be either discrete (e.g., "Confident," "Non-Confident") or continuous (represented through a real number).

A short description of reputation models follows. Furthermore, we present our point of view related to their advantages and disadvantages.

Simple Mathematical Models

They are the simplest models. In these models simple calculations are used in order to compute reputation values. For example, the system may store the number of positive and negative opinions for agents and compute the final score. If the positive opinions are $P=\{p_1, p_2, \ldots, p_k\}$ and the negative are $N=\{n_1, n_2, \ldots, n_m\}$ then the final score is:

$$Score = |P| - |N| \quad (3)$$

where $|x|$ denotes the cardinality of set x. The higher the value of *Score* is the more reliable the agent is considered. For example, if an agent has received 10 positive and 2 negative referrals, it has a reputation degree of 8. This agent is more reliable than another that has a reputation degree of 5. However, these models do not take into consideration the initial numbers from which the final result is computed. Let us examine an agent that may have received 100 positive and 90 negatives opinions. This means that the agent has approximately 47% negative opinions in the community. Nevertheless, this agent is more reliable than another with 10 positives and 1 negative referral (approximately 9% negative opinions).

In order to cover these disadvantages, advanced mechanisms use a weighted sum to compute an average which shows the reputation level. These mechanisms use the information related to the ratings such as the age of each rating, the distance between rating and current reputation value, and so forth (Josang, Roslam, & Colin, 2006).

It should be noted that such systems do not take into consideration critical issues concerning the selfish and dynamic nature of agents. It is possible that agents may form coalitions in order to exchange positive marks thus achieving better reputation scores. Furthermore, the simple mathematical models do not examine in depth the referrals retrieved from the community members in order to extract useful information about the behaviour of an agent.

Bayesian Reputation Systems

Bayesian systems are based on statistics. They compute reputation using the Beta probability density function. This function can be used to describe probability distributions of binary events. For simplicity, we give only a short description of such systems.

A Bayesian reputation system takes binary ratings and uses the *a priori* reputation score and the current ratings to compute the *a posteriori* result (Josang & Ismail, 2002; Mui, Mohtashemi, Ang, Szolovtis, & Halberstadt, 2001; Whitby, Josang, & Indulska, 2004). In agent systems, we can describe the behaviour of an agent as "Honest" vs "Dishonest," or as "Reliable" vs "Unreliable" which constitute binary events. If, for one agent, there are x positive and y negative observations, then the reputation score can be computed as follows:

$$\alpha=x+1,\ \beta=y+1 \text{ with } x,y \geq 0 \quad (4)$$

the probability expectation value is:

$$E(p) = \frac{\alpha}{\alpha+\beta} \quad (5)$$

for the Beta distribution, which can be expressed as:

$$B(p|\alpha,\beta) = \frac{\Gamma(\alpha+\beta)}{\Gamma(\alpha)\cdot\Gamma(\beta)} p^{\alpha-1}\cdot(1-p)^{\beta-1} \quad (6)$$

where $p \in [0..1]$, $\alpha,\beta>0$, $p \neq 1$ when $b<1$ and $p \neq 0$ when $a<1$.

When the a priori probability does not exist, then we consider $\alpha=1$ and $\beta=1$. From this function, each agent can compute the possibility that a potential partner is reliable based on previous values of reputation. For example, if the expectation value E(p) has a score of 0.9 means that the most likely value of positive outcomes in the future is 0.9, but the actual outcomes are uncertain.

Bayesian models give a computational theoretical framework for the reputation score good for autonomous computational entities as agents are. It is an efficient mechanism to combine evidences. An entity must keep track of the outcomes of others and compute the reliability possibility through the above referred functions. A full description of a system representative of this kind of model can be found in Josang and Ismail (2002). However, these models are complicated due to the calculations that must be performed in order to derive a final reputation value. Also, the definition of a priori probability used for the calculations is necessary. The probability value is important for these models and must be derived by a subjective method.

Social Networks

Social networks are originated in sociology. Social networks can be represented as graphs that depict relations between members of a community. Social networks analysis emerged as a set of methods for the analysis of social structures (Sabater & Sierra, 2002). In MAS, agents must retrieve data concerning the relation among the members of the system in order to decide the reputation level of a potential partner. However, it is difficult to use methods taken from sociology in order to poll information for the network architecture. For example, sociologists use methods as the opinion poll or interviews. In cases where autonomous computational components are the nodes of a social network, more "computational" ways must be found to extract the necessary information.

The procedure that agents adopt for building the network is critical for the success of such systems. They must describe as much relational data as they can in order to have a significant view of the system. We must also note, that these networks change dynamically due to the open nature of MAS. Agents may enter or leave at every time and, moreover, may alter their goals, behaviours and intentions.

Generally speaking, in systems based on social networks there is a set $A=\{a_1, a_2, a_3, \ldots, a_n\}$ of agents that want to obtain information from others in order to complete their goals. Each agent builds a social network $G=\{d_1, d_2, \ldots, d_n\}$, with nodes d_i representing the members of the system. Edges that connect nodes show a relation between them. For example, edge $e_{i,j}$ denotes that there is a relation among nodes i and j. Next, it finds its potential partners and tries to collect information about them. It may be based on referrals or on the system or on a combination of them. Referrals are taken from other members that have interaction history with the target agents, and after a careful selection. These referrals must be aggregated in order to extract a final value through which reputation is computed.

Three critical factors must be taken into account: the possibility that agents may tell lies, the possibility that agents may conceal information about others and the careful selection of referral agents. In these cases, social networks may be constructed based on false values as for the reputation degree of each member. Also, there are cases in which agents ally with others and for this reason may hide the bad reputation that an examined agent acquired. A method to alleviate the problem of lying in MAS is presented in Schillo, Funk, and Rovatsos (2000).

As mentioned above, in MAS social networks agents appear as nodes and their relationships as edges. Each edge has a value which represents the weight of the relationship between the two connected agents. After the graph creation, the construction of the network and the computation of reputation follow. Such computation is based on the weights assigned to edges. An extensive survey on reputation mechanisms based on social networks can be found in Ramchourn et al. (2004).

In this point, we present two simple examples of social networks models. A first example of such model is presented in Pujol and Sanguesa (2002). Authors describe an algorithm which is named *NodeRanking* and is used to extract a reputation value for members in a community. Its main idea is that every node and, respectively, agents have an authority degree which is an importance measure. For example, the authority of a node x is computed as a function of the total measure of authority presented in the network and the authority of the nodes pointing in x. The main rationale is that if a node has a lot of edges pointing to it, this means that the node is important in the community because this means that it cooperates with many other members. Another critical issue is that every authority value is propagated through the out-edges. The reputation value is computed based on the authority value of each node, taking into consideration the importance and hence the number of agents that are related to the examined entity.

Another system that uses social information in order to compute reputation is REGRET (Sabater & Sierra, 2002). Reputation in REGRET has three dimensions, the *individual*, the *social*—according to the source of the information used to extract a reputation value—and the *ontological* dimension, which helps to transfer the reputation between related contexts. While the individual dimension takes into consideration the results of direct interactions between potential partners, the social dimension utilizes information taken from the other members of the community. In the second dimension of reputation, agents may use witnesses from others or consider neighbourhood reputation. Furthermore, the system assigns a reputation level to every role defined in it. Hence, agents that have a specific role in the system inherit the reputation level assigned to the role.

System reputation is the easiest to compute but is dangerous because a role held by an agent does not convey information about its intentions. In the REGRET system, reputation is combined with a domain and calculated using a table in which rows are the potential roles and columns are the reputation types. More complicated are the remaining two methods. Witnesses reputation uses the referrals of others in order to establish a reputation level. REGRET gives the opportunity to an agent to define a set of witnesses and aggregate their referrals based on fuzzy rules. Of course, witnesses are entities that have interacted in the past with the target agent and they are taken into account based on the same event, if it is possible. Neighbourhood reputation is not related to the physical location of the agents but to the links created by interactions. These interactions and the relations between agents are very useful to compute reputation level for a target agent. Fuzzy rules are also used in this case.

Belief Theory Models

Belief theory characterizes the remaining of the subtraction between 1 and the summary of the possibilities of the all possible outcomes, as uncertainty. In these models, agents use their beliefs about the behaviour of another entity. In Yu and Singh (2002), authors propose the use of Dempster-Shafer theory (Dempster, 1968; Shafer, 1976) for the computation of reputation degree. There are two kinds of belief in their model: *Local* and *Total*. Local belief is obtained from direct interactions with the target agent and Total belief is extracted from the opinions of others combined with local belief. Witnesses from others are necessary when interactions are not available. Each agent models the information from others using belief functions. There are two outcomes related to the reliability of an agent: *Trustworthy* or *Not Trustworthy,* each of them has a belief value m(T) and m($\neg$ T) respectively, taken from the correspondent belief function. The reputation score for an agent A is:

$$\Gamma(A)=\beta_A(\{T_A\})-\beta_A(\{\neg T_A\}) \quad (7)$$

where β_A is the cumulative belief result computed using the testimonies from a set of L neighborhoods. When no testimonies are available, then the reputation score is 0. Also, authors present the reputation value of a set of K agents, which is:

$$\text{Group\_Reputation} = \frac{1}{K}\sum_{i\in[1..K]}\Gamma(A_i) \quad (8)$$

These models are able to exhibit the beliefs of agents, accumulated from past experiences or others, in functions that combine them and produce the final result. However, the definition of a threshold value is critical. This threshold defines when an agent may be characterized as trustworthy or not. Moreover, the assumption of only two possible outcomes limits the model.

Fuzzy Models

Fuzzy models try to catch the subjective point of view of an agent related to another member of a community. In these models, reputation is presented through linguistic fuzzy values in contrast to other models in which common reputation levels are defined by means of real numbers. For example an agent may be characterised as "reliable" or "not reliable." Fuzzy logic (Zadeh, 1989) is very important because it provides reasoning techniques for the extraction procedure. Rules may have the next form:

IF andecent THEN consequent

where *andecent* is represented with fuzzy sets.

Fuzzy logic techniques depend on subjective criteria that may lead an agent to cooperate based on its thoughts about others. This means that if an agent has very optimistic views of the community he may rely on others that have bad intentions.

In Rubiera, Molina Lopez, and Muro (2001) a method for the computation of reputation based on a fuzzy model is presented. Each agent retrieves opinions only from entities that are highly appreciated. Based on their answers it computes a value that is extracted from a fuzzy set according to its point of view. The result is the weight of an agent's opinion. Furthemore, there is interest on combination of the new and the old reputation values. The old reputation score of the candidate partner is taken into consideration and, hence, the final result is the average of two fuzzy values: the old and the new one. The agent is responsible to decide on defection. Usually, if a threshold is reached, cooperation is held.

Another system that relies on fuzzy rules is REGRET, which was briefly described in the previous section.

Role-Based Reputation

In some models, reputation can be seen as a value of a role fulfilment. A role is a set of obligations and actions that an entity has in the community. If an agent acts and behaves as a role dictates, then it has the reputation level that this role offers. In Carter and Ghorbani (2004) a framework for the role fulfilment measurement is presented. Three roles are investigated: The Assistant, the Provider and the Citizen. A general overview of how measurements take place in each role is given by the authors. In order to compute the final value of reputation, authors examine the satisfaction degree of each role for a specific entity and combine these partial results. The reputation is a weighted sum of each value that reflects the fulfilment of each role.

The main problem with these models is that they do not examine in depth the intentions that agents have. Meeting the requirements of a role by an agent does not mean that the agent will not change its behaviour.

Unfair Ratings: Deception

As mentioned above in distributed reputation models, when a central authority is absent each agent that wants to cooperate with others must collect ratings from the environment in order to decide the reputation degree of an entity. In these cases important issues are:

- The possibility that some agents provide unfair ratings for others. These ratings may be unfairly positive or unfairly negative (Dellarocas, 2000).
- The possibility that an agent deludes others.

According to Whitby et al. (2004), methods of avoiding unfair ratings are divided into *endogenous* and *exogenous*.

Endogenous methods are based on the statistical analysis of the rating values. They can give or exclude ratings that are possible to be unfair. Classical examples of this kind of systems are Bayesian reputation systems. Authors describe an algorithm that filters unfair ratings in a Bayesian model. Exogenous methods are based on factors that are related to external elements such as the reputation of the witness. The main idea is that an entity with low reputation is likely to give unfair ratings and vice versa. In the relevant literature, one can find a lot of works falling in the aforementioned categories.

Another algorithm for the detection of deceptive agents is proposed in Yu and Singh (2003). This method uses exogenous characteristics of the witnesses as we presented above. The algorithm assigns weights to witnesses and makes a prediction based on the weighted sum of their ratings. The second idea is to tune these weights when a prediction fails. In this case, the weight of successful witnesses is increased and the weight for the unsuccessful is decreased. We must note that the ratings that an agent takes are belief functions and for this reason the algorithm maps belief functions to probabilities in order to be able to compute and update the weights of each witness. Moreover, authors study the number of witnesses and its effect to the system's prediction values.

ISSUES CONCERNING TRUST

Trust is usually researched in the security domain. The main reason is that these two concepts are related, but they have different orientations. However, trust and security provide protection against malicious components. In this sense, trust can be considered as a *soft security* mechanism. This term first appeared in Rasmusson and Jansson (1996). Authors discern *hard* and *soft* security mechanisms. Hard tools are authentication, cryptography, and so forth, and soft tools are those that take into consideration social control issues, as they are trust and reputation.

In MAS, trust plays an important role because agents need to cooperate with other members of the community. The importance of trust in MAS is shown in Castelfranchi and Falcone (1998). Critical questions arise such as: When can I trust another entity? Which entity is trustworthy? What are the elements that can be used in order to conclude a trust level? What methods should be used to conclude trust level? Such questions are addressed in the fourth section of our chapter.

Discussion

The main differences between trust and reputation are:

a. Usually, trust is a score that reflects the subjective view of an entity from another, whereas reputation is a score that reflects the view of the community.
b. In trust systems, transitivity is considered explicitly while in reputation systems is seen implicitly (Wang, Hori, & Sakurai, 2006).

The common element between the two concepts is that both of them try to help someone that wants to find trustworthy partners to achieve its goals through cooperation. However, trust is more complicated concept that involves many parameters. For this reason, it is very difficult to assign a strict definition to trust.

As mentioned above, trust is a subjective view of an entity. It is based on some beliefs that an entity has for another, but it is not clear where such beliefs originated. This means that an agent may be reliable only for a set of other agents and not for all of them. The level of trust is also depended on the context in which it is being studied. For example, an agent may be trustworthy when providing information but it is nontrustworthy when selling products. These two factors are basic to open systems and must be taken into consideration. Moreover, trust is dynamic. An agent may consider another entity as reliable in a specific time but its opinion may change accordingly based on the behaviour of the target entity.

The simplest form of trust is centralized. In such systems, there is a central authority that keeps the trust level of entities who rate each other after every transaction. Soft mathematical calculations are held to provide the final result. It is a scheme that must take into account issues concerning lying entities or unfair ratings. Also, there must be a high security level to prevent violations in central database where ratings are kept. On the other hand, decentralized trust models are complex and require effort from the side of each entity that tries to find partners. In such systems, critical issues are the storage of trust values, the location of witnesses and the inference procedure.

In general, a trust function has the following parameters: the beliefs of the examiner (B), the reputation of the examinee (R), previous trust values (P) and the context (C).

$$Trust\_value=f(B, R, P, C) \quad (9)$$

An interesting value is B. B may be extracted from direct experiences through communication or past experiences with the target entity. It may be a positive or a negative belief. Relation (9) concerns a general function form. As we discuss in the following paragraphs all the described models use a function that follows this general form and takes into consideration one or more values from B, R, P or C.

The result of the referred function may be discrete or continuous values. For example, in discrete models, as fuzzy models are, a trustworthy behaviour may be characterized as "Very Trustworthy," "Trustworthy," "Untrustworthy," or "Very untrustworthy," for direct trust between two entities, or "Very good," "Good," "Bad," or "Very bad" for recommender trust (Abdul-Rahman & Hailes, 2000). Either discrete or continuous, the final trust value reflects a confidence over the knowledge we have about an entity.

Trust mechanisms vary from these that use simple computations to those that use more complicated characteristics of the entities involved in such situations. However, the common procedure among them is that they map a set of features to trust information. In the following paragraphs, we give a description of some important categories of trust and examples of each one.

A key issue concerning trust is its dynamic nature. Trust evolves over time as entities cooperate with others. For this reason, it is critical to define a trust update procedure. Especially in open environments like MAS, where goals, intentions and beliefs of each agent change continually, there is a need for dynamic adaptation of the trust level. This means that in every model that is used to compute trust, developers must take into consideration how trust levels evolve over time and transactions. The evolution of trust may be based on the experiences of the trustor or on new information taken from other members of the community.

In conclusion, in the computing trust procedure, the phases that an agent may handle are:

a. Trust Discovery Phase (TDP);
b. Trust Aggregation Phase (TAP); and
c. Trust Evolution Phase (TEP).

In TDP, each agent tries to find the appropriate sources for referrals and may communicate with the target agents in order to elicit useful information about their behaviour. In this phase it is important to have mechanisms to identify if a group of agents have formed a coalition and share good referrals among them. In TAP, the most important issue is to use an appropriate aggregation function in order to derive the final value of trust. This function may take into consideration the results of direct experiences and of course the referrals of peers. Finally, the TEP is a continuous procedure through which an initial trust level is evolving over time based on observations. Its great importance relies on the dynamic nature of MAS. If an agent is trusted in a specific time this does not mean it is to be trusted forever. Agents are selfish and may change their behaviour without warning or may ally with others.

Trust Propagation

MAS can be viewed as graphs where their agents are represented by nodes. Edges consist of their relationship in the community and weights represent the trust value between the two connected nodes. Graphs may be used to transfer trust information among members. Every agent trusts some others in the network. *Estimated trust belief* is derived through the trust network based on inferences while *expected trust belief* is the ideal target (Ding, Kolari, Ganjugante, Finin, & Joshi, 2004). It is very difficult to achieve the expected trust belief due to the lack of global knowledge of the community and its members. As the trust network evolves over time and more information is gathered from the agents, the ultimate goal is gradually approached.

Propagation of trust is very important in such networks because it gives the opportunity to derive beliefs about agents through the combination of values taken from multiple sources. The most common method for trust propagation is referrals. Referrals have been investigated in reputation mechanisms (see Section "Issues Concerning Reputation"). An agent, having collected opinions from a set of peers, needs an aggregation method in order to define the final estimated value of trust. Through this procedure, trust information can be propagated over the network. Additionally, we must take into consideration that agents may ally with others and give positive recommendations for their allies. However, when an agent establishes trust based on recommendations from others, this trust value cannot be greater than the trust value between the agent and the recommender, and neither the trust value between the recommender and the target agent (Lindsay, Yu, Han, & Ray Liu, 2006). Another effort on trust propagation is discussed in Guha, Kumar, Raghavan, and Tomkins (2004).

Simple Trust Models

Simple trust models try to determine relations that depict the behaviour of an entity as a function of positive and negative opinions. The simplest form, found in Deriaz (2006), is:

$$Trust\_score = \frac{|Positive\ Ratings|}{|Positive\ Ratings| + |Negative\ Ratings|} \quad (10)$$

where |Positive Ratings| and |Negative Ratings| represent the number of positive and negative ratings, respectively.

Whenever an certain agent has only positive ratings and no negatives, then the trust score is equal to 1. Therefore, *Trust_score* can take values ranging from 0 to 1. An extension to this model that take into consideration the time in which these ratings are provided gives more efficient trust computation because it can exclude obsolete ratings. Furthermore, trust can be computed if we scale recent events. Accordingly, if in Deriaz's model we set a=b=c=1, which are the default values of parameters a, b, c, then the following holds:

$$Trust\_score = \frac{|PosRatings| \cdot (1+\delta_p) + A}{|PosRatings| \cdot (1+\delta_p) + |NegRatings| \cdot (1+\delta_n) + B} \quad (11)$$

where

$$A = \frac{1}{T} \sum_{i \in [1..|PosRatings|]} t(po_i) \quad (12)$$

$$B = \frac{1}{T} \left(\sum_{i \in [1..|PosRatings|]} t(po_i) + \sum_{i \in [1..|NegRatings|]} t(pn_i) \right) \quad (13)$$

and δ_p, δ_n are the statistical variance of the positive and negative outcomes, T is the current time, po_i, pn_i are positive and negative rating received at time i, |PosRatings| and |NegRatings| are the cardinality of positive and negative ratings, respectively.

The disadvantage of this mechanism is that it does not take into account the context in which these ratings were made. The above forms deal with the positive and the negative opinions without separating them into the context fields for which ratings are formulated. Furthermore, the model does not notice cases where entities may ally with others in order to elicit positive outcomes or the case that an entity is neutral to another.

Entropy-Based & Probability-Based Trust Model

In this section, we present two models of trust based on the work reported in Lindsay et al. (2006). Authors, influenced by information theory, present *Entropy-based* and *probability-based* trust. The trust relationship between two entities is represented by T(S,A,AC), where S is the subject which examines the trust level of the agent A for an action AC. Similarly, the probability that and agent A will perform the action AC in the subject's S point of view is represented by P(S,A,AC). The entropy based value of trust is calculated as follows:

$$T(S,A,AC) = \tag{14}$$

$$\begin{cases} 1+p\cdot\log_2(p)+(1-p)\cdot\log_2(1-p) & 0.5\le p\le 1 \\ -p\cdot\log_2(p)-(1-p)\cdot\log_2(1-p)+1 & 0\le p<0.5 \end{cases}$$

where

$$p = P(S,A,AC) \tag{15}$$

The final trust value is a real number in the interval [-1,1]. Some important examples that show the trust level of an agent are:

$$C = \begin{cases} \textit{Subject trusts the agent} & p=1 \textit{ and } T=1 \\ \textit{Subject distrusts the agent the most} & p=0 \textit{ and } T=-1 \\ \textit{Subject has no trust} & p=0.5 \textit{ and } T=0 \end{cases} \tag{16}$$

In general, the following holds:

$$T(S,A,AC) = \begin{cases} <0 & when\ p\in[0..0,5) \\ >0 & when\ p\in(0,5..1] \\ =0 & when\ p=0,5 \end{cases} \tag{17}$$

We should note that the probability P(S,A,AC) represents the view of a specific subject which means that different agents have different opinions about a target agent.

The entropy-based model depends on the trust value described above. Especially for the propagation of trust, a simple product is used where the two factors are the recommendation value of another agent multiplied by the trust value of the recommender. For multipath recommendations, the final result is the weighted sum of each recommendation. In the probability-based model, the probability that an agent will perform the specific tasks combined with the probability that a recommender make correct recommendations is adjusted.

Reputation-Based Trust Models

Reputation based trust models are used in distributed systems where there is little information on the overall network. If an entity has a high reputation level in a community then others may trust it more easily than another that has lower reputation value. For the computation of trust, an agent depends on opinions of a set of community members. Important issues in these cases are the collection method of ratings and the aggregation procedure. It should be reminded that trust is a concept derived from direct interactions between two entities, while reputation is the view of a member from the community side.

Reputation-based models rely on methods that give the opportunity to an entity to gather referrals from other members and apply an aggregation function in order to calculate the final value of trust. It is wiser to combine these results with values obtained from direct experiences. In Ramchourn et al. (2004), authors give an extensive review describing methods for the retrieval of ratings and the aggregation procedure from a social network point of view.

Reputation models are discussed in previous section.

Bayesian Network Trust Models

A Bayesian network is a network where probability relationships of some entities are presented (Ben-Gal, 2007). The final trust value is represented by the root of the network and leafs are the sources in which the beliefs of the examiner are based. A formalization of Bayesian trust networks is given in Melaye and Demazeau (2005). Each agent forms its basic beliefs investigating a number of belief sources. We have:

$$Basic\_Belief_i = f(B_{i1}, B_{i2}, \ldots, B_{iN}) \tag{18}$$

where $i \in [1..number\_of\_basic\_beliefs]$ and N are the number of belief sources. An important point is that the function f is depended on conditional probabilities, which means that a tree node may be more influential than another. Each trust component is associated with a probability of satisfaction. Accordingly, the final trust value is calculated as follows:

$$\text{Trust\_value}=g(BB_1, BB_2, \ldots, BB_N) \quad (19)$$

where BB_i is the i-th basic belief taken from (18). It is obvious that in such cases a bottom-up approach is adopted, starting from the belief sources and concluding with the final result.

Another representative example is shown in Wang and Vassileva (2003). Each agent builds a Bayesian network for every potential partner. Each network has a root with two values. The first represents the satisfaction level while the second the nonsatisfaction degree. The satisfaction level is derived from the number of the successful interactions divided by the total number of interactions. The leaf nodes in the network represent the different capabilities that potential partners have. In this model, the recommendations of other agents are taken into consideration.

Belief and Fuzzy Models

In belief models, we meet methods that use the beliefs of an agent that another entity is trustworthy or not. In belief theory, each opinion may be represented as a triplet (belief, disbelief, uncertainty). The sum of probabilities of these three values is 1:

$$\text{belief}+\text{disbelief}+\text{uncertainty}=1 \quad (20)$$

Agents use their and others' beliefs in order to extract the final score. This score is computed through the use of belief theory and consists of a subjective certainty of the pertinent beliefs. An example of a belief trust model one can be found in Josang (1999, 2001). Josang names his trust model "subjective logic" and combines belief theory with Bayesian probabilities. The forms used for this purpose are:

$$\text{belief}=\frac{p}{p+n+2} \quad (21)$$

$$\text{disbelief}=\frac{n}{p+n+2} \quad (22)$$

$$\text{uncertainty}=\frac{2}{p+n+2} \quad (23)$$

where p and n are the positive and negative ratings, respectively. These two parameters are also used in the beta probability density function (see Section titled "Bayesian Reputation Systems"). For the combination of beliefs, external or internal, an aggregation function is used. "Majority consensus" functions are well-known for handling beliefs with discrete values while numerical functions are more appropriate for handing beliefs with continuous values (Ding et al., 2004). The authors in Josang (1999, 2001) use operators for the combination of opinions that are not based on Dempster-Shaffer theory as Yu and Singh do (2002).

In fuzzy models trust and reputation are described with linguistic fuzzy values. Reasoning is used in order to achieve the definition of a trust level. The most important and completed example is the system REGRET, which is described in the "Social Networks" section.

Role-Based Trust

These models are based on the notion of role and its assigned permissions to operate in a system. Hence, credentials are used to define access to a system such as identity, authentication, and so forth. Thus, an entity can be uniquely identified by the system and can obtain a specific role. Roles are used to give information about an entity. The key is the trust management mechanism that employs different languages and engines for reasoning on rules for trust establishment. It is a model used in access control systems and tries to determine the trust level of an entity based on credentials and security policies. A framework based on roles is presented in Li and Mitchell (2003).

REPUTATION AND TRUST ENGINEERING

Modeling trust requirements is the most important issue in developing efficient systems. Especially in cases where open systems are examined (e.g., MAS), this feature receives more attention. This section of our work aims to show the basic elements in which a requirements engineering procedure must be based.

As mentioned, MAS are open systems with members that change their behaviours continually. They are characterized by openness, heterogeneity, and dynamic character. Selfish agents try to locate partners in order to achieve their goals. Main issues that must be taken into consideration in MAS are:

- **Agents' identity.** Each autonomous component that interacts in a system should have a unique name and should be able to prove its identity. Identity is a requirement when agents communicate with others, because it shows to the potential partner that the component is a registered user of the system. Of course, a critical issue is the administration of the names. For example, an agent having bad reputation in a community may change its name in order to avoid the consequences.
- **Agents' communication.** Agents' communication is also important. Developers of MAS should introduce standards for communication. A security policy is necessary in order to avoid problems in the exchange of messages. Corrupted messages should be recognised by the recipients. Mechanisms that can be use for safe communication are authentication, cryptography, and so forth.
- **Agents' context.** The context in which each agent is activated should be defined. Mechanisms that take the context into consideration should be developed. This could provide efficiency in the cooperation procedure.
- **Agents' behaviour.** Agents' behaviour should be observed by the interested members and from the system. It is imperative to have the opportunity to observe and recognize bad or good behaviours in the system. Based on behaviour, trust is developed and members obtain a high reputation value in the community. Constructive trust must be promoted through the observation of interactions of an agent in the society. Also, from the systems' point of view, mechanisms that "punish" bad behaviours should be developed.

In Wong and Sycara (1999), the authors present a number of possible threats in MAS and their potential solutions. The basic threats that MAS may meet are related to corrupted naming of agents, insecure communication channels, insecure delegation and lack of accountability. Synoptically, the proposed solutions are:

- The use of trusted agent name servers and matchmakers.
- The provision of methods for the unique identification of each agent and proofs for their identifications.
- Secure the communication channels through the authentication of messages.
- Force agents to prove their owners.
- The provision of methods through which owners of the agents may be liable for their actions.

Authors give a full description of these solutions and explain the mechanisms with which these goals will be achievable.

A methodology for agent-based software development is described in Giorgini, Massacci, and Zannone (2005) and Giorgini, Mouratidis, and Zannone (2007). In TROPOS there are phases through which the trust establishment is feasible. The first phase is the requirement phase. In this stage, the functional and the nonfunctional requirements are determined in two subphases:

the *early requirements phase* and the *late requirements phase*. The key concepts in secure TROPOS are:

- **Actor.** It is an entity that represents a physical or software agent as well as a role or position, having its goals and intentions.
- **Goal.** Represents actors' interests that they wish to accomplish.
- **Plan.** It concerns a number of steps targeting to achieve a goal.
- **Resource.** It is a physical or an information entity.
- **Dependency.** Indicates that an actor depends on some other entity in order to complete their goals.

Accordingly, in TROPOS four new relationships are defined, which are: **Ownership** which indicates that an actor has a goal, **provisioning** which is the capability of an actor to achieve a goal or to have a plan or to provide some information, **trust** which indicates the belief of an actor that another entity will perform some task according to their goals and plans and **delegation** which shows that an actor delegates to some other to achieve its goals.

In TROPOS methodology, basic operations are:

- The definition of the actor's model and the dependency model. The essential actors should be recognised as well as the dependencies among them from the point of view of achieving their goals.
- The trust and delegation models. They determine the relationships between actors.
- The goal and plan models. Such models identify, from the viewpoint of actors, goals and plans that are necessary to achieve (sub)goals. Moreover, the resources needed to this procedure are recognised.

An extension to classic TROPOS is the security constraint modeling, which involves security constraints posed by the actors and the system. The architectural design development process for this extension relies on the following:

1. Secure Architectural style model.
2. Actor model, Goal/plan model, and security constraint model.
3. Capability and secure capability model.
4. Agent model.

In this chapter, due to the space constraints, we have presented only a short description of the TROPOS methodology. For a full discussion the interested reader should refer to the Giorgini et al. (2005, 2007).

FUTURE DIRECTIONS

Reputation and trust are important concepts in today's dynamic systems. However, there are some open issues that must be addressed in order to develop efficient methods for adoption in MAS.

First of all, for the construction of a theoretical framework that covers all the aspects of reputation and trust generation, manipulation and propagation is necessary. This model will set basic specifications in which developers may be based on, in order to construct efficient and productive systems. Such a framework will set the essentials that will allow the comparison of existing models. Today, this comparison is very difficult to accomplish, because the existing models come from specific domains and they are not based on a common theoretical framework.

The social network dimension in MAS presents new opportunities in reputation and trust management. However, it must be validated in real applications in order to discover its advantages and disadvantages. In social networks, there is an extensive need to deal with problems related to strategic lying and strategic coalition formation. This domain must be further studied in order to produce effective methods to deal with. Moreover, there is a need to define basic mechanisms through which opinions of members can be stored and secured in order to provide a higher security level.

Interactions of agents are held in a system under specific context. Context must be taken into consideration when defining the trust and reputation level of an entity. To the best of our knowledge, only a few works deal with this issue. Additionally, concern should be posed in trust propagation in specific contexts. Propagation allows the building of relations between all the agents communicating in a system. New methods for propagation should be developed with regard to the combination of the aforementioned models such as statistical functions, belief and fuzzy theory.

CONCLUSION

This chapter introduces the reader to the domain of reputation and trust in Multi-agent systems. It presents the existing reputation and trust quantification methods that are used in commercial and research applications. We show the importance of these two concepts especially in open systems where control over the actions of agents is limited. Also, the environment, where agents act, may change at any time due to the nature of the involved entities, which are autonomous components trying to serve their owners. For this reason, they are selfish and change their behavior subject to new conditions. A lot of models have been proposed for reputation and trust extraction in specific domains. We shortly present the most important of them, giving their basic characteristics. Certain models are very simple, while others are more sophisticated and utilize statistical functions, belief or fuzzy theory. Finally, we discuss key contribution in the domain of trust and reputation engineering. It is a critical field that leads to more efficient and productive systems. The engineering process must drive developers to construct systems that pay special attention to issues that ensure secure communication and fair ratings for all.

Reputation and trust will play an important role in future systems. Such concepts will be extensively adopted and used for gaining access to information sources in open environments like MAS and the Semantic Web.

REFERENCES

Abdul-Rahman, A., & Hailes, S. (2000). Supporting trust in virtual communities. In *Proceedings of the Hawaii International Conference on System Services* (Vol. 6, pp. 6007).

Artz, D., & Gil, Y. (2006). *Survey of trust in computer science and the Semantic Web.* Submitted for publication, Information Sciences Institute, University of Southern California. Retrieved April 3, 2008, from http://www.isi.edu/~dono/pdf/artz06survey.pdf

Ben-Gal, I. (2007). Bayesian networks. In F. Ruggeri, F. Faltin, & R. Kenett (Eds.), *Encyclopedia of statistics in quality and reliability.* John Wiley & Sons.

Boella, G., & Van Der Torre, L. (2005). Normative multiagent systems and trust dynamics. *Trusting agents for trusting electronic societies, LNAI 3577,* (pp. 1-17).

Boella, G., Van Der Torret, L., & Verhagen, H. (2006). Introduction to normative multiagent systems. *Computational & Mathematical Organisation Theory, 12*(2-3), 71-79.

Carter, J., & Ghorbani, A. A. (2004). Value centric trust in multiagent systems. In *Proceedings of the IEEE/WIC International Conference on Web Intelligence*, (pp. 3-9).

Castelfranchi, C., & Falcone, R. (1998). Social trust: Cognitive anatomy, social importance, quantification, and dynamics. In *Proceedings of the First International Workshop on Trust,* Paris, France, (pp. 72-79).

Dellarocas, C. (2000). Immunizing online reputation reporting systems against unfair ratings and discriminatory behavior. In *Proceedings of the ACM Conference of Electronic Commerce,* (pp. 150-157).

Dempster, A. P. (1968). A generalisation of BaysianBayesian inference. *Journal of the Royal Statistical Society, Series B, 30,* 205-247.

Deriaz, M. (2006). *What is trust? My own point of view.* University of Geneva. Retrieved April 3, 2008, from http://cui.unige.ch/ASG/publications/TR2006/

Ding, L., Kolari., P., Ganjugunte, S., Finin, T., & Joshi, A. (2004). Modeling and evaluating trust network inference. In *Proceedings of the 7th International Workshop on Trust in Agent Societies at AAMAS 2004,* New York.

Doran, J. E., Franklin, S., Jennings, N. R., & Norman, T. J. (1997). On cooperation in multi-agent systems. *The Knowledge Engineering Review, 12*(3), 309-314.

Durfee, E. H., & Lesser, V. (1989). Negotiating task decomposition and allocation using partial global planning. In L. Gasser & M. Huhns (Eds.), *Distributed artificial intelligence* (Vol. 2, pp. 229-244). San Francisco: Morgan Kaufmann.

Giorgini, P., Massacci, F., & Zannone, N. (2005). Security and trust requirements engineering. *Foundations of security analysis and design III—tutorial lectures, LNCS 3655* (pp. 237-272). Springer-Verlag.

Giorgini, P., Mouratidis, H., & Zannone, N. (2007). Modelling security and trust with secure TROPOS. *Integrating security and software engineering: Advances and future vision.* Hershey, PA: Idea Group.

Grandison, T., & Sloman, M. (2000). A survey of trust in Internet applications. *IEEE Communications Surveys and Tutorials, 4th Quarter, 3*(4), 2-16.

Guha, R., Kumar, R., Raghavan, P., & Tomkins, A. (2004). Propagation of trust and distrust. *In Proceedings of the 13th International Conference on World Wide Web (WWW 2004),* New York, (pp. 403-412).

Huynh, D., Jennings, N. R., & Shadbolt, N. R. (2006). An integrated trust and reputation model for open multi-agent systems. *Autonomous Agents and Multi-Agent Systems, 13*(2), 119-154.

Josang, A. (1999). Trust-based decision making for electronic transactions. In L. Yngstrm & T. Scensson (Eds.), *Proceedings of the 4th Nordic Workshop on Secure Computer Systems.* Stockholm, Sweden: Stockholm University Report 99-005.

Josang, A. (2001). A logic for uncertain probabilities. *International Journal of Uncertainty, Fuzziness and Knowledge-Based Systems, 9*(3), 279-311.

Josang, A., & Ismail, R. (2002). The Beta reputation system. In *Proceedings of the 15th Bled Conference on Electronic Commerce,* Slovenia, (pp. 324-337).

Josang, A., Roslam, I., & Colin, B. (2006). A survey of trust and reputation systems for online service provision. *Decision support systems.*

Li, N., & Mitchell, J. C. (2003). RT: A role-based trust-management framework. In *Proceedings of the 3rd DARPA Information Survivability Conference and Exposition (DISCEX III).*

Lindsay, Y., Yu, W., Han, Z., & Ray Liu, K. J. (2006). Information theoretic framework of trust modelling and evaluation for the ad hoc networks. *IEEE International Journal on Selected Areas in Communications, 24*(2), 305-317.

McKnight, D. H., & Chervany, N. L. (1996). The meanings of trust (Tech. Rep.). University of Misessota, Management Information Systems Research Center. Retrieved April 3, 2008, from http://www.misrc.umn.edu/workingpapers/

Melaye, D., & Demazeau, Y. (2005). Bayesian dynamic trust model. In *Proceedings of Multi-agent Systems and Applications IV: 4th International Central and Eastern European Conference on Multi-agent Systems, CEEMAS 2005*, (pp. 480-489). Springer-Verlag, LNCS 3690.

Mui, L., Halberstadt, A., & Mohtashemi, M. (2002). Notions of reputation in multi-agent systems: A review. In *Proceedings of the 1st International Joint Conference on Autonomous Agents and Multiagent Systems,* Bologna, Italy, (pp. 280-287).

Mui, L., Mohtashemi, M., Ang, C., Szolovtis, P., & Halberstadt, A. (2001). Ratings in distributed systems: A bayesian approach. In *Proceedings of the Workshop on Information Technologies and Systems (WITS)*, Miami, Fl.

Nwana, H. S. (1996). Software agents: An overview. *The Knowledge Engineering Review, 11*(3), 205-244.

Osman, N. (2006). *Formal specification and verification of trust in multi-agent systems.* School of Informatics, University of Edinburgh. Retrieved April 3, 2008, from http://homepages.inf.ed.ac.uk/s0233771/trust.pdf

Poslad, S., Calisti, M., & Charlton, P. (2002). Specifying standard security mechanisms in multi-agent systems. In *Proceedings of the Workshop on Deception, Fraud and Trust in Agent Societies, AAMAS 2002*, Bologna, Italy, (pp. 122-127).

Pujol, J. M., & Sanguesa, R. (2002). Reputation measures based on social networks metrics for multi agent systems. In *Proceedings of the 4th Catalan Conference on Artificial Intelligence CCIA-01,* Barcelona, Spain, (pp. 205-213).

Ramchourn, S. D., Huynh, D., & Jennings, N. R. (2004). Trust in multi-agent systems. *The Knowledge Engineering Review, 19*(1), 1-25.

Rasmusson, L., & Jansson, S. (1996). Simulated social control for secure Internet commerce. In C. Meadows (Ed.), *Proceedings of the 1996 New Security Paradigms Workshop,* (pp. 18-26).

Rubiera, J. C., Molina Lopez, M. J., & Muro D. J. (2001). A fuzzy model of reputation in multi-agent systems. In *Proceedings of the 5th International Conference on Autonomous Agents*, Montreal, Quebec, Canada, (pp. 25-26).

Sabater, J., & Sierra, C. (2002). Reputation and social network analysis in multi-agent systems. In *Proceedings of the 1st International Joint Conference on Autonomous Agents and Multiagent Systems*, Bologna, Italy, (pp. 475-482).

Schillo, M., Funk, P., & Rovatsos, M. (2000). Using trust for detecting deceitful agents in artificial societies. *Applied Artificial Intelligence, Special Issue on Trust, Deception and Fraud in Agent Societies, 14*(8), 825-848.

Shafer, G. (1976). *A mathematical theory of evidence.* Princeton University Press.

Sycara, K. P. (1998). Multiagent systems. *Artificial Intelligence Magazine, 19*(2), 79-92.

Wang, Y., Hori, Y., & Sakurai, K. (2006). On securing open networks through trust and reputation–architecture, challenges and solutions. In *Proceedings of the 1st Joint Workshop on Information Security*, Seoul, Korea.

Wang, Y., & Vassileva, J. (2003). Bayesian network-based trust model. In *Proceedings of IEEE International Conference on Web Intelligence,* Hallifax, Canada.

Whitby, A., Josang, A., & Indulska, J. (2004). Filtering out unfair ratings in bayesian reputation systems. In *Proceedings of the AAMAS 2004,* New York.

Wong, H. C., & Sycara, K. (1999). Adding security and trust to multi-agent systems. In *Proceedings of Autonomous Agents '99 Workshop on Deception, Fraud, and Trust in Agent Societies*, (pp. 149-161).

Yu, B., & Singh, P. M. (2002). An evidential model of distributed reputation management. In *Proceedings of the 1st International Joint Conference on Autonomous Agents and Multiagent Systems*, Bologna, Italy, (pp. 294-301).

Yu, B., & Singh, P. M. (2003). Detecting deception in reputation management. In *Proceedings of the 2nd International Joint Conference on Autonomous Agents and Multiagent Systems*, Melbourne, Australia, (pp. 73-80).

Zadeh, L. A. (1989). Knowledge representation in fuzzy logic. *IEEE Transactions on Knowledge and Data Engineering, 1*(1), 89-100.

Chapter IX
The Concept of Autonomy in Distributed Computation and Multi-Agent Systems

Mariusz Nowostawski
University of Otago, New Zealand

ABSTRACT

The concept of autonomy is one of the central concepts in distributed computational systems, and in multi-agent systems in particular. With diverse implications in philosophy, social sciences and the theory of computation, autonomy is a rather complicated and somewhat vague notion. Most researchers do not discuss the details of this concept, but rather assume a general, common-sense understanding of autonomy in the context of computational multi-agent systems. In this chapter, we will review the existing definitions and formalisms related to the notion of autonomy. We re-introduce two concepts: relative autonomy and absolute autonomy. We argue that even though the concept of absolute autonomy does not make sense in computational settings, it is useful if treated as an assumed property of computational units. For example, the concept of autonomous agents facilitates more flexible and robust architectures. We adopt and discuss a new formalism based on results from the study of massively parallel multi-agent systems in the context of Evolvable Virtual Machines. We also present the architecture for building such architectures based on our multi-agent system KEA, where we use an extended notion of dynamic and flexibly linking. We augment our work with theoretical results from chemical abstract machine algebra for concurrent and asynchronous information processing systems. We argue that for open distributed systems, entities must be connected by multiple computational dependencies and a system as a whole must be subjected to influence from external sources. However, the exact linkages are not directly known to the computational entities themselves. This provides a useful notion and the necessary means to establish an autonomy in such open distributed systems.

INTRODUCTION

This work concentrates on the general notion of autonomy in multi-agent systems. We will initially define an abstract concept of relative and absolute autonomy in the context of a computational agent. We think that the concept of autonomy must be always linked with the context and with the reference to what a given notion is applied to. Autonomy means different things to various researchers and it seems necessary to provide appropriate context and qualification of the term. Based on the notion of relative autonomy, we review some of the existing multi-agent systems. We will then discuss the objectives of the research community and the motivations regarding the concept of autonomy of a given computational unit (an agent) in the context of open multi-agent systems, adaptability and complexity growth. We argue that, to build an open and adaptable multi-agent system, agents must be subjected to constant external influences. These influences must (possibly indirectly) affect and control a given agent's behaviour, and therefore negate the generally accepted requirement of agent's absolute autonomy. Based on our results with experimental Evolvable Virtual Machines (EVM) (Nowostawski, Epiny, & Purvis, 2005a) framework, we draw conclusions that computational agents can never be truly autonomous or else the applicability of multi-agent systems in solving complex problems in an open environment would be constrained or even impossible. That means that restrictions on autonomy imposed by the multi-agent system designer are not only based on the pragmatic needs to limit and manage general complexity of the system. These restrictions come directly from an inherent property of the dynamics of the MAS as a distributed asynchronous computational system.

To demonstrate and discuss the issues related to autonomy, we build an experimental framework called Evolvable Virtual Machines (EVM). This framework has been used for modeling and analysis of metacomputational architectures, metalearning, self-organisation and adaptive computing. We will present results related to a contemporary model of computation for massively parallel open-ended evolutionary computations based on EVM (Nowostawski, Purvis, & Cranefield, 2004; Nowostawski, Epiney, & Purvis, 2005b; Nowostawski et al., 2005a). The model has been used to investigate properties of asynchronously communicating agents in a massively parallel multi-agent system. In this context, we discuss the concept of computational complexity, evolutionary learning and adaptability. We will show that with our computational evolutionary system a constant flux of external information is necessary to provide an open-ended increase of complexity of generated (discovered) computational programs. Our results suggest that any closed (or fully autonomous) collection of computational agents would be limited in their ability to learn and adapt to new circumstances. We claim that the notion of autonomy should be revisited and used in a clearly specified context. Computational agents should be subjected to direct or indirect external influences to allow continuous learning and adaptation by the system as a whole.

AUTONOMY IN MAS

Autonomy, from the Greek Auto-nomos (*auto* meaning *self*, and *nomos* meaning *law*), refers to an entity that gives oneself its own laws. In other words, autonomy means self-governance, and freedom from external influence or authority.

Generally, in multi-agent systems there are two basic attempts and formalisations of the concept of autonomy: internal and external views. The internal notion applies the above definition of autonomy to the agent itself, and specifies a set of principles or architectural constraints that are claimed for an autonomous operation of a given agent. The external view takes a different approach. It does not prescribe anything about internals of the agent or agent architecture itself.

It is, rather, an assumption that other agents are autonomous in an abstract sense and cannot be controlled/influenced directly. Agent's behaviour cannot be imposed by any other agent; hence the interactions and agents collaboration must take into account various aspects of the assumed participants' autonomy. We discuss briefly these two notions below.

The general notion of autonomy is invariant of the usual architectural or behavioural interpretations. In the next subsections, we review proposals for the definition from the internal and the external point of view.

Internal Autonomy

From a simple engineering perspective, the concept of autonomy has been used as one of the distinguishing features between traditional object-oriented and agent-driven systems. See, for example, discussion in Franklin and Graesser (1996) or Castelfranchi (1995). It is important to note that the notion of autonomy in MAS is often confused with the notion of automatic or independent operation. We want to stress that autonomy does not collapse to a mere independent operation. In complex software systems, it is a simple truism that many complex inter-dependencies and influences must exist between various computational units. However, there is always an element of choice. Indeterminacy is essential, from the external observer point of view, to be able to talk of autonomous computing. As an example of internal view of autonomy, consider the work of Luck and d'Inverno (1995), who have postulated that agent's motivation and the ability to create own goals is essential for autonomy. Using the *Z* specification language, they described a three-tiered hierarchy comprising objects, agents and autonomous agents where agents are viewed as objects with goals, and autonomous agents are agents with motivations. The ability to create goals according to some internal hidden and changeable agenda/motives is, according to their classification, essential for achieving true autonomy.

External Autonomy

Compared to internal autonomy, we can turn the roles around. Instead of concentrating on our own agent and its autonomy, we can insist on the assumption that all entities and agents that our software agent interacts with are autonomous in the abstract sense. How this is achieved, or if it is possible at all, is not our primary concern. What is important is the fact that no fixed assumptions can be made regarding the interactions, agents, goals delegation, motives, environment, and so forth. The research community is somewhat divided into two independent groups. One follows a strict internal view of autonomy, and proposes ways to enhance and promote autonomy in various agent architectures. The other group moved away from the initial strict internal requirements on agents' autonomy, toward more open, distributed systems that are driven by interactions, dialogues, negotiations and collaborations of multiple individual participants, that are assumed autonomous from the external point of view. The role of autonomy for individual agents became an external assumption, rather than architectural requirement. The best discussion on this is presented in the work of Weigand and Dignum (2003). In their work, they have argued that architectural requirements of autonomy on agents are not as important as the expectations of autonomy on behalf of other agents. The agents that a given software agent interacts with must be assumed to be autonomous. Agents must be prepared to deal with other agents' autonomy, and participate and collaborate with supposedly autonomous participants. This somewhat inverts the original requirements from those that support autonomy directly through elaborated architectures, into those that support features that work with autonomous agents.

COMPUTATIONAL AUTONOMY IN MAS

Most researchers base the definition of autonomy on two primitives: self-governance and independence (e.g., Gouaich, 2003; Carabelea, Boissier, & Florea, 2003). Self-governance refers exclusively to the internals of the agent and its architecture. As we pointed out, this is not necessary in general discussion or in practical agent-oriented software engineering directly. Both notions, however, seem relevant when trying to formalise the concept of autonomy. One of the attempts in providing comprehensive definition is provided in Carabelea et al. (2003):

An agent *A* is autonomous with respect to *B* for *p* in the context *P*, if, in context *P*, its behaviour regarding *p* is not imposed by *B*.

The symbols used are: p represents a property, A and B represent the active entities, agents, and P represents the context. The property *p* in the above definition relates to the object of autonomy, and emphasis is placed on the relational nature of the concept of autonomy. There are, however, two main problems with the above definition. The first problem lies in the fact that multiple vague concepts are being used: *context, property* (or autonomy object) *p* and the notion of *imposed.* The precise and formal meaning of these terms in the above definition is not clear. Nevertheless, the above definition is useful and conveys the common-sense understanding of the concept of autonomy.

Formal Definition

To make the above definition less ambiguous, we propose to base the definition on a formal notion of computation. Let us assume computation *C* to mean the universal Turing machine transformation of input data from the input tape into output data on an output tape (we assume here a two-tape setup, with read-only input tape and write-only output tape). For more details and formal introduction to Turing-machine computational models, see for example, Lynch and Tuttle (1989) or Hopcroft and Ullman (1979). We will denote computation from input X into output Y as: $X \rightarrow Y$. Let us assume data *D* to be a particular mapping of symbols into an input tape for the universal Turing machine. Let us assume that a computational agent *A* has access to a particular collection of data sources $D_i \in E$, where *E* stands for *environment*, or context. In other words, agent *A* is capable of performing universal Turing machine computation on a set of data accessible from its environment E. The data can be represented as a sequence of symbols from a particular alphabet, for example, [0,1] as in the original work of Turing (1936, 1937). Without any loss of generality, let us assume the following properties:

- Data decomposition: $\exists D_m = \emptyset, \forall D = D_i + D_j$;
- Data composition: $\forall D_i, D_j + D_k = D_i : D_i - D_j = D_k$ and $D_i - D_k = D_j$; and
- Computational composition, $\forall D_i, D_i \rightarrow D_{i+1}$ and $D_{i+1} \rightarrow D_{i+2} : D_i \rightarrow D_{i+2}$.

Data composition and decomposition simply captures the fact that data can be combined or split, without any loss of information. The computational composition ensures that computations do not have any side effects. Note that data can be read from or written to by various agents, and there is no distinction for input or output data. Only during the actual computation can a single data source be used as input or output (exclusive or).

Agent *A* is not autonomous in respect to agent *B* in the context E_s, and Agent *B* is said to *control* agent *A*, if:

$$\forall D_i \in E_s \exists D_y \xrightarrow{agent_B} D_i : D_i \xrightarrow{agent_A} D_x. \quad (1)$$

If no such agent B exists, then we say that Agent A is relatively autonomous in context E_s:

$$\exists D_i \in E_s : \forall D_i \overset{agent_A}{\rightarrow} D_x, \forall D_y \overset{agent_B}{\rightarrow} D_k, D_k \neq D_i. \quad (2)$$

The agent A is absolutely autonomous in the context E_s, if:

$$\exists D_i \in E_s : \forall D_i \overset{agent_A}{\rightarrow} D_x, \forall D_y \rightarrow D_k, D_i \neq D_k. \quad (3)$$

Discussion of the Definition

The above equations provide the following intuitive interpretations.

1. If Agent A uses a particular subset of its environment $E_s \in E$ with data sources $D_i \in E_s$ to perform its computation C_{E_s}, and there exists an agent B that can output into all of D_i sources, we say that agent A is not autonomous in respect to agent B in the context E_s. Agent B is said to *control* agent A.
2. If no such agent B exists, then we say that Agent A is relatively autonomous in context E_s.
3. If there is no set of agents that can collectively output to all of the $D_i \in E_s$, then we say that agent A is absolutely autonomous in the context of E_s.

Based on the above definition, we propose the following general autonomy classes in MAS. The three general classes below are often informally discussed in MAS literature, and these are now easy to define formally.

- **User autonomy.** Agent A is said to be autonomous in respect to a user, if the user does not provide all the data inputs that control agent A. In such a case, users cannot impose agent's behaviour directly; hence, we talk about agent's relative autonomy in respect to the user.
- **Interactions autonomy (social autonomy).** Agent A is autonomous socially, if it not only takes its inputs from other agents through interactions, but uses other sources of input at the same time. These sources are not bound to social interactions (e.g., user input). This means that agents cannot simply impose any goals or behaviour directly on other agents, because interactions are not enough to "drive" agent's computations.
- **Organisational autonomy (norm autonomy).** Organisational and institutional norms modeled as data sources cannot be used to impose behaviour of agents directly. Agents use various data-sources that influence their behaviour and computational choices.

Some authors, in particular Carabelea et al. (2003), postulate also a notion of *environmental autonomy*. In our definition of computational agents, *environment* encompasses all the possible input data sources for a given agent: user input, other agents, static data, norms, and any other. Therefore, there is no possibility of an agent performing any other computational mapping than $E_{input} \rightarrow E_{output}$. Agent is, by definition, just a computational function from the input environment, to the output environment. The concept of environmental autonomy, in our setup, does not make sense. To discuss environmental autonomy, we would need to establish a partitioning of E into subenvironments, one exclusively called *environment*, and other subsets labeled differently. We feel that partitioning of E into such disjoint classes is questionable in a general sense, although it might be useful for certain aspects of MAS, namely user interactions, social interactions and organisational interactions. If we model a closed system, where all data sources are in some way dependent upon agent's interactions and computations, then each single agent cannot be absolutely autonomous. To have a meaningful concept of absolute autonomy we have to deal with open systems, where some of the data sources are beyond the scope of the MAS itself. In that case, we talk about a stream of randomness (or indeterminacy) that comes from outside of the system itself.

INDETERMINACY AS AUTONOMY

Our definition of autonomy as presented above rests entirely on the formal and intuitive notions of indeterminacy. Let us consider a case of two simple homoeostatic processes: one performed by a thermostat, and one performed by a bacterium. We intuitively feel that there is some difference between these two in respect to how autonomous they are. In the case of the thermostat, even though it operates completely automatically and independently and we do not know or cannot predict exactly when it will switch from state to state, the degrees of freedom are quite limited. A thermostat is usually embedded in a well-insulated environment, where the temperature reacts almost exclusively to the heater/cooler system controlled by the thermostat itself. In the case of bacterium, even though the performed functions are sometimes as simple as those of the thermostat, the actual degrees of freedom seem to be larger. This is mostly due to the fact that in the case of a thermostat, the environment is almost exclusively controlled by the thermostat itself (the thermostat can make the ambient temperature to go up or down). The environment, to a certain extent, is simple and reactive. In the case of bacterium, there is no direct control over the environment as such. The interactions with the environment are of different types. Bacterium must deal continuously in a highly unpredictable environment. We will not argue if there is any categorical distinction between these two autonomy classes. We just want to point out, that the main distinguishing feature from autonomous and nonautonomous process lies in the indeterminacy and predictability of the environment. If there is a process, that is entirely deterministic and predictable from a given observer's point of view, then we say that there is no autonomy within that process. The process is simply determined as a function of its environment. If the process is not entirely predictable, then we talk about autonomy, and about a *choice* (the process can *choose* one or the other trajectory for its evolution).

Let us consider a multi-agent system within a formalism of Chemical Abstract Machine (cham) (Berry & Boudol, 1989). Cham has been successfully used as a modeling formalism for other process calculi and process algebras, most notably for Milner's CCS (Milner, 1989), and Nicola's and Hennessy's TCCS (Nicola & Hennessy, 1987). It is possible to model any asynchronous computational system within cham formalism, and therefore some observations within cham can be extended to any other process calculi. In cham the states of a machine are modeled as solutions consisting of data-structures floating and interacting in an abstract space. These data-structures can be of any type: primitive, such as numbers and strings, complex objects, or agents. Assuming that the data-structures are individual agents, and the interactions are equivalent to reactions in cham, we can talk about two aspects in respect to autonomy (and indeterminacy):

- Interactions are random. They are not preordered or prespecified by the system design, and
- Reaction rules may or may not be followed by the individual agents. Note that in the original cham formalism all the reaction rules must be strictly followed by the system.

Now, let us consider a system (example inspired from Banâtre et al., 1988), consisting of *n* agents named *2 . . . n+1* and a reaction rule (interaction) between agents, such as if A_i, A_{i*j} where $i, j \in [2, n+1]$, then A_{i*j} annihilates itself. That means that if two agents meet, and one of the agents is a multiple of another, the multiple will annihilate itself. From the initial solution of all *n* agents, after some time, there will be only agents named with prime numbers left. This is assuming both, autonomy in the interaction choices and autonomy in the adoption of the general annihilation, rule.

The above example demonstrates that in some circumstances, global coherent behaviour can be obtained in systems where autonomy is present on some of the underlying levels of abstraction. However, this is not always the case with all the systems. In some systems, autonomy must be restricted for the system to achieve a desirable stable point. For example, in the case of cham it is not easy to advise autonomous rules that would lead the system to calculate factorial. We will discuss this in more detail in the context of our EVM model. In the next section, we will briefly introduce the notion of autonomy in our multi-agent system KEA (Nowostawski, Purvis, & Cranefield, 2001).

MULTI-AGENT SYSTEM KEA

The aim of the KEA project (Nowostawski et al., 2001) is to provide a modular agent platform with an enterprise-level backend. The architecture supports the use of agent-oriented ideas at multiple levels of abstraction. At the lowest level are micro-agents, which are robust and efficient implementations of agents that can be used for many conventional programming tasks. Agents with more sophisticated functionality can be constructed by combining these micro-agents into more complicated agents. Consequently, the system supports the consistent use of agent-based ideas throughout the software engineering process, because higher level agents may be hierarchically refined into more detailed agent implementations. This enables scalability, flexibility and robustness of the platform, providing at the same time uniform modeling and programming paradigm.

The main distinguishing feature of KEA architecture as compared with traditional software engineering techniques is the autonomous dynamic linking facility. In traditional statically linked code, the function call is statically linked with appropriate library during compilation time. In dynamic linking, the function call is not linked with an appropriate implementation until the runtime. Then, the code is dynamically linked. The dynamic linkage with the library is unconditional (the library cannot refuse the linkage). Once the linkage has been made, it (usually) lasts until the end of the execution of the runtime system.

In KEA, the concept of dynamic linking has been extended further. The association between agents (or function calls if using the traditional programming nomenclature) is postponed until the very time when it is needed. At that time, the linkage is initiated, and may or may not be established. The participating party may refuse participation, in which case the caller will have to deal with this situation by trying alternatives. In case of successful association (when the linkage has been established), it will only last until the end of the current task (or function). After that, a new dynamic linkage must be initiated and established again.

Such a model promotes high-levels of autonomy because no fixed assumption can be made upon available participants. Agents must be prepared to deal with situations where given functionality may not be immediately available, and alternative means of achieving one's goals must be undertaken. More details about the KEA platform can be found in Nowostawski et al. (2001).

EVOLVABLE VIRTUAL MACHINES (EVM)

Overview

There has been research conducted regarding autonomous asynchronously-interacting computations pursued in diverse areas of theoretical computer science. Certain properties investigated in those settings have been found to be invariant and shared between different complex systems. Our original desire was to integrate the recent advances from various fields onto a single coherent theoretical model, together with an experimental computational framework which could be used for practical investigations on massively parallel computational framework. Originally designed as an artificial evolution modeling tool (Nowostawski et al., 2005b), the EVM architecture is a model for autonomously interacting, evolving, complex and hierarchically organised software systems. The EVM architecture stems from recent advances in evolutionary biology and utilises notions such as specialisation, *symbiogenesis* (Margulis, 1981), and *exaptation* (Gould & Vrba, 1982). From the computational perspective, it is a massively distributed asynchronous collection of interactive agents that utilises computational reflection. The EVM framework has been used for multitask learning and metalearning. Hence, computational reflection and reification, on one hand, provide a compact and expressive way to deal with complex computations, and on the other hand, provide ways of expanding a computation on a given level via the metalevels and metacomputations.

Symbiogenesis researchers argue that symbiosis and cooperation are primary sources of biological variation, and that acquisition and accumulation of random mutations alone is not sufficient to develop high levels of complexity (Margulis, 1970, 1981). Other opponents of the traditional biological gradualism suggest that evolutionary change may happen in different ways, most notably through exaptation (Gould & Vrba, 1982), that is, a process whereby a structure evolved for one purpose that has come to be used for another, unrelated purpose (or function).

The EVM architecture follows the biological models of: symbiogenesis, exaptation and specialisation. EVM allows independent computing elements to engage in symbiotic relationships, same as in cham, where independent agents are engaged in relationships through reaction rules and the concept of a *membrane*, that limits interactions only to local data within a membrane. In the case of EVM the interactions are not only 2-way, but they may involve an arbitrary number of participants. EVM allows a given agent to specialise in specific tasks, or to evolve toward new, more complex, tasks, similarly to the specialisation principle from biology. EVM also allows agents to be used in different contexts than originally designed for, similar to the exaptation principle.

The EVM architecture can be also seen as a computational model that combines the features of a trial-and-error machine (Bringsjord & Zenzen (2003) and the multi asynchronously-interacting machines paradigm. The trial-and-error behaviour is achieved through continuous looping of different hypotheses and their re-evaluation until the desired precision of the hypothesis is achieved.

The EVM model is similar to the one of cham. There are, however, some main differences. In cham, reaction rules are (typically) written between two agents in the solution. In EVM the interactions can happen between more than a pair of agents. Also, in cham, the reaction rules are written beforehand, and not changed during the abstract machine execution. This is not the case for EVM. In EVM, the initial machines executed can modify the rules. It is beyond the scope of this article to analyse the exact formal equivalence and relationship between these two models, and we leave it for future work.

In the following subsection, we will present the details of the EVM implementation, and discuss the experimental.

Implementation

Our current implementation of the EVM architecture is based on a stack-machine. With small differences, the EVM implementation is comparable to an integer-based subset of the Java Virtual Machine (JVM). There are two independent but compliant implementations: one is written entirely in Java and the second one in C. Developers and researchers can obtain the sources from CVS http://www.sf.net/projects/cirrus. The basic data unit for processing in our current implementation is a 64-bit signed integer. This somewhat arbitrary constraint is dictated by practical and efficient implementation on contemporary computing devices. The basic input/output and argument-passing capabilities are provided by the operand stack, called *the data stack*, which is a normal integer stack. At the moment, only integer-based computations are supported. All the operands for all the instructions are passed via the stack. The only exception is the instruction push, which takes its operand from the *program list* itself. Unlike other virtual machines (such as the JVM), our virtual machine does not provide any operations for creating and manipulating arrays. Instead, the architecture facilitates operations on lists. There is a special stack, called *the list stack,* for storing integer-based lists.

Execution frames are managed in a similar way to the JVM, via a special execution frames stack. There is a lower-level machine handle attached to each of the execution frames. Machine is a list

Figure 1. Schematic view of a single EVM processor: Program and Program Counter, BaseMachine, data (operand) stack and list stack. Below, a concrete instantiation during the execution of swap instruction

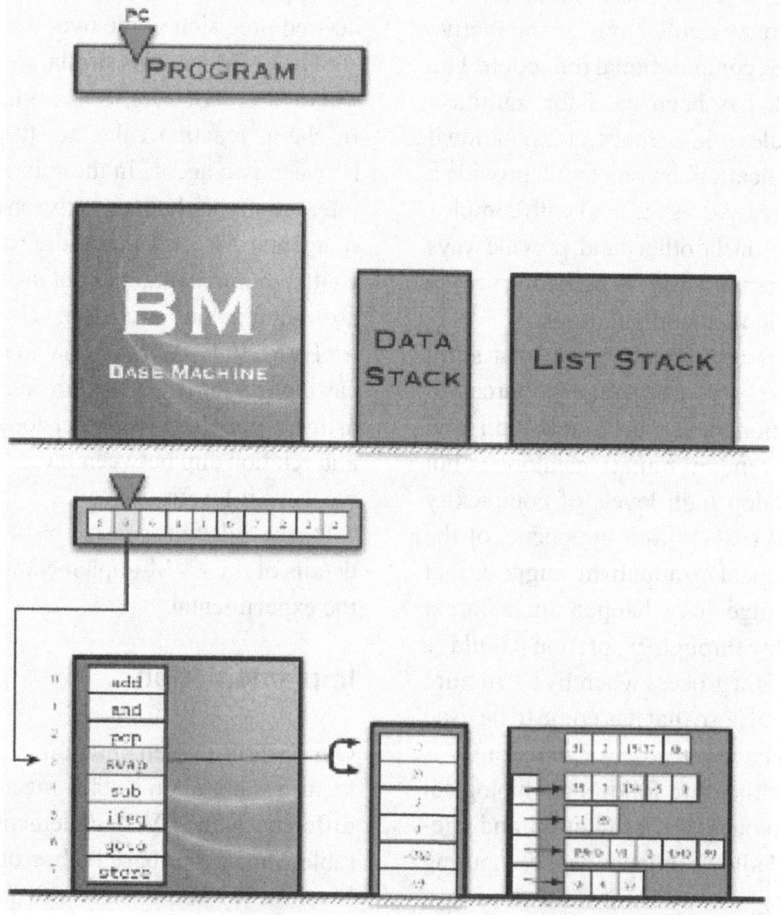

of lists, where each individual list represents an implementation of a single instruction for the given machine. In other words, the machine is a list of lists of instructions, each of which implements a given machine instruction. If a given instruction is not one of the primitive Base Machine units, that is, primitive instructions for that machine, then the instruction sequence must be executed on another, lower-level machine. The Base Machine implements all the primitive instructions that are not reified further into more primitive units. To distinguish those primitive instructions that are executed on the Base Machine, we refer to them as *operations*.

Potentially, EVM programs can run indefinitely, and therefore, for practical reasons, each thread of execution has a special limit to constrain the number of instructions each program can execute. This is especially crucial in a multi-EVM environment. Once the limit is reached, a given program will unconditionally halt.

The EVM offers rich reflection and reification mechanisms. The computing model is relatively fixed at the lowest-level, but it does provide the machines with multiple computing architectures to choose from. The model allows the programs to reify the virtual machine on the lowest level. For example, programs are free to modify, add, and remove instructions from or to the lowest level virtual machine, as well as any other level. Also, programs can construct higher-level machines and execute themselves on these newly created levels. In addition, a running program can switch the context of the machine, to execute some commands on the lower-level, or on the higher-level machine. Altogether, the EVM provides limitless flexibility and capabilities for reifying individual EVM executions. Due to this high level of flexibility, there have been no attempts to formalise the full EVM model in any of the existing process calculi or other computational algebras. We have only attempted partial formalisations of the model.

Each individual EVM is a computational agent that can reference any other machine in the multi-EVM environment (the EVM Universe). This is achieved by using the first 32 bytes of the instruction to address any computer in the Internet, and the second 32 bytes for the index of the instruction on that machine. That mean that the theoretical limits of the EVM universe are bound by 2^{32} number of hosts with 2^{32} instructions on each of the hosts.

One possible way of instantiating part of the computational environment for the architectural framework is by adapting bias-optimal search primitives (Levin, 1973), or the incremental search methods (Schmidhuber, 2004). To narrow the search, one can combine several methods together. For example, it is possible to construct a generator of problem solver generators, and employ multiple metalearning strategies for a given computational task at hand. A more detailed description of the abstract EVM architecture is given in Nowostawski et al. (2004). The experimental results are described in detail in Nowostawski et al. (2005a).

EVM Search Process

The EVM Universe is composed of a spatially distributed grid of EVM agents (cells) each of which is trying to solve one (or many) tasks provided to the system through the special *Task Manager* that is part of *the environment*. Each individual cell works independently of other cells, and it can perform a finite number of operations. We have implemented and deployed several different search algorithms that the EVM can use: random search, stochastic search, genetic algorithms, exhaustive bias-optimal search. The actual search performed by the individual cell is decided by the cell itself. In this research, we concentrated our attention on the interactions and dependencies between the cells. A single EVM agent can use other agents. This is typically done in such a way that a solution for a complex problem is decomposed into subtask, each of which is delegated to other agents. It is common for any of the search methods to quickly solve subtasks by delegating the processing to the other agents. Each cell can specialise in a given subtask by reusing another cell for this subtask. Such a cell will be called a *first-level parasite*. If a parasite uses another parasite, we call it *higher-order parasite*. Parasites appear often within certain classes of problems, as it becomes exponentially easier for a cell to parasite an existing solution than to come up with a solution on its own. This is due to the interactions between the spatial aspects of the asynchronous interactions between agents and internal processing capabilities of the computational agents themselves.

The dependencies and *bonding* between cells has been achieved through random search process (trial-and-error search). For certain classes of arithmetic problems with which this model has been tested, random search was sufficient to find solutions to tasks consisting of up to 3 subtasks (the neighbourhood of a single cell was restricted to only four cells). For problems with higher order of combinatorial search space the random search was insufficient, and must be augmented with other heuristics to establish proper bonding between computational units.

Simple Grid Experiments

In one of our early experiments (Nowostawski et al., 2005a) we have used regular toroidal grid, with only one program per cell. We used four neighbours models, and locality played an important role in the dynamical evolution of the cells' interactions. Our multi-agent system exhibits properties found in other artificial life systems such as Tierra (Ray, 1991) (for instance, knowledge diffusion, parasitism, self-assembly). The experiments were conducted on a single level of EVM, with the aim of discovering a good compact machine suitable for a given sequence of tasks. We have used two probability-based learning methods: one based on individual probabilities of instructions, and the second based on conditional probabilities of sequences of instructions. We have introduced, investigated and compared five different specialisation mechanisms:

1. Random search and exhaustive search;
2. Classic genetic algorithm;
3. Ad-hoc stochastic search for a fixed-size program;
4. Improved tree search based on successful patterns (building-blocks); and
5. Universal reinforcement acceleration mechanism.

All these search mechanisms operate at the low, cellular level, and dictate, in interaction between agents within the multi-agent system and the macroscopic behaviour of the system. A typical run, exhibiting many features of our self-organising, self-adaptable cellular system is presented in Figure 2.

In order to be efficient in a multitask context, we have assumed that the model must fulfill these requirements:

1. All tasks must be eventually solved.
2. Solving difficult tasks should lead to greater rewards than solving easy ones.
3. Computational resources should focus on unsolved tasks.
4. Solutions must not be forgotten, as long as they are useful.
5. Knowledge diffusion should be facilitated (previous solutions must be accessible).
6. Dynamic environments should be supported: tasks can be added or removed at any time, dynamically.

For arithmetical problems, cell's programs are short (typically four instructions). The reason is that we want to focus on the way these programs will collaborate and reuse other existing subexpressions from neighbours. Some of the tasks tackled are enumerated in Table 1. Note that, for

Figure 2. Typical evolution of the cellular system. Note that the topology is toroïdal. For example, right neighbours of the cells on the right border of the grid are the cells along the left border of the grid.

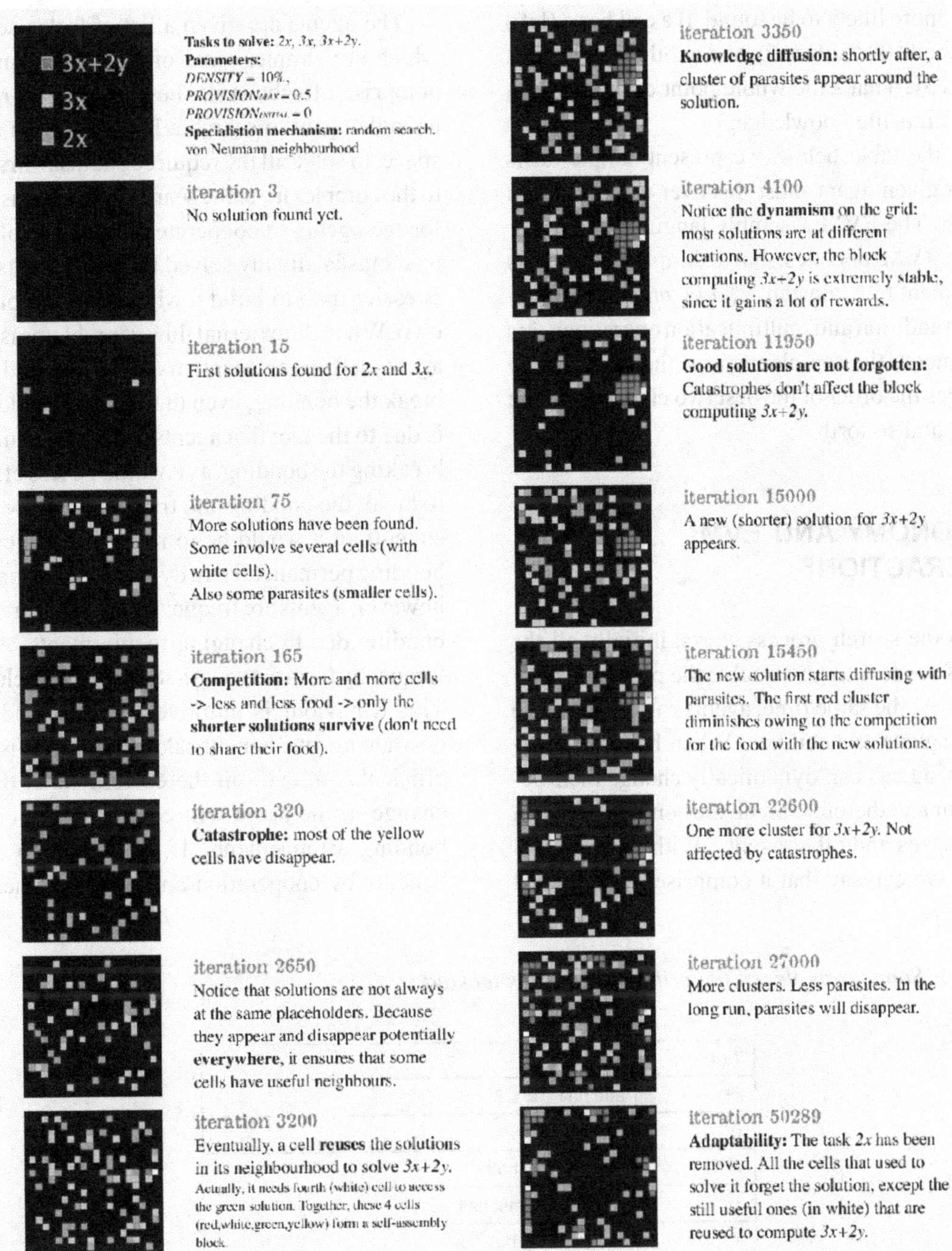
3x+2y
3x
2x
Tasks to solve: 2x, 3x, 3x+2y.
Parameters:
DENSITY = 10%,
PROVISIONmax = 0.5
PROVISIONinitial = 0
Specialistion mechanism: random search.
von Neumann neighbourhood
iteration 3
No solution found yet.
iteration 15
First solutions found for 2x and 3x.
iteration 75
More solutions have been found.
Some involve several cells (with white cells).
Also some parasites (smaller cells).
iteration 165
Competition: More and more cells -> less and less food -> only the shorter solutions survive (don't need to share their food).
iteration 320
Catastrophe: most of the yellow cells have disappear.
iteration 2650
Notice that solutions are not always at the same placeholders. Because they appear and disappear potentially everywhere, it ensures that some cells have useful neighbours.
iteration 3200
Eventually, a cell reuses the solutions in its neighbourhood to solve 3x+2y.
Actually, it needs fourth (white) cell to access the green solution. Together, these 4 cells (red,white,green,yellow) form a self-assembly block.
iteration 3350
Knowledge diffusion: shortly after, a cluster of parasites appear around the solution.
iteration 4100
Notice the dynamism on the grid: most solutions are at different locations. However, the block computing 3x+2y is extremely stable, since it gains a lot of rewards.
iteration 11950
Good solutions are not forgotten: Catastrophes don't affect the block computing 3x+2y.
iteration 15000
A new (shorter) solution for 3x+2y appears.
iteration 15450
The new solution starts diffusing with parasites. The first red cluster diminishes owing to the competition for the food with the new solutions.
iteration 22600
One more cluster for 3x+2y. Not affected by catastrophes.
iteration 27000
More clusters. Less parasites. In the long run, parasites will disappear.
iteration 50289
Adaptability: The task 2x has been removed. All the cells that used to solve it forget the solution, except the still useful ones (in white) that are reused to compute 3x+2y.

instance, the solution to $2x + 3y$ can be only five instruction long *(leftNeighbourProgram* swap *rightNeighbourProgram* add halt), and thus much more likely to be found, if a cell has a (left) neighbour that solves $2x$ and another (right) that solves $3x$. That's the whole point of our system, that of reusing knowledge.

In the table below, we present simple tasks that a given agent must discover using random search. The EVM assembly language is similar to the JVM opcodes. For example, *const_2* is equivalent to a constant 2, *add*, *mul* are equivalent to addition and multiplication operations, *inc* increments the top element on the stack, *swap* changes the order of the first two elements on the stack, and so forth.

AUTONOMY AND EVM INTERACTIONS

When the search process starts, initially all the agents are free to try to solve the problem themselves. At the same time, agents can try to re-use other agents' capabilities. When being used by others, agents can dynamically change their behaviour and therefore break the bonding between themselves and other agents. With such a MAS setup, we can say that it comprises a high-level of relative autonomy, because agents are free in their choices to conduct or not any interactions with other agents.

The agents are given a list of tasks, some of which are simple, some of which are hard and comprise of subtasks. The harder tasks require a single agent to search a large program search space, to solve all the required subtasks first. Due to the complexity of the search process, it is easier for the agents to cooperate, and reuse solutions to subtasks already solved by other agents. This is easier than to build a whole solution on their own. When the external flow of problems is fixed, agents self-organise into fixed clusters and rarely break the bonding, even though they could. This is due to the fact that agents have no incentive of breaking the bonding, as it would be less efficient to break the bonding and try to find a new stable state, than it would be to remain in the current bonding permanently. In dynamic environments, however, agents are frequently forced to break the bonding due to changing requirements. Agents frequently form new quasi-stable bonding clusters. This behaviour is intuitively simple, and boils down to agents' benefit calculations: if it is more efficient to remain in the current state, than to change it, the agent will continue with current bonding arrangements. If it is faster to obtain benefits by cooperation and bonding, then this

Table 1. Some examples of the arithmetical tasks tackled

Task	Solution
$x+y$	add halt
xy	mul halt
$2x$	const_2 mul halt
$3x$	const_2 inc mul halt
X	ifge nop neg halt
$2x+y$	const_2 mul add halt
$2x-y$	const_2 mul sub halt
$2x+3y$	const_2 mul swap const_2 inc mul add halt
7	const_2 const_2 inc add halt

Figure 3. Self-assembly. Bottom. Left: first, some cells discover the solutions to the easy tasks (2x and 3x). These solutions can be reused by their neighbour to compute a more difficult, but related task (3x + 2y). These three cells live in symbiosis together (middle). At the same time, two other cells (green and grey) manage to discover together a solution to (49 – x). The green cell computes that solution with the help of the grey cell. The grey cell solves no task alone, but still gets rewards and survives because it contributes to the computation of the green cell. Right: Eventually, a cell between these two blocks of cells connects them to solve a more complex task (49–(3x+2y)). Top: Details of the cells' programs.

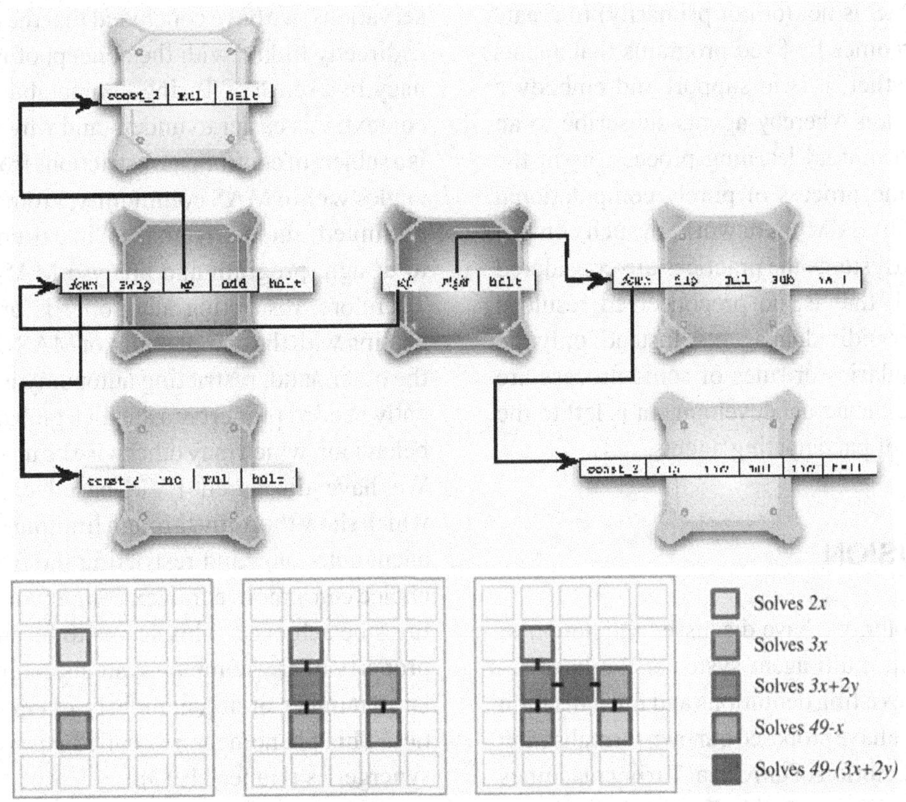

is observed. If simple tasks are more efficiently solved by individual agents alone, then this is a predominant behaviour. This is very similar to other phenomena in complex asynchronous systems in physics, chemistry and biology. This phenomenon of restricting degrees of freedom of a given entity is called *enslavement* (Haken, 1983). This means, that a certain stable energy level has been obtained. The amount of participation in a particular arrangement beyond a certain threshold locks more and more participation. A process of enslavement is being observed.

Some researchers designing and developing MAS have also noticed that the autonomy and independence are not always the most desirable features. Some constraints are often necessary for the overall goals of the MAS system to exceed capabilities of individual agents. Based on our experiments, we can conclude that full autonomy and complete freedom is not always desirable. A form of *enslavement* must be present when designing and building MAS. Enslavement helps the system to settle in certain configurations or dynamical patterns, which would be not attainable otherwise.

Another related question is how to make MAS (absolutely) independent from the actual programmer. This requires that agents have the abilities to learn new activities, adopt new goals and devise new learning strategies themselves (see also Witkowski & Stathis, 2003). The aim in open MAS is not (or not primarily) to create certain outcomes by fixed programs that agents execute. Rather, it is to support and embody a transformation whereby agents subscribe to an open-ended mutual learning process, as in the programming process of purely computational agents within EVM framework. In such computational ecosystems the practices are considered open-ended, that is, no preconceived result is intended for individual agents. Instead, only the initial boundaries or rules of some process are defined and the actual development is left to the interaction of participating agents.

CONCLUSION

In this chapter, we have discussed the notion of autonomy in multi-agent systems. We have reviewed the existing definitions and formalisation attempts. We have proposed our own formalisation based on the notion of universal Turing machines computational agents, with the abstract notion of data sources and data transformations. Based on the assumed notions of computation, the concept of relative and absolute autonomy for a given computational agents have been presented. We compared our definition to existing intuitive definitions in multi-agent literature. We have provided also a comparison of general autonomy classes in MAS, with intuitive and formal notions of autonomy.

In the context of autonomy in MAS, we have presented details of two multi-agent systems: KEA and EVM. These frameworks tackle the challenges of autonomous computing in various ways. In both the emphasis is placed on the central notion of unsecured and unreliable interprocess (or interagent) communication. The KEA framework is using the notion of autonomous dynamic linking between agents. The EVM system is using the notion of unstructured self-assembly and dynamic aggregation of computational components.

Based on the literature review and our own observations, we have concluded that the autonomy is directly linked with the concept of indeterminacy in a sense of Turing-computability. In that context, it is easier to understand why autonomy is a subject of continuous restrictions from various angles within MAS community. From one hand, unlimited autonomy makes it extremely hard to design, program and analyse MAS systems. Therefore, restricting autonomy is one way of dealing with the complexities of MAS design. On the other hand, restricting autonomy is an inherently needed property to achieve global coherent behaviour, which may otherwise be unattainable. We have discussed EVM-based experiments which show that only through limiting individual agent autonomy and restricting the freedoms of choice can a more complex computational structures be achieved. This seems to be an inherent property of any complex systems composed of a large number of autonomously interacting entities. This phenomenon is called *enslavement* in synergetics (Haken, 1983).

REFERENCES

Berry, G., & Boudol, G. (1989). *The chemical abstract machine*. New York: ACM Press.

Bringsjord, S., & Zenzen, M. (2003). Superminds: People harness hypercomputation, and more. *Studies in cognitive systems* (Vol. 29). Kluwer Academic. Cen BF311 B 4867.

Carabelea, C., Boissier, O., & Florea, A. (2003). Autonomy in multi-agent systems: A classification attempt. In Nickles et al. (Eds.), (pp. 103-113).

Castelfranchi, C. (1995). Guarantees for autonomy in cognitive agent architecture. In *Proceedings of the Workshop on Agent Theories, Architectures, and Languages, ATAL'94*, (Vol. 890 of LNAI, pp. 56-70). New York: Springer-Verlag.

Franklin, S., & Graesser, A. (1996). Is it an agent, or just a program?: A taxonomy for autonomous agents. In *Proceedings of the Third International Workshop on Agent Theories, Architectures, and Languages*, (pp. 21-36).

Gouaich, A. (2003). Requirements for achieving software agents autonomy and defining their responsibility. In Nickles et al. (Eds.), (pp. 128-139).

Gould, S. J., & Vrba, E. (1982). Exaptation—a missing term in the science of form. *Paleobiology, 8*, 4-15.

Haken, H. (1983). *Synergetics, an introduction: Nonequilibrium phase transitions and self-organization in physics, chemistry, and biology* (3rd rev. ed.). Berlin: Springer-Verlag.

Hopcroft, J. E., & Ullman, J. D. (1979). *Introduction to automata theory, languages, and computation*. Reading, MA: Addison-Wesley.

Levin, L. A. (1973). Universal sequential search problems. *Problems of information transmission, 9*(3), 265-266.

Luck, M., & d'Inverno, M. (1995). A formal framework for agency and autonomy. In *Proceedings of the First International Conference on Multi-agent Systems (ICMAS)*, (pp. 254-260).

Lynch, N., & Tuttle, M. R. (1989). An introduction to input/output automata. *CWI Quarterly, 2*(3), 219-246.

Margulis, L. (1970). *Origin of eukaryotic cells*. New Haven, CT: University Press.

Margulis, L. (1981). *Symbiosis in cell evolution*. San Francisco: Freeman.

Milner, R. (1989). *Communication and concurrency*. Upper Saddle River, NJ: Prentice Hall.

Nickles, M., Rovatsos, M., & Weiß, G. (Eds.). (2004). Agents and computational autonomy—potential, risks, and solutions. In *Postproceedings of the 1st International Workshop on Computational Autonomy—Potential, Risks, Solutions (AUTONOMY 2003), held at the 2nd International Joint Conference on Autonomous Agents and Multi-agent Systems (AAMAS 2003)*, July 14, 2003, Melbourne, Australia. Lecture Notes in Computer Science (Vol. 2969). Springer-Verlag.

Nicola, R. D., & Hennessy, M. (1987). CCS without tau's. In *Proceedings of the International Joint Conference on Theory and Practice of Software Development: Advanced Seminar on Foundations of Innovative Software Development I and Colloquium on Trees in Algebra and Programming*, (Vol. 1, pp. 138-152).

Nowostawski, M., Epiney, L., & Purvis, M. (2005a). Self-adaptation and dynamic environment experiments with evolvable virtual machines. In S. Brueckner, G. M. Serugendo, D. Hales, & F. Zambonelli (Eds.), *Proceedings of the Third International Workshop on Engineering Self-organizing Applications (ESOA 2005)*, (pp. 46-60). Springer-Verlag.

Nowostawski, M., Epiney, L., & Purvis, M. (2005b). Self-adaptation and dynamic environment experiments with evolvable virtual machines. In *Proceedings of the Third International Workshop on Engineering Self-organizing Applications (ESOA 2005)*, (pages 46-60). Utrech, The Netherlands: Fourth International Joint Conference on Autonomous Agents & Multi Agent Systems.

Nowostawski, M., Purvis, M., & Cranefield, S. (2001). Kea—multi-level agent infrastructure. In *Proceedings of the 2nd International Workshop of Central and Eastern Europe on Multi-agent Systems (CEEMAS 2001)*, (pp. 355-362). Krak'ow,

Poland: Department of Computer Science, University of Mining and Metallurgy.

Nowostawski, M., Purvis, M., & Cranefield, S. (2004). An architecture for self-organising evolvable virtual machines. In S. Brueckner, G. D. M. Serugendo, A. Karageorgos, & R. Nagpal (Eds.), *Engineering self organising sytems: Methodologies and applications*. Lecture Notes in Artificial Intelligence (No. 3464). Springer-Verlag.

Ray, T. S. (1991). An approach to the synthesis of life. In C. Langton, C. Taylor, J. D. Farmer, & S. Rasmussen, (Eds.), *Artificial life II: Santa Fe Institute Studies in the Sciences of Complexity* (Vol. XI, pp. 371-408). Redwood City, CA: Addison-Wesley.

Schmidhuber, J. (2004). Optimal ordered problem solver. *Machine Learning, 54*, 211-254.

Turing, A. M. (1936-1937). On computable numbers with an application to the entscheidungsproblem. *Proceedings of the London Mathematical Society, 42*(2), 230-265, also *43*, 544-546.

Weigand, H., & Dignum, V. (2003). I am autonomous, you are autonomous. In Nickles et al. (Eds.), (pp. 227-236).

Witkowski, M., & Stathis, K. (2003). A dialectic architecture for computational autonomy. In Nickles et al. (Eds.), (pp. 261-274).

Chapter X
An Agent-Based Library Management System Using RFID Technology

Maryam Purvis
University of Otago, New Zealand

Toktam Ebadi
University of Otago, New Zealand

Bastin Tony Roy Savarimuthu
University of Otago, New Zealand

ABSTRACT

The objective of this research is to describe a mechanism to provide an improved library management system using RFID and agent technologies. One of the major issues in large libraries is to track misplaced items. By moving from conventional technologies such as barcode-based systems to RFID-based systems and using software agents that continuously monitor and track the items in the library, we believe an effective library system can be designed. Due to constant monitoring, the up-to-date location information of the library items can be easily obtained.

INTRODUCTION

One of the primary objectives of a library is to provide a collection of information artefacts and enable easy and fast access to those artefacts. Most modern libraries provide open stack access for browsing and retrieving of the items available. This open access may lead to misplacement of items in various sections of large libraries. When an item is misplaced it cannot be reached by its potential users. It is tedious for the library staff to find and track a misplaced book that is needed by another user. In addition, it can be costly to locate the item, and possibly replace the item (when it is not possible to locate the item at the time that is needed). In this chapter, we describe an approach that can reduce the effort associated with finding such items.

RFID is an upcoming technology that facilitates easy object identification, in particular, when voluminous entities have to be tracked and monitored (such as products in the supply chain context, library items in a library). An item that is marked with an RFID tag can be read by a RFID reader. This information can be used in tracking and managing the tagged items. The cost of RFID tags (in particular, the passive ones) are low enough to make it feasible to be used for the identification of large quantities of items. Currently, more than 20 million books worldwide are embedded with RFID tags (Research Information, 2007) in more than 300 libraries (RFID Gazette, 2007).

Software agent systems are one of the well studied areas of artificial intelligence, as agents can be embedded with intelligent decision-making capabilities. Robots are physical embodiments of software agents. Software agents when embedded in a robot can be used for a variety of purposes such as planet exploration, handling nuclear wastes, and fire rescue. The study of collaboration using agents is important because they are indispensable for carrying out tasks in unmanned zones and industrial automation.

In our approach, the agents interact with each other in order to ensure up-to-date information in the central library database. They read the tag in the environment using a RFID reader, undertake appropriate processing and communicate the information to another agent. To provide inter-agent communication they can use languages such as FIPA (The Foundation for Intelligent Physical Agents (FIPA), 2007) ACL over WI-FI network. In this project, an agent is used to identify and obtain the location of a misplaced book.

BACKGROUND

Some researchers have worked in integrating agent-based systems with RFID technology for tracking and monitoring purposes (Mamei & Zambonelli, 2005). Our work is inspired by their approach in adopting the RFID technology with agent-based systems.

Related Work in the Context of Library Environment

In the previous works (Choi, et al., 2006; Molnar & Wagner, 2004) that have used RFIDs for library management system, most of the focus has been on automating the process of check-in and check-outs carried out at the circulation desks, automation of inventory management process and sorting returned items (RFID Sorting, 2007). The RFID technology has also been used in enabling antitheft functionality by requiring the gate sensors to check whether an item has been issued or not.

The authors of R-LIM system (Choi et al., 2006) describe how the position of tagged items in the library can be identified within a shelf, based on the shelf locator tags that indicate the relative position of the books in a particular rack of the bookshelf. In their approach, manual scanning (using a hand-held scanner) was employed to read the tags of the library items in a shelf. It was assumed that the library items are placed in their correct location. This may not be easily assumed in an open library stack where numerous patrons interact with the library artefacts. To ensure consistency, the library staffs need to periodically check the shelves for possible misplaced items. This is a tedious and time consuming operation. To our knowledge, not much work has been done that identifies the location of misplaced items in an automated manner.

In our system we have incorporated the idea of continuous monitoring of the library items which facilitates easier identification of misplaced items and their locations.

HIGH LEVEL DESCRIPTIONS OF APPLICATION AND ARCHITECTURE

We describe the design of an agent-based system that can be used for library book tracking. One of the common problems in a large library is that the books are often moved around and misplaced in different sections of a library. This problem can be solved by placing RFID tags on each book and using robotic agents to locate and track the books.

Assume that the library is made up of different floors. Each floor is partitioned into different reading zones. Each zone contains a certain number of bookshelves. Each shelf is made up of a number of racks where tagged books are kept.

Figure 1. Architecture of the RFID-based library system

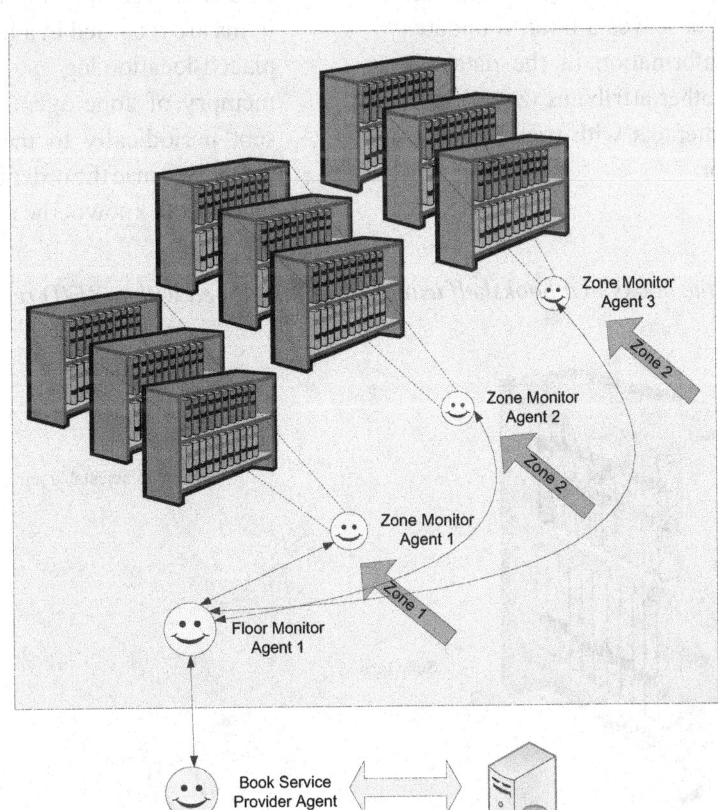

The tag embedded in each book contains information such as unique id, floor, zone, shelf, rack and availability details using a simple encryption mechanism.

In our system, there are different types of agents (shown in Figure 1), such as library service provider agent, floor agent, zone monitor agent, and tracker agent.

The book service provider agent is the agent to whom book tracking requests can be submitted. It performs the following tasks:

- Initializes the library items with appropriate location information
- Maintains the changes made to the location of library items
- Provides status information of the library items

The floor agent resides at appropriate entry/exit points of a floor. The floor agent monitors when a book enters or leaves a floor. It updates the current floor information in the database while resetting the other attributes (zone, shelf, and rack). It also interacts with tracker agents assigned to that floor.

The zone monitor agent is responsible for monitoring the library items placed in shelves assigned to it. The zone monitor agent performs the following tasks:

a. Periodically takes a snapshot of the tagged items within its reach.
b. It finds the discrepancies between the currently read books and the expected book list for its zone. This includes the items that have been removed and the items that have been added which do not belong to the current zone.
c. The database is updated to indicate that a particular item is not in its correct place. In addition, the approximate position of the misplaced item is recorded (the current position). This includes the information with regard to the zone. For the removed items all the current position attributes are reset except the floor information. The misplaced items are recorded in a log file called "misplaced-location.log" stored locally in the memory of zone agent. The log files are sent periodically to the service provider agent. Because the order in which the reader reads is not known, the zone agent can only

Figure 2. Scanning the books in a bookshelf using a robot equipped with a RFID reader

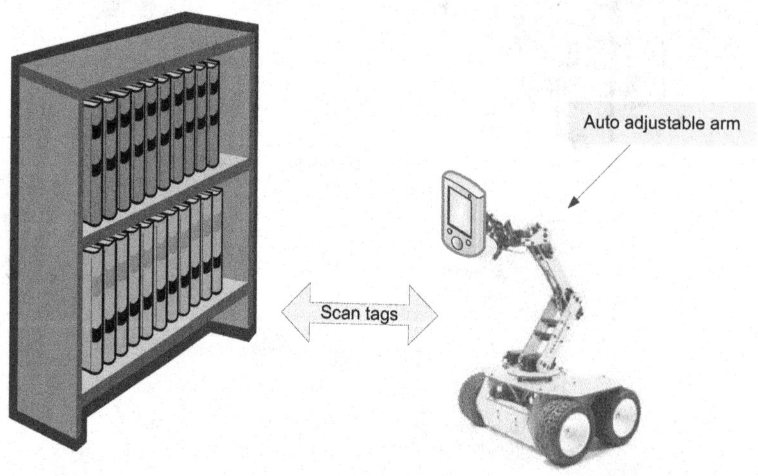

indicate that an item belonging to another zone is present within its zone and obtain the corresponding tag values (which indicate the correct location of the misplaced item). To find the exact location of the misplaced book, we use a tracker agent.

Finding the Location of the Misplaced Item

In this scenario, we know only the existence of a misplaced item within a zone but not the current location of the item. To find this information, we need to use RFID readers with lower range of readability. In our approach, we use robotic agents that are equipped with the RFID readers and they can be used to scan the tags (shown in Figure 2). Based on the log file entries, a particular misplaced item can be identified by the robot. The robot is capable of moving back and forth across a shelf and it is equipped with an automatic adjustable arm which can read items in different (higher) racks. Shelves will be equipped with the beginning of the shelf and end of the shelf tags. The end shelf tags will have directional information which is used by the robot to locate the next shelf within a zone.

The tracker agent is capable of finding items misplaced across zones as well as within its current zone. The tracker agent locates an item that belongs to another zone, by reading each tag in its range and comparing it with the tag code of the target item (misplaced book recorded in the log file). After locating the item, it derives the location of the misplaced item by obtaining the location information from its neighbouring items. The current location of the misplaced item is stored in another log file called "found-location.log" and the database is updated accordingly.

In this process, the tracker agent is also checking the correct relative order of items that are being read. Whenever it finds an item that is out of order, it identifies it as a misplaced item and derives its location information based on its neighbourhood and stores it in the "found-location.log" file. This process ensures identification of items that are misplaced both within and across zones.

The library staffs periodically check the log files updated by the tracker agent and place the misplaced items in their correct location.

Operational Scenarios

Initial Configuration

All the library items are labeled (tagged) appropriately. All the items are recorded in the database. Whenever new items are added to the library, some adjustments to the neighbouring items may be required.

The library database consists of the following details associated with each item:

a. Call number
b. Unique identifier
c. Availability
d. Correct position (original location, as specified by the administrator)
    i. Floor, Zone, Shelf, Rack
e. Current position (as indicated by the floor and the zone agents)
    i. Floor, Zone, Shelf, Rack

When the items are initialized, the correct position and current position of an item are the same. When an item is moved from one location to another, the current position is updated. The correct position of an item remains the same (unless the administrator resets it to accommodate the growth of the library).

The database would have more details other than the above information such as due date and reserve status. The unique code, current position (in the encoded format) and the availability information are placed on the tag belonging to each library item.

Requesting a Book Scenario

When a request for locating the current position of an item is made, it may be an item that is in its original correct place. In this case, the user is informed of the item location details.

If the requested item is identified as a misplaced item, then the current zone is known. In this case, there could be two possibilities:

1. It can be found in the "found-location.log" file. In that case, the staff can fetch the item from the location and update the database, and the log-file.
2. Otherwise, the item details are found in the "misplaced-location.log" file. In this case the staff can use a hand-held reader to locate the item and update the relevant information such as database and log file. Alternatively, the tracker agent can be assigned to look for the location of the misplaced item.

It is possible that a misplaced book is in an unzoned area of the library floor such as a reading area. In this case, it is assumed that the library staff will collect all these books at the end of the day and place them in a designated shelf for further processing (to be placed in their correct location).

DESIGN CONSIDERATION AND IMPLEMENTATION

RFID Infrastructure

We are planning to use the RFID-chips conforming to ISO 15693 and avoid any proprietary tags belonging to one particular vendor. Our system uses two kinds of RFID readers. The long range RFID reader covers 3-5 meters while the short range reader used by the robotic agent covers 10-50 centimetres. A RFID tag can only be read if the reader has the appropriate authorizations.

Implementation Details of the Robotic Agent

We are using Garcia robot (Acroname Robotics, 2007) which is embedded with a RFID reader. Each robot has an onboard processor called Stargate (Crossbow Technology Inc, 2006). The brainstem C development kit (Acroname Robotics, 2007) installed in the robot provides the API for the control of Garcia robot movement (moving forward, turning left and right). We are currently working on implementing a controller for the automatic arm adjustment (moving up and down).

Otago Agent Platform (OPAL) has been used to support multi-agent cooperation (Purvis et al., 2002). OPAL is a FIPA-compliant agent platform. Tracker agent is an OPAL agent which is made up of two components, namely Garcia robot controller and RFID reader. The instructions for the robot to perform certain operations can be issued using the FIPA ACL (The Foundation for Intelligent Physical Agents (FIPA), 2007) standard.

In our system, when a particular request is made for an item that is misplaced, then the service provider agent communicates this information to the tracker agent in order to find the current location of the item. For communication between zone monitor agent and tracker agent, we are using WIFI protocol.

Upon receiving a request for finding a particular book from a service provider agent, the tracker agent (which is an OPAL agent) instructs the Garcia robot to initiate a search using its RFID reader. In this process, when the end of rack tag is read, the robot agent is instructed to adjust the arm to reach to the next rack and also turns the robot around in order to be able to read the next rack. If the end of shelf tag is read, then the robot agent is directed to adjust the arm to its lower position and move to the next shelf using the directional information that is placed on the end of the shelf tag. When the requested book tag is found, the position of the book is calculated as described earlier.

Figure 3. Psuedocode for a tracker agent locating the position of a misplaced book

```
Find-tag(tag-code)
{
        Read till the-end-of-zone-tag
        {
                Read till the-end-of-shelf-tag
                {
                        Read till end-of-rack-tag
                        {
                                MessageToReaderAgent(readNextItem());
                                if (book found)
                                {
                                        ProcessInformation();
                                        Exit();
                                }
                        }
                        MessageToRobotAgent(goToNextRack());
                }
                MessageToRobotAgent(goToNextShelf());
        }
}
```

Figure 3 shows the pseudo code that indicates the sequence of steps taken by a tracker agent when it tries to locate a misplaced book within a zone. In this code, the processInformation() method corresponds to the calculation of the current book position based on the neighbourhood and the update of the database and corresponding log files.

Communication Between Agents

The agents in our system can communicate with each other using FIPA ACL messages. The following interactions can take place in our system:

a. Library service provider agent can send a request for searching a book to a tracker agent;

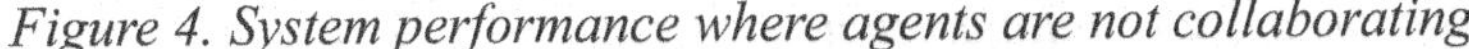

Figure 4. System performance where agents are not collaborating

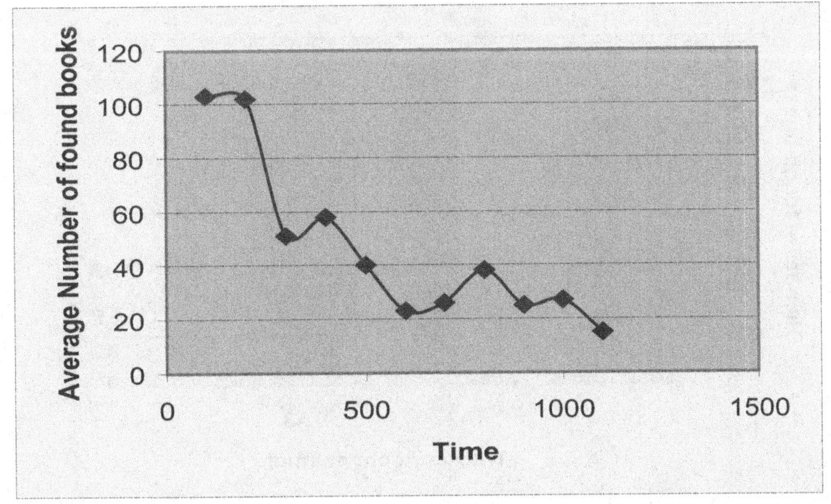

Figure 5. System performance where agents are collaborating

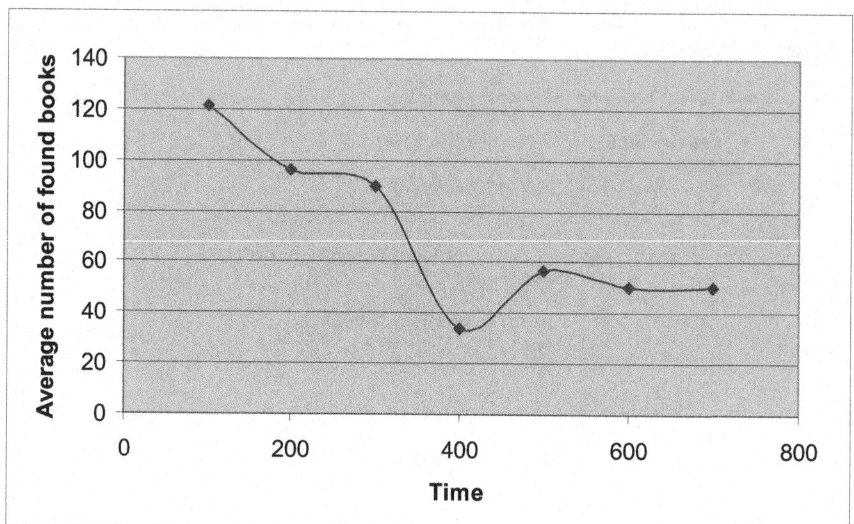

b. In case a particular tracker agent is not available (or busy or unavailable due to charging or maintenance), the library service provider agent will send a request to the floor agent to find a replacement;

c. When a zone agent wants to update the database, it sends all the position data to the library service provider agent, which then updates the repository. Similarly, floor agent and tracker agent send the data to the library service provider agent.

Simulation and Testing of the Prototype System

In order to verify the operational correctness of the system, we are currently implementing a simulation system which is populated with a large

Figure 6. Agents participation where agents are not collaborating

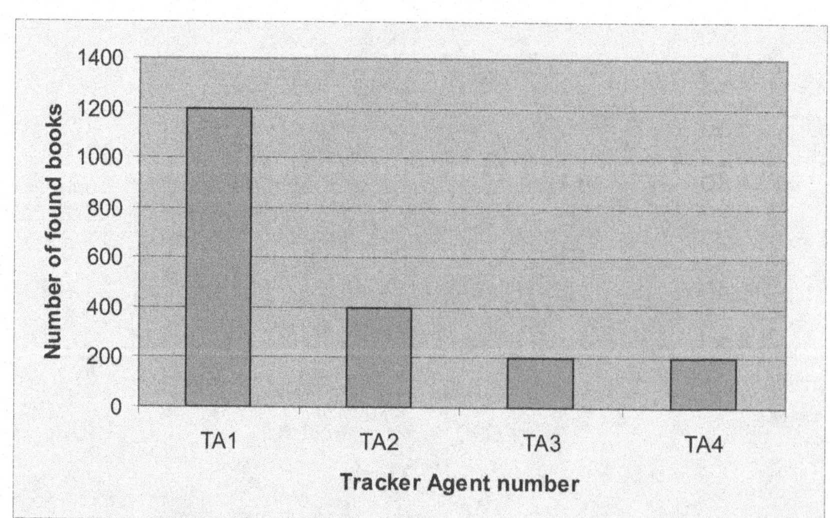

Figure 7. Agents participation where agents are collaborating

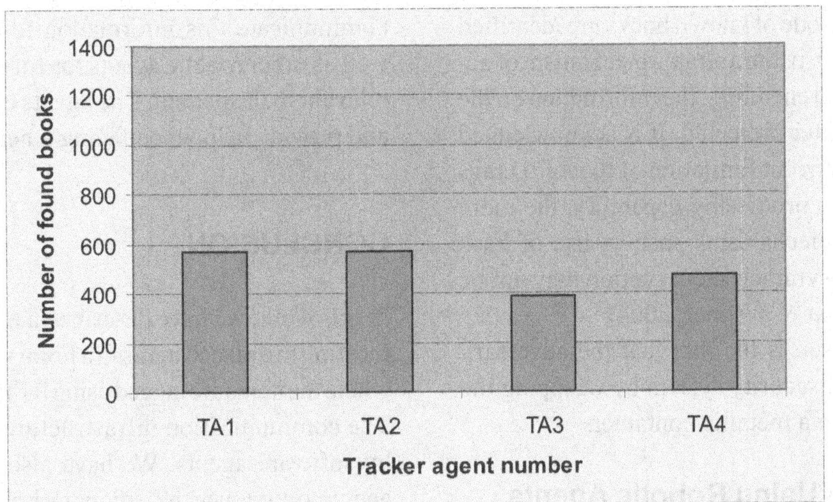

number of books. We have parameterized the number of floors, zones, shelves and racks. The user requests are modeled based on the anonymous historic data of our local library. Based on this information, we measure the performance of the system by calculating the time it takes to find the misplaced books identified by the zone agents.

In Figures 4 and 5 we show the performance of the system in finding the misplaced books in two different scenarios. Figure 4 shows the time it takes to find the misplaced books where various tracker agents are assigned to locate books in specific zones (in this example, four tracker agents are assigned respectively to four different zones). Figure 5 shows the system performance where the tracker agents cooperate with each other (an idle tracker agent may help another agent with higher work load). It is clear that when the agents work together the time that it takes to complete the same set of requests is shorter.

Figures 6 and 7 show the outcome of the same experiments from the individual tracker agent point of view. In Figure 7 the tracker agents' participations in finding the books are more evenly distributed, which resulted in a better overall system performance.

More simulation experiments will be designed to examine the system performance when different priorities are assigned for performing different tasks. In particular, we would like to explore an optimum time that has to be spent by the tracker agents to locate the misplaced books identified by the zone agents (using the long range RFID readers) as opposed to the time spent on locating the misplaced books within each zone (using the short range RFID readers by the tracker agents).

DISCUSSIONS

Issues with Use of RFID

Privacy is one of the important concerns of RFID systems. An adversary can attempt to read the library book details (such as the title, author of the book) which might reveal some personal information without the owner's consent. Molnar and Wagner discuss various methods that an adversary may use the tag information to reveal details about a person associated with certain tagged items. They describe how this can be achieved even when a unique identifier (such as barcode) is used by the process of association (linking different people reading the same book) or book hotlisting (where the barcode of known books are identified and tracked). By using a simple mechanism of access control and encoding, the information on the tag may be better protected. It is acknowledged that with the current limitation of the RFID tags in terms of the processing capability, the more sophisticated mechanisms such as use of hash functions and symmetric encryption may not be feasible (Molnar & Wagner, 2004).

Another issue is the fact that the adversary can bypass the security system by wrapping the library items in a metallic container.

Issues with Using Robotic Agents in the Library

The use of robotic agents in a library environment can interfere with the movements of the patrons of the library. This can be addressed if the robotic tasks are performed after the closing of the library. But this will increase the latency of the information that can be made available to the users. Alternatively, the robots can be assigned to designated paths which can be made known to the patrons.

The robotic agents may run out of power to operate. A mechanism needs to be provided so that it can be recharged at appropriate time intervals.

Other Applications

Our approach can be applied to a variety of applications that have tracking and monitoring requirements. For example, our robotic agents can be used for patient monitoring in hospitals. Assume that each patient is identified using an RFID tag. A robotic agent can read a patient's RFID tag and find the appropriate information about that patient (names of possible medication and the corresponding timing information for taking the medicine). The agent can then provide specialized service (such as dialysis) based on the information obtained. The agent can then communicate this information to the server as well as other robotic agents for further treatment relevant to the patient. The agents can request for and provide help when facing a heavy load.

CONCLUSION

In this work, we have described a mechanism for locating misplaced items in a library environment where the items are tagged using RFID technology. The communication infrastructure is facilitated by software agents. We have also used robotic agents to automate the tedious task associated with locating the position of the misplaced items. Our approach is promising, as it makes the misplaced items easily known due to the continuous monitoring. The proposed architecture enables the library administrative person to make more informed decision on allocation of the library resources (staff as well as robot trackers) based on the list of misplaced books identified by the RFID readers. We are currently working on the implementation details associated with our approach.

REFERENCES

Acroname Robotics. (2007). *Garcia manual.* Retrieved April 3, 2008, from http://www.acroname.com/garcia/man/man.html

Choi, J. W., Oh, D. I., & Song, I. Y. (2006). R-LIM: An affordable library search system based on RFID. In *Proceedings of the International Conference on Hybrid Information Technology (ICHIT'06)*, (pp. 103-108).

Crossbow Technology, Inc. (2006). *Stargate developers guide.* Retrieved April 3, 2008, from http://www.xbow.com/Support/Support_pdf_files/Stargate_Manual.pdf

Mamei, M., & Zambonelli, F. (2005). Spreading pheromones in everyday environments through RFID technology. In *Proceedings of the 2nd IEEE Symposium on Swarm Intelligence, (pp. 281-288).* IEEE Press.

Molnar, D., & Wagner, D. (2004). *Privacy and security in library RFID: Issues, practices, and architectures*. New York: ACM Press.

Purvis, M., Cranefield, S., Nowostawski, M., & Carter, D. (2002). Opal: A multi-level infrastructure for agent-oriented software development. *Autonomous agents and multi-agent systems.* Bologna, Italy: ACM Press.

Research Information. (2007). *Radio-tagged books*. Retrieved April 3, 2008, from http://www.researchinformation.info/rimayjun04radio-tagged.html

RFID Gazette. (2007). *RFID applications for libraries*. Retrieved April 3, 2008, from http://www.rfidgazette.org/libraries/

The Foundation for Intelligent Physical Agents (FIPA). (2007). *Agent communication language specification*. Retrieved April 3, 2008, from http://www.fipa.org

Chapter XI
Mechanisms to Restrict Exploitation and Improve Societal Performance in Multi-Agent Systems

Sharmila Savarimuthu
University of Otago, New Zealand

Martin Purvis
University of Otago, New Zealand

Maryam Purvis
University of Otago, New Zealand

Mariusz Nowostawski
University of Otago, New Zealand

ABSTRACT

Societies are made of different kinds of agents, some cooperative and uncooperative. Uncooperative agents tend to reduce the overall performance of the society, due to exploitation practices. In the real world, it is not possible to decimate all the uncooperative agents; thus the objective of this research is to design and implement mechanisms that will improve the overall benefit of the society without excluding uncooperative agents. The mechanisms that we have designed include referrals and resource restrictions. A referral scheme is used to identify and distinguish noncooperators and cooperators. Resource restriction mechanisms are used to restrict noncooperators from selfish resource utilization. Experimental results are presented describing how these mechanisms operate.

INTRODUCTION

By nature's design different kinds of people exist in a society. Every society has cooperative and uncooperative members. In the real world, it is not possible to get rid of all uncooperative members in the society. It is an injustice to exclude people with certain behaviour from the society. However, it is possible to restrict the performance of uncooperative members and prevent the cooperative members from being exploited. The uncooperative members take advantage of cooperative members by making suckers out of them and also causing damage to the common good. Special mechanisms need to be designed and deployed to control the behaviour of such particular groups, especially in electronic societies such as P2P file sharing and so forth.

BACKGROUND

In previous work (Purvis, Savarimuthu, Oliveira, & Purvis, 2006) we used simple tags and showed the self-organization of cooperative and uncooperative groups. We have now included the referral mechanism into our approach. In this work, we investigate how effective the referral mechanism is in reducing the performance of uncooperative members and increasing the overall society benefit. Additionally, we also show that resource restriction for uncooperative members improves the society benefit.

This chapter is organised as follows. First, we discuss some concepts which are related to our experiments. Then, we explain our experiments which use tags and referral mechanisms. Next, we explain about the experiments which use resource restriction mechanism. Finally, we present our conclusion and future work.

COOPERATIVE BEHAVIOUR IN MULTI-AGENT SOCIETY

For a society to operate effectively, agents within the society must obey certain social rules and norms. So far, much of the focus in this area has been on work devoted to the identification of malevolent agents, where the goal is to identify a noncooperator and exclude it from the society. However, in the real world, it is not going to be applicable in all situations. Our focus is on situations where society members are behaving in an uncooperative manner, but are not necessarily "evil" and deserving of expulsion. This is the issue of the "Tragedy of the Commons" (Hardin, 1968).

Tragedy of the Commons

In Hardin's classic paper (Hardin, 1968), "Tragedy of the Commons," he outlines the "tragedy." A common pasture is open to herders, each of which tries to maintain as many cattle as possible on the commons. A herder will reckon that the positive benefits of adding one additional animal will all go to him, alone, whereas the negative effects from overgrazing of that one additional animal will be shared borne by all the herders. Accordingly, self-interested herders may continue adding one more animal to their herds, even if they know that collectively this is destroying the commons. The question is: how are restrict selfish herders to avoid the tragedy?

The Tragedy of the Commons can be related to the "Prisoner's Dilemma" situation (Axelrod, 1984). Two collaborating criminals are imprisoned and questioned separately. Each criminal may cooperate with his fellow criminal by refusing to divulge details of the crime or defect by ratting on his colleague. It is possible to establish a reward structure (see Figure 1) such that:

- If both criminals cooperate they get a reward, R,
- If they both defect, they are punished (punishment, P),
- If one player defects and the other cooperates, then the defector gets high reward (temptation, T) and the other gets a severe punishment (sucker, S)
- And $T > R > P > S$, and $2R > T + S$
  Under these reward conditions, each individual criminal will reason that if the other
  - Cooperates, he does better by defecting, and if the other
  - Defects, he also does better by defecting.

Thus, the Nash equilibrium situation for this game is for both players to defect, even though they would collectively get a higher reward if they were both to cooperate. The Tragedy of the Commons can be likened to a situation in which the individual herder is playing the Prisoner's Dilemma game against the collection of all the other herders: his selfish interests lead him to defect, even though they are all better off if they cooperate.

Another cooperation game that is discussed in the literature is the Stag Hunt game. The metaphor here is two hunters who may cooperate to hunt a stag (high reward, S). If they operate by themselves, they each can only catch a rabbit (lower reward, R). A hunter seeking to hunt a stag without cooperation gets nothing. But there is no sucker's reward here. The reward structure is shown below (see Figure 2).

Figure 1. Payoff matrix for prisoner's dilemma

Criminals 1 / 2	Cooperate	Defect
Cooperate	R/R	S/T
Defect	T/S	P/P

Power Laws in Network Behaviour

A related issue in networked situations is associated with "power law" behaviour (Barabási, 2002; Huberman, 2001; Shirky, 2003b). Globalised economic environments in which previous barriers to resource access have been greatly reduced exhibit power law properties. For example, Shirky (2003b) has observed that Web logs ranked by number of inbound links closely follows a power law distribution. The spread of telecommunications and the resulting easy media access has meant that a great proportion of earnings made in connection with professional musicians, actors, writers, and other entertainers goes to a small group of people. Whereas it was once the case that almost every town had some musicians that people eagerly wanted to hear, now most people get their music listening pleasure from the electronic media, and traditional skills in instrumental music are increasingly unrewarded and disappearing.

In some circumstances, this power-law situation may be a good thing, but there may be times when it is desirable to achieve a more equable distribution. In particular, a more equable distribution of networked links may lead to a more robust and dynamically adaptable society. It may be the case that preferred nodes "hubs" in a network become single points of failure and risk being choked with the amount of traffic that they must handle. As with the Tragedy of the Commons situation, we may find that there are situations in which there are advantages in introducing some sort of regulations in the traffic, just as we find it necessary to regulate and sometimes restrict vehicular traffic in urban environments.

Figure 2. Payoff matrix for Stag Hunt

Hunters 1 / 2	Cooperate	Defect
Cooperate	S/S	0/R
Defect	R/0	R/R

In our view, this is what needs to be done in connection with agent societies. Mechanisms are needed not only to uncover malefactors, but also to guide and sometimes restrict agent behaviour so that a more cooperative environment is fostered.

P2P File Sharing

P2P file sharing, using Napster, Gnutella, Kazaa, or BitTorrent, is widely engaged in, but uncooperative behaviour is frequently observed. BitTorrent is currently particularly popular, and Hales and Patarin's analysis of BitTorrent's workings (Hales & Patarin, 2005) is of interest. With BitTorrent, groups of users "swarm" with an interest in a specific media file coordinate to speed-up the process. A given file is partitioned into pieces, and each peer is responsible for obtaining and sharing with the other peers some of the pieces. Each swarm is managed by a "tracker," which keeps track of the peers interested in a file or group of files. Peers may query the tracker for a random list of other peers in the swarm, and once obtained, the peers can exchange their piece lists so that they may determine which peers may have pieces that they need. Because a peer may not be able to service at once all the peers that need its piece, it only services up to a limited number of other peers, with remaining peers being left out "choked." Presumably, peers will choose to cooperate with those peers which, in turn, have cooperated with it, and so cooperative behaviour is presumed to be induced by an implicit "tit-for-tat" strategy. But Hales determined that it would be easily possible to cheat under these arrangements and be a "free rider" the bane of all P2P file sharing systems; yet such cheating is not observed in connection with BitTorrent, but is observed on other systems. Why?

Hales's suggested answer to this question (Hales & Patarin, 2005) concerns the way that file metadata is handled with BitTorrent. BitTorrent does not provide a central distribution for metadata; instead the acquisition of metadata is left to the users. To download a file using BitTorrent, one must supply information which can be found in a special .torrent file, but the user must use his or her own devices, such as user-run Web sites, to find this file. This means that the connectivity of this "network" of interested users is not complete: separate, possibly somewhat isolated, groups of users will form and share the metadata. Although this is sometimes thought to be a weakness of BitTorrent, Hales suggests that this may be an advantage, because separate swarms with their individual trackers can be formed for the same file. This can lead to a swarm selection process, whereby higher performing swarms (with more cooperative members) are selected and poorly performing swarms (with more free-riders) are deselected and eventually die off.

The suggested mechanism at work here is that, by means of probably unintended limitations in terms of metadata access, there is an arrangement in BitTorrent that can lead self-interested peers to generate multiple groups and a group-swarm-selection process that ultimately yields more cooperative (and hence higher overall performing) groups. Thus, by having some restrictions in a group, the Tragedy of the Commons can be avoided.

Cooperation Using Tags

Advocates of the Semantic Web envision an IT future in which intelligent agents achieve effective collaboration by employing automated reasoning facilities in connection with rich online ontological information. Others (Doctorow, 2001; McCool, 2005, 2006), have expressed doubts that a realisation of this vision can be practically achieved in the foreseeable future, because the Semantic Web requires too many new tools and constant new data encoding to respond to ever-changing contexts in order to be able to achieve the required take-up and that simpler, lower threshold structures and mechanisms are needed (Shirky,

2003a). One simpler idea is to use simple tags that do not define their semantics, but which are interpreted by application agents for their own particular circumstances (Hales & Patarin, 2005). Since Holland (1993) invented Tags, it has been interest of so many researchers.

Research investigating how cooperation has arisen in biological and social groups, for example, has suggested that simple tagging may provide a better account for the evolution of cooperation than do notions of "tit-for-tat" reciprocity and the "shadow of the future" (Riolo, Cohen, & Axelrod, 2001). In these scenarios, tagging offers a simple mechanism that can facilitate cooperative behaviour on the part of selfish individuals. Individuals just need to like or feel comfortable interacting with other individuals who are readily observed to be like them because they have the appropriate visual tag. This is certainly a natural phenomenon in ordinary human social intercourse.

Feedbacks/Referrals

Information about other members of the society is important and helpful. Before selecting the strategy, getting feedback about the opponent is a rational thing to do. In the model described in Purvis et al. (2006), agents were playing the Prisoner's Dilemma game. They chose a strategy to cooperate or defect by their value of cooperativeness which was assigned randomly when they were created. They did not change their strategy based on their knowledge of the past behaviour. The agents were playing with the nonchanging strategy. The purpose was to simplify the experiments and show the self-organization of cooperative and noncooperative groups achieved by using simple tags.

In this chapter, we have adopted a hybrid approach that uses the concept of tags and recorded history of agent interactions. We change the aspect of nonchanging strategy by allowing the player to ask for feedback about the opponent. By getting the feedback about the opponent, the player can decide whether to cooperate or defect. Here, the feedback about an agent is called the referral. In our approach, there is no lying in the referrals because it is happening within the group. Agents give feedback based on their own observation about the other agent. We do not associate a cost for referrals as we use referrals as a mechanism to improve overall societal benefit. And also, if the referral is positive about the opponent, the agent cooperates. Otherwise, the agent defects to avoid getting a sucker reward.

In the following section, we outline our experiments that take advantage of some of these concepts.

EXPERIMENTS USING TAGS AND REFERRALS

For our experiments, an artificial agent simulation environment has been set up with a society of 100 agents divided into 5 subgroups of 20 agents each. In each subgroup, each agent played 10 games with 19 other agents in its group. Among these 10 games, we call the first five games as first half and the next five games as second half. In the first half, agents play using their value of cooperativeness assigned to them. Every agent keeps the history of the first half which can be referred for the second half, so they know who cooperated or defected with them in the past five games. But they know nothing about how much the other agents might have cooperated with each other. In the second half, each agent asks for referral about the opponent to its best five cooperators of the first half. Among five of them, if at least three of them say that the opponent is a cooperator, the agent will cooperate; otherwise it defects. So each agent plays 190 (19 *10) games in a round. Then, the subgroup monitor will conduct a survey among the subgroup members to determine which is the most cooperative and least cooperative member of each subgroup.

When a player plays with every other member of its subgroup, it computes a cooperation score for each player it played with. The performance of an individual agent is measured by its individual score (not the cooperation score) and is denoted as P.

The score of an individual member is different from its Degree of Cooperation denoted as DC, as voted by its fellow members. For instance, in the Prisoner's Dilemma game, an uncooperative group member may make suckers out of its fellow members and achieve a high score (P), even though being considered least cooperative by its fellow group members (low value of DC). At the end of each round, the voting is performed. Voting is the process of ranking of subgroup members based on their degree of cooperativeness (DC) for that round.

The agents' vote is based on their individual playing experience with other agents. The votes are tallied by the subgroup monitor. Thus, after surveying each subgroup member, the monitor will know its most cooperative (highest DC) and least cooperative (lowest DC) member for that round. The monitor uses this information to kick out the least cooperative member and promote the most cooperative member to other subgroups.

In addition, for each round, the five subgroups are themselves ranked in terms of their Overall Performance (OP), which is the sum of the individual scores of all of its members in the games as given by the formula below.

$$OP = \sum_{p=1}^{p=20} P$$

where P , the score of a player (performance) is given by

$$P = \sum_{g=1}^{g=10} S_g$$

where Sg is the score per game.

To determine movement between subgroups, the procedure given below is followed:

- The highest ranked subgroup in terms of performance (P) kicks out its least cooperative member.
- The 2nd, 3rd, and 4th subgroups in terms of performance (P) also kick out their least cooperative members, but also promote their most cooperative members for movement to a new group.
- The lowest ranked subgroup in terms of performance (P) promotes its most cooperative member for movement to a new group.

There are, thus, eight agents that have been placed into a separated pool for moving to another group: four promoted from the 2nd, 3rd, 4th, and 5th ranked subgroups and four kicked out of the 1st, 2nd, 3rd, and 4th ranked subgroups. Now tagging is employed. Commonly in Hales's (Hales, 2003a, 2003b, 2004a, 2004b; Hales & Edmonds, 2004) and Riolo's(Riolo, 1997; Riolo et al., 2001) work, all the agents are tagged and the tags serve the purpose of showing the identity of the agent and also specifying which group the agent belongs to. Here, our purpose of using tags is just to represent the status of an agent which is currently in the pool. The status could be high (promoted) or low (kicked out). The four promoted agents are given blue tags, signifying promotion, and the four kicked out agents are given red tags, signifying demotion. The monitor agent chooses players with blue tags in preference to players with red tags without knowing the performance scores of the players in the pool. The monitor agent takes players with blue tags if they are still available when it comes to its turn to choose.

- At this stage, the highest performance ranked subgroup gets one agent among the pool members in order to replace the member that has been kicked out (1st group gets 1 blue tagged agent).
- Then, the 2nd, 3rd, and 4th ranked subgroups get two agents to replace the two agents that they have lost (2nd group gets 2 blue tagged

agents, 3rd group gets a blue tagged and a red tagged agent and the 4th group gets 2 red tagged agents).

- Finally, the lowest ranked subgroup winds up with the remaining agent of the pool (5th group gets a red tagged agent).

With the newly created subgroups, another round of play is initiated. As mentioned before, individual game-playing agents are programmed to cooperate or defect with a tendency determined by a constant cooperation threshold parameter value of cooperativeness, CT that was randomly initialised to have a value between 0 and 1. Then, when each game was played, a random number was selected, and if it was less than CT, the player defected on that occasion, while if it was equal to or above CT, the player cooperated. At the outset of the game, each of the five subgroups populated with a random collection of 20 players having various tendencies to cooperate or defect within the group. Remember, only in the first half of every game, all the agents use the assigned constant cooperation threshold value to select the strategy. In the second half, they use referral scheme to select the strategy. The goal of the experiments is to show how well the performance of noncooperators can be restricted and society benefit can be improved. An agent is considered to be cooperative if DC is at least over a threshold value CT. In our case we set CT = 0.5. That corresponds to having a degree of cooperation (DC) such that: DC $\geq$ 0.5. For instance, if an agent has DC=0.4, it cooperates 4 times out of 10 times. Because its DC value is less than CT, it is considered to be uncooperative.

When the Prisoner's Dilemma game was played over 100 rounds, initially the groups were approximately having equal number of cooperators and defectors in every group. But after about 15 rounds, we could clearly distinguish different groups:

- Two groups had almost all cooperators;
- Two other groups had mostly defectors; and
- A middle group that had about half cooperators and half defectors.

Thus, these groups self organise themselves. This kind of emergence falls on the category of Type 3 according to Fromm (2005). In his chapter, Formm has explained comprehensive classification of the major types and forms of emergence in Multi-agent Systems. According to his classification, the result we got shows multiple emergence because it is influenced by several factors like tagging, referrals and assigned value of cooperativeness. This is shown in Figure 3.

With the separation of groups (based on their behaviour) we compared the scores (see Table 1).

Our general observation when using the referral scheme is that the groups who have more cooperators in the population gain a higher score than what it scored in the first half. And the uncooperative group (full of defectors) scores less than what it scored in the first half. In such groups, by

Figure 3. Multiple emergence influenced by tagging, referrals and assigned value of cooperativeness

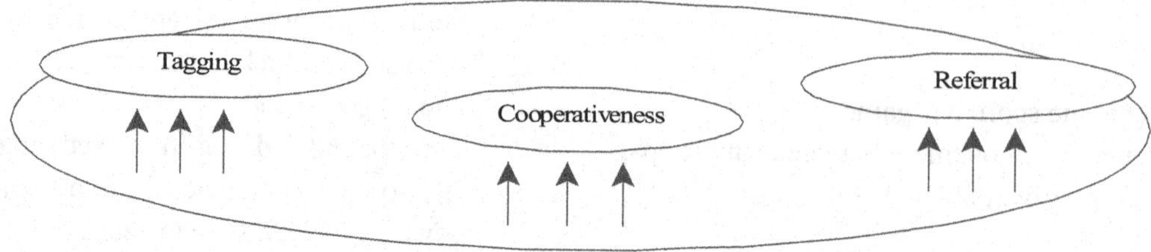

Table 1. Different groups and their performance in PD game

Groups (ranked by OP)	*1*	*2*	*3*	*4*	*5*
C : D	*18:2*	*16:4*	*10:10*	*4:16*	*0:20*
P in Firsthalf	*3397*	*3293*	*2961*	*2688*	*2247*
P in Secondhalf	*3531*	*3374*	*3147*	*2636*	*2174*

OP–Overall Performance C-Cooperators
D-Defectors P-Performance

using referrals, every agent comes to know that the opponent is a defector, so they all defect each other and get a low score. Because of this, their group score also goes low. The performance of a group which has an equal number of cooperators and defectors is determined by its overall tendency to cooperate/defect. If the tendency to cooperate is higher than defection, the group increases its score from its first half score. This is also the same for the groups who have a very small difference in the number of cooperators and defectors. For instance, a group with 8 defectors and 12 cooperators can get a lesser score in the second half, if the groups overall tendency to defect is more than to cooperate. For the first three groups, the score increased in the second half. But for the last two groups, the score decreased because they are mostly defectors (see Figure 4).

Also, we observed that the sum of scores of all groups in the second half is always higher than that of the first half's (see Figure 5). This shows that the referral scheme works well to improve the scores which is good for the overall society.

As the second experiment, the Stag Hunt game was played (S = 9, R = 7),over 100 rounds in the same manner as the previous PD game experiment. This also showed similar results as PD game in score increase while using referral (similar to Figure 5).

With the separation of groups, we compared the scores for Stag Hunt (see Table 2). Here, all the groups showed a score increase in the second

Figure 4. Performance of groups for PD game

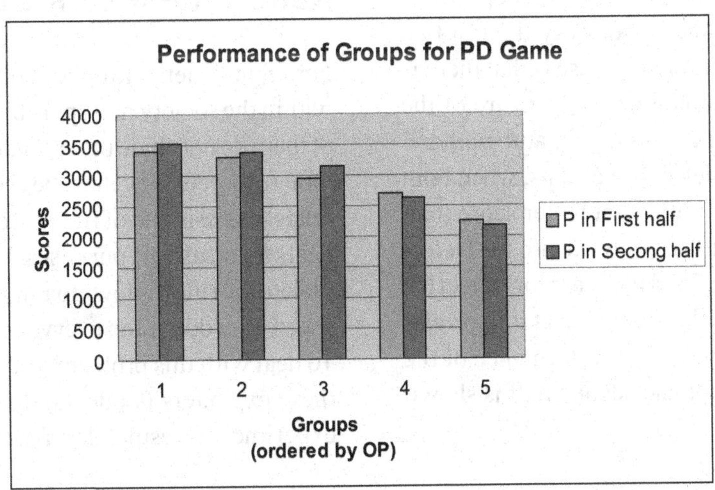

Table 2. Different groups and their performance in Stag Hunt game

Groups (ranked by OP)	1	2	3	4	5
C : D	18:2	15:5	10:10	1:19	3:17
P in First half	13777	12126	11156	11022	10665
P in Second half	15172	12225	11504	11322	11398

OP–Overall Performance C-Cooperators
D-Defectors P-Performance

Figure 5. Comparison of scores for first half (without referral) and second half (with referral) is shown for 10 runs

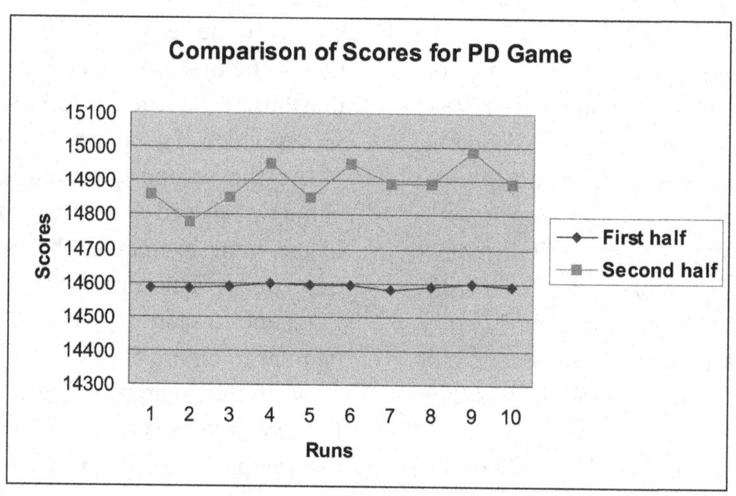

half. It is normal that the first 3 groups show score increase. But the reason why the last two groups also showed score increase is that there is no sucker reward in the reward structure of the Stag Hunt game. One cannot gain at the others' expense. So, in the defectors' groups, when both select to defect they both get a better score than being one of them defected (the score for Defect (7) + Cooperate (0) = 7 and the score for Defect (7) + Defect (7) = 14).When both defect it improves the group score as well. The performance of the groups in the first half and second half is shown (see Figure 6).

RESOURCE RESTRICTION

For agent societies to operate effectively, all agents within the society must play by rules. When some of them do not abide the rules and exploit the common resource, they must be restricted to do so. A related issue is that of Internet access given to students for academic purposes. When some of them violate the rule by downloading music/videos, the speed goes down and everyone is affected by this. To deal with this problem, the Internet access for these exploiters (students) must be limited. Our experimental results describe this.

Figure 6. Performance of groups for Stag Hunt game

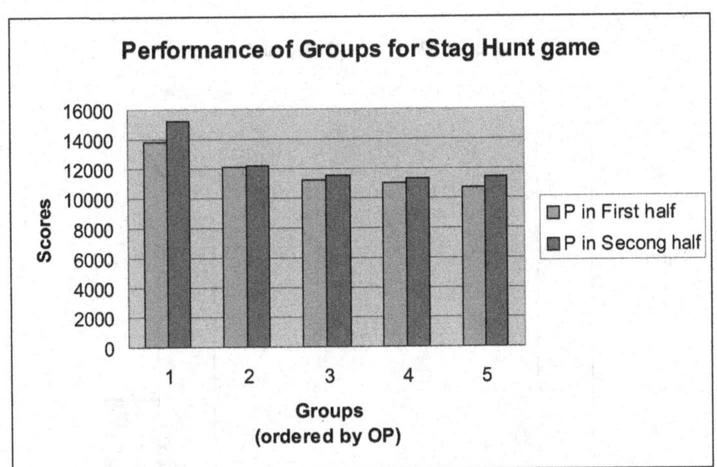

Resource Restriction for Groups

We tried this experiment in the same environment as the previous PD game experiment. There are 100 agents divided into 5 groups each having 20 agents. We made the agents in each group play five games per round (just like the first half of the previous PD game experiment, but no history is stored here). They played 100 rounds. They played according to how they were randomly assigned at the beginning.

After few rounds (about 15 rounds) the groups started separating. Out of five groups, they separated two groups full of mostly cooperators, two groups full of mostly defectors and a middle group having half of both. Now the OP is calculated for each group and the groups were ranked based on that. And also the overall average (sum of the average score of all five groups) is calculated. We call this average *the average without resource restriction.*

The number of games each agent plays is treated as the resource. We want to limit the resource (number of games) to the groups who are not performing up to the standards. The limitation for using the resource varies to groups according to their performance. The 1st ranked group is allowed to play five games as usual, because it is the best performer. The 2nd ranked group is restricted to play four games, the 3rd ranked group to play three games, the 4th ranked group to play two games and the 5th ranked group to play just one game, which is the worst performer among all. Now, again, the overall average is calculated. We call this *the average with resource restriction.*

When we compared the performance (see Figure 7), the total score with resource restriction is less than without resource restriction, because of less number of games. But the overall average is higher. This shows that resource restriction for exploiters is good for society benefit.

Resource Restriction for Individual

In another experiment using the same experimental set up as the previous one, we applied resource restriction for individuals who are uncooperative. Unlike the previous experiment, here the past history is stored. A player can decide whether to play or not with a particular opponent, based on past history. The player calculates how many times a particular opponent cooperated and defected. If the opponent cooperated more times, then the player plays with the opponent; otherwise the player opts not to play the game.

Figure 7. Comparison of performance

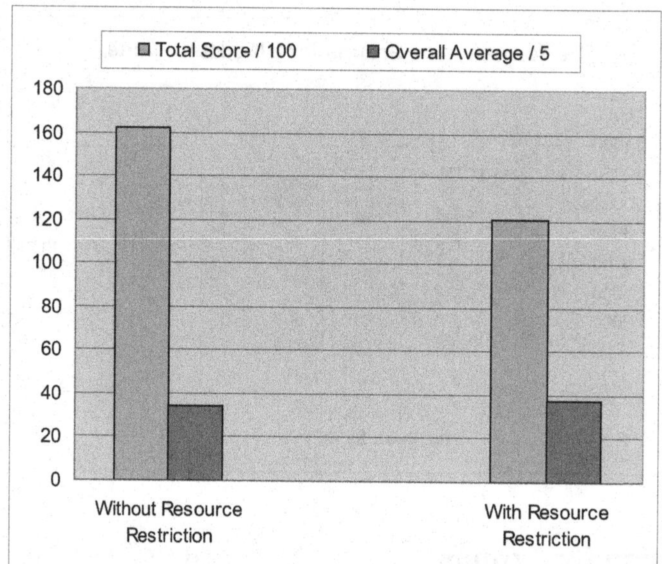

Figure 8. Self-organization of groups based on individual resource restriction is shown for 25 runs

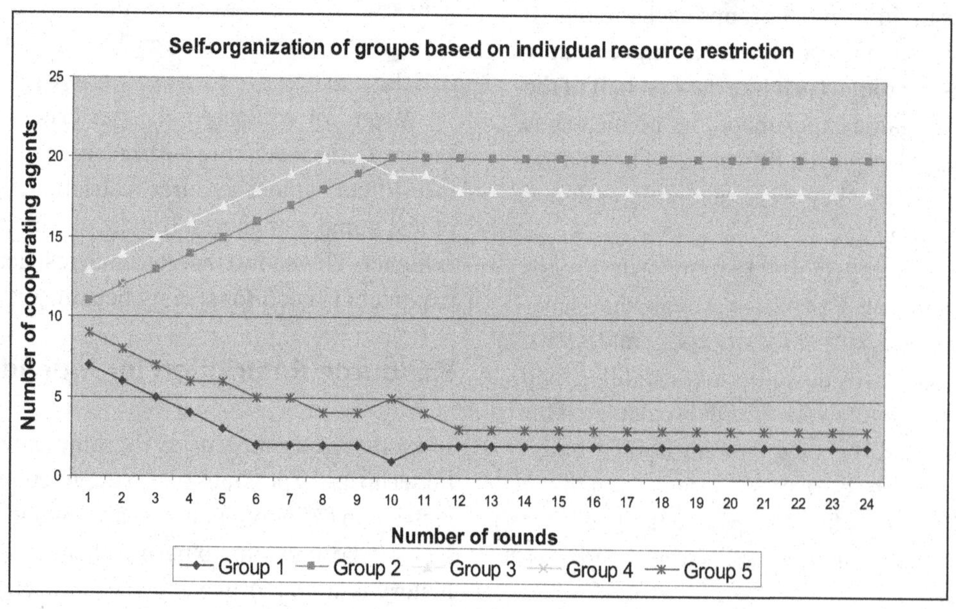

By this mechanism, the groups with a greater number of cooperators score more and the groups with a greater number of uncooperators score less.

In the very first experiment of this chapter, we observed the convergence around 15 rounds. In this experiment, the convergence takes place faster. The groups started separating around 7-10 rounds (see Figure 8).

The emergence of self-organization when using individual resource restriction leads to faster convergence than the mechanism that used referral because this mechanism uses personal history of all the past interactions with other agents, while the referral mechanism used the history of only five games.

CONCLUSION AND FUTURE WORK

In this work, we have described mechanisms that help to resist exploitation and improve overall societal performance in a multi-agent society that has cooperative and uncooperative agents. Our first mechanism uses tags and referrals. We have demonstrated that our proposed mechanism helps in improving the society benefit when the agents play the PD game as well as the Stag hunt game. We have also proposed resource restriction mechanisms and have demonstrated that the group level resource restriction and the individual level resource restriction can be used to improve the overall societal performance.

In the future, we are planning to use the concepts of tags and referrals and propose mechanisms for real life electronic societies (similar to orkut), so that cooperative members can be encouraged and uncooperative members can be restricted from social interactions.

REFERENCES

Axelrod, R. (1984). *The evolution of cooperation.* New York: Basic Books.

Barabási, A. L. (2002). *Linked: The new science of networks.* PA: Perseus Books Group.

Doctorow, C. (2001). *Metacrap: Putting the torch to seven straw-men of the meta-utopia* [Electronic Version]. Version 1.3. Retrieved April 3, 2008, from http://www.well.com/~doctorow/metacrap.htm

Fromm, J. (2005). *Types and forms of emergence* [Electronic Version]. Adaptation and self-organizing systems. Retrieved April 3, 2008, from http://www.citebase.org/abstract?id=oai:arXiv.org:nlin/0506028

Hales, D. (2003a). *Evolving specialisation, altruism and group-level optimisation using tags* (Vol. 2581/2003, Lecture notes in computer science). Berlin/Heidelberg: Springer-Verlag.

Hales, D. (2003b, September 16-19). Understanding tag systems by comparing tag models. In *Paper presented at the to-Model Workshop (M2M2) at the Second European Social Simulation AssociationConference (ESSA'04)*, Valladolid, Spain.

Hales, D. (2004a). Change your tags fast!--a necessary condition for cooperation?. In *Paper presented at the The Multi-agent-based Simulation (MABS 2004)*, (pp. 89-98).

Hales, D. (2004b). Self-organising, open and cooperative P2P Societies--from tags to networks. In *Paper presented at the Engineering Self-Organising Applications (ESOA 2004)*, (pp. 123-137).

Hales, D., & Edmonds, B. (2004). Can tags build working systems?—From MABS to ESOA. *Lecture notes in computer science* (pp. 186-194).

Hales, D., & Patarin, S. (2005). *How to cheat BitTorrent and why nobody does. UBLCS* (Tech. Rep. No. UBLCS-2005-12).

Hardin, G. (1968). The tragedy of the commons. *Science, 162*, 1243-1248).

Holland, J. H. (1993). *The effect of labels (tags) on social interactions* (Vol. SFI Working Paper 93-10-064). NM: Santa Fe Institute.

Huberman, B. A. (2001). *The laws of the Web: Patterns in the ecology of information.* Cambridge: MIT Press.

McCool, R. (2005). Rethinking the Semantic Web, part 1. *IEEE Internet Computing, 9*(6), 86-88.

McCool, R. (2006). Rethinking the Semantic Web, part 2. *IEEE Internet Computing, 10*(11), 93-96.

Purvis, M. K., Savarimuthu, S., Oliveira, M. D., & Purvis, M. A. (2006). Mechanisms for cooperative behaviour in agent institutions. In *Paper presented at the Intelligent Agent Technology (IAT 2006)*, Hong Kong.

Riolo, R. L. (1997). *The effects of tag-mediated selection of partners in evolving populations playing the iterated prisoner's dilemma* (Paper No. 97-02-016). NM: Santa Fe Institute.

Riolo, R. L., Cohen, M. D., & Axelrod, R. (2001). Cooperation without reciprocity. *Nature, 414*, 441-443.

Shirky, C. (2003a). *Permanet, nearlynet, and wireless data* [Electronic Version]. Economics & Culture, Media & Community, Open Source. Retrieved April 3, 2008, from http://www.shirky.com/writings/permanet.html

Shirky, C. (2003b). *Power laws, Web logs, and inequality* [Electronic Version]. Economics & Culture, Media & Community, Open Source, Version 1.1. Retrieved April 3, 2008, from http://shirky.com/writings/powerlaw_weblog.html

Chapter XII
Norm Emergence in Multi-Agent Societies

Bastin Tony Roy Savarimuthu
University of Otago, New Zealand

Maryam Purvis
University of Otago, New Zealand

Stephen Cranefield
University of Otago, New Zealand

ABSTRACT

Norms are shared expectations of behaviours that exist in human societies. Norms help societies by increasing the predictability of individual behaviours and by improving cooperation and collaboration among members. Norms have been of interest to multi-agent system researchers, as software agents intend to follow certain norms. But, owing to their autonomy, agents sometimes violate norms, which needs monitoring. In order to build robust MAS that are norm compliant and systems that evolve and adapt norms dynamically, the study of norms is crucial. Our objective in this chapter is to propose a mechanism for norm emergence in artificial agent societies and provide experimental results. We also study the role of autonomy and visibility threshold of an agent in the context of norm emergence.

INTRODUCTION

Norms are behaviours that are expected by the members of a particular society. These expected behaviours are common in human societies and sometimes even in animal societies (Clutton-Brock & Parker, 1995). The human society follows norms such as tipping in restaurants, exchange of gifts at Christmas, dinner table etiquette and driving vehicles on the left or right hand side of the road. Some of the well-established norms may become laws. The norms are of interest to researchers because they help to improve the predictability of the society. Norm adherence enhances coordination and cooperation among the members of the society (Axelrod, 1986; Shoham & Tennenholtz, 1995). Norms have been of interest in different areas of research such as sociology, economics, psychology and computer science (Elster, 1989). Sociologists and economists are divided on their view of norms based on the theories of *homo economicus* and *homo sociologicus* (Elster, 1989). Sociologists consider that the norms are always used for the overall benefit of the society. Economists, on the other hand, state that the norms exist because they cater to the self-interest of every member of the society and each member is thought to be rational (Gintis, 2003). A more integrated view of norms from sociology and economics point of view is provided by Conte and Castelfranchi (1999). Applying social theories in multi-agents is synergetic, as agents are modeled using some of the social concepts such as autonomy and speech act theory. Both disciplines complement each other as agents serve as a platform to design, test and validate social theories. Some researchers (Boman, 1999; Verhagen, 2000, 2001) have undertaken agent-based simulations of social theories. Even though researchers in different fields have been trying to answer questions such as why agents follow certain norms and the implications of not following these norms, there has been limited work on mechanisms that propose the emergence of these norms. In this chapter, we explain a mechanism for norm emergence and discuss the role of autonomy and visibility threshold of an agent in an agent society.

BACKGROUND

In this section, we describe different types of norms and the treatment of norms in multi-agent systems. We also describe the work related to norm emergence.

Types of Norms

Due to multidisciplinary interest in norms, several definitions for norms exist. Habermas (1985), one of the renowned sociologists, identified norm regulated actions as one of the four action patterns in human behaviour. A norm to him means *fulfilling a generalized expectation of behaviour*, which is a widely accepted definition for social norms. Researchers have divided norms into different categories. Tuomela (1995) has categorized norms into the following categories.

- r-norms (rule norms)
- s-norms (social norms)
- m-norms (moral norms)
- p-norms (prudential norms)

Rule norms are imposed by an authority based on an agreement between the members (e.g., one has to pay taxes). Social norms apply to large groups such as a whole society (e.g., one should not litter). Moral norms appeal to one's conscience (e.g., one should not steal or accept bribe). Prudential norms are based on rationality (e.g., one ought to maximize one's expected utility). When members of a society violate the societal norms, they may be punished. Many social scientists have studied why norms are adhered. Some of the reasons for norm adherence include:

- Fear of authority;
- Rational appeal of the norms; and
- Feelings such as shame, embarrassment and guilt that arise because of nonadherence.

Elster (1989) categorizes norms into consumption norms (e.g., manners of dress), behaviour norms (e.g., norm against cannibalism), norms of reciprocity (e.g., gift-giving norm), norms of cooperation (e.g., voting and tax compliance) and so forth.

Normative Multi-Agent Systems

The research of norms in multi-agent systems is recent (Boman, 1999; Conte, Falcone, & Sartor, 1999; Shoham & Tennenholtz, 1995). Norms in multi-agent systems are treated as constraints on behaviour, goals to be achieved or as obligations (Castelfranchi, 1995). There are two main research branches in normative multi-agent systems. The first branch focuses on normative system architectures, norm representations and norm adherence and the associated punitive or incentive measures. The second branch of research is related to emergence of norms.

Lopez and Marquez (2004) have designed an architecture for normative BDI agents and Boella and Torre (2006) have proposed a distributed architecture for normative agents. Some researchers are working on using deontic logic to define and represent norms (Boella & Torre, 2006; Garcia-Camino, Rodriguez-Aguilar, Sierra, & Vasconcelos, 2006). Several researchers have worked on mechanisms for norm compliance and enforcement (Aldewereld et al., 2006; Axelrod, 1986; Lopez, Luck, & Inverno, 2002). A recent development is the research on emotion-based mechanism for norm enforcement by Fix, Scheve, and Moldt (2006).

Related Work on Emergence of Norms

The second branch focuses on two main issues. The first issue is on norm propagation within a particular society. According to Boyd and Richerson (1985), there are three ways by which a social norm can be propagated from one member of the society to another. They are:

- Vertical transmission (from parents to offspring);
- Oblique transmission (from a leader of a society to the followers); and
- Horizontal transmission (from peer to peer interactions).

Norm propagation is achieved by spreading and internalization of norms. Boman and Verhagen (Boman, 1999; Verhagen, 2000, 2001) have used the concept of normative advice (advise from the leader of a society) as one of the mechanisms for spreading and internalizing norms in an agent society. Their work focuses on norm spreading within one particular society and does not address how norms emerge when multiple societies interact with each other. The concept of normative advice in their context assumes that the norm has been accepted by the top-level enforcer, the Normative Advisor, and the norm does not change. But, this context cannot be assumed for scenarios where norms are being formed (when the norms undergo changes). So, the issue that has not received much attention is the emergence of norms in multi-agent societies. But, there are lots of literature in the area of sociology on why norms are accepted in agent societies and how they might be passed on. Karl-Dieter Opp (Opp, 2001) has proposed a theory of norm emergence based on sociological concepts. Epstein (2001) has proposed a model of emergence based on the argument that the norms reduce individual computations and has provided some results. Our objective in this chapter is to propose a mechanism for norm emergence based on the concept of oblique norm transmission in artificial agent societies. We also provide our experimental results.

PROPOSED MECHANISMS

In this section, we will describe the mechanisms that help norm emergence when different agent societies with different norms interact with each other. Assume that two agent societies with different norms inhabit a particular geographical location. When these societies are co-located, interactions between them are inevitable. When they interact with each other, their individual societal norms might change. The norms may tend to emerge in such a way that it might be beneficial to the societies involved. Our working hypothesis is *Interactions between agent societies with different norms in a social environment (with a shared context), results in the convergence of norms. Norm convergence might result in the improvement of the average performance of the societies.* To demonstrate our hypothesis, we have experimented with agents that play the Ultimatum game (Slembeck, 1999). The shared context of interaction is the knowledge of the rules of the game. This game has been chosen because it is claimed to be sociologists' counter argument to the economists' view on rationality (Elster, 1989).

Ultimatum Game

The Ultimatum game (Slembeck, 1999) is an experimental economics game in which two parties interact anonymously with each other. The game is played for a fixed sum of money (say x dollars). The first player proposes how to divide the money with the second player. Say, the first player proposes y dollars to the second player. If the second player rejects this division, neither gets anything. If the second accepts, the first gets (x-y) dollars and the second gets y dollars. For example, assume that each game is played for a sum of 100 dollars by two agents, A and B. Assume that A offers 40 dollars to B. If B accepts the offer, then A gets 60 dollars and B gets 40 dollars. If B rejects the offer both of them do not get any money.

Description of the Multi-Agent Environment

An agent society is made up of a fixed number of agents. For our experiments we have designed two kinds of societies, namely selfish and benevolent societies, as shown in Figure 1. Society 1 and Society 2 correspond to selfish and benevolent societies, respectively. Society 1 is modeled after the materialistic world where agents try to maximize their personal income. Selfish agents propose the least amount of money and accept any non-zero amount. The second kind of society

Figure 1. Architecture of the experimental framework

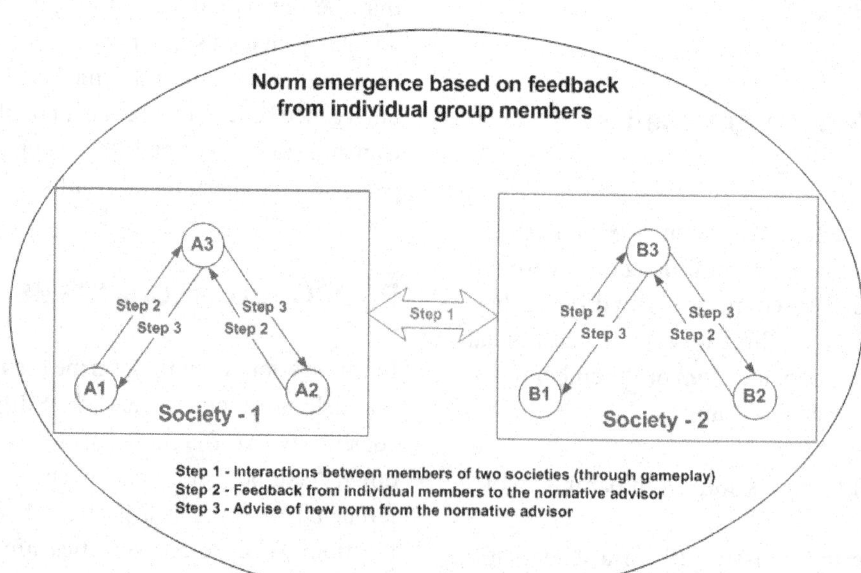

is the benevolent society such as the Ika tribe of Ethiopia (Elster, 1989). The benevolent agents are generous agents. They propose more than the fair share. But, they expect nothing less than the fair share. They also reject high offers. Each agent has two types of norms:

- Group norm (G norm); and
- Personal norm (P norm).

The G norm is shared by all the members of the society. The P norm is internal to the agent and it is not known to any other member. Autonomy is an important concept associated with choosing either a G norm or a P norm when an agent interacts with another agent. When an agent is created, it has an autonomy value uniformly distributed between 0 and 1. Depending upon the autonomy value, an agent chooses either the G norm or the P norm. For example, if the autonomy of an agent is .4, it chooses P norm 4 times and the G norm 6 times out of 10 games. Normative Advisor is one of the agents in the society, which is responsible for collecting the feedback from the individual agents. It modifies the G norm of the society and advises the change to all the members of the society. As shown in Figure 1, the Normative Advisor agents of the two societies are A3 and B3, respectively.

Experimental Parameters

The G norm and P norm are made up of two sub norms, namely the proposal norm and the acceptance norm. The proposal norm corresponds to the range of values (minimum and maximum values) that an agent is willing to propose to other agents. The acceptance norm corresponds to the range of values that an agent is willing to accept from other agents. A sample G norm for a selfish agent looks like the following where min and max are the minimum and maximum values when the game is played for a sum of 100 dollars.

- G-Proposal norm (min=1, max=30)
- G-Acceptance norm (min=1, max=100)

The representations given above indicate that the group proposal norm of the selfish agent ranges from 1 to 30 and the group acceptance norm of the agent ranges from 1 to 100. A sample P norm for a selfish agent might look like the following.

- P-Proposal norm (min=10, max=40)
- P-Acceptance norm (min=20, max=100)

Initially, the G norm of a society is assigned with a particular value, which will be shared by all the members of the society. The personal norms will vary from one agent to another. An agent can accept or reject a proposal based on the norm it chooses (which is based on its autonomy).

Collective Feedback Mechanism for Norm Emergence

In this section, we describe our mechanism for norm emergence that is based on collective feedback of individual agent experiences when playing the Ultimatum game against agents in the other society. The agents have a common G norm to start with. They also have an internal P norm. Both norms continuously evolve based on social learning to maximize the benefit of the society. In the context of the Ultimatum game, the goal is to improve the performance of the overall society while maximizing their own benefit. In one iteration, every agent in a society plays an equal number of games against all the agents in the other society. After the end of each game the agents record the history of interactions (both successes and failures). At the end of each iteration, all the agents submit their successful proposal and acceptance values to the Normative Advisor Agent of their society.

The Normative Advisor Agent uses the average successful values submitted by all the agents in a society and derives the new G norm value for the group. In each iteration the Normative Advisor Agent fractionally increases or decreases G norm values for a society so that it can accommodate the norms of the other society. This mechanism will reduce the overall losses and increase the overall income. After each iteration, the group norm will be propagated to all the agents in the society. Similar to the G norm, P norm of an agent will also change continuously. While G norm changes only at the end of each iteration, P norm changes within each iteration. When an agent chooses P norm over G norm, the outcome of that game determines whether the P norm will change or not. For example, when an agent's proposal that is based on a P norm is rejected *n* consecutive times, the agent modifies its P norm. The agent modifies its P norm fractionally so that it moves closer to the G norm.

EXPERIMENTATION AND RESULTS

The agents in our experiments are built on Otago Agent Platform (Purvis et al., 2002) and they communicate using FIPA ACL messages ("Foundation for Intelligent Physical Agents (FIPA)," 2007). Our experimental set up is made up of two societies with fixed number of agents in each society. In each iteration an agent plays the ultimatum game with all the players in the other group. The games were played over a fixed number of iterations (5 to 5000). In the first experiment, the agents do not use the designed mechanisms. In the second experiment, the agents use designed mechanism. At the end of each experiment, we observe whether norms emerge (whether the proposal norms stabilize or not). In the third and the fourth experiments, we explore the role of autonomy and the visibility threshold, respectively, on norm emergence.

The initial G norms associated with the three experiments are given below.

- G-Proposal norm for selfish society (min=1, max=30)
- G-Acceptance norm for selfish society (min=1, max=100)
- G-Proposal norm for benevolent society (min=55, max=70)
- G-Acceptance norm for benevolent society (min=45, max=55)

In our experimental setup the minimum and maximum values are parameterized and can be changed easily. We have chosen these sample values to demonstrate the results that we obtained.

Experiment 1: Societies that Resist Changes

Assume that the two societies that play the Ultimatum game resist changes to their G norms and P norms. In this scenario, the G norms are the same across all agents in one society. The P norms will be different from one agent to another. The agents do not change their G or P norms over all iterations.

The results of the average game money won by both societies in this scenario are shown in Figure 2. It can be observed that the performance of both societies are well below what could be achieved by both groups if they were rational such as the Utopian Society. Utopian Society, in its most common and general meaning, refers to a hypothetical perfect society. It is synonymous to a fair society where the average income for the Ultimatum game will be 50. When sociologists conducted Ultimatum game experiments in modern societies, many of the societies proposed the fair 50-50 split. This indicates that the norm of fairness had evolved in these societies (Elster, 1989). The performance of the selfish society in this experiment is better than the benevolent society because the selfish agents accept any non-zero proposal.

Figure 2. Performance of societies based on initial societal norms

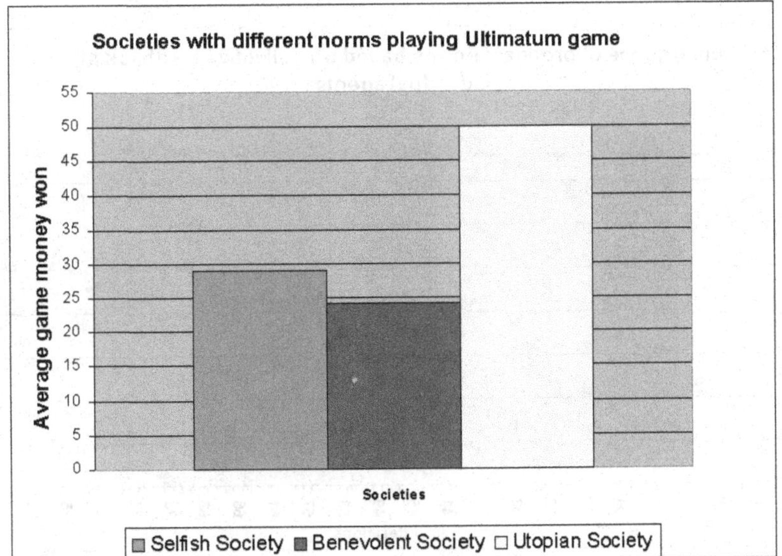

Experiment 2: Societies that Use Collective Feedback from Agents

In this experiment, both societies use the collective feedback mechanism. Figure 3 shows the G-Proposal norm changes of the benevolent as well as the selfish societies over 100 iterations. It can be observed that both groups are continuously changing their G-Proposal norm to accommodate the G-Proposal norm of the other group. Initially, the G-Proposal norm values for the benevolent group decrease because the Normative Advisor Agent changes the norm closer to the selfish societies' G-Proposal norm (based on the collective feedback). For the same reason the G-Proposal norm values for the selfish society increase (until iteration 32). Then, the norms in both societies oscillate to move closer to each other. When, one societies' maximum and minimum values are closer to the other, the G proposal norms start to converge (around iteration 80). These experiments show that the overall performance of the societies have improved as a result of norm emergence, as shown in Figure 3. It can also be observed that the ideal values are not reached as the agents are autonomous and may choose to ignore the G norm, particularly when the autonomy values are high. But, when the number of iterations increased to 5000, the outcomes were closer to the norm of fairness.

Experiment 3: Effect of Autonomy on Norm Convergence in an Agent Society

Unlike previous experiments where norm emergence was observed when two societies come together, in this experiment we observe the effect of autonomy on norm emergence in a single agent society.

The objective of the experiment was to study the effect of autonomy on norm emergence. There were 20 agents in a society and the agents played the Ultimatum game. The experiments were conducted over 20, 50 and 100 iterations. These experiments were carried out using two values of autonomy for all agents (0.2 and 0.8) representing lower and higher autonomy values.

It can be observed from Figure 4 that, when the autonomy of an agent is high (0.8), the con-

Figure 3. Emergence of norms based on collective feedback mechanism

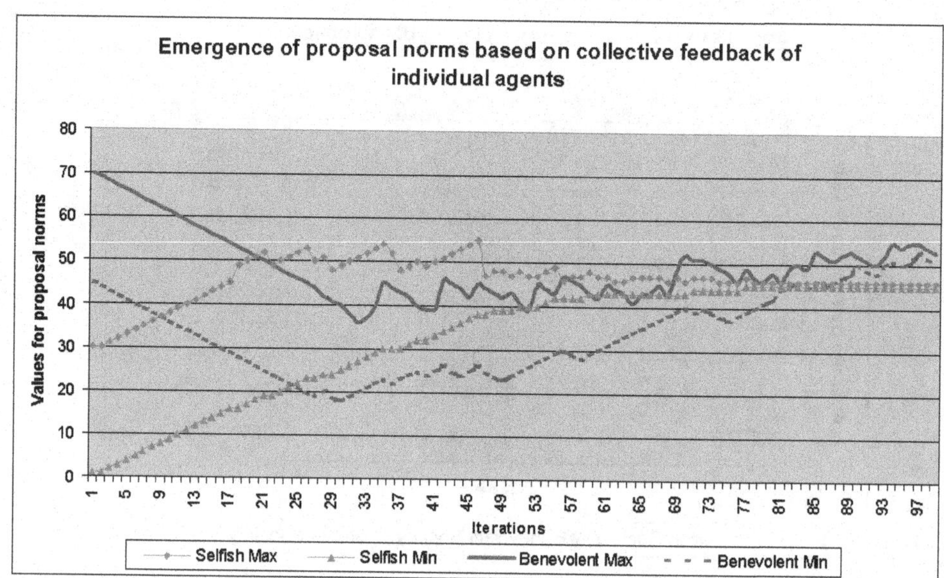

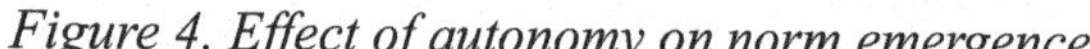

Figure 4. Effect of autonomy on norm emergence

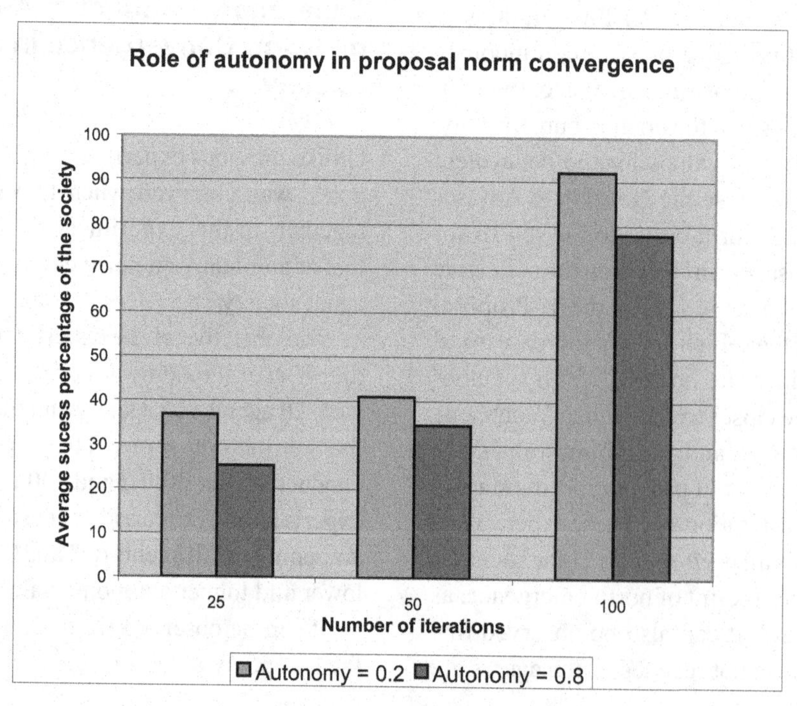

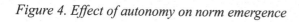

Figure 5. Effect of visibility threshold on norm emergence

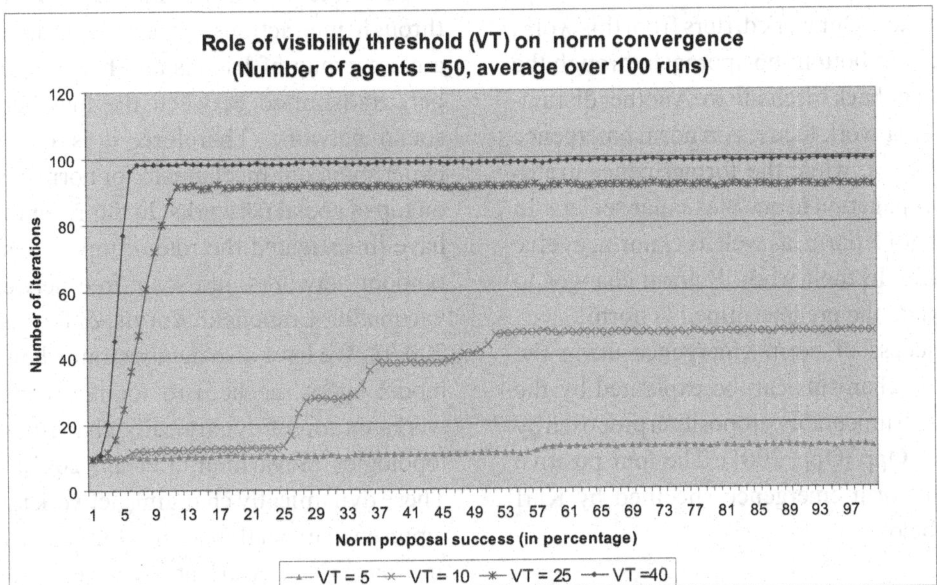

vergence of the norm is low. This indicates the negative effect of autonomy on the system. This result indicates that societies that have more autonomous agents will adopt or evolve norms slower than agent societies that have less autonomous or cooperative agents. This is because the agents that have higher autonomy tend to resist changes to their norms. After obtaining the feedback from the Normative Advisor agent, they move close to the advisor's norm depending upon their autonomy. If the autonomy is higher they do not readily adopt the recommendations provided through normative advice.

Experiment 4: Effect of Visibility Threshold on Norm Emergence

Assume that the collective feedback mechanism is modified in such a way that an agent can choose to seek advice from a local normative advisor agent as opposed to a centralized normative advisor agent. In this modified mechanism, an agent can choose another agent as its normative advisor whose successful proposal norm is within a limit represented by Visibility Threshold (VT). For example, if VT = 5 and an agent's successful proposal average is 80%, then the agent can choose another agent whose successful proposal average is between 80 and 85%.

We have conducted experiments using a society of 50 agents and varying the values for VT (5, 10, 25, 50). It can be observed from Figure 5 that, as the visibility threshold increases, the rate of norm emergence increases. When VT increases, an agent gets to choose a normative advisor within a broader spectrum and the probability of choosing a highly successful role model is high. So, convergence is faster for larger values of VT.

DISCUSSION

The experiments described in this chapter are our initial efforts in the area of norm emergence. Verhagen's thesis (Verhagen, 2000) focuses on the spreading and internalizing of norms. This assumes that a norm is agreed or chosen by a top-level entity (say, a Normative Advisor) and this G norm does not change. The G norm is spread to the agents through the normative advice using a top-down approach. Our work differs from this work, as we employ a bottom-up approach through the collective feedback mechanism. Another distinction is that our work focuses on norm emergence across societies, while the former concentrates on norm propagation in one particular society. In our work both P norm, as well as G norm, evolve continuously. In their work, P norm changes to accommodate the predetermined G norm.

The success of norm emergence using the proposed mechanisms can be explained by the theory of instrumentality proposition proposed by Karl-Dieter Opp (Opp, 2001). The four positive criteria for norm emergence specified by Karl are given below.

1. ***Homogeneity of goals G*** - In our experiments, the goal of an agent was to maximize its personal and societal income.
2. ***Knowledge that a norm N leads to G*** - The agents in our system worked toward establishing a norm that leads to an increase in overall score of the society.
3. ***Knowledge that behaviour B leads to N*** - The agents are aware that by reporting their experience to the Normative Advisor Agent, they can help to achieve the group goal.
4. ***Incentives to perform B*** - The agents know that they can increase their own personal score by providing feedback and receiving the advice. Another incentive for an agent to report experiences is its eagerness to predict other agents' behaviour (e.g., knowing the acceptance range of the other agent).

This emerging area of research on norm emergence offers interesting avenues for further research. In the real world, people are not related to each other by chance. They are related to each other through the social groups that they are in, such as the work group, church group, ethnic group and the hobby group. Information tends to percolate among the members of the group through interactions. People seek advice from a close group of friends and hence information gets transmitted between the members of the social network. Therefore, it is important to experiment our mechanism for norm emergence on top of social networks. In our recent work, we have investigated the role of topologies such as random networks and scale-free networks (Savarimuthu, Cranefield, Purvis, & Purvis, 2007a, 2007b). We have also demonstrated how the role model agent mechanism for norm emergence works on top of dynamically changing network topologies (Savarimuthu et al., 2007a, 2007b). These dynamically changing network topologies represent the social space in which agents can join and leave the network at any time.

An interesting problem in the context of norm emergence mechanism is to experiment with attaching weights to the advice provided by others. The weights of the edges (links) should be considered when the agent makes a decision on whom to choose as advisor agents. We plan to incorporate these ideas in our future experiments.

CONCLUSION

We have explained a mechanism for norm emergence in artificial agent societies. The mechanism used collective feedback of individual agent experiences. We have demonstrated the use of oblique norm transmission in these mechanisms for norm emergence. Through the experimental results, we have shown that norms emerge in agent societies when two different societies are brought together, and this norm might be beneficial to the societies as a whole. We have demonstrated the role of autonomy and visibility threshold of an agent on norm emergence. We have also discussed our future work.

REFERENCES

Aldewereld, H., Dignum, F., Garcia-Camino, A., Noriega, P., Rodriguez-Aguilar, J. A., & Sierra, C. (2006). Operationalisation of norms for usage in electronic institutions. In *Paper presented at the AAMAS*, Hakodate, Japan.

Axelrod, R. (1986). An evolutionary approach to norms. *The American Political Science Review, 80*(4), 1095-1111.

Boella, G., & Torre, L. v. d. (2006). An architecture of a normative system: Counts-as conditionals, obligations and permissions. In *Paper presented at the AAMAS*, New York.

Boman, M. (1999). Norms in artificial decision making. *Artificial Intelligence and Law, 7*(1), 17-35.

Boyd, R., & Richerson, P. J. (1985). *Culture and the evolutionary process.* Chicago: University of Chicago Press.

Castelfranchi, R. C. C. (1995). *Cognitive and social action.* London: UCL Press.

Clutton-Brock, T. H., & Parker, G. A. (1995). Punishment in animal societies. *Nature, 373*, 209-216.

Conte, R., & Castelfranchi, C. (1999). From conventions to prescriptions—Towards an integrated view of norms. *Artificial Intelligence and Law, 7*(4), 323-340.

Conte, R., Falcone, R., & Sartor, G. (1999). Agents and norms: How to fill the gap?. *Artificial Intelligence and Law, 7*(1), 1-15.

Elster, J. (1989). Social norms and economic Theory. *The Journal of Economic Perspectives, 3*(4), 99-117.

Epstein, J. M. (2001). Learning to be thoughtless: Social norms and individual computation. *Computer Economy, 18*(1), 9-24.

Fix, J., Scheve, C. v., & Moldt, D. (2006). Emotion-based norm enforcement and maintenance in multi-agent systems: Foundations and petri net modeling. In *Paper presented at the AAMAS.*

Foundation for Intelligent Physical Agents (FIPA). (2007). Retrieved April 3, 2008, from http://www.fipa.org

Garcia-Camino, A., Rodriguez-Aguilar, J. A., Sierra, C., & Vasconcelos, W. (2006). Norm-oriented programming of electronic institutions. In *Paper presented at the AAMAS*, New York.

Gintis, H. (2003). Solving the puzzle of prosociality. *Rationality and Society, 15*(2), 155-187.

Habermas, J. (1985). *The theory of communicative action: Reason and the rationalization of society* (Vol. 1). Beacon Press.

Lopez, F., Luck, M., & Inverno, M. (2002). Constraining autonomy through norms. In *Paper presented at the Proceedings of the First International Joint Conference on Autonomous Agents and Multi Agent Systems AAMAS'02.*

Lopez, F. L. Y., & Marquez, A. A. (2004). An architecture for autonomous normative agents. In *Paper presented at the Fifth Mexican International Conference in Computer Science (ENC'04)*, Los Alamitos, CA, USA.

Opp, K.-D. (2001). How do norms emerge? An outline of a theory. *Mind and Society, 2*(1), 101-128.

Purvis, M., Cranefield, S., Nowostawski, M., Ward, R., Carter, D., & Oliveira, M. (2002). Agent cities interaction using the opal platform. In *Paper presented at the Proceedings of the Workshop on Challenges in Open Agent Systems, AAMAS.*

Savarimuthu, B. T. R., Cranefield, S., Purvis, M., *& Purvis, M. (2007a). Mechanisms for norm emergence in multi-agent societies.* In *Paper presented at the Sixth International Joint Conference on Autonomous Agents and Multiagent Systems (AAMAS)*, Honolulu, HI, USA.

Savarimuthu, B. T. R., Cranefield, S., Purvis, M., & Purvis, M. (2007b). Role model based mechanism for norm emergence in artificial agent societies. In *Paper presented at the International Workshop on Coordination, Organization, Institutions and Norms (COIN) at AAMAS 2007*, Honolulu, HI, USA.

Shoham, Y., & Tennenholtz, M. (1995). On social laws for artificial agent societies: Off-line design. *Artificial Intelligence, 73*(1-2), 231-252.

Slembeck, T. (1999). *Reputations and fairness in bargaining--experimental evidence from a repeated ultimatum game with fixed opponents* (Experimental). EconWPA.

Tuomela, R. (1995). *The importance of us: A philosophical study of basic social notions.* Stanford, CA: Stanford Series in Philosophy, Stanford University Press.

Verhagen, H. (2000). *Norm autonomous agents.* Department of Computer Science, Stockholm University.

Verhagen, H. (2001). Simulation of the learning of norms. *Social Science Computer Review, 19*(3), 296-306.

Chapter XIII
Multi-Agent Systems Engineering:
An Overview and Case Study

Scott A. DeLoach
Kansas State University, USA

Madhukar Kumar
Software Engineer, USA

ABSTRACT

This chapter provides an overview of the Multi-agent Systems Engineering (MaSE) methodology for analyzing and designing multi-agent systems. MaSE consists of two main phases that result in the creation of a set of complementary models that get successively closer to implementation. MaSE has been used to design systems ranging from a heterogeneous database integration system to a biologically based, computer virus-immune system to cooperative robotics systems. The authors also provide a case study of an actual system developed using MaSE in an effort to help demonstrate the practical aspects of developing systems using MaSE.

INTRODUCTION

This chapter describes the Multi-agent Systems Engineering (MaSE) methodology for analyzing and designing multi-agent systems. MaSE was originally designed to develop closed, general purpose, heterogeneous multi-agent systems. MaSE has been used to design systems ranging from a heterogeneous database integration system to a biologically based, computer virus-immune system to cooperative robotics systems. While the multi-agent systems designed by MaSE are typically closed (the number and type of all agents are known a priori), the number of agents is unlimited, although, practically, the number of types of different agents is limited to something less than 50.

MaSE uses the abstraction provided by multi-agent systems to help designers develop intelligent, distributed software systems. MaSE views agents as a further abstraction of the object-oriented paradigm where agents are a specialization of objects. Instead of simple objects, with methods that can be invoked by other objects, agents coordinate with each other via conversations and act proactively to accomplish individual and system-wide goals. Agents are a convenient abstraction that allows designers to handle intelligent and non-intelligent system components equally within the same framework.

MaSE builds on existing object-oriented techniques and applies them to the specification and design of multi-agent systems. Many of the models developed with MaSE are similar to models defined in the Unified Modeling Language. However, the semantics of the models are often specialized for the multi-agent setting.

MaSE was designed to be used to analyze, design, and implement multi-agent systems by proceeding in an orderly fashion through the development lifecycle (DeLoach, Wood, & Sparkman, 2001). MaSE has been automated via an analysis and design environment called agentTool, which is a tool that supports MaSE and helps guide the system designer through a series of models, from high-level goal definition to automatic verification, semi-automated design generation, and finally to code generation.

The MaSE methodology consists of two main phases that result in the creation of a set of complementary models. The phases and the respective models that result at the end of each phase are listed below. While presented sequentially, the methodology is, in practice, iterative. The intent is to free the designer to move between steps and phases such that with each successive pass, additional detail is added and, eventually, a complete and consistent system design is produced.

ANALYSIS PHASE

The first phase in developing a multi-agent system using the MaSE methodology is the analysis phase. The goal of the MaSE analysis phase is to define a set of roles that can be used to achieve the system-level goals. These roles are defined explicitly via a set of tasks, which are described by finite state models. This process is captured in three steps: capturing goals, applying use cases, and refining roles.

Capturing Goals

The purpose of the first step in the analysis phase is to capture goals of the system by extracting the goals from a set of system requirements. The initial system requirements may exist in many forms including informal text and tell the designer about how the system should function based on specific inputs and the system state. The MaSE methodology uses these requirements to define goals in two specific sub-steps: Identifying goals and Structuring goals.

Phases	Models
1. Analysis Phase a. Capturing Goals b. Applying Use Cases c. Refining Roles	Goal Hierarchy Use Cases, Sequence Diagrams Concurrent Tasks, Role Model
2. Design Phase a. Creating Agent Classes b. Constructing Conversations c. Assembling Agent Classes d. System Design	Agent Class Diagrams Conversation Diagrams Agent Architecture Diagrams Deployment Diagrams

Identifying Goals

The main purpose of this step is to derive the overall system goal and its subgoals from the initial set of requirements. This is done by first extracting scenarios from the requirements and then identifying the goals of the scenarios. These initial scenarios are usually abstract in nature and are critical to the entire system. Therefore, the goals identified from these scenarios are at a very high level. These high-level goals then serve as the basis of analysis of the entire system. The roles defined later in the analysis phase must support one of these goals. Later, if the analyst defines a role that does not support one of these goals, either the role is not needed or the initial set of goals was incomplete and a new goal must be added.

Structuring Goals

After the goals have been identified, the second step is to categorize and structure them into a goal tree. The result is a Goal Hierarchy Diagram whose nodes represent goals and arcs define goal/subgoal relationships. The Goal Hierarchy Diagram is acyclic; however, there some subgoals that may have more that one parent goal.

To structure the goals, the analyst first identifies the main goal of the system. In the case where there is more than one main goal, those goals must be summarized as one high-level goal that is decomposed into a set of subgoals that are easier to manage and understand. To decompose a goal into subgoals, the developer must analyze *what* must be done to achieve the parent goals. A subgoal should support its parent goal by describing a subgoal that must be achieved in order to achieve the parent goal.

Although superficially similar, goal decomposition is different from function decomposition since goals define *what* tasks must be done instead of *how* a task is achieved, which is functional decomposition. Thus, goal decomposition should stop when the designer thinks that any further decomposition will result in functions and not subgoals. MaSE goal decomposition is similar to the KAOS approach (van Lamsweerde & Letier, 2000) except that MaSE goals do not have to be strictly AND-refined or OR-refined.

There are four types of goals in a Goal Hierarchical Diagram: summary goals, partitioned goals, combined goals, and non-functional goals. Any goal or subgoal can take on the attributes of any one, or more, of these types of goals. The four types of goals are described below.

1. **Summary Goal.** A summary goal encapsulates a set of existing "peer" goals to provide a common parent goal for the set. This often happens at the highest level of the Goal Hierarchical Diagram when a goal may be needed to support multiple high-level goals.
2. **Non-Functional Goal.** As the name suggests, non-functional goals are derived from non-functional requirements of the system, such as maintaining reliability or response times of the system. These goals need not directly support the overall functional goals of the system. When a non-functional goal is discovered, a new branch is generally created under the overall system goal, which can then be decomposed into either functional or non-functional sub-goals.
3. **Combined Goal.** While analyzing the goals of a system, often a number of subgoals are discovered in a hierarchy that are identical or very similar and can be grouped into a combined goal. This often results when the same basic goal is a subgoal of two different goals. In this case, the combined goal becomes a subgoal of both the goals.
4. **Partitioned Goal.** A partitioned goal is one of a set of goals that collectively meet a parent goal. This is identical to the notion of a KAOS *conjunctive goal.*

Once the goals have been identified and structured, the developer is ready to move to the next step of the MaSE analysis phase, applying use cases.

Applying Use Cases

In this step, the goals and subgoals are translated into use cases. These use cases typically capture the scenarios discovered in the previous step by providing a textual description and a set of sequence diagrams that are similar to the UML sequence diagrams. The main difference between MaSE sequence diagrams and UML is that in MaSE they are used to represent sequences of events between roles instead of objects. The events sent between roles are used in later steps to help define the communications between the agents that will be eventually playing these roles.

The use case at this stage helps the developer in representing desired system behaviors and sequences of events. When the use cases are converted to sequence diagrams, the roles that are identified become the initial set of roles that will be used in the next step of refining roles.

While not all requirements can be captured as use cases, the developer should try to represent the critical requirements as either positive or negative use cases. Positive use cases define the desired system behaviors, and negative use cases describe a breakdown or an error in the system. Both are useful in defining roles that must be played in the system.

Refining Roles

With the Goal Hierarchy Diagram and use cases in place, the analyst is ready to move to the next step, Refining Roles. This step involves further defining roles by associating them with specific tasks. The roles produced from this step are defined in such a way as to ensure that each system goal is accounted for and form the building blocks for the agents that will eventually populate the system.

MaSE is built on the assumption that the system goals will be satisfied if each goal maps to a role, and every role is played by at least one agent class. In general, the mapping of goals to roles involves a one-to-one mapping. However, the developer may choose to allow a role to be responsible for multiple goals for the sake of convenience or efficiency. At this stage, the developer may also choose to combine several roles; although this will most certainly increase the complexity of the individual roles, it can significantly simplify the overall design.

In MaSE, goal refinement is captured in a Role Model (Kendall, 1998). In this model, the roles are represented by a rectangle, while a role's tasks are represented by ovals attached to them. The arrows between tasks designate communication protocols, with arrows pointing from the initiator of the protocol toward the responder. Solid lines represent external communication (role-to-role), while dashed lines indicate internal communication between tasks belonging to the same role instance.

Once the roles are decomposed into a set of tasks, the individual tasks are designed to achieve the goals for which the role is responsible. It is important to note here that roles should not share tasks with other roles. Sharing a task among different roles indicates improper role decomposition. If the analyst believes that a task needs to be shared, then a separate role should be created for that task. This will allow the task to be incorporated into different agent classes, thus being effectively shared.

Concurrent Task Model

After the roles are defined, the analyst must define the details of each task in the role model. Task definition is performed via a Concurrent Task Diagram, which is based on finite state automata. Semantically, each task is assumed to run concurrently and may communicate with other tasks either internally or externally. Taken collectively, the set of tasks for a specific role should define the behavior required for that role.

A concurrent task consists of a set of states and transitions. The states in the concurrent tasks represent the internal functioning of an agent while transitions define the communication between tasks. Every transition in the model has a source state, destination state, trigger, guard condition, and transmissions. The transitions use the syntax

trigger [guard] ^ transmission(s)

If there are multiple transmissions required, they can be concatenated using a semicolon (;) as a separator; however, no ordering is implied. In general, events sent as triggers or transmissions are associated with events sent to tasks within the same role instance, thus allowing for internal task coordination. To represent messages sent between agents, however, two special events—send and receive —are used.

The *send* event is used to represent a message sent to another agent and is denoted by send(message, agent) while a *receive* event, denoted by receive(message, agent), is used to define a message received from another agent. The message itself consists of a *performative*, the intent of the message along with a set of *parameters*. It is also possible to send the same message to several agents at the same time using multicasting by using a group name of the agents as compared to the name of a single agent.

Task states may contain *activities* that represent internal reasoning, reading a percept from sensors, or performing actions via actuators. More than one activity may be included in a single state and they are performed in an uninterruptible sequence, which, when combined with states and transitions, gives a general computational model. Once inside a state, the task remains there until the activity sequence is complete. Variables used in activity and event definitions are visible within the task, but not outside of the task or within activities. All messages sent between roles and events sent between tasks are queued to ensure that all messages are received even if the agent or task is not in the appropriate state to handle the message or event immediately.

Once a transition is enabled, it is executed instantaneously. If multiple transitions are enabled, then internal events are handled first, external messages (the send/receive events) are next, and the transitions with guard conditions only are last (DeLoach, 2000).

To reason about time, the Concurrent Task Model provides a built-in timer activity. An agent can define a timer using the setTimer activity, t = setTimer(time). The *setTimer* activity takes a time as input and returns a timer that will timeout in exactly the time specified. The timer that can then be tested via the timeout activity, timeout(t), which returns a Boolean value, to see if it has "timed out."

DESIGN PHASE

In the analysis phase, a set of goals was derived and used to create a set of use cases and sequence diagrams that described basic system behavior. These models were then used to develop a set of roles and tasks that showed how the goals should be achieved. The purpose of the design phase is to take those roles and tasks and to convert them into a form that is more amenable to implementation, namely, agents and conversations. The MaSE design phase consists of four steps. These steps include designing agent classes, developing conversation between the agents, assembling agents, and finally deploying the agents at system-level design.

Construction of Agent Classes

The first step in the design phase involves designing the individual agent classes, which is documented in an Agent Class Diagram. In this step, the designer maps each role defined in the analysis phase to at least one agent class. Since roles are derived from the system goals and are responsible for achieving them, enforcing the constraint that each role is assigned to at least one agent class in the system helps to ensure that the goals are actually implemented in the system. In general, an agent class can be thought of as a template for creating the actual agent instances that will be part of the multi-agent system. These templates are defined in terms of the roles they play and the protocols they use to coordinate with other agents.

The first step in constructing agent classes is to assign roles to the agent classes. If the designer chooses to assign more than one role to the same agent class, the roles may be performed either concurrently or sequentially. The assignment of roles to agents allows the multi-agent organization to be easily modified, since the roles can be manipulated modularly. This allows the designer to manipulate the design to account for various software engineering principles, such as functional or temporal cohesion.

Once the agents are created by identifying the roles they will be playing, the conversations between agents are designed accordingly. For example, if two roles, R1 and R2, that shared a communication protocol were assigned to agent classes A1 and A2 respectively, then A1 and A2 would require a conversation (to implement the protocol) between them as well.

The Agent Class Diagram that results from this step is similar to object-oriented class diagrams. They are different in that (1) agent classes are defined by the roles they play instead of their attributes and methods, and (2) the relationships between agent classes are always conversations.

Constructing Conversations

Once the agent classes have been defined and the required conversations identified, the detailed design of the conversations is undertaken. These details are extracted from the communications protocols identified in the analysis phase.

Conversations are modeled using two different Conversation Class Diagrams, one for the initiator and the other for the responder. These diagrams are based on finite state automata and use states and transitions to define the inter-agent communication, similar to concurrent tasks. The transitions in the conversation diagrams use a slightly different syntax

rec-mess(args1) [cond] / action ^ trans-mess(args2)

This means that that if the message *rec-mess* is received with the arguments *args1* and the condition *cond* holds true, then the method *action* is called and the message *trans-mess* is sent with arguments *args2*.

Conversations are derived from the concurrent tasks of the analysis phase, based on the roles the agents are required to play. Thus, each task that defines an external conversation (outside the role) ends up becoming one or more conversation between agents. However, if all task communication is internal (within the same role) or with roles that are performed by the same agent, then the communication translates into internal function or method calls. Generally, however, concurrent tasks translate into multiple conversations, as they require communication with more than one agent class.

During this stage, the designer also needs to take into account other factors besides the basic protocols defined in the concurrent tasks. For example, what should an agent do if it does not receive the message it was expecting? Perhaps the communication medium was disabled or the other agent failed. Therefore, the designer should attempt to make conversations robust enough to handle potential run-time errors.

Assembling Agent Classes and Deployment Design

The last two stages in MaSE involve the internal design of the agent classes and the system-level design. The first of these stages, Assembling Agent Classes, involves two steps, defining the agents' architecture and defining the individual components of the architecture. MaSE does not assume any particular agent architecture and attempts to allow a wide variety of existing and new architectures to be used. Thus, the designer has the choice of either using pre-existing agent architecture like Beliefs, Desires, and Intentions (BDI) or creating a new architecture from scratch. The same goes for the architecture components. The step of assembling agents result in an Agent Architecture Diagram in which the components are represented by rectangular boxes connected to either inner or outer agent connectors. The inner-agent connectors, represented by thin arrows define visibility between components, while the outer agent connectors, represented by dashed arrows, define external connections to resources like other agents effectors, databases, and so on. A more detailed discussion of this step can be found in Robinson (2000).

The last step in building a multi-agent system using the MaSE methodology is to decide on the actual configuration of the system, which consists of deciding the number and types of agents in the system and the platforms on which they should be deployed. These decisions are documented in a Deployment Diagram, which is very similar to a UML Deployment Diagram and is used for much the same purpose. In a Deployment Diagram, agents are represented by three-dimensional boxes, while rectangles with dashed lines represent physical computing platforms. The lines between agents represent the actual lines of communication between the agents. In a dynamic multi-agent system in which agents move or are created and destroyed, the Deployment Diagrams are used to show snapshots of possible system configurations.

EXAMPLE CASE STUDY: MULTI-AGENT RESEARCH TOOL (MART)

To show how to use the MaSE methodology outlined above, this section presents an example of using MaSE to develop an actual multi-agent system. The Multi-agent Research Tool (MART) was developed as part of a MS project at Kansas State University and is being considered for distribution by a private company. The analysis and design was performed using the agentTool development environment, with the implementation being done in Java.

Overview

Writing articles is an important part of work for a researcher at a university or a content provider working for a media company. While writing research or news articles, the author often conducts searches on the World Wide Web (WWW) to unearth relevant information that can be used to write the article. However, when an author is writing an article, it is often a distraction to stop writing, visit a few search engines, conduct keyword searches, retrieve relevant information, and then incorporate it into the article. This is not very efficient, especially when the author has to deliver the article by a specific deadline.

The motivation for developing a Multi-agent Research Tool (MART) was to develop a tool that helps authors to research while writing an article without wasting valuable time. This means that the research tool should not only be smart and efficient in conducting searches, but that it should also be able to work in the background and, at the same time, be non-intrusive to the user. Moreover, since use of the Internet has become commonplace, it is assumed that a person using MART has access to the Internet. It would be more useful if the research tool could use distributed computing to retrieve research material and present it to the user whenever he/she decides to view or use them.

Based on the nature of the original motivation, it was decided to build MART as a multi-agent system since the location and numbers of the various components within the local network would not be known in advance. Since MART was developed using the MaSE methodology, the decision to use the agentTool development environment—a software engineering tool that directly supports MaSE analysis and design—was straightforward.

Developing Goals of the System

In the first step of the Analysis Phase, the following goals were defined based on the requirements for the MART system, as presented above. As shown in Figure 1, the overall goal of the system is to produce the results of a search for keywords from the user's article. This goal is partitioned into four subgoals: ranking and refining the keywords used in the search, searching the Web for results, producing and presenting the result to the user, and managing the entire system.

The *rank and refine search keywords* goal is partitioned into two subgoals: reading user keywords and ranking the keywords. The goal, *search the Web*, is also partitioned; however, it has only one subgoal, namely, *search Web sites*. Although not technically required, this goal structure was adopted so that future versions could add additional goals that could include searching other types of information sources such as databases located on the host computer and/or local network. Finally, the goal *produce results* is partitioned into three subgoals that allow for reading raw results, refining the raw results, and producing the final results that will be presented to the user.

Applying Use Case

After defining the goals, three primary use cases were generated based on the three main subgoals (1.1, 1.2, and 1.3 in Figure 1). These use cases are *Refine and Rank Keywords*, *Search the Web*, and *Generate Results*. Each is presented in detail below.

Refine and Rank Keywords

The Refine and Rank Keywords use case defines how the system should behave when it is initially asked to perform as search. As shown in Figure 2, the manager of the search process asks the reader to read the predefined user preferences and keywords and then asks the ranker to rank the keywords that were returned. The user preferences define exactly how the user prefers the search to be conducted while the keywords are the specific words on which the user wants the search to be conducted. These keywords are then ranked in terms of relevance to the article the user is currently writing.

Figure 1. MART system goals

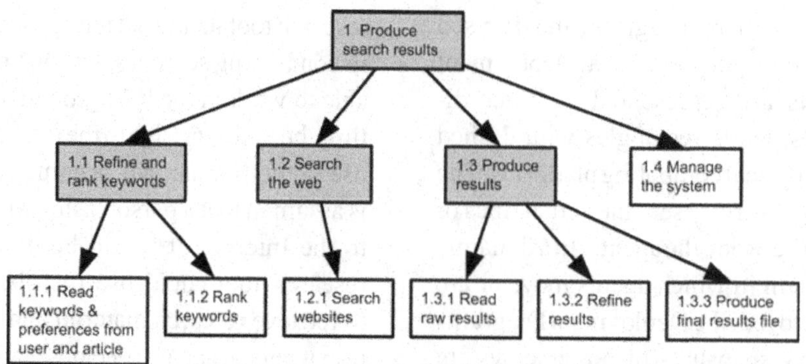

Figure 2. Sequence diagram for "Refine and Rank Keywords" use case

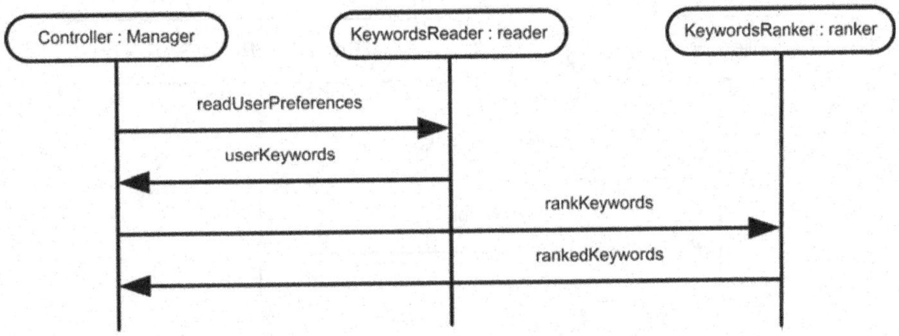

Search the Web

As shown in Figure 3, the Search the Web use case defines the basic search process of the system once a set of keywords has been developed. Each searcher agent is asked to search its known Web sites for a specific set of keywords. Exactly where and how each searcher conducts its search is not known to the manager. However, once results are received back by the searcher agents, the results are returned to the controller, who tabulates the results for a variety of searchers.

Generate Results

The sequence diagram in Figure 4 shows that the manager, once it has the raw results, sends a message to the result generator along with the raw results. The result generator refines the results by extracting duplicates and providing proper formatting and then sends back the finished product back to the manager.

Refining Roles

The role diagram depicts how the different goals are mapped to the roles of the system. Figure 5 shows that the controller has many tasks that collaborate with the other roles in order to read keywords, perform a search, and generate the finished product. The numbers inside rectangles (roles) indicate the goals for which they are responsible. For example, the Controller role is responsible for goal 1.4 from Figure 1, which is the goal of managing the system.

As discussed above, the solid lines connecting the different roles represent the communication between the roles. The dotted line between tasks

Figure 3. Sequence diagram for the use case "Search the Web"

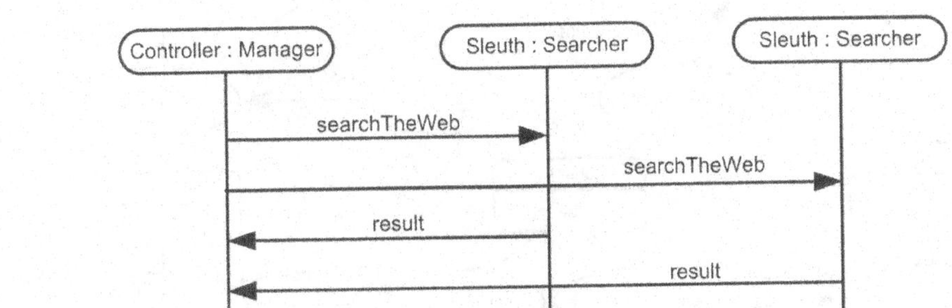

Figure 4. Sequence diagram for the use case "Generate results"

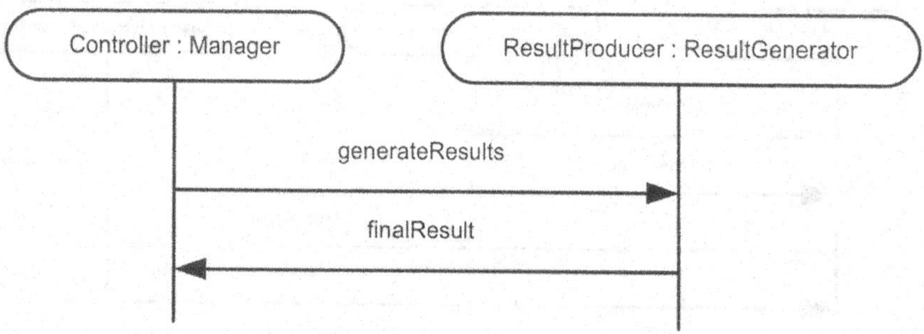

in the Sleuth role (the makeRaw protocol) shows that it is an internal communication between tasks within the same role instance (agent). The makeRawResults task is invoked by the rawResults task of the Sleuth once it receives the *searchTheWeb* request from the controller.

Concurrent Task Model

Once the Role Model has been constructed, a concurrent task model was defined for each task in the role model. For example, *showKeywords* is a task for the KeywordsReader role. An example of a concurrent task model for *showKeywords* is shown in Figure 6. The task starts when a *readPrefs* message is received from an agent named controller. After receiving the message, the user preferences are read via the activity readPreferences() in the readPreferences state. Upon completion of the activity, the task enter the readKeywords state where it gets the keywords from the user via the readKeywords() activity. If the keywords list is empty (null), then the task ends without sending a response. Otherwise, the task sends the set of keywords back to the controller.

Constructing Agent Classes

After all the tasks from the Role Model have been defined via concurrent task diagrams, the analysis phase ended and the design phase commenced. The first step of the design phase is to define the basic system architecture using an Agent Class Diagram. The initial task was to create agent classes and assign them specific roles to play.

Figure 5. MART role model

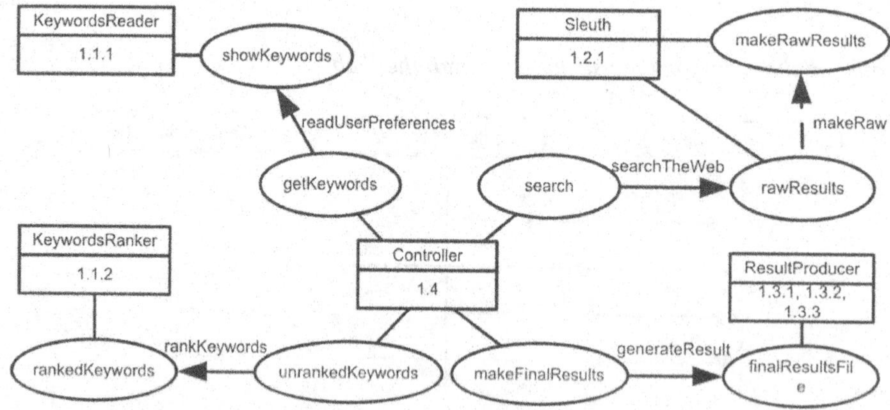

Figure 6. The showKeywords task for role KeywordsReader

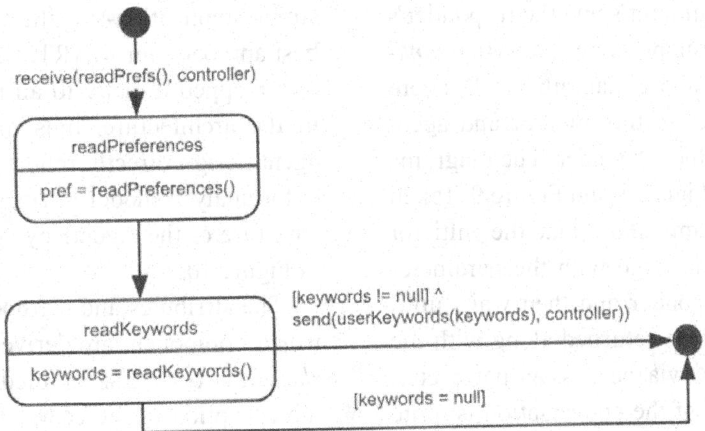

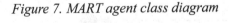

The Agent Class Diagram shown in Figure 7 shows that MART has five different agents: AgentManager, AgentKey, AgentProducer, AgentGoogle, and AgentTeoma. The lines connecting the agents represent the conversations between the agents. For example, *searchTheWeb* is a conversation that is initiated by AgentManager. *searchTheWeb* is unique in that it exists between the AgentManager and two different agent types: AgentGoogle and AgentTeoma. Actually, both of these agent types implement the Sleuth role, and the conversation can be directed from the AgentManager to either agent type requesting them to conduct a search and return raw results. The agent classes in Figure 7 represent independent processes operating in their own thread of control. These agents could be placed on different machines and still be able to talk to each other using the conversations defined in the system.

Constructing Conversations

After creating the agent classes and documenting them via the Agent Class Diagram, the individual communication between the agents was defined, based on the protocols between the appropriate roles from which they were derived. Each resulting conversation was documented using a pair of Conversation Diagrams, which are similar to and can be derived from the concurrent tasks models

Figure 7. MART agent class diagram

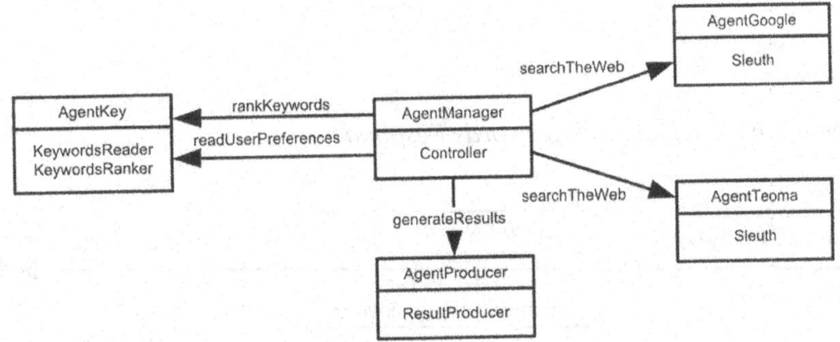

developed during the analysis phase (Sparkman, DeLoach, & Self, 2001). Each conversation is represented from the initiator's and the responder's points of view. For example, the conversation *rankKeywords* from the above diagram has the agent class AgentManager as the initiator and agent class AgentKey as the responder. The diagrams are show below in Figure 8 and Figure 9. Taken together, the diagrams show that the initiator sends the rankKey message with the parameter keywords to the responder and then waits until a rankedKey message is returned along with the ranked set of keywords via the message parameter. The responder side of the conversation is quite similar, with the messages sent by the initiator being received and the messages received by the initiator being sent. The obvious difference between the two is that the responder side includes an activity, rankKeywords(), that is called to actually perform the ranking process.

Assembling Agents and Deployment

After developing the conversations required for MART, the next step was constructing the individual components of the agent classes. As discussed earlier, there is a choice of either using either pre-existing agent architectures or creating an application-specific architecture. Because the MART agents were simple, it was decided that a simple application-specific architecture was the best approach for MART. Each concurrent task was mapped directly to an internal component in the architecture, thus making the internal agent design directly reflect the roles and tasks of the analysis model. An example of component structure of the AgentKey agent class is shown in Figure 10.

The attributes and methods of the *showKeywords* component are derived directly from the *showKeywords* task defined in Figure 6, with the exception of the conv_r_readUserPreferences method. The conv_r_readUserPreferences method was created to initiate the *readUserPreferences* conversation. When the agent wants to start the *readUserPreferences* conversation, it calls the method, which contains all the implementation dependent code for handling the conversation. The *rankedKeywords* component was derived similarly. Because these two components do not communicate directly (they are derived from the *showKeywords* and *rankedKeywords* tasks in Figure 5), there is not a visibility connection between them.

The last step in the design of the MART system was to develop the overall system deployment

Figure 8. Conversation model for rankKeywords Initiator

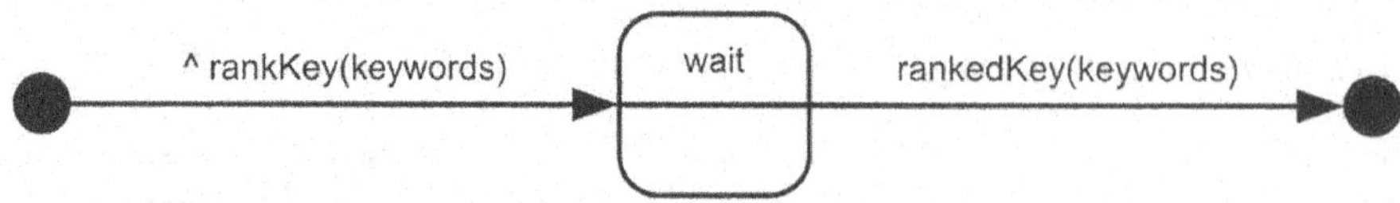

Figure 9. Conversation model for rankKeywords Responder

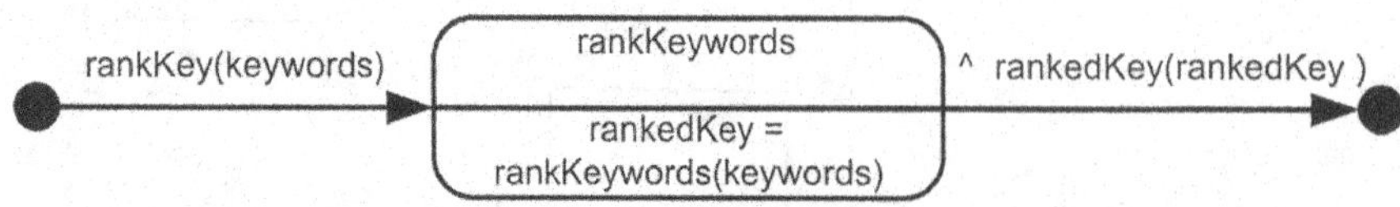

Figure 10. Components of AgentKey

showKeywords
+pref : type +keywords : type +controller : type +keyword : type
#readPreferences() : Object #readKeywords() : Object #conv_r_readUserPreferences(in controller : Object, in keyword : Object) : void

rankedKeywords
+keywords : Set(type) +controller : type +rank : type
+conv_r_rankKeywords(in controller : Object) : void +rankKeywords(in keywords : Object) : Object

Figure 11. Deployment diagram for MART

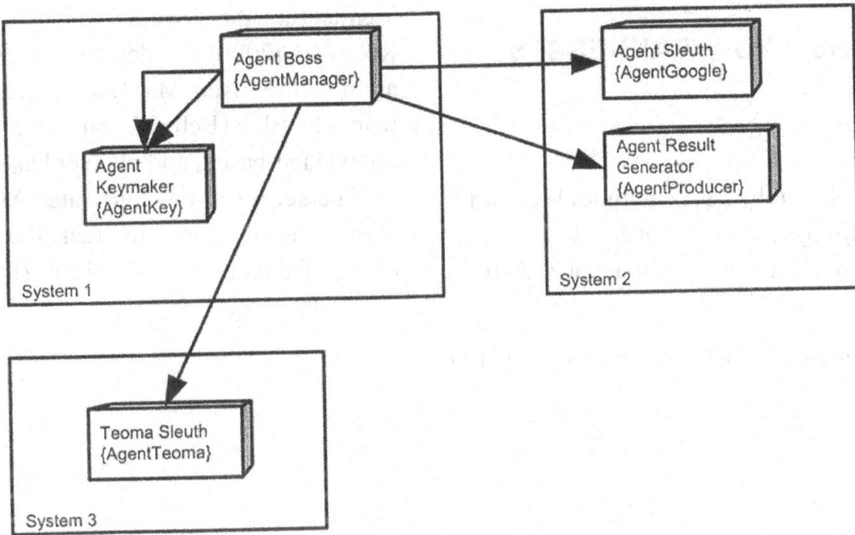

design. The MART Deployment Diagram, as shown in Figure 11, was created as an example of how the MART system could be deployed. As the MART system is designed to be deployed in a number of settings, this deployment is notional. Typically, the AgentKey and AgentManager agents are started on the user's computer while the AgentGoogle and AgentTeoma agents must be pre-deployed on local network computers (the AgentProducer agent may be deployed anywhere on the local network as well).

Refining the Object Model

After completing the analysis and design for MART, the implementation of the system began. As the system analysis and design was performed using agentTool, the first step was to verify that the conversations were correct and deadlock free. After this step was completed, the code generation capabilities of agentTool were employed to generate the initial code, which included stubs for each agent, each component, and each side of the conversations. Generally, each agent component is implemented as a single class. However, due to the simplicity of the components defined during the Assembling agents step, each agent class design was implemented by integrating all the agent components into a single class (one class for each agent type). This resulted in the simplified class structure shown in Figure 12.

The system architecture is shown in Figure 13, where each agent class is represented as a package. Obviously, multiple versions of each agent may exist. The KeyObjectPackage, which is accessed by each of the agents, includes shared definitions of data objects passed between agents. The diagram also includes a user-defined stereotype «conversation» to denote the existence of conversations between the various agent packages. The system was implemented using the agentMom agent-oriented middleware system, which inherently supports the concept of conversations as defined in MaSE (Mekprasertvit, 2004).

STRENGTHS AND WEAKNESSES

Strengths

MaSE is a comprehensive methodology for building multi-agent systems. It has been used to develop both software multi-agent systems and multi-agent cooperative robotic systems (DeLoach, Matson, & Li, 2003). One of the major strengths of MaSE is that it provides guidance throughout the entire software development life-cycle—from requirements elicitation through to implementation.

Firstly, MaSE is independent of any particular agent architecture or underlying environment. While the example above does not use pre-existing agent architectures, MaSE does allow the task and conversation behavior to be implemented in any architecture the designer wishes. For example, Robinson (2000) has defined a variety of agent architectures using MaSE components including reactive, BDI (Belief Desires and Intentions), knowledge-based, and planner based.

The sequence of interrelated MaSE models allows the developer to track the mapping of entities from one model to the next. This is most

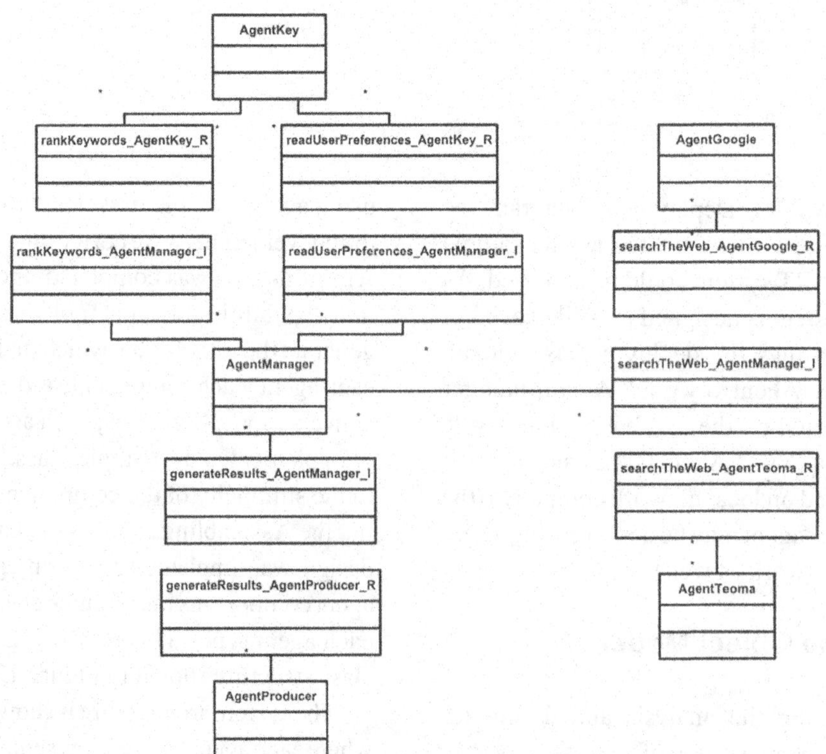

Figure 12. Simplified UML class model for MART

Figure 13. MART packages

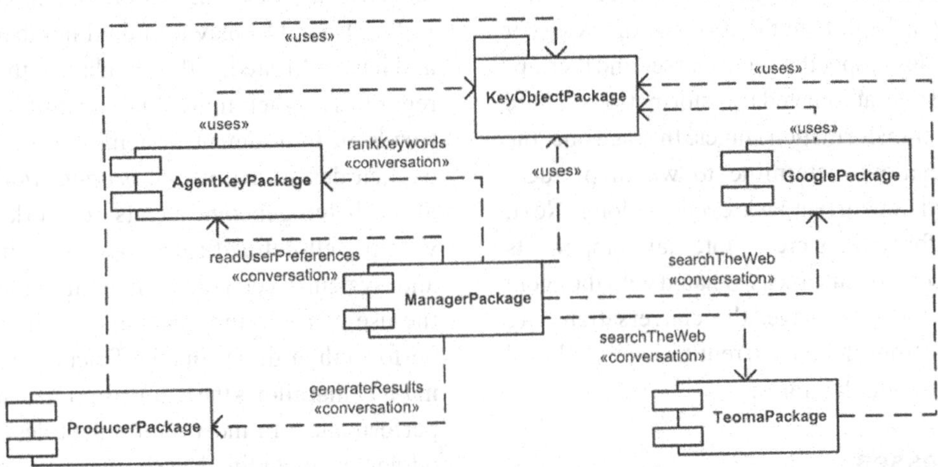

readily apparent in the mapping of goals to roles, roles to agents, and communications protocols to conversations. This mapping allows the developer to move between models, showing that the entities defined in previous models are implemented successfully in the current model. It also provides an excellent tool for tracking down system errors. If a particular goal is not being successfully achieved, the developer can track that goal directly to the responsible role and then on the implementing agent classes.

Often in multi-agent approaches, developers are allowed to specify behavior and agent communications protocols; however, the relationship between the two is not always clear. MaSE provides a way of directly defining the relationship between agent communication protocols and the internal behavior of the agent. This relationship is captured in the concurrent task diagrams and is carried over to agent conversation diagrams. By studying a set of concurrent tasks, it becomes evident how the communications between roles, and eventually agents, directly affects and is affected by the results of the communication. For instance, in Figure 6, it is clear that the computation (the *readPreferences* activity) starts after receiving the *readPrefs* message, and the results of the *readKeywords* activity determine whether the *userKeywords* message is even sent.

MaSE is also supported by the agentTool development environment. AgentTool is a software engineering tool built to help designers create multi-agent systems using the MaSE methodology. Using agentTool, a multi-agent system can be developed by following the MaSE steps in both the analysis and design phases. Since agentTool is a graphical-based tool, all the diagrams and the models described in the MaSE methodology are created using the tool. During each step of system development, the various analysis and design diagrams are available via agentTool and the developer is allowed to move freely back and forth between models in the various MaSE steps. A developer may also use agentTool to verify a conversation at any point by using the conversation verification capability (Lacey & DeLoach, 2000), which uses the Spin model checker (Holzmann, 1997) to check for deadlocks, as well as non-progress loops, syntax errors, unused messages, or unused states. If an error exists, the verification results are presented textually to the developer as well as by directly highlighting the offending conversation diagram. AgentTool includes developing support for semi-automatic transformations that convert a set of analysis models into the appropriate design models (Sparkman, DeLoach, & Self, 2001). To initiate the process, the designer assigns roles to specific agent classes and then applies the semi-automated transformations. There are three transformation stages. In stage one, the transformations determine to which protocol individual concurrent task events belong. Next, the transformations create internal components for each concurrent task associated with the agent class. In the final stage, the conversations are extracted from the concurrent tasks and placed in conversation diagrams.

Weaknesses

While MaSE provides many advantages for building multi-agent systems, it is not perfect. It is based on a strong top-down software engineering mindset, which makes it difficult to use in some application areas. First, MaSE is not currently appropriate for the development of *open* multi-agent systems. Since MaSE predefines the communications protocols between agent classes, the resulting system assumes that any agents trying to participate in the system implicitly know what those protocols are. In addition, MaSE does not inherently support the use of different ontologies, although an extension to MaSE by DiLeo, Jacobs, and DeLoach (2002) does incorporate the notion of ontologies into MaSE and agentTool. In general, however, MaSE implicitly defines an ontology that is embedded in the task communication protocols and is implemented within each agent.

The MaSE notion of conversations can also be somewhat bothersome, as they tend to decompose the protocols defined in the analysis phase into small, often extremely simple pieces when the original protocol involves more than two agents. This often results in conversations with only a single message. This makes comprehending how the individual conversations fit together more difficult.

MaSE also tends to produce multi-agent systems with a fixed organization. Agents developed in MaSE tend to play a limited number of roles and have a limited ability to change those roles, regardless of their individual capabilities. Recent trends in multi-agent systems are towards the explicit design and use of *organizations*, which allow heterogeneous agents to work together within well-defined roles to achieve individual and system-level goals. In multi-agent teams, the use of roles and goals allows the agents to perform their duties in an efficient and effective manner that allows the team to optimize its overall performance. In most multi-agent design methodologies, including MaSE, the system designer *analyzes* the possible organizational structure and then *designs* one organization that will suffice for most anticipated scenarios. Unfortunately, in dynamic applications—where the environment as well as the agents may change—a designer can rarely account for or even consider all possible situations.

Ideally, a multi-agent team would be able to design its own organization at runtime. To accomplish this, MaSE would have to be extended to be able to analyze and design multi-agent organizations. While MaSE already incorporates many of the required organizational concepts such as goals, roles, and the relations between these entities, it cannot currently be used to define a true multi-agent organization.

CONCLUSION

MaSE provides a detailed approach to the analysis and design of multi-agent systems. MaSE combines several established models into a comprehensive methodology. It also provides a set of transformation steps that shows how to derive new models from the existing models thus guiding the developer through the analysis and design process.

MaSE has been successfully applied in many graduate-level projects as well as several research projects. The Multi-agent Distributed Goal Satisfaction Project used MaSE to design the collaborative agent framework to integrate different constraint satisfaction and planning systems. The Agent-Based Mixed-Initiative Collaboration Project also used MaSE to design a multi-agent system focused on distributed human and machine planning. MaSE has been used successfully to design an agent-based heterogeneous database system as well as a multi-agent approach to a biologically based computer virus-immune system. More recently, we applied MaSE to a team of autonomous, heterogeneous search and rescue robots (DeLoach, Matson, & Li, 2003). The MaSE approach and models worked very well. The concurrent tasks mapped nicely to the typical behaviors in robot architectures. MaSE also provided the high-level, top-down approach missing in many cooperative robot applications.

Future work on MaSE will focus on specializing it for use in adaptive multi-agent and cooperative robotic systems based on an organizational theoretic approach. We are currently developing an organizational model that will provide the knowledge required for a team of software or hardware agents to adapt to changes in their environment and to organize and re-organize to accomplish team goals. Much of the information needed in this organizational model—goals, roles and agents—is already captured in MaSE. However, we will have to extend MaSE analysis to capture more detail on roles, including the capabilities required to play roles.

REFERENCES

DeLoach, S.A. (2000). *Specifying agent behavior as concurrent tasks: Defining the behavior of social agents.* Technical Report, U.S. Air Force Institute of Technology, AFIT/EN-TR-00-03.

DeLoach, S.A., Matson, E.T., & Li, Y. (2003). Exploiting agent oriented software engineering in cooperative robotics search and rescue. *The International Journal of Pattern Recognition and Artificial Intelligence, 17*(5), 817-835.

DeLoach, S.A., Wood, M.F., & Sparkman, C.H. (2001). Multiagent systems engineering. *The International Journal of Software Engineering and Knowledge Engineering, 11*(3), 231-258.

DiLeo, J., Jacobs, T., & DeLoach, S.A. (2002). Integrating ontologies into multiagent systems engineering. In *Proceedings of the Fourth International Bi-Conference Workshop on Agent-Oriented Information Systems (AOIS 2002) at AAMAS '02,* Bologna, Italy, July 16. Available from *http://CEUR-WS.org/Vol-59*

Holzmann, G.J. (1997). The model checker spin. *IEEE Transactions on Software Engineering, 23*(5), 279-295.

Kendall, E. (1998). Agent roles and role models: New abstractions for multiagent system analysis and design. In *Proceedings of the International Workshop on Intelligent Agents in Information and Process Management*, Bremen, Germany, September.

Lacey, T.H. & DeLoach, S.A. (2000). Automatic verification of multiagent conversations. In *Proceedings of the 11th Annual Midwest Artificial Intelligence and Cognitive Science Conference,* University of Arkansas, Fayetteville, April 15-16 (pp. 93-100). AAAI Press.

Mekprasertvit, C. (2004). AgentMom user's manual. Kansas State University. Retrieved June 30, 2004, from *http://www.cis.ksu.edu/~sdeloach/ai/software/agentMom_2.0/*

Robinson, D. (2000). A *component based approach to agent specification.* Master's Thesis. AFIT/GCS/ENG/00M-22. School of Engineering, Air Force Institute of Technology, Wright-Patterson AFB.

Sparkman, C.H., DeLoach, S.A., & Self, A.L. (2001). Automated derivation of complex agent architectures from analysis specifications. In *Proceedings of the Second International Workshop on Agent-Oriented Software Engineering, (AOSE 2001), Springer Lecture Notes in Computer Science* (Vol. 2222, pp. 188-205).

van Lamsweerde, A. & Letier, E. (2000). Handling obstacles in goal-oriented requirements engineering. *IEEE Transactions on Software Engineering, 26*(10), 978-1005.

Section III
Fuzzy-Based and Other Methods for Integration

Chapter XIV
Modeling, Analysing, and Control of Agents Behaviour

František Čapkovič
Institute of Informatics, Slovak Academy of Sciences, Slovak Republic

ABSTRACT

An alternative approach to modeling and analysis of agents' behaviour is presented in this chapter. The agents and agent systems are understood here to be discrete-event systems (DES). The approach is based on the place/transition Petri nets (P/T PN) that yield both the suitable graphical or mathematical description of DES and the applicable means for testing the DES properties as well as for the synthesis of the agents' behaviour. The reachability graph (RG) of the ***P/T PN-based model*** *of the agent system and the space of feasible states are found. The RG adjacency matrix helps to form an auxiliary hyper-model in the space of the feasible states. State trajectories representing the actual interaction processes among agents are computed by means of the mutual intersection of both the straight-lined reachability tree (developed from a given initial state toward a prescribed terminal one) and the backtracking reachability tree (developed from the desired terminal state toward the initial one; however, oriented toward the terminal state). Control interferences are obtained on the base of the most suitable trajectory chosen from the set of feasible ones.*

INTRODUCTION

Behaviour of an agent in surroundings as well as among other agents in **multi-agent systems (MAS)** is one of the most important parts of the research in the intelligence integration. Agents are usually understood (Fonseca, Griss, & Letsinger, 2001) to be persistent (especially software, but not only software) entities that can perceive, reason, and act in their environment and communicate with other agents. Hence, MAS can be apprehended as a composition of collaborative agents working in shared environment. The agents together perform a more complex functionality.

Communication enables the agents in MAS to exchange information. Thus, the agents can coordinate their actions and cooperate with each other. However, an important question arises here, namely: What communication mechanisms enhance the **cooperation between communicating agents**?

In general, the agent interaction is a specialized kind of the behaviour. Roughly speaking, the agent behaviour has both internal and external attributes. From the external point of view the agent is (Demazeau, 2003) a real or virtual entity that (i) evolves in an environment; (ii) is able to perceive this environment; (iii) is able to act in this environment; (iv) is able to communicate with other agents; and (v) exhibits an autonomous behaviour. On the other hand, from the internal point of view, the agent is a real or virtual entity that encompasses some local control in some of its perception, communication, knowledge acquisition, reasoning, decision, execution, and action processes. While the internal attributes characterize, rather, the agent inherent abilities, different external attributes of agents manifest themselves in different measures in a rather wide spectrum of MAS applications, like for example, computer-aided design, decision support, manufacturing systems, robotics and control, traffic management, network monitoring, telecommunications, e-commerce, enterprise modeling, society simulation, office and home automation, and so forth. Even (Demazeau, 2003), the applications in computer vision, natural language processing, spatial data handling and so forth, are known as well.

It is necessary to distinguish two groups of agents or agent societies, namely, human and artificial. The principle difference among them consists especially in the different internal abilities. These abilities are studied by many branches of sciences including those finding themselves out of the technical branches; for example, economy, sociology, psychology, and so forth. This chapter does not set itself these abilities as a goal of studies. It takes no account of the causes of them. Simply said, the internal behaviour happens and it is practically idle to consider how it happens. Here, the appearance of the internal abilities in the form of discrete events is important only. However, on the other hand, the external (i.e., inter-agent) behaviour is very important as to the quality of the **communication or cooperation process in MAS**. At the **cooperation in MAS** two principle characteristics of the agents are usually distinguished. Namely, either each agent is able to solve the whole problem but the use of many agents in parallel speeds up the problem solving, or the agents are specialized to solve different subproblems. While, in the former case, the cooperation consists of the purely physical (i.e., spatial or temporal) decomposition of the work between the agents, for example, each agent either solves a part of the problem or works for a given time, in the latter case each agent solves the problem for which it is specialized. However, a mix both of them seems to be more effective. Namely, it is very useful when an agent being free is able to substitute (at least partially) the activities of another agent in case of a failure or to help another agent asking for help (e.g., in case when it is not able to solve a problem).

As to the agent abilities, we can speak about cognitive and reactive agents. The cognitive agents are those that can form plans for their behaviours, whereas reactive agents are those that just have reflexes. Ferber (1999) showed how both approaches could converge in the end. Namely, one kind of research focuses on the building of individual intelligences whose communication is organised, whereas the other imagines very simple entities whose coordination emerges in time without the agents being conscious of it. However, in fact, a number of different schools of MAS persist, all coming from different theoretical backgrounds.

During the **agent cooperation,** different kinds of conflicts and synergies occur. The agent interactions in MAS occur in order to achieve some desired global goals. The goals are external to each individual agent and they must be reached by the interaction of agents. Desired behaviour of MAS is external to the agents. However, the behaviour of individual agents is motivated from their own goals and capabilities, that is, agents bring in their own ways into the MAS. The **MAS behaviour** emerges (Weigand, Dignum, Meyer, & Dignum, 2003) from the goal-pursuing behaviour of the individual agents within the constraints set by the organizational model (OM). OM describes the global behaviour and structure (it is not dependent on the agents themselves). OM also describes the desired or intended behaviour and overall structure of the society from the perspective of the organization in terms of roles, interaction scripts and social norms. Emergence means (Walker & Wooldridge, 1995) that conventions develop from within a group of agents.

MAS are also used in intelligent control, especially for a cooperative problem solving (Yen, Yin, Ioerger, Miller, Xu, & Volz, 2001). The **negotiation** belongs to the most important interactions in such a case. It is the process of multilateral bargaining for mutual profit. In other words (Brams & Kilgour, 2001; Thompson, 1998), the **negotiation** is a decision process where two or more participants make individual decisions and interact with each other in order to reach a compromise.

The agent behaviour can be understood as driven by discrete events. Consequently, it can be modeled by the **discrete event systems (DES)**. We can even speak about the **discrete event dynamic systems (DEDS)** because the events occur in intrinsic time instants. An approach to modeling and analysing **interactions among agents in MAS** is presented in this chapter and the possibility of their control is pointed out. In order to model the DES or DEDS different kinds of **Petri nets (PN)** (Petersen, 1981) are frequently used. The place/transition PN (P/T PN) are utilised here in order to model the behaviour of the agents. Namely, PN can be seen as a graphical modeling tool having the well understanding mathematical background (Murata, 1989), the formalism based on discrete mathematics. Consequently, many important static and dynamic properties are proved in PN theory. Existing automated tools make it possible to prevent or avoid erroneous situations, for example, deadlocks. Particular modules of a complex system can be modeled separately by simple PN modules. The PN modules can cooperate by means of the communication modules. In a simple case, the communication modules consist either of PN transitions or of PN places. In a complex case the communication modules can be formed as the PN modules (subnets) with a suitable structure.

The reachability graph of the **PN-based model of MAS** and the space of feasible states are found. The set of feasible trajectories representing the feasible ways of the system dynamics development is computed by means of the mutual intersection of both the straight-lined reachability tree (SLRT) and the backtracking reachability tree (BTRT). The SLRT is developed from a given initial state toward the prescribed terminal one. The BTRT is developed from the terminal state toward the initial one. However, it is oriented toward the terminal state. The most suitable trajectory can be chosen from the set of feasible ones on the base of a criterion (if any). Namely, the criterion is usually given only verbally or in nonanalytical terms. Sometimes several criteria (even contradictory) can be formulated. The chosen trajectory yields the control interferences applicable to the real system.

PETRI NET-BASED MODELING THE DEDS

DEDS (Cassandras & Lafortune, 1999) are driven by discrete events. PN in general are frequently used for modeling systems that are characterized as being DEDS (Murata, 1989; Petersen, 1981). Hence, PN are able to describe and study the behaviour of such systems because they are able to competently express the systems being concurrent, parallel, asynchronous, distributed, and so forth. Especially, PN are characterized by their ability to handle operation sequence, concurrency, conflict and mutual exclusion in systems, and the like. These features predetermine them to be a suitable tool for describing and analysing concurrent and real-time systems such as flexible manufacturing systems (FMS), communication systems, transport systems, and so forth.

As a graphical tool, PN can be used as a visual-communication aid. In addition, tokens are used in these nets to simulate the dynamic and concurrent activities of systems. As a mathematical tool, it is possible to set up the PN state equation as a **mathematical model** governing the behaviour of the systems in question.

Thus, the Petri net is a graphical and mathematical modeling tool. It consists of the set of places $P = \{p_1, p_2, ..., p_n\}$, the set of transitions $T = \{t_1, t_2, ..., t_m\}$, $P \cap T = \emptyset$, and arcs that connect them. The set of oriented arcs $F \subseteq P \times T$ connects places with transitions (the arcs start at a place and end at a transition), while the set of oriented arcs $G \subseteq T \times P$, $F \cap G = \emptyset$, connects the transitions and places (the arcs start at a transition and end at a place). Formally, the quadruplet $\langle P, T, F, G \rangle$ represents the PN structure. It can be said that PN are the bipartite directed graphs with two kinds of nodes (places and transitions) and two kinds of edges (arcs emerging from places and entering transitions, and arcs emerging from transitions and entering places). The places usually model the elementary activities and they can contain tokens. The tokens symbolize the activities. Mathematically, they are expressed by integers. The absence of tokens in a place represents the passivity while the presence of a different number of tokens in the place represents a kind of activity. For example, when the place models a resource, existence of tokens in the place indicates the availability of the resource. The current state of the modeled system (the marking) is given by the number of tokens in each place. Transitions model discrete events (starting or ending activities), which can occur (the transition fires) in the system and change the state of the system (the marking of the Petri net). In such a way, firing of a transition represents occurrence of the event or execution of the activity. Transitions are only allowed to fire if they are enabled, which means that all the preconditions for the activity must be fulfilled (there are enough tokens available in the input places). Thus, transition is enabled if each of its input places contains a number of tokens that is greater than or equal to the weight of the arrow (expressing the number of tokens which can simultaneously pass throughout this oriented way) connecting the input place to the transition. An enabled transition may fire. Firing of a transition t is the action that removes from each its input place p_i the number of tokens equal to the weight of arrow from p_i to t and then inserts into each output place p_j the number of tokens equal to the weight of arrow from t to p_j.

The marking development (the PN dynamics) can be expressed by another quadruplet $\langle X, U, \delta, \mathbf{x}_0 \rangle$. Here, $\mathbf{x}_0$ is an initial state vector, $X = \{X_1, X_2, ..., X_N\}$ is the set of feasible state vectors (i.e., the vectors representing the mutually different feasible states of marking all of the PN places in different steps of the system dynamics development) where $X_1 = \mathbf{x}_0$ and the states X_2, ..., X_N are reachable from X_1 by means of firing certain sequences of enabled transitions, $U = \{U_1, U_2, ..., U_M\}$ is the set of control vectors (i.e., the vectors representing the states all of the PN transitions in different steps, as to enabling or firing), and finally, $\delta : X \times U \rightarrow X$ is the transition function of the PN. The transition function formally expresses that the new state of the system depends on the previous state and the firing of transitions representing the occurrence of discrete events. This function can be formed as the linear discrete equation

$$\mathbf{x}_{k+1} = \mathbf{x}_k + \mathbf{B}.\mathbf{u}_k,\ k = 0, 1, ... \qquad (1)$$

Here, $\mathbf{B} = \mathbf{G}^T - \mathbf{F}$ and $\mathbf{F}.\mathbf{u}_k \leq \mathbf{x}_k$, $k = 0, 1, \ldots$ with $\mathbf{x}_k = (\sigma^k_{p_1}, \sigma^k_{p_2}, \ldots, \sigma^k_{p_n})^T$ being the state vector of the PN places in the step k where $\sigma^k_{p_i} \in \{0,1,\ldots,c_{p_i}\}$ is the state of the elementary place p_i, $i = 1, 2, \ldots, n$, and c_{p_i} is the capacity (as to the number of tokens). The capacity of the place p_i can be finite or infinite. The state $\sigma^k_{p_i} = 0$ expresses the passivity of the place p_i in the step k and $\sigma^k_{p_i} \geq 1$ expresses a measure of the activity of the place p_i in the step k. The vector $\mathbf{u}_k = (\gamma^k_{t_1}, \gamma^k_{t_2}, \ldots, \gamma^k_{t_m})^T$ is the state vector of the PN transitions in the step k with $\gamma^k_{t_j} \in \{0,1\}$ being the state of the elementary transition t_j, $j = 1, 2, \ldots, m$, where $\gamma^k_{t_j} = 0$ expresses that the transition t_j is disabled in the step k and $\gamma^k_{t_j} = 1$ expresses that the transition t_j is enabled in the step k. Because the transitions represent the discrete events able to change PN marking (the states of the PN places) $\mathbf{u}_k$ can be named as the control vector. $\mathbf{F}$ is the $(n \times m)$-dimensional incidence matrix corresponding to the set F and $\mathbf{G}$ is the $(m \times n)$ -dimensional incidence matrix corresponding to the set G. The elements of the matrices **F**, **G** are, respectively, integers expressing existence as well as weights of the arcs oriented from places to transitions and those oriented from transitions to places. Hence, the element $f_{ij} \in \{0, 1, \ldots, M_{f_{ij}}\}$, $i = 1, 2, \ldots, n$; $j = 1, 2, \ldots, m$ of the matrix $\mathbf{F}$ expresses the weight (i.e., multiplicity) of the arc oriented from p_i to t_j while the element $g_{ij} \in \{0, 1, \ldots, M_{g_{ij}}\}$, $i = 1, 2, \ldots, m$; $j = 1, 2, \ldots, n$ of the matrix $\mathbf{G}$ expresses the weight (i.e., multiplicity) of the arc oriented from t_i to p_j. The superscript $(.)^T$ symbolizes transpose of matrices and vectors, while k represent the discrete step of the system dynamics development.

Reachability of States

Reachability is very important term in PN theory as well as in the **DEDS control synthesis**. It determines whether a system can reach a specific state or exhibit a particular functional behaviour. The PN **reachability tree (RT)** or the **reachability graph (RG)** are very important as to the reachability property. Exact definitions of the RT are introduced in PN theory basic sources, for example, by Petersen (1981) and Murata (1989). To have an idea about the RT it is sufficient to introduce here only a short description of it. The PN reachability tree $G_{rt} = (V_{rt}, E_{rt})$ is the tree where the set of nodes $V_{rt} = \{v_0, v_1, \ldots, v_{N_r}\}$ is represented by the set of PN states, that is, v_i, $i = 0, 1, \ldots, N_r$ represent the state vectors $\mathbf{x}_i$, $i = 0, 1, \ldots, N_r$. Thus, $V_{rt} = \{\mathbf{x}_0, \mathbf{x}_1, \ldots, \mathbf{x}_{N_r}\}$ with the initial state $\mathbf{x}_0$ being the root of the RT. The set of edges $E_{rt} = \{e_1, e_2, \ldots, e_M\}$ consists of edges marked by the PN transitions $t_j \in T$; $j = 1, \ldots, m$. Namely, two nodes $v_i, v_j \in V$ are connected by the oriented arc $e = e_{v_i \to v_j} \in E$ directed from v_i toward v_j. The arc is marked by the PN transition $t = t_{v_i \to v_j} = t_{\mathbf{x}_i \to \mathbf{x}_j} \in T$ just when it is enabled in the state $\mathbf{x}_i$ and the new state $\mathbf{x}_j$ will be reached by means of its firing. The RT has to involve a corresponding node for every PN state and a corresponding edge for any PN transition enabled in the given state. To avoid complications (especially the infiniteness of the generated RT) the so-called duplicity nodes are defined as the leaf nodes (the graph leaves). In such a way subtrees, which already were included into the RT, are eliminated. Namely, the node $v_i \in V_{rt}$ for which exists a node $v_j \in V_{rt}$ such that $v_j \prec v_i$ is named as the duplicity node. The operator $\prec$ represents here the binary antireflexive transitive antisymmetric relation expressing the ordering of nodes in the set V_{rt}. Consequently, from the duplicity node no edges emerge. Connecting all of the duplicity nodes of the node $v_j \in V_{rt}$ together and also with the node v_j itself we obtain (after doing this for $j = 1, \ldots, N_r$) the RG from the RT. It is important that both the RG and the RT have the same adjacency matrix.

Another descriptive definition of the RG is presented by Heljanko (2006). What is distinctive on it is that it starts by defining two functions: (i) *enabled(v)*: given a state v, this function returns the set of transitions t that are enabled in v; and

(ii) *fire(v, t)*: given a state v, and a transition $t \in enabled(v)$, this function returns the state v_0 reached from v by firing t and consequently, the RG is defined as $G_{rt} = (V_{rt}, T, E_{rt}, v_0)$ being the graph with the smallest sets of nodes V_{rt}, transitions T, and edges E_{rt} such that (i) $v_0 \in V_{rt}$, where v_0 is the initial state of the system, and (ii) if $v \in V_{rt}$, then for all $t \in enabled(v)$ it holds that $t \in T, fire(v, t) \in V_{rt}$, and $(v, t, fire(v, t)) \in E_{rt}$. In PN theory, the RT or RG are very important, especially for testing the PN properties. In this chapter, they will be utilized at the DEDS control synthesis.

The Matlab procedure for computing the G_{rt} was developed by Čapkovič (2003). Its entries are the PN incidence matrices **F**, **G** and the initial state $\mathbf{x}_0$. The procedure computes the adjacency matrix $\mathbf{A}_{rt}$ of G_{rt} and the set of the reachable states given as columns of the matrix $\mathbf{X}_{reach}$. The matrix $\mathbf{A}_{rt}$ has a quasi-functional form $\mathbf{A}_{rt}(k)$, because the nonzero elements of this matrix represent the indices of the transitions. The matrices $\mathbf{A}_{rt}$ and $\mathbf{X}_{reach}$ fully characterize the RT or RG of the PN. The functional adjacency matrix $\mathbf{A}_k$ corresponding to the quasi-functional matrix $\mathbf{A}_{rt}(k)$ can be constructed when the nonzero integer elements of $\mathbf{A}_{rt}$ representing the indices of PN transitions $t_j \in T; j = 1, \ldots, m$ are replaced by the corresponding transition functions $g^k_{t_j}, j = 1, \ldots, m; k = 0, 1, \ldots, K$. The matrix $\mathbf{A}_k$ represents the PN causality. It can be said that when DEDS is modeled by PN, the strictness of the DEDS causality (Čapkovič, 2007) is rigorously adhered by the PN RT or RG. Therefore, the RG cannot be avoided at the DEDS control synthesis, of course.

The PN RG can be utilized at modeling, analysing and control of interactions among agents. However, it is necessary to keep in mind that the number of feasible states $N = N_r + 1$ because also the initial state is one of them. The attention will be focused especially on the negotiation process.

PN-Based Approach in the Light of other Approaches

PN are widely used for modeling DES or DEDS in general (in flexible manufacturing systems, communication systems of different kinds, transport systems, etc.). In spite of the fact that the main aim of this chapter is to point out their applicability to modeling, analysing and control of **agents and agent systems** understood to be a kind of DES or DEDS, it is necessary to make a mention of using PN in the context of a specific application. Consider workflow modeling to be such an application, because it is frequently used in different areas. The workflow is (Eshuis & Wieringa, 2003) a set of activities (production, business, etc.) that are ordered according to a set of procedural rules to deliver a service. It is the computerized facilitation or automation of a process. The workflow is defined by a model. In general, there are two important dimensions of workflows (Leymann & Roller, 2000; van der Aalst, 2000), the control-flow dimension and the resource one.

Some authors have argued that PN are suitable for workflow modeling, for example, van der Aalst (1998) and Salimifard and Wright (2001). They use arguments, (i) that PN yields both the graphical model and the mathematical one; (ii) that PN have a formal semantics; (iii) that they are able to express most of the desirable routing constructs; (iv) that there are many techniques for proving their basic properties (Murata, 1989; Petersen, 1981) like reachability, liveness, boundedness, conservativeness, reversibility, coverability, persistence, fairness, and so forth.; and (v) that PN represent the enough general means to be able to model a wide class of systems. These arguments are the same as those used in **DES or DEDS modeling** in general in order to prefer PN to other approaches. In addition, there were many methods developed in PN-theory that are very useful at model checking, for example, the methods of the deadlocks avoidance, methods for computing P (place)-invariants and T (transition)-invariants, and so forth. Moreover, PN-based models dispose of the possibility to express not only the event causality, but also of the possibility to express analytically the current states expressing the system dynamics development. Even linear algebra and matrix calculus can be utilised in this way. It is very important, especially at the **DES or DEDS control synthesis**. The fact that most PN properties can be tested by means of methods based on the reachability tree and invariants is indispensable too. Thus, the reachability tree and invariants are very important in **PN-based modeling DES**.

On the other hand, there are some authors (Eshuis & Wieringa, 2003; Eshuis & Dehnert, 2003) that have argued that most of the introduced arguments that seem to be in general an advantage of PN did not refer to the domain of workflow modeling (only the routing argument does). Namely, after van der Aalst, ter Hofstede, Kiepuszewski, and Barros (2000) identified workflow patterns (see also van der Aalst, ter Hofstede, Kiepuszewski, & Barros, 2003), it has been shown that they can be modeled using activity diagrams (White, 2004). There have been efforts for defining semantics for activity diagram, so that execution of the workflow models can be done (Eshuis & Wieringa, 2001a, 2001b, 2002). The PN critics have affirmed that PN semantics is not specifically intended for workflow modeling, because the analysis of the PN-based workflow model presupposes that the real workflow is modelled by PN faithfully. They have argued that from an unrealistic model, no reliable analysis results can be inferred, that the PN semantics models closed, active systems that are nonreactive, whereas the semantics of UML (unified modeling language) activity diagrams models open, reactive systems. Namely, the activity diagrams describing the workflow behaviour of a system are also useful (Fowler & Scott, 2000) for analysing a use case by describing what actions need to take place and when they should occur, for describing a complicated sequential algorithm, and for modeling applications with parallel processes. However, activity diagrams do not give details (Fowler & Scott, 2000) about how objects behave or how objects collaborate.

Because workflow systems are open, reactive systems, the PN critics conclude that PN are not entirely suitable for workflow modeling. Therefore, they have defined a reactive PN semantics (Eshuis & Dehnert, 2003). Namely, a reactive system runs in parallel with its environment (Harel & Pnueli, 1985) and responds or reacts to input events by creating certain desirable effects in the environment. The reactive PN semantics can model behaviour of a reactive system and its environment. Eshuis and Dehnert (2003) compared this semantics with the token-game semantics of nonreactive PN and proved that under some conditions the reactive semantics and the token-game semantics induce similar behaviour. Next, they applied the reactive semantics to workflow modeling and showed how a workflow net can be transformed into a reactive workflow net. They proved that under some conditions the soundness property of a workflow net is preserved when the workflow net is transformed into a reactive workflow net. This result shows that to analyse soundness, the token-game semantics can safely be used, even though that semantics is not reactive.

However, some reasons for preferring PN modeling in connection with workflow modeling to other notations are presented also in newer papers (see e.g., Purvis, Purvis, Haidar, & Savarimuthu, 2005). The authors emphasize the following facts: (1) PN have formal semantics. It makes the execution and simulation of PN models unambiguous. (2) It is shown that PN can be used to model workflow primitives. (3) Typical process modeling notations, such as dataflow diagrams, are event-based, but PN can model both states and events. (4) There are many analysis techniques associated with Petri nets, which make it possible to identify "dangling" tasks, deadlocks, and safety issues. (5) Other standardization protocols do not cater to expressiveness, simplicity and formal semantics.

There are even authors (Meena, Saha, Mondal, & Prabhakar, 2005) that transform activity diagrams to PN. They have proposed a new way of looking at analysis of workflows. They model workflows by activity diagrams (they understand the activity diagrams to be the language, which is easy and more intuitive to work with). However, afterward they analyse the model using PN. They have mentioned three properties of PN, which are useful in commenting on workflow models: boundedness, safeness, and deadlock. Here, the verbal definitions of these properties are (Murata, 1989; Peterson, 1981; Murata) as follows. A given PN with initial marking M0 (given by the initial state vector $X_1 = \mathbf{x}_0$ of the PN-based model) is said to be bounded, if for any reachable marking M (i.e., any reachable state vector $X_i \in X$) the number of tokens in each place does not exceed a finite value. A given PN with initial marking M0 is safe, if it is bounded and maximum allowable token in each place is one. A given PN net with initial marking M0 is said to be live, if from any reachable marking it is possible to fire any transition after some firing sequence. A transition is said to be dead if it can never be fired. If, in a firing sequence, we reach a point where a particular transition cannot be fired, then the net is in a potential deadlock.

The absence of boundedness indicates (Meena et al., 2005) that a particular place has infinite number of tokens. However, this indicates that the end place can never be reached without having left some tokens in other places. In workflow domain, this implies that an activity can never be ended without leaving some reference to it. Safeness property in workflow domain will ensure that we don't have more than one reference to an object to be processed. This makes sense because there is no need of processing two of the same objects when one is needed. Deadlock property is very useful from workflow point of view. Namely, it indicates that the corresponding workflow has some activity which cannot be reached; hence the design has some flaws. More such properties can be looked for (Meena et al., 2005) and at the same time, more constructs from activity diagrams can be added.

Symbolically said, nothing is either white or black. Everything has some advantages on one hand and some disadvantages on other hand. Whenever we have to model a system, we have a possibility to decide what approach is the most suitable for our needs. The author of this chapter attaches oneself to (or endorses) the PN-based approach. The main reason of this is the fact that PN allows the expression of the state of the modeled object at any step of the system dynamic development. Especially, the P/T PN approach enables using the linear algebra and matrix calculus, that is, exact and in practice verified approaches. This makes the analysis of the systems in analytical terms possible, especially computing reachability graph, invariants, testing properties, model checking and **model-based control synthesis**. Moreover, PN can be used not only for handling software agents, but also for "material" agents, like robots and other technical devices. PN are suitable also at modeling, analysing and control of any **modular DES or DEDS** and they are able to deal with any problem in that way. Consequently, in this chapter, devoted to modeling, analysing and control of agents or agent systems, the P/T PN-based approach is used only and no attention is paid to other approaches (activity diagrams, reactive PN, etc.). **Mutual interactions of agents** are considered within the framework of the global model. Such an approach is sufficiently general in order to allow us to create the model that yields the possibility to analyse any situation. Even the environment behaviour can be modeled as an agent of the agent system too. In such a case, it can acquire arbitrary structure and consequently, generate different situations.

The big advantage of this approach is that the model is given in analytical terms, as a discrete system described by the vector linear difference equation (1). The same holds also for modules expressing the agents. Consequently, the model can be utilized at mathematical dealing with problems of different kinds (testing properties, reachability analysis, computing invariants, control synthesis, etc.). Of course, PN-based models cannot change their structure dynamically, during the operation, that is, online. However, recently the research was initiated (Llorens & Olivier, 2004, 2006) in the area of reconfigurable PN.

PN-Based Modeling Agents

Consider the agent A having (on a chosen level of abstraction) the following elementary activities represented by the PN places p_i, $i=1, 2,..., 12$, that is, $p_i \in P = \{p_1, p_2,...,p_{12}\}$, where the meaning of the places is the following: p_1 means the agent A is free; p_2 means a problem P_A has to be solved by A; p_3 means A is able to solve P_A ; p_4 means A is not able to solve P_A ; p_5 means P_A is solved; p_6 means P_A cannot be solved by A and another agent(s) should be contacted; p_7 means A asks another agent(s) to help him to solve P_A ; p_8 means A is asked by another agent(s) to solve a problem P_B; p_9 means A refuses the help; p_{10} means A accepts the request of another agent(s) for help; p_{11} means A is not able to solve P_B; and p_{12} means A is able to solve P_B. However, it is necessary to say that another concept of both the agent structure and the activities is not excluded. The **PN model** of A is given in Figure 1. The PN transitions t_j, $j=1, 2,..., 7$, that is, $t_j \in T = \{t_1, t_2,..., t_7\}$ represent the discrete events expressing the starting or ending of the activities.

The RG corresponding to the initial state $\mathbf{x}_0 = (1,1,1,0,0,0,0,0,0,0,0,0)^T$ is in Figure 2a. It means that A is able to solve P_A. The RG corresponding to $\mathbf{x}_0 = (1,1,0,0,1,0,0,0,0,0,0,0)^T$ is in Figure 2b. It means that A is not able to solve P_A. Finally, the RG corresponding to $\mathbf{x}_0 = (1,0,0,0,0,0,0,1,0,0,0,0)^T$ is in Figure 2c. It means that A is asked by another agent(s) for help. Because the initial state $X_1 = \mathbf{x}_0$ is different for each of these tree cases, it is also clear that the states reachable from X_1 will be different. In case a) $X_2 = (0,0,0,0,1,0,0,0,0,0,0,0)^T$ that is, when the agent is able to solve the problem

Figure 1. Two identical versions of the graphical PN-based model of the agent A; in the version on the right the input/output possibilities of the interconnection with another agent(s) are emphasized

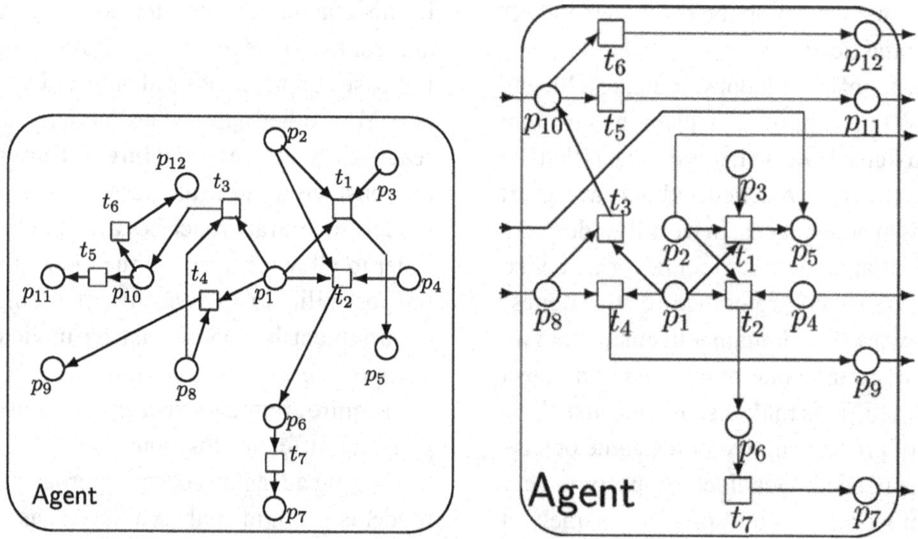

(the state X_1) the problem is solved by it (the state X_2). In case b) $X_2 = (0,0,0,0,0,1,0,0,0,0,0,0)^T$, $X_3 = (0,0,0,0,0,0,1,0,0,0,0,0)^T$ that is, when the agent is not able to solve the problem (the state X_1) the problem cannot be solved by the agent (the state X_2) and the agent has to ask another agent(s) for help (the state X_3). Finally, in case c) $X_2 = (0,0,0,0,0,0,0,0,1,0,0)^T$, $X_3 = (0,0,0,0,0,0,0,0,0,1,0,0,0)^T$, $X_4 = (0,0,0,0,0,0,0,0,0,0,0,1,0)^T$, $X_5 = (0,0,0,0,0,0,0,0,0,0,0,1)^T$ that is, when the free agent is asked for help by another agent (the state X_1) it can either refuse the request (the state X_3) or accept the request (the state X_2) however, in case of accepting the request it may be either able to solve the problem of the applicant (the state X_5) or unable to solve the problem (the state X_4). All of the states can be computed by means of the equation (1) starting from the corresponding initial state and the system parameters given by the matrices **F**, **G** given as follows

$$\mathbf{F} = \begin{pmatrix} 1&1&1&1&0&0&0 \\ 1&1&0&0&0&0&0 \\ 1&0&0&0&0&0&0 \\ 0&1&0&0&0&0&0 \\ 0&0&0&0&0&0&0 \\ 0&0&0&0&0&0&1 \\ 0&0&0&0&0&0&0 \\ 0&0&1&1&0&0&0 \\ 0&0&0&0&0&0&0 \\ 0&0&0&0&1&1&0 \\ 0&0&0&0&0&0&0 \\ 0&0&0&0&0&0&0 \end{pmatrix};$$

$$\mathbf{G}^T = \begin{pmatrix} 0&0&0&0&0&0&0 \\ 0&0&0&0&0&0&0 \\ 0&0&0&0&0&0&0 \\ 0&0&0&0&0&0&0 \\ 1&0&0&0&0&0&0 \\ 0&1&0&0&0&0&0 \\ 0&0&0&0&0&0&1 \\ 0&0&0&0&0&0&0 \\ 0&0&0&1&0&0&0 \\ 0&0&1&0&0&0&0 \\ 0&0&0&0&1&0&0 \\ 0&0&0&0&0&1&0 \end{pmatrix}. \qquad (2)$$

Moreover, by means of a simple Matlab program (Čapkovič, 2003, 2005) the quasi-functional adjacency matrix $\mathbf{A}_{rt}$ of the RT (its nonzero elements represent the indices of the PN transitions) as well as the space of reachable states represented by the matrix $\mathbf{X}_{reach}$ can be computed. For example, in case c) the matrices are the following

$$\mathbf{A}_{rt} = \begin{pmatrix} 0&3&4&0&0 \\ 0&0&0&5&6 \\ 0&0&0&0&0 \\ 0&0&0&0&0 \\ 0&0&0&0&0 \end{pmatrix};$$

$$\mathbf{X}_{reach} = \begin{pmatrix} 1&0&0&0&0 \\ 0&0&0&0&0 \\ 0&0&0&0&0 \\ 0&0&0&0&0 \\ 0&0&0&0&0 \\ 0&0&0&0&0 \\ 0&0&0&0&0 \\ 1&0&0&0&0 \\ 0&0&1&0&0 \\ 0&1&0&0&0 \\ 0&0&0&1&0 \\ 0&0&0&0&1 \end{pmatrix} \qquad (3)$$

It is necessary to say that **PN-based modeling tools** (graphical as well as mathematical) allow us to express the agent having practically an arbitrary internal structure and arbitrary numbers of places and transitions.

MODELING THE COOPERATION OF AGENTS IN MAS

As it was mentioned above, there exist several kinds of **cooperation among agents in MAS**. PN are used for **e-negotiations activities,** for example, by Hung and Mao (2002). The following five principle properties of e-negotiation are defined there: 1. interactivity (it involves the agents to participate and communicate with each

Figure 2. The RG of the PN-based model of the agent in different situations given by different initial states: a) $X_1 = \mathbf{x}_0 = (1,1,1,0,0,0,0,0,0,0,0,0)^T$, b) $X_1 = \mathbf{x}_0 = (1,1,0,0,1,0,0,0,0,0,0,0)^T$, c) $X_1 = \mathbf{x}_0 = (1,0,0,0,0,0,0,1,0,0,0,0)^T$

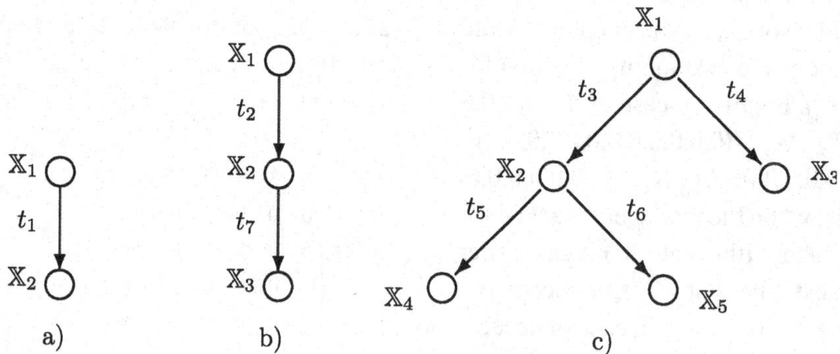

other); 2. informativeness (it generates, transmits and stores information); 3. irregularity (it behaves differently according to the combination of agents, strategies, events, tasks, issues, alternatives, preferences and criteria); 4. integrity (it affords speed, consistency and absence of errors through efficient and effective mechanisms); and 5. inexpensiveness (it automates or semi-automates negotiation activities to save time and cost).

The **negotiation process** itself consists of several principle activities (Hung & Mao, 2002). The following ones are especially the most important: defining the negotiation environment, initial contact of agents, offer(s) and counter offer(s) among them, evaluation of proposals, and outcomes of the negotiation process. In general, the coordination plan of the negotiation process can be formally described by DEDS modeled by PN. The PN places represent the activities and the PN transitions represent the discrete events. In addition, PN can effectively help to express the properties to be satisfied, especially the first 3 that are generic.

PN were chosen to model MAS in the whole, for example, by Nowostawski, Purvis, and Cranefield (2001), Purvis,Cranefield, Nowostawski, Ward, Carter, and Oliveira (2002) and many other authors. However, PN-based models are suitable also for the analysis of MAS. In order to analyse complicated interactions among agents in MAS, modeling of them can be used too. On the base of previous experience (Čapkovič, 2002, 2003; Čapkovič & Čapkovič, 2003) with PN-based modeling and control synthesis of the DEDS and the **agent cooperation** (Čapkovič, 2004a, 2004b, 2005, 2006) a new approach to modeling, analysis and control of the **cooperation process** is proposed here. The cooperative process of the agents in MAS (the negotiation process as well) can be understood to be DEDS. It seems to be natural, because such a process is discrete in nature and simultaneously it is causal. The approach consists of: (1) creating the **PN-based mathematical model** of the cooperation process; (2) generating the space of feasible states which are reachable from the given initial state; and (3) utilizing the RG in order to find the feasible state trajectories to a prescribed feasible terminal state. After a thorough analysis of the set of possibilities, the most suitable strategy (the control trajectory) can be chosen.

The PN-based Model of Two Agents Negotiation

The collaboration/negotiation of two agents A_1 and A_2 with the same structure is given in Figure 3.

Figure 3. The PN-based model of two agents' negotiation

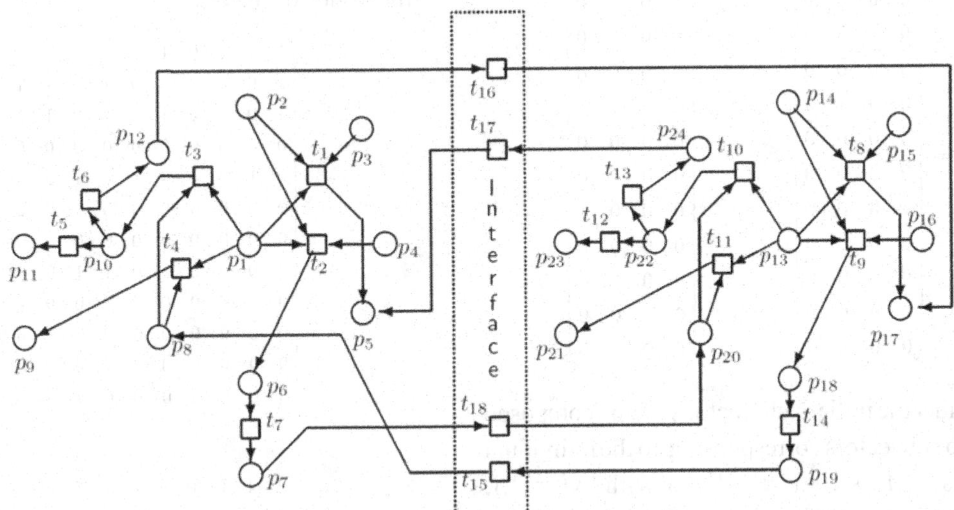

In case of more agents (e.g., N_A) the place numbering of the agent $k = 1, ..., N_A$ is p_{i+12j}, $i = 1, ..., 12, j = k-1 = 0, ..., N_A-1$. In case of several agents, both the PN model and the RG will be more intricate. However, the **model structure** can be expressed in the block form as follows:

$$\mathbf{F} = \begin{pmatrix} \mathbf{F}_1 & \mathbf{0} & \cdots & \mathbf{0} & \mathbf{0} & | & \mathbf{F}_{c_1} \\ \mathbf{0} & \mathbf{F}_2 & \cdots & \mathbf{0} & \mathbf{0} & | & \mathbf{F}_{c_2} \\ \vdots & \vdots & \ddots & \vdots & \vdots & | & \vdots \\ \mathbf{0} & \mathbf{0} & \cdots & \mathbf{F}_{N_A-1} & \mathbf{0} & | & \mathbf{F}_{c_{N_A-1}} \\ \mathbf{0} & \mathbf{0} & \cdots & \mathbf{0} & \mathbf{F}_{N_A} & | & \mathbf{F}_{c_{N_A}} \end{pmatrix};$$

$$\mathbf{G} = \begin{pmatrix} \mathbf{G}_1 & \mathbf{0} & \cdots & \mathbf{0} & \mathbf{0} \\ \mathbf{0} & \mathbf{G}_2 & \cdots & \mathbf{0} & \mathbf{0} \\ \vdots & \vdots & \ddots & \vdots & \vdots \\ \mathbf{0} & \mathbf{0} & \cdots & \mathbf{G}_{N_A-1} & \mathbf{0} \\ \mathbf{0} & \mathbf{0} & \cdots & \mathbf{0} & \mathbf{G}_{N_A} \\ --- & --- & --- & --- & --- \\ \mathbf{G}_{c_1} & \mathbf{G}_{c_2} & \cdots & \mathbf{G}_{c_{N_A-1}} & \mathbf{G}_{c_{N_A}} \end{pmatrix}. \quad (4)$$

Thus, in case of two agents A_1, A_2 ($N_A = 2$) with general structure ($A_1 \neq A_2$) we have:

$$\mathbf{F} = \begin{pmatrix} \mathbf{F}_1 & \mathbf{0} & \mathbf{F}_{c_1} \\ \mathbf{0} & \mathbf{F}_2 & \mathbf{F}_{c_2} \end{pmatrix}; \ \mathbf{G} = \begin{pmatrix} \mathbf{G}_1 & \mathbf{0} \\ \mathbf{0} & \mathbf{G}_2 \\ \mathbf{G}_{c_1} & \mathbf{G}_{c_2} \end{pmatrix}.$$

However, when the agents have the same structure like the agent A ($A_1 = A_2 = A$) with the parameters (2), the structural matrices of the agents cooperation are as follows:

$$\mathbf{F}_{AA} = \begin{pmatrix} \mathbf{F} & \mathbf{0} & \mathbf{F}_{c_1} \\ \mathbf{0} & \mathbf{F} & \mathbf{F}_{c_2} \end{pmatrix}; \mathbf{G}_{AA} = \begin{pmatrix} \mathbf{G} & \mathbf{0} \\ \mathbf{0} & \mathbf{G} \\ \mathbf{G}_{c_1} & \mathbf{G}_{c_2} \end{pmatrix} \quad (5)$$

$$\mathbf{F}_{c_1} = \begin{pmatrix} 0&0&0&0 \\ 0&0&0&0 \\ 0&0&0&0 \\ 0&0&0&0 \\ 0&0&0&0 \\ 0&0&0&0 \\ 0&0&0&1 \\ 0&0&0&0 \\ 0&0&0&0 \\ 0&0&0&0 \\ 0&0&0&0 \\ 0&1&0&0 \end{pmatrix} \quad \mathbf{F}_{c_2} = \begin{pmatrix} 0&0&0&0 \\ 0&0&0&0 \\ 0&0&0&0 \\ 0&0&0&0 \\ 0&0&0&0 \\ 0&0&0&0 \\ 1&0&0&0 \\ 0&0&0&0 \\ 0&0&0&0 \\ 0&0&0&0 \\ 0&0&0&0 \\ 0&0&1&0 \end{pmatrix}$$

$$\mathbf{G}^T{}_{c_1} = \begin{pmatrix} 0&0&0&0\\ 0&0&0&0\\ 0&0&0&0\\ 0&0&0&0\\ 0&0&1&0\\ 0&0&0&0\\ 0&0&0&0\\ 1&0&0&0\\ 0&0&0&0\\ 0&0&0&0\\ 0&0&0&0\\ 0&0&0&0 \end{pmatrix} \quad \mathbf{G}^T{}_{c_2} = \begin{pmatrix} 0&0&0&0\\ 0&0&0&0\\ 0&0&0&0\\ 0&0&0&0\\ 1&0&0&0\\ 0&0&0&0\\ 0&0&0&0\\ 0&0&0&1\\ 0&0&0&0\\ 0&0&0&0\\ 0&0&0&0\\ 0&0&0&0 \end{pmatrix}.$$

Consider the initial state vector $\mathbf{x}_0$ being composed of two subvectors corresponding to the individual agents A_1, A_2 as $\mathbf{x}_0 = ({}^{A1}\mathbf{x}^T{}_0, {}^{A2}\mathbf{x}^T{}_0)^T$ with ${}^{A1}\mathbf{x}^T{}_0 = (1,1,1,0,0,0,0,0,0,0,0,0)^T$, ${}^{A2}\mathbf{x}^T{}_0 = (1,1,0,1,0,0,0,0,0,0,0,0)^T$, that is, the situation when A_1 is able to solve its own problem P_{A1}; however, A_2 is not able to solve its own problem P_{A2}. Therefore, A_2 has to ask A_1 for help. We can compute the adjacency matrix $\mathbf{A}_{RG}$ of the RG ($\mathbf{A}_{RG} = \mathbf{A}_{rt}$) and the matrix $\mathbf{X}_{reach}$ representing the space of reachable states. The RG of the negotiations is given in Figure 4.

$$\mathbf{A}_{RG} = \begin{pmatrix} 0&1&9&0&0&0&0&0&0&0&0&0&0\\ 0&0&0&9&0&0&0&0&0&0&0&0&0\\ 0&0&0&1&14&0&0&0&0&0&0&0&0\\ 0&0&0&0&0&14&0&0&0&0&0&0&0\\ 0&0&0&0&0&1&15&0&0&0&0&0&0\\ 0&0&0&0&0&0&0&15&0&0&0&0&0\\ 0&0&0&0&0&0&0&1&3&4&0&0&0\\ 0&0&0&0&0&0&0&0&0&0&0&0&0\\ 0&0&0&0&0&0&0&0&0&0&5&6&0\\ 0&0&0&0&0&0&0&0&0&0&0&0&0\\ 0&0&0&0&0&0&0&0&0&0&0&0&0\\ 0&0&0&0&0&0&0&0&0&0&0&0&16\\ 0&0&0&0&0&0&0&0&0&0&0&0&0 \end{pmatrix};$$

$$\mathbf{X}_{reach} = \begin{pmatrix} {}^1\mathbf{X}_{reach} \\ {}^2\mathbf{X}_{reach} \end{pmatrix}.$$

The blocks of the matrix $\mathbf{X}_{reach}$ are the following. The columns of ${}^1\mathbf{X}_{reach}$, ${}^2\mathbf{X}_{reach}$ contain, respectively, the states of the agents A_1, A_2 being the nodes of the RG.

$${}^1\mathbf{X}_{reach} = \begin{pmatrix} 1&0&1&0&1&0&1&0&0&0&0&0&0\\ 1&0&1&0&1&0&1&0&1&1&1&1&1\\ 1&0&1&0&1&0&1&0&1&1&1&1&1\\ 0&0&0&0&0&0&0&0&0&0&0&0&0\\ 0&1&0&1&0&1&0&1&0&0&0&0&0\\ 0&0&0&0&0&0&0&0&0&0&0&0&0\\ 0&0&0&0&0&0&0&0&0&0&0&0&0\\ 0&0&0&0&0&0&1&1&0&0&0&0&0\\ 0&0&0&0&0&0&0&0&0&1&0&0&0\\ 0&0&0&0&0&0&0&0&1&0&0&0&0\\ 0&0&0&0&0&0&0&0&0&0&1&0&0\\ 0&0&0&0&0&0&0&0&0&0&0&1&0 \end{pmatrix}$$

$${}^2\mathbf{X}_{reach} = \begin{pmatrix} 1&1&0&0&0&0&0&0&0&0&0&0&0\\ 1&1&0&0&0&0&0&0&0&0&0&0&0\\ 0&0&0&0&0&0&0&0&0&0&0&0&0\\ 1&1&0&0&0&0&0&0&0&0&0&0&0\\ 0&0&0&0&0&0&0&0&0&0&0&0&1\\ 0&0&1&1&0&0&0&0&0&0&0&0&0\\ 0&0&0&0&1&1&0&0&0&0&0&0&0\\ 0&0&0&0&0&0&0&0&0&0&0&0&0\\ 0&0&0&0&0&0&0&0&0&0&0&0&0\\ 0&0&0&0&0&0&0&0&0&0&0&0&0\\ 0&0&0&0&0&0&0&0&0&0&0&0&0\\ 0&0&0&0&0&0&0&0&0&0&0&0&0 \end{pmatrix}$$

In the case when $N_A > 2$ we can use **interconnections among the agents in MAS** by means of the Interface with different structure. A symbolic schema of a general **cooperation Interface** among three agents with the same structure (2) is given in Figure 5. The numbers at inputs and outputs of the blocks A_1, A_2, A_3, and representing the agents are the numbers of the places occurring in the agents' PN-based model. The details of the three **agents' negotiation** as well as the internal structure of the interface are introduced in Figure 6.

The cooperation structure need not be fossil (rigid), of course. Among the interface blocks also the mutually exclusion can occur. For example, in the case when an agent asks all of the agents for help, several of them can be ready to help it in the same time. However, only one of them has to be chosen because help from two or more agents

Figure 4. The RG of the negotiation process between A_1, A_2

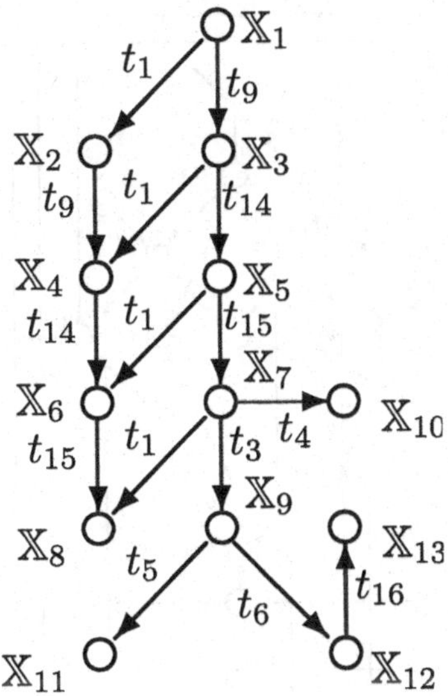

cannot be accepted simultaneously. In Figure 7, the three agents' negotiation is roughly outlined and the structure of the **PN-based model of the general MEX** (mutual exclusion) block is presented. On the left side of Figure 7, the case of the three agents' negotiation is displayed. The agent A_3 asked the agents A_1 and A_2 for help. The transitions t_{3-1} and t_{3-2} symbolize the discrete events expressing the start of the requests. Both of the agents offer the help. However, A_3 can accept the offer from one of them only, that is, from A_1 or from A_2. On the right side of the Figure 7, the PN-based model of the general structure of the MEX block is introduced.

Other Forms of the Agent Structure and the Agents' Cooperation

The approach to modeling the agents in MAS is not rigid. It is suitable for modeling and analysing the agents with different structure as well as for the agent cooperation in general. Namely, the PN subnets modeling the structure of agents can be built arbitrarily (according to demands of the model creator). The cooperating agents can have the mutually different structure. Even, any interface among the agents in MAS needs not to be modeled only by the additional PN transitions. On the contrary, in general the interface can be modeled by a PN subnet consisting of the additional

Figure 5. The three agents cooperation—a symbolic schema of a general cooperation of three agents with the same structure given by the relationship (2)

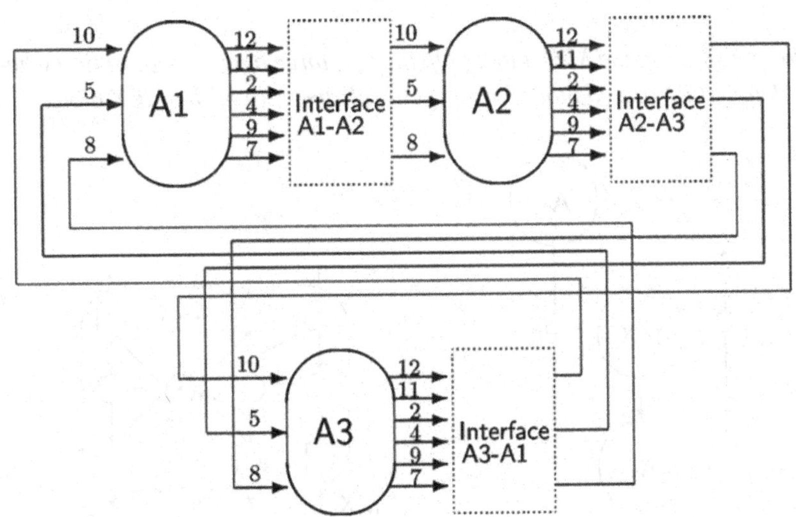

Figure 6. The three agents' cooperation—the detailed PN-based model of the three agents' negotiation

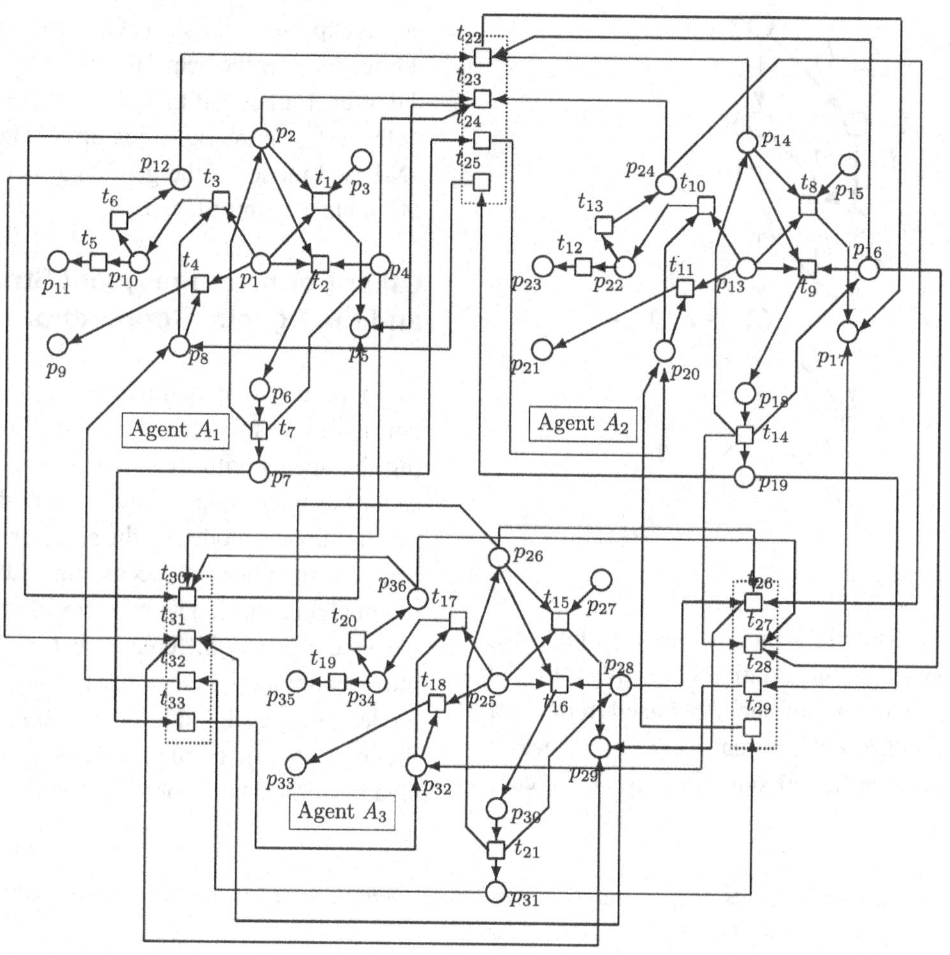

Figure 7. The example of using the MEX block: On the left, three agents' negotiation when MEX is used; on the right, the detailed PN-based model of the general structure of the MEX block

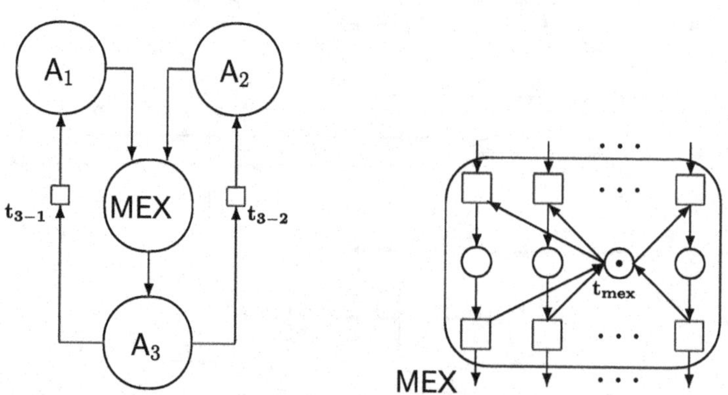

Figure 8. The PN-based model of the two agents' cooperation in general

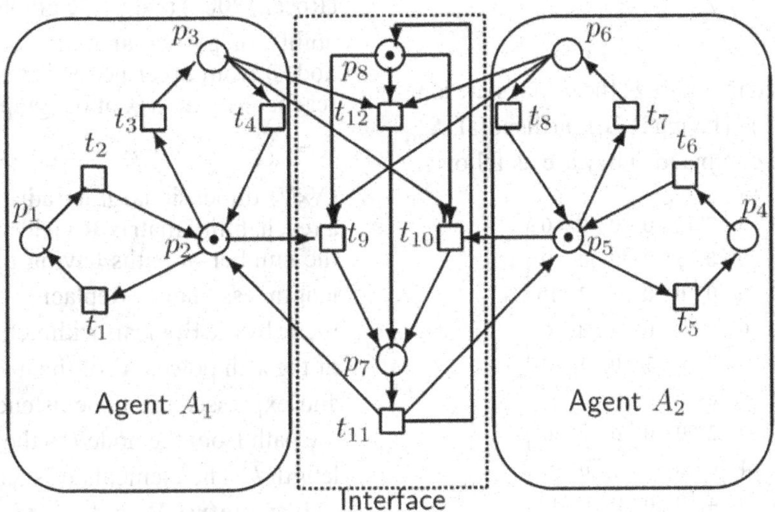

PN transitions and the additional PN places. The structure of such an interface can be created without restrictions at pleasure of the model creator. To illustrate this, consider the simple structure of the agent defined by Saint-Voirin, Lang, and Zerhouni (2003). The interpretation of the places in the PN model presented in Figure 8 is the following: p_1 – A_1 does not want to communicate; p_2 – A_1 is available; p_3 – A_1 wants to communicate; p_4 – A_2 does not want to communicate; p_5 – A_2 is available; p_6 – A_2 wants to communicate; p_7 - communication; and p_8 - availability of the communication channel(s) Ch (representing the interface). The PN transition t_9 fires the communication when A_1 is available and A_2 wants to communicate with A_1, t_{10} fires the communication when A_2 is available and A_1 wants to communicate with A_2, and t_{12} fires the communication when both A_1 and A_2 wants to communicate each other. Here, it is clear that the interface—the communication channel—has a form of the PN module (PN subnet) consisting of both the places and transitions.

To use the above described approach to PN-based modeling, it is sufficient to take and apply the following model parameters and to choose an initial state. After doing this, there are no restrictions as with using the approach.

$$\mathbf{F} = \left(\begin{array}{cccc|cccc|cccc} 0 & 1 & 0 & 0 & 0 & 0 & 0 & 0 & 0 & 0 & 0 & 0 \\ 1 & 0 & 1 & 0 & 0 & 0 & 0 & 0 & 1 & 0 & 0 & 0 \\ 0 & 0 & 0 & 1 & 0 & 0 & 0 & 0 & 0 & 1 & 0 & 1 \\ \hline 0 & 0 & 0 & 0 & 0 & 1 & 0 & 0 & 0 & 0 & 0 & 0 \\ 0 & 0 & 0 & 0 & 1 & 0 & 1 & 0 & 0 & 1 & 0 & 0 \\ 0 & 0 & 0 & 0 & 0 & 0 & 0 & 1 & 1 & 0 & 0 & 1 \\ \hline 0 & 0 & 0 & 0 & 0 & 0 & 0 & 0 & 0 & 0 & 1 & 0 \\ 0 & 0 & 0 & 0 & 0 & 0 & 0 & 0 & 1 & 1 & 0 & 1 \end{array}\right)$$

$$= \begin{pmatrix} \mathbf{F}_{A_1} & \mathbf{0}_{3x4} & \mathbf{F}_{A_1-A_2} \\ \mathbf{0}_{3x4} & \mathbf{F}_{A_2} & \mathbf{F}_{A_2-A_1} \\ \mathbf{0}_{2x4} & \mathbf{0}_{2x4} & \mathbf{F}_{Ch} \end{pmatrix}$$

$$\mathbf{G}^T = \left(\begin{array}{cccc|cccc|cccc} 1 & 0 & 0 & 0 & 0 & 0 & 0 & 0 & 0 & 0 & 0 & 0 \\ 0 & 1 & 0 & 1 & 0 & 0 & 0 & 0 & 0 & 0 & 1 & 0 \\ 0 & 0 & 1 & 0 & 0 & 0 & 0 & 0 & 0 & 0 & 0 & 0 \\ \hline 0 & 0 & 0 & 0 & 1 & 0 & 0 & 0 & 0 & 0 & 0 & 0 \\ 0 & 0 & 0 & 0 & 0 & 1 & 0 & 1 & 0 & 0 & 1 & 0 \\ 0 & 0 & 0 & 0 & 0 & 0 & 1 & 0 & 0 & 0 & 0 & 0 \\ \hline 0 & 0 & 0 & 0 & 0 & 0 & 0 & 0 & 1 & 1 & 0 & 1 \\ 0 & 0 & 0 & 0 & 0 & 0 & 0 & 0 & 0 & 0 & 1 & 0 \end{array}\right)$$

$$= \begin{pmatrix} \mathbf{G}^T{}_{A_1} & \mathbf{0}_{3x4} & \mathbf{G}^T{}_{A_1-A_2} \\ \mathbf{0}_{3x4} & \mathbf{G}^T{}_{A_2} & \mathbf{G}^T{}_{A_2-A_1} \\ \mathbf{0}_{2x4} & \mathbf{0}_{2x4} & \mathbf{G}^T{}_{Ch} \end{pmatrix}$$

As it is displayed in Figure 8, the initial state is $\mathbf{x}_0 = (0,1,0,0,1,0,0,1)^T$. Having $\mathbf{F}, \mathbf{G}, \mathbf{x}_0$ at disposal, $\mathbf{A}_{RG}$ an $\mathbf{X}_{reach}$ can be computed. They are as follows.

$$\mathbf{A}_{RG} = \begin{pmatrix} 0 & 1 & 3 & 5 & 7 & 0 & 0 & 0 & 0 & 0 \\ 2 & 0 & 0 & 0 & 0 & 5 & 7 & 0 & 0 & 0 \\ 4 & 0 & 0 & 0 & 0 & 0 & 0 & 5 & 7 & 10 \\ 6 & 0 & 0 & 0 & 0 & 1 & 0 & 3 & 0 & 0 \\ 8 & 0 & 0 & 0 & 0 & 0 & 1 & 0 & 3 & 9 \\ 0 & 6 & 0 & 2 & 0 & 0 & 0 & 0 & 0 & 0 \\ 0 & 8 & 0 & 0 & 2 & 0 & 0 & 0 & 0 & 0 \\ 0 & 0 & 6 & 4 & 0 & 0 & 0 & 0 & 0 & 0 \\ 0 & 0 & 8 & 0 & 4 & 0 & 0 & 0 & 0 & 12 \\ 11 & 0 & 0 & 0 & 0 & 0 & 0 & 0 & 0 & 0 \end{pmatrix};$$

$$\mathbf{X}_{reach} = \begin{pmatrix} 0 & 1 & 0 & 0 & 0 & 1 & 1 & 0 & 0 & 0 \\ 1 & 0 & 0 & 1 & 1 & 0 & 0 & 0 & 0 & 0 \\ 0 & 0 & 1 & 0 & 0 & 0 & 0 & 1 & 1 & 0 \\ 0 & 0 & 0 & 1 & 0 & 1 & 0 & 1 & 0 & 0 \\ 1 & 1 & 1 & 0 & 0 & 0 & 0 & 0 & 0 & 0 \\ 0 & 0 & 0 & 0 & 1 & 0 & 1 & 0 & 1 & 0 \\ 0 & 0 & 0 & 0 & 0 & 0 & 0 & 0 & 0 & 1 \\ 1 & 1 & 1 & 1 & 1 & 1 & 1 & 1 & 1 & 0 \end{pmatrix}$$

THE ANALYSIS AND CONTROL OF THE AGENTS' BEHAVIOUR

The process of modeling yields the mathematical or graphical description of different kinds of the agent activities. Subsequently, the model can be utilized for analysing the properties as well as for the control synthesis. Basic properties of PN, that is, findings whether PN are live, safety, bounded, reversible, and so forth, are performed especially (Murata, 1989; Petersen, 1981) by means of the PN RG and PN invariants (P-invariants and T-invariants). Because here, in this chapter, the RG will be utilized in order to analyse the control possibilities, we are interested only in **reachability** and its consequences. In discrete mathematics (Rosen, 2002) and in the graph theory the reachability of particular nodes of a graph (with N nodes) from other nodes is tested by means the reachability matrix of the graph having the form $\mathbf{R} = \sum_{i=1}^{N} \mathbf{A}^i = \mathbf{A} + \mathbf{A}^2 + \ldots + \mathbf{A}^{N}$, where $\mathbf{A}$ is the $(N \times N)$-dimensional graph adjacency matrix. Such a reachability matrix $\mathbf{R}$ yields information about the number of paths having the length N or the length less than N. Replacing the ordinary arithmetic by the Boolean arithmetic, the element $a^{(k)}_{ij}$ of the k-th power $\mathbf{A}^k$ of the matrix $\mathbf{A}$ is Boolean and expresses the nonexistence or existence of the path from the node i to the node j having the length k. The elements of such a logical **reachability matrix** $\mathbf{R}_L = \mathbf{A} \vee \mathbf{A}^2 \vee \ldots \vee \mathbf{A}^N$ yield information about the reachability in itself, that is, they decide the reachability.

To avoid computing the powers of the RG adjacency matrix, we will use another approach at the control synthesis. We will define the **RG-based dynamic model** working with vicarious vectors $\mathbf{X}_k$ being something like *hyper-state* vectors corresponding to the original state vectors $\mathbf{x}_k$ in the PN-based model (1).

The Graph-Based Model and Vicarious Vectors

The RT or RG of PN expresses the straightforward causality among the discrete events and the states of the modeled system. Namely, they represent relations between the cause (occurred discrete event at the actual state) and the consequence (the new state of system due to the cause), that is, *the actual state* → *an event* → *the new state.* However, also *backward* causality can be tested by the *backward* RT or RG. They express the relation between the actual state on one hand and the causes of why the system is in this state (i.e., from which states and by means of which discrete events the system passed to this state) on the other hand, that is, *which states?* ➔ *which events?* ➔ *the actual state*. Knowledge of both the straightforward causality and the backward one give us very useful information about the feasible states of the modeled system as well as about the feasible trajectories among them. These trajectories represent or characterize the reachability of the states and they are very useful at the analysing of the system behaviour as well as at the control synthesis.

Define the functional (k-variant) adjacency matrix $\mathbf{A}_k$ of the RG. As a matter of fact, it is the matrix $\mathbf{A}_{rt}(k)$. Such an unusual term like functional or k-variant matrix is a consequence of the fact that elements of $\mathbf{A}_k$ are the transition functions $g^k_{t_{i\to j}}$, $i=1,2,\ldots,N$; $j=1,2,\ldots,N$. Here, the transition $t_{i\to j}$ is the transition between two feasible states X_i and X_j, directed from X_i to X_j as follows: $X_i \xrightarrow{t_{i\to j}} X_j$. It means that $\mathbf{A}_k$ expresses a dynamic structure of the RG, as well as the causal relations among corresponding feasible states X_i, $i=1,2,\ldots,N$, from the set of feasible states $X = \{X_1, X_2, \ldots, X_N\}$. The matrix $\mathbf{A}_k$ can be created from the quasi-functional adjacency matrix $\mathbf{A}_{rt} = \mathbf{A}_{RG}$ in such a way that the nonzero integer elements of $\mathbf{A}_{RG}$ representing the indices of the transitions are replaced by the transition functions of these transitions. In order to work with $\mathbf{A}_k$, let us define the N-dimensional vicarious vectors $\mathbf{X}_k$, $k = 0, 1, \ldots$ as follows. $\mathbf{X}_k = ({}^kX_1, {}^kX_2, \ldots, {}^kX_N)^T$ where

$$
{}^kX_i = \begin{cases} 1 & \text{if } i = k+1 \\ 0 & \text{otherwise} \end{cases}; \; i = 1, 2, \ldots, N
$$

In the **mathematical model based on the RG** the vicarious vectors act as the PN state vectors in the model (1). Note this fact as $\mathbf{X}_k \cong \mathbf{x}_k$, $k = 0, 1, \ldots$ Namely, while the PN state vectors $\mathbf{x}_k$ appearing in the PN-based model (1) represent the system dynamics development in the steps $k = 0, 1, \ldots$ the vicarious vectors appear in the following model

$$
\mathbf{X}_{k+1} = \mathbf{A}^T_k . \mathbf{X}_k \, ; \; k = 0, 1, \ldots \tag{6}
$$

which yields $\mathbf{X}_k = \mathbf{A}^T_{k-1} . \mathbf{A}^T_{k-2} \ldots \mathbf{A}^T_0 . \mathbf{X}_0$. The model (6) is a ***hypermodel*** corresponding to the model (1). It is useful to mention also the **backward model** of such a kind, namely

$$
\mathbf{X}_{k-1} = \mathbf{A}_{k-1} . \mathbf{X}_k \, ; \; k = K, K-1, \ldots \tag{7}
$$

where K is an integer denoting the terminal state to be reached. Such a model represents the backward development of the system dynamics when we are interested in finding how (i.e., from which previous states, predecessors) the present state can be reached. The backward model yields $\mathbf{X}_0 = \mathbf{A}_0 . \mathbf{A}_1 \ldots \mathbf{A}_{K-2} . \mathbf{A}_{K-1} . \mathbf{X}_K$.

Feasible Trajectories and Control Synthesis

The models (6), (7) express, respectively, the straight-lined and the backward causality. They can be utilized at finding the system of feasible trajectories from a given initial state $\mathbf{x}_0$ represented by the vicarious vector $\mathbf{X}_0$ to a prescribed terminal state $\mathbf{x}_t = \mathbf{x}_K$ represented by the vicarious vector $\mathbf{X}_K$. To avoid symbolic computations on this way the matrix $\mathbf{A}_k$ is replaced by the constant matrix $\mathbf{A}$ containing all of the nonzero elements equal to 1. It corresponds to the situation when all of the transitions are enabled and fired in any step k.

Hence, we are able to compute the **straight-lined reachability tree (SLRT)** starting from $\mathbf{X}_0$ and developed toward $\mathbf{X}_K$ as well as the **backtracking reachability tree (BTRT)** starting from $\mathbf{X}_K$ and developed toward $\mathbf{X}_0$, however, oriented from $\mathbf{X}_0$ to $\mathbf{X}_K$. SLRT is computed as ${}^{sl}\{\mathbf{X}_1\} = \mathbf{A}^T . \mathbf{X}_0$, ${}^{sl}\{\mathbf{X}_2\} = \mathbf{A}^T . {}^{sl}\{\mathbf{X}_1\} = \mathbf{A}^T . \mathbf{A}^T . \mathbf{X}_0$, …, ${}^{sl}\{\mathbf{X}_K\} = \mathbf{A}^T . {}^{sl}\{\mathbf{X}_{K-1}\} = \underbrace{\mathbf{A}^T \cdots \mathbf{A}^T}_{K-factors} . \mathbf{X}_0$.

Here, ${}^{sl}\{\mathbf{X}_i\}$ means the aggregated states due to the fact that all transitions are fired in the step i.

Analogically, the BTRT is computed as ${}^{bt}\{\mathbf{X}_{K-1}\} = \mathbf{A}.\mathbf{X}_K, {}^{bt}\{\mathbf{X}_{K-2}\} = \mathbf{A}.{}^{bt}\{\mathbf{X}_{K-1}\} = \mathbf{A}.\mathbf{A}.\mathbf{X}_K, \ldots, {}^{bt}\{\mathbf{X}_0\} = \mathbf{A}.{}^{bt}\{\mathbf{X}_1\} = \underbrace{\mathbf{A} \cdots \mathbf{A}}_{K-factors}.\mathbf{X}_K$.

Store the SLRT as the columns of the matrix $\mathbf{M}_1 = (\mathbf{X}_0, {}^{sl}\{\mathbf{X}_1\}, \ldots, {}^{sl}\{\mathbf{X}_{K-1}\}, {}^{sl}\{\mathbf{X}_K\})$ and the BTRT as the columns of the matrix $\mathbf{M}_2 = ({}^{bt}\{\mathbf{X}_0\}, {}^{bt}\{\mathbf{X}_1\}, \ldots, {}^{bt}\{\mathbf{X}_{K-1}\}, \mathbf{X}_K)$. Let us perform the column-to-column intersection $\mathbf{M} = \mathbf{M}_1 \cap \mathbf{M}_2 = (\mathbf{X}_0 \cap {}^{bt}\{\mathbf{X}_0\}, {}^{sl}\{\mathbf{X}_1\} \cap {}^{bt}\{\mathbf{X}_1\}, \ldots, {}^{sl}\{\mathbf{X}_{K-1}\} \cap {}^{bt}\{\mathbf{X}_{K-1}\}, {}^{sl}\{\mathbf{X}_K\} \cap \mathbf{X}_K) = (\mathbf{X}_0, \{\mathbf{X}_1\}, \ldots, \{\mathbf{X}_{K-1}\}, \mathbf{X}_K)$ with $\{\mathbf{X}_i\} = {}^{sl}\{\mathbf{X}_i\} \cap {}^{bt}\{\mathbf{X}_i\} = min({}^{sl}\{\mathbf{X}_i\}, {}^{bt}\{\mathbf{X}_i\})$, $i = 0, 1, \ldots, K$ where ${}^{sl}\{\mathbf{X}_0\} = \mathbf{X}_0$ and ${}^{bt}\{\mathbf{X}_K\} = \mathbf{X}_K$. The matrix **M** stores the graph being the intersection of the SLRT and BTRT, that is, it stores the feasible trajectories of the system. Namely, it is the consequence of the fact that the element $a^{(k)}_{ij}$ of the k-th power $\mathbf{A}^k$ of the adjacency matrix **A** represents (Preparata & Yeh, 1974; Rosen, 2002) the number of the paths having the length k from the node i of the graph represented by **A** to the node j of this graph.

What is very important and interesting is that the principle of causality allows us to find shorter trajectories when the longer ones were already computed. Namely, having at disposal the matrices $\mathbf{M}_1$, $\mathbf{M}_2$, we can compute not only the trajectories of the corresponding length but also the trajectories shorter for *1, 2, ..., j* steps in such a way that before the intersection of these matrices we shift the matrix $\mathbf{M}_2$ to the left for *1, 2, ..., j* columns as follows.

$${}^{-1}\mathbf{M} = \mathbf{M}_1 \cap {}^{-1}\mathbf{M}_2;\ \text{where } {}^{-1}\mathbf{M}_2 = ({}^{bt}\{\mathbf{X}_1\}, \ldots, {}^{bt}\{\mathbf{X}_{k-1}\}, {}^{bt}\mathbf{X}_k, \mathbf{0})$$

$${}^{-2}\mathbf{M} = \mathbf{M}_1 \cap {}^{-2}\mathbf{M}_2;\ \text{where } {}^{-2}\mathbf{M}_2 = ({}^{bt}\{\mathbf{X}_2\}, \ldots, {}^{bt}\{\mathbf{X}_{k-1}\}, {}^{bt}\mathbf{X}_k, \mathbf{0}, \mathbf{0})$$

$$\vdots \qquad \vdots$$

$${}^{-j}\mathbf{M} = \mathbf{M}_1 \cap {}^{-j}\mathbf{M}_2;\ \text{where } {}^{-j}\mathbf{M}_2 = ({}^{bt}\{\mathbf{X}_j\}, \ldots, {}^{bt}\{\mathbf{X}_{k-1}\}, {}^{bt}\mathbf{X}_k, \underbrace{\mathbf{0}, \ldots, \mathbf{0}}_{j\text{-zero column vectors}})$$

Here, **0** is the zero column vector of the corresponding dimensionality. When an intersection $\mathbf{M}_1 \cap {}^{-j}\mathbf{M}_2$ does not exist, the matrix ${}^{-j}\mathbf{M}$ is the zero-matrix with the corresponding dimensionality. It means that no trajectory shorter for j steps exists in such a case.

It is necessary to say that the number of steps K in which the system reaches the terminal state $\mathbf{x}_t$ is not predetermined. It is bounded only by the relation $K \leq (N - 1)$ which results from graph theory (where $(N - 1)$ is the maximal length of the paths in the graph with N nodes) as well as by the relation ${}^{sl}\{\mathbf{X}_K\} \geq \mathbf{X}_K$ (to be sure that $\mathbf{X}_K$ is comprehended in ${}^{sl}\{\mathbf{X}_K\}$). When such a K is found, the number of the columns of $\mathbf{M}_1$ is determined and the backtracking development starting from $\mathbf{X}_K \cong \mathbf{x}_t = \mathbf{x}_K$ can be performed in order to obtain the matrix $\mathbf{M}_2$.

To illustrate the approach, consider the two agents' negotiation process introduced above. Then, for the given initial state $\mathbf{x}_0 = X_1$ and the prescribed terminal state $\mathbf{x}_t = X_{13}$, we have only one trajectory. The trajectory can be expressed also in the form $X_1 \xrightarrow{t_9} X_3 \xrightarrow{t_{14}} X_5 \xrightarrow{t_{15}} X_7 \xrightarrow{t_3} X_9 \xrightarrow{t_6} X_{12} \xrightarrow{t_{16}} X_{13}$. It is displayed in Figure 9. The trajectory follows from the below introduced intersection **M** of the matrices $\mathbf{M}_1$, $\mathbf{M}_2$ and it corresponds to the trajectory $\mathbf{X}_0 \xrightarrow{t_9} \mathbf{X}_1 \xrightarrow{t_{14}} \mathbf{X}_2 \xrightarrow{t_{15}} \mathbf{X}_3 \xrightarrow{t_3} \mathbf{X}_4 \xrightarrow{t_6} \mathbf{X}_5 \xrightarrow{t_{15}} \mathbf{X}_6$.

$$\mathbf{M}_1 = \begin{pmatrix} 1 & 0 & 0 & 0 & 0 & 0 & 0 \\ 0 & 1 & 0 & 0 & 0 & 0 & 0 \\ 0 & 1 & 0 & 0 & 0 & 0 & 0 \\ 0 & 0 & 2 & 0 & 0 & 0 & 0 \\ 0 & 0 & 1 & 0 & 0 & 0 & 0 \\ 0 & 0 & 0 & 3 & 0 & 0 & 0 \\ 0 & 0 & 0 & 1 & 0 & 0 & 0 \\ 0 & 0 & 0 & 0 & 4 & 0 & 0 \\ 0 & 0 & 0 & 0 & 1 & 0 & 0 \\ 0 & 0 & 0 & 0 & 1 & 0 & 0 \\ 0 & 0 & 0 & 0 & 0 & 1 & 0 \\ 0 & 0 & 0 & 0 & 0 & 1 & 0 \\ 0 & 0 & 0 & 0 & 0 & 0 & 1 \end{pmatrix};$$

$$\mathbf{M}_2 = \begin{pmatrix} 1 & 0 & 0 & 0 & 0 & 0 & 0 \\ 0 & 0 & 0 & 0 & 0 & 0 & 0 \\ 0 & 1 & 0 & 0 & 0 & 0 & 0 \\ 0 & 0 & 0 & 0 & 0 & 0 & 0 \\ 0 & 0 & 1 & 0 & 0 & 0 & 0 \\ 0 & 0 & 0 & 0 & 0 & 0 & 0 \\ 0 & 0 & 0 & 1 & 0 & 0 & 0 \\ 0 & 0 & 0 & 0 & 0 & 0 & 0 \\ 0 & 0 & 0 & 0 & 1 & 0 & 0 \\ 0 & 0 & 0 & 0 & 0 & 0 & 0 \\ 0 & 0 & 0 & 0 & 0 & 0 & 0 \\ 0 & 0 & 0 & 0 & 0 & 1 & 0 \\ 0 & 0 & 0 & 0 & 0 & 0 & 1 \end{pmatrix};$$

$$\mathbf{M} = \begin{pmatrix} 1 & 0 & 0 & 0 & 0 & 0 & 0 \\ 0 & 0 & 0 & 0 & 0 & 0 & 0 \\ 0 & 1 & 0 & 0 & 0 & 0 & 0 \\ 0 & 0 & 0 & 0 & 0 & 0 & 0 \\ 0 & 0 & 1 & 0 & 0 & 0 & 0 \\ 0 & 0 & 0 & 0 & 0 & 0 & 0 \\ 0 & 0 & 0 & 1 & 0 & 0 & 0 \\ 0 & 0 & 0 & 0 & 0 & 0 & 0 \\ 0 & 0 & 0 & 0 & 1 & 0 & 0 \\ 0 & 0 & 0 & 0 & 0 & 0 & 0 \\ 0 & 0 & 0 & 0 & 0 & 0 & 0 \\ 0 & 0 & 0 & 0 & 0 & 1 & 0 \\ 0 & 0 & 0 & 0 & 0 & 0 & 1 \end{pmatrix}$$

Consequently, it answers to the trajectory $\mathbf{x}_0 \xrightarrow{t_9} \mathbf{x}_1 \xrightarrow{t_{14}} \mathbf{x}_2 \xrightarrow{t_{15}} \mathbf{x}_3 \xrightarrow{t_3} \mathbf{x}_4 \xrightarrow{t_6} \mathbf{x}_5 \xrightarrow{t_{16}} \mathbf{x}_6$ representing the real states of the system passed through at during the system dynamics development.

The transitions between the states in questions are unambiguously given by the elements of the matrix $\mathbf{A}_{RG}$. The sequence of transitions $T_c = \{t_9, t_{14}, t_{15}, t_3, t_6, t_{16}\}$ represents the control sequence of discrete events realizing the transition of the system from the initial state to the terminal one. In other words, the sequence T_c is the product of the **control synthesis process** and can be realized in real process. Analogically, we can compute and analyse the ways of how to reach the arbitrary state $X_i \in \{X_2, X_3, \ldots, X_N\}$ from X_1.

Figure 9. The feasible state trajectory

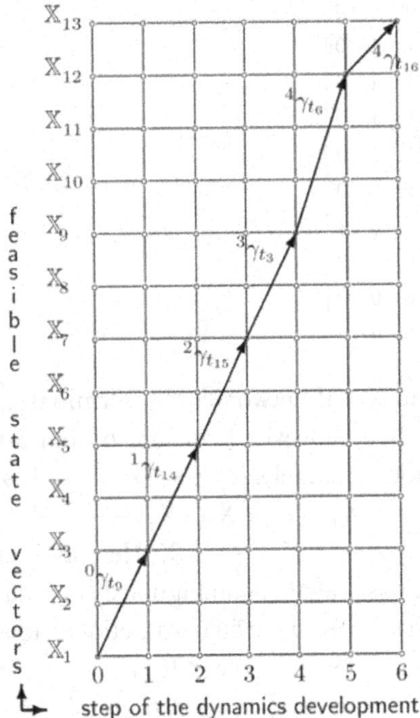

In general, several trajectories (and consequently several sequences of transitions) can occur as a product of the control synthesis, of course. For example, consider the system displayed in Figure 8. When we are interested in passing the system from the initial state X_1 being the real state $\mathbf{x}_0 = (0,1,0,0,1,0,0,1)^T$ to the terminal state being X_1 again that is, the state $\mathbf{x}_t = (0,1,0,0,1,0,0,1)^T$ we have the following matrices $\mathbf{M}_1$, $\mathbf{M}_2$, $\mathbf{M}$ storing, respectively, the SLBT, BTRT and the feasible trajectories

$$\mathbf{M}_1 \begin{pmatrix} 1 & 0 & 4 \\ 0 & 1 & 0 \\ 0 & 1 & 0 \\ 0 & 1 & 0 \\ 0 & 1 & 0 \\ 0 & 0 & 2 \\ 0 & 0 & 2 \\ 0 & 0 & 2 \\ 0 & 0 & 2 \\ 0 & 0 & 2 \end{pmatrix} \quad \mathbf{M}_2 = \begin{pmatrix} 4 & 0 & 1 \\ 0 & 1 & 0 \\ 1 & 1 & 0 \\ 0 & 1 & 0 \\ 1 & 1 & 0 \\ 2 & 0 & 0 \\ 2 & 0 & 0 \\ 2 & 0 & 0 \\ 3 & 0 & 0 \\ 0 & 1 & 0 \end{pmatrix}$$

$$\mathbf{M} = \begin{pmatrix} 1 & 0 & 0 \\ 0 & 1 & 0 \\ 0 & 1 & 0 \\ 0 & 1 & 0 \\ 0 & 1 & 0 \\ 0 & 0 & 0 \\ 0 & 0 & 1 \\ 0 & 0 & 0 \\ 0 & 0 & 0 \\ 0 & 0 & 0 \end{pmatrix} \quad (9)$$

It means that after two steps the terminal state can be reached, however, by means of four different trajectories, namely, $X_1 \xrightarrow{t_1} X_2 \xrightarrow{t_2} X_1$ or $X_1 \xrightarrow{t_3} X_3 \xrightarrow{t_4} X_1$ or $X_1 \xrightarrow{t_5} X_4 \xrightarrow{t_6} X_1$ or $X_1 \xrightarrow{t_7} X_5 \xrightarrow{t_8} X_1$. Here, assigning the transitions is again unambiguously obvious from $\mathbf{A}_{RG}$, that is, from the quasi-functional adjacency matrix $\mathbf{A}_k$. The sequences $\{t_1, t_2\}$, $\{t_3, t_4\}$, $\{t_5, t_6\}$, $\{t_7, t_8\}$ are concerning *hesitations* of agents as to the cooperation (cooperate or not to cooperate). In such a case—in case of two steps—the most suitable trajectory can be chosen relatively simply. There exist two longer (three steps) trajectories computed by means of the following matrices expressing SLRT, BTRT and their intersection

$$\mathbf{M}_1 = \begin{pmatrix} 1 & 0 & 4 & 2 \\ 0 & 1 & 0 & 8 \\ 0 & 1 & 0 & 8 \\ 0 & 1 & 0 & 8 \\ 0 & 1 & 0 & 8 \\ 0 & 0 & 2 & 0 \\ 0 & 0 & 2 & 0 \\ 0 & 0 & 2 & 0 \\ 0 & 0 & 2 & 0 \\ 0 & 0 & 2 & 2 \end{pmatrix} \mathbf{M}_2 = \begin{pmatrix} 2 & 4 & 0 & 1 \\ 8 & 0 & 1 & 0 \\ 9 & 1 & 1 & 0 \\ 8 & 0 & 1 & 0 \\ 9 & 1 & 1 & 0 \\ 0 & 2 & 0 & 0 \\ 1 & 2 & 0 & 0 \\ 1 & 2 & 0 & 0 \\ 2 & 3 & 0 & 0 \\ 4 & 0 & 1 & 0 \end{pmatrix}$$

$$\mathbf{M} = \begin{pmatrix} 1 & 0 & 0 & 1 \\ 0 & 0 & 0 & 0 \\ 0 & 1 & 0 & 0 \\ 0 & 0 & 0 & 0 \\ 0 & 1 & 0 & 0 \\ 0 & 0 & 0 & 0 \\ 0 & 0 & 0 & 0 \\ 0 & 0 & 0 & 0 \\ 0 & 0 & 0 & 0 \\ 0 & 0 & 1 & 0 \end{pmatrix} \quad (10)$$

The trajectories realize the mutual communication in the case when one of the agents wants to communicate and another one is available. They are the following $X_1 \xrightarrow{t_3} X_3 \xrightarrow{t_{10}} X_{10} \xrightarrow{t_{11}} X_1$ and $X_1 \xrightarrow{t_7} X_5 \xrightarrow{t_9} X_{10} \xrightarrow{t_{11}} X_1$. It means that the product of the control synthesis is one of the sets of transitions $\{t_3, t_{10}, t_{11}\}$, $\{t_7, t_9, t_{11}\}$. To illustrate the process of finding shorter trajectories by means of shifting $\mathbf{M}_2$ before intersection with $\mathbf{M}_1$, let us perform such an intersection when $\mathbf{M}_2$ in (10) is shifted for one column to the left and then the matrix ${}^{-1}\mathbf{M}_2$ is intersected with $\mathbf{M}_1$. The result of such a procedure is the final matrix ${}^{-1}\mathbf{M}$ which is equal to the matrix $\mathbf{M}$ given in (9). It is the consequence of the causality. This fact is very important, because the goal of the control synthesis process need not be always the shortest trajectory, as usually are optimisation based on graphs.

In addition to this, when we have at our disposal the matrices $\mathbf{M}_1$, $\mathbf{M}_2$ for an arbitrary number K of steps (long trajectories), even for $K > N$, we are able to find all of the shorter trajectories by means of shifting $\mathbf{M}_2$ for $1, 2, \ldots, J, J < K$, columns. This can be done when the shifted matrices ${}^{-1}\mathbf{M}_2$, ${}^{-2}\mathbf{M}_2$, …, ${}^{-J}\mathbf{M}_2$ are used in the process of intersection with $\mathbf{M}_1$. In the case when shorter trajectories do not exist, the intersection gives the zero matrix, that is, ${}^{-J}\mathbf{M} = \mathbf{0}$.

Figure 10. The view on the screen of the PN-editor handling the model. The model is on the left while the corresponding reachability tree is on the right.

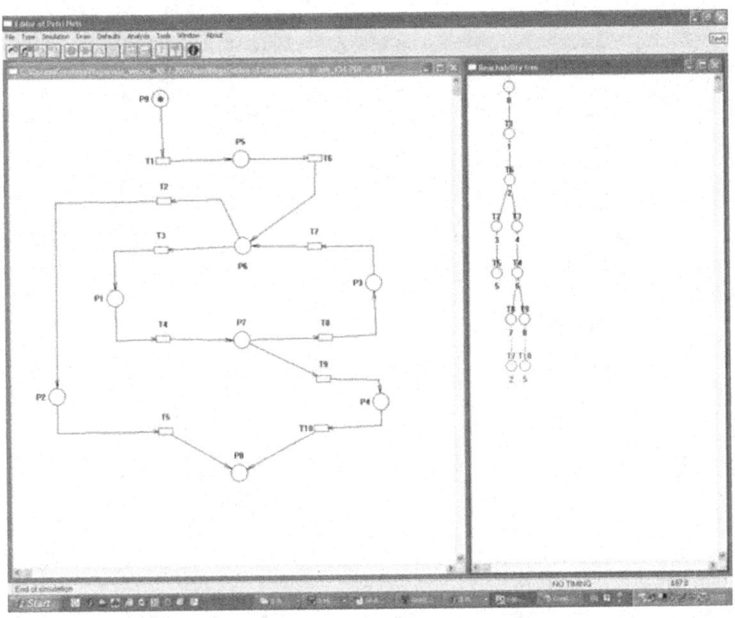

In the case when several feasible trajectories occur, the most suitable trajectory has to be chosen on base of a criterion or on base on a set of criteria. The situation is easier when a state or several states to be passed through are predefined. The more states to be passed through that are predefined, the better. In any case, to form concrete quantified criteria is not any simple task. Namely, the demands are usually formulated only verbally or in other nonanalytical terms. Consequently, the most suitable approach seems to be a form of rule-based representation of knowledge about the control task specifications. The approaches based on the so-called soft computing—especially those utilizing fuzzy logic—can be used in this way. To represent logical or fuzzy-logical rules, the special kinds of PN can be used too, for example, the logical PN or fuzzy-logical PN (Čapkovič, 1995, 1996/1997, 1999; Tzafestas & Čapkovič, 1997).

THE EXAMPLE OF THE APPLICATION NEGOTIATION SCENARIO

Consider two virtual companies (Lenz, Oberweis, & Schneider, 2001) company A and the company B.

There are different phases in the process of setting the cooperation between two organizations up. The following **negotiation scenario** takes place after the potential partners have found each other. In the displayed case, company A creates an information document containing the issues of the project (e.g., which the mutual software agents already agreed on) and those that are still unclear.

The PN-based model created by means of the PN editor is given on the left part in Figure 10, while the corresponding reachability tree is on the right side. The interpretations of the PN places and the PN transitions are the following.

Figure 11. The view on the screen of the GraSim tool for the control synthesis based on the reachability graphs

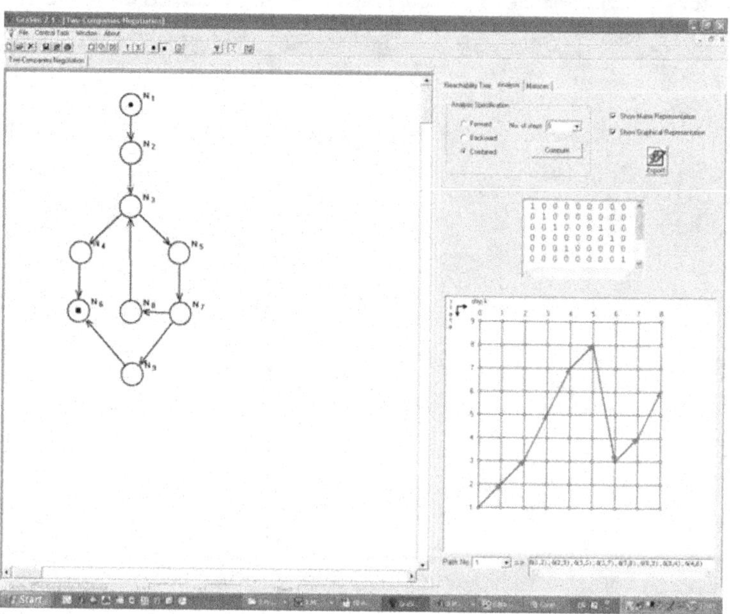

The places are: pl, p3 – the updated proposal; p2, p4 – the unchanged proposal; p5 – the information document; p6 – the proposal to A; p7 – the proposal to B; p8 – the contract; and p9 – the start. The transitions are: tl – creating the information document; t2, t9 – checking the proposal and the agreement with it; t3, t8 – checking the proposal and asking changes; t4 – sending the updated proposal; t5, t10 – accepting the unchanged proposal; t6 – preparing the proposal; and t7 – sending the updated proposal. The nodes of the reachability tree are the states (state vectors) reachable from the initial state $\mathbf{x}_0$. In Figure 11, the **RG-based control synthesis**, realized by the GraSim tool, is displayed. The reachability graph of the modeled agent system is on the left side, while one of the state trajectories from the initial state $\mathbf{x}_0$ (node N1) to the terminal state (in the given flash it is the node N6) is on the right side. The state vectors can be seen (in the real tool, because in Figure 11 only a flash of its operation is displayed) by means of the roller as the columns of the matrix situated in the small window over the trajectory. The control interferences are displayed in the lower right corner of the screen.

The example simultaneously illustrates that PN places can also create the Interface among agents. Namely, the places pl, p2 and the transitions tl, …, t5 together with their interconnections represent company A (agent A) and the places p3, p4 and the transitions t6, …, t10 together with their interconnections represent company B (agent B). Finally, the places p5, …, p9 concern the model of the Interface among the agents.

CONCLUSION

The alternative approach to PN-based modeling of agents and MAS was presented in this chapter. Using RT or RG the agents' behaviour can be efficiently analysed. In addition, the control synthesis can be comfortably performed. Both the straightforward and backward causality of the system dynamics development were utilized in this way. The approach yields the results in analytical terms (computed by means of Matlab) as well as graphically (obtained by means of the graphical tools, the PN simulator and the GraSim tool). The RT or RG, as well as the space of reachable states, were generated on the base of both the model parameters and the given initial state of the system. The coherence between the causality and the RG was pointed out. Finally, the feasible state trajectories were found by means of the mutual intersection of both the SLRT representing the straightforward causal development of the system and BTRT representing the backward causal development of the system. The possibility of the wider utilization of the proposed approach was pointed out.

ACKNOWLEDGMENT

The research was partially supported by the Slovak Grant Agency for Science VEGA under grant 2/6102/26. The author thanks VEGA for the support.

REFERENCES

Brams, S. J., & Kilgour, D. M. (2001). Fallback bargaining. *Group Decision and Negotiation, 10*, 287-316.

Čapkovič, F. (1995). Using fuzzy logic for knowledge representation at control synthesis. *Busefal, 63*, 4-9.

Čapkovič, F. (1996/1997). Petri nets and oriented graphs in fuzzy knowledge representation for DEDS control purposes. *Busefal, 69*, 21-30.

Čapkovič, F. (1999). Knowledge-based control synthesis of discrete event dynamic systems. In S.G. Tzafestas (Ed.), *Advances in manufacturing systems: Decision, control and information technology* (pp. 195-206). London, UK: Springer-Verlag.

Čapkovič, F. (2002). Control synthesis of a class of DEDS. *Kybernetes, 31*(9/10), 1274-1281.

Čapkovič, F. (2003). The generalised method for solving problems of DEDS control synthesis. In P. W. H. Chung, C. Hinde, & M. Ali (Eds.), *Developments in applied artificial intelligence: Lecture notes in artificial intelligence* (Vol. 2718, pp. 702-711). New York-Heidelberg-London: Springer-Verlag.

Čapkovič, F. (2004a). An application of the DEDS control synthesis method. In M. Bubak, G. D. van Albada, P. M. A. Sloot, & J. J. Dongarra (Eds.), *Computational science—ICCS 2004, Part III: Lecture notes in computer science* (Vol. 3038, pp. 528-536). New York-Heidelberg-London: Springer-Verlag.

Čapkovič, F. (2004b). DEDS control synthesis problem solving. In R. Yager & V. Sgurev (Eds.), *Proceedings of the 2nd IEEE Conference on Intelligent Systems*, Varna, Bulgaria (pp. 299-304.). Piscataway, NJ, USA: IEEE Press.

Čapkovič, F. (2005). An application of the DEDS control synthesis method. *Journal of Universal Computer Science, 11*(2), 303-326.

Čapkovič, F. (2006). Modeling, analysing and control of interactions among agents in MAS. In V. N. Alexandrov, G. D. van Albada, P. M. A. Sloot, & J. J. Dongarra (Eds.), *Computational science - ICCS 2006, Part III: Lecture notes in computer science* (Vol. 3993, pp. 176-183). Berlin-Heidelberg: Springer-Verlag.

Čapkovič, F. (2007). Modelling, analysing and control of interactions among agents in MAS. *Computing and Informatics, 26*(5), 507-541.

Čapkovič, F., & Čapkovič, P. (2003). Petri net-based automated control synthesis for a class of DEDS. In L. Gomes, R. Zurawski, C. Couro, & J. Fuertes (Eds.), *Proceedings of the 2003 IEEE*

Conference on Emerging Technologies and Factory Automation—ETFA 2003, Lisbon, Portugal (Vol. 2, pp. 297-304). Piscataway, NJ, USA: IEEE Electronic Society Press.

Cassandras, C.G., & Lafortune, S. (1999). *Introduction to discrete event systems*. Boston: Kluwer Academic.

Demazeau, Y. (2003, September 29-October 4). MAS methodology. In *Proceedings of the Tutorial at the 2nd French-Mexican School of the Repartee Cooperative Systems - ESRC 2003*. Rennes, France: IRISA.

Eshuis, R., & Dehnert, J. (2003). Reactive Petri nets for workflow modelling. In W.M.P. van der Aalst, & E. Best (Eds.), *Applications and theory of Petri nets: Lecture notes in computer science* (Vol. 2679, pp. 296-315). New York-Heidelberg-London: Springer-Verlag.

Eshuis, R., & Wieringa, R. (2001a). *A formal semantics for UML Activity Diagrams—normalising workflow models* (Tech. Rep. No. CTIT-01-04). University of Twente, The Netherlands: Department of Computer Science.

Eshuis, R., & Wieringa, R. (2001b). A real time execution semantics for UML activity diagrams. In H. Hussmann (Ed.), *Fundamental approaches to software engineering: Lecture notes in computer science* (Vol. 2029, pp. 76-90). Berlin-Heidelberg: Springer-Verlag.

Eshuis, R., & Wieringa, R. (2002, May 19-25). Verification support for workflow design with UML activity graphs. In *Proceedings of the 24th International Conference on Software Engineering (ICSE'02)*, Orlando, FL, USA, (pp. 166-176). New York: ACM Press.

Eshuis, R., & Wieringa, R. (2003). Comparing Petri net and activity diagram variants for workflow modelling—a quest for reactive Petri nets. In H. Ehrig, W. Reisig, G. Rozenberg, & H. Weber (Eds.), *Petri net technology for communication-based systems: Lecture notes in computer science* (Vol. 2472, pp. 321-351). New York-Heidelberg-London: Springer-Verlag.

Ferber, J. (1999). *Multi-agent systems: An introduction to distributed artificial intelligence*. Harlow, UK: Addison-Wesley Longman.

Fonseca, S., Griss, M., & Letsinger, R. (2001). *Agent behavior architectures—a MAS framework comparison* (HP Labs Tech. Rep. No. HPL-2001-332). Palo Alto, CA, USA: HP Laboratories. Retrieved April 3, 2008, from http://www.hpl.hp.com/techreports/2001/HPL-2001-332.pdf

Fowler, M., & Scott, K. (2000). *UML distilled: A brief guide to the standard object modeling language*. Indianapolis, IN: Addison-Wesley.

Harel, D., & Pnueli, A. (1985). On the development of reactive systems. In K.R. Apt (Ed.), *Logics and models of concurrent systems, NATO ASI Series F: Computer and systems sciences* (pp. 477-498). New York: Springer-Verlag.

Heljanko, K. (2006). *T79.4301. Parallel and distributed systems (4 ECTS), Lecture 5* (pp. 1-24). Department of Computer Science, University of Helsinki, Finland. Retrieved April 3, 2008, from http://www.tcs.hut.fi/Studies/T-79.4301/2006SPR/lecture5.pdf

Hung, P. C. K., & Mao, J. Y. (2002). Modeling e-negotiation activities with Petri nets. In R. H. Spraguer, Jr. (Ed.), *Proceedings of the 35th Hawaii International Conference on System Sciences HICSS 2002*, Big Island, HI, (Vol. 1, p. 26), CD ROM. Piscataway, NJ, USA: IEEE Computer Society Press.

Lenz, K., Oberweis, A., & Schneider, S. (2001). Trust based contracting in virtual organizations: A concept based on contract workflow management systems. In B. Schmid, K. Stanoevska-Slabeva, & V. Tschammer (Eds.), *Towards the e-society—E-commerce, e-business, and e-government* (pp. 3-16). Boston: Kluwer Academic.

Leymann, F., & Roller, D. (2000). *Production workflow—Concepts and techniques*. New York: Prentice-Hall.

Llorens, M., & Oliver, J. (2004). Structural and dynamic changes in concurrent systems: Reconfigurable Petri nets. *IEEE Transactions on Computers, 53*(9), 1147-1158.

Llorens, M., & Oliver, J. (2006). A basic tool for the modelling of Marked-Controlled Reconfigurable Petri Nets. In *Proceedings of the Workshop on Petri Nets and Graph Transformation (PNGT 2006): Electronic Communications of the EASST: Petri Nets and Graph Transformations* (Vol. 2, pp. 1-13). Retrieved April 3, 2008, from http://eceasst.cs.tu-berlin.de/index.php/eceasst/article/viewFile/23/28 or http://eceasst.cs.tu-berlin.de/index.php/eceasst/issue/view/10

Meena, H.K., Saha, I., Mondal, K.K., & Prabhakar, T.V. (2005, October 16-17). An approach to workflow modeling and analysis. In *Proceedings of the 20th Annual ACM SIGPLAN Conference on Object-oriented Programming, Systems, Languages, and Applications (OOPSLA'05): The OOPSLA Workshop on Eclipse technology eXchange*, San Diego, CA, (pp. 85-89). New York: ACM Press.

Murata, T. (1989). Petri nets: Properties, analysis and applications. *Proceedings of the IEEE, 77*(4), 541-588.

Nowostawski, M., Purvis, M., & Cranefield, S. (2001, May 28-June1). A layered approach for modelling agent conversations. In T. Wagner & O. F. Rana (Eds.), *Proceedings of 2nd International Workshop on Infrastructure for Agents, MAS, and Scalable MAS, 5th International Conference on Autonomous Agents - AA 2001*, Montreal, Canada, (pp. 163-170). Menlo Park, CA, USA: AAAI Press.

Peterson, J.L. (1981). *Petri nets theory and the modelling of systems*. Englewood Cliffs, NJ, New York: Prentice Hall.

Preparata, F.P., & Yeh, R.T. (1974). *Introduction to discrete structures*. Reading, MA, USA: Addison-Wesley.

Purvis, M., Cranefield, S., Nowostawski, M., Ward, R., Carter, D., & Oliveira, M. A. (2002). Agentcities interaction using the Opal platform. In B. Burg, J. Dale, T. Finin, H. Nakashima, L. Padgham, C. Sierra, & S. Willmott (Eds.), *Proceedings of Workshop on Agentcities: Research in Large-scale Open Agents Environments, First International Joint Conference on Autonomous Agents and Multi-agent Systems–AAMAS 2002*, Bologna, Italy. CD/ROM. New York: ACM Press. Retrieved April 3, 2008, from http://www.agentcities.org/Challenge02/Proc/Papers/ch02_56_purvis.pdf

Purvis, M., Purvis, M., Haidar, A., & Savarimuthu, B.T.R. (2005). A distributed workflow system with autonomous components. In M.W.B., & N. Kasabov (Eds.), *Intelligent agents and multi-agent systems: Lecture notes in computer science* (Vol. 3371, pp. 193-205). New York-Heidelberg-London: Springer-Verlag.

Rosen, K. (2002). *Discrete mathematics and its applications*. New York: McGraw-Hill.

Saint-Voirin, D., Lang, C., & Zerhouni, N. (2003, July 16-20). Distributed cooperation modelling for maintenance using Petri nets and multi-agents systems. In *Proceedings of the 5th IEEE International Symposium on Computational Intelligence in Robotics and Automation, CIRA'03, Kobe, Japan*, (Vol. 1, pp. 366-371). Piscataway, NJ, USA: IEEE Press.

Salimifard, K., & Wright, M. (2001). Petri net-based modelling of workflow systems: An overview. *European Journal of Operational Research, 134*(3), 664-676.

Thompson, L. (1998). *The mind and heart of the negotiator*. Upper Saddle River, NJ: Prentice Hall.

Tzafestas, S.G., & Čapkovič, F. (1997). Petri net-based approach to synthesis of intelligent control systems for DEDS. In S.G. Tzafestas (Ed.), *Computer assisted management and control of manufacturing systems* (pp. 325-351). London, UK: Springer-Verlag.

van der Aalst, W.M.P. (1998). The application of Petri nets to workflow management. *The Journal of Circuits, Systems and Computers, 8*(1), 21-66.

van der Aalst, W.M.P. (2000). Workflow verification: Finding control-flow errors using Petri net-based techniques. In W. van der Aalst, J. Desel, & A. Oberweis (Eds.), *Business process management: Models, techniques, and empirical studies, Lecture notes in computer sciences* (Vol. 1806, pp. 161-183). New York-Heidelberg-London: Springer-Verlag.

van der Aalst, W.M.P., ter Hofstede, A.H.M., Kiepuszewski, B., & Barros, A.P. (2000). *Workflow patterns, BETA working paper series, WP 47,* Eindhoven University of Technology, Eindhoven, The Netherlands.

van der Aalst, W.M.P., ter Hofstede, A.H.M., Kiepuszewski, B., & Barros, A.P. (2003). Workflow patterns. *Distributed and Parallel Databases, 14*(1), 5-51.

Walker, A., & Wooldridge, M. (1995, June 12-14). Understanding the emergence of conventions in multi-agent systems. In V. R. Lesser & L. Gasser (Eds.), *Proceedings of the First International Conference on Multi-agent Systems,* San Francisco, (pp. 384-389). Boston: The MIT Press.

Weigand, H., Dignum, M. V., Meyer, J.J., & Dignum, F. (2003, September 16-17). Specification by refinement and agreement: Designing agent interaction using landmarks and contracts. In P. Petta, R. Tolksdorf, & F. Zambonelli (Eds.), *Engineering Societies in the Agents World III: Third International Workshop, ESAW 2002: Lecture Notes in Computer Science,* Madrid, Spain, (Vol. 2577, pp. 257-269). Berlin-Heidelberg: Springer-Verlag.

White, S.A. (2004, March). *Process modelling notations and workflow patterns.* BP Trends (pp. 1-24). IBM Corporation. Retrieved April 3, 2008, from http://www.omg.org/bp-corner/bp-files/Process_Modeling_Notations.pdf

Yen, J., Yin, J., Ioerger, T. R., Miller, M. S., Xu, E., & Volz, R. A. (2001). CAST: Collaborative agents for simulating teamwork. In B. Nebel (Ed.), *Proceedings of 17th International Joint Conference on Artificial Intelligence—IJCAI' 2001,*Seattle, WA, USA, (Vol. 2, pp. 1135-1142). San Francisco: Morgan Kaufmann.

Chapter XV
Using Fuzzy Segmentation for Colour Image Enhancement of Computed Tomography Perfusion Images

Martin Tabakov
Wroclaw University of Technology, Poland

ABSTRACT

This chapter presents a methodology for an image enhancement process of computed tomography perfusion images by means of partition generated with appropriately defined fuzzy relation. The proposed image processing is used to improve the radiological analysis of the brain perfusion. Colour image segmentation is a process of dividing the pixels of an image in several homogenously- coloured and topologically connected groups, called regions. As the concept of homogeneity in a colour space is imprecise, a measure of dependency between the elements of such a space is introduced. The proposed measure is based on a pixel metric defined in the HSV colour space. By this measure a fuzzy similarity relation is defined, which next is used to introduce a clustering method that generates a partition, and so a segmentation. The achieved segmentation results are used to enhance the considered computed tomography perfusion images with the purpose of improving the corresponding radiological recognition.

INTRODUCTION

Data clustering is a popular technique for statistical data analysis, which is used in many fields, including machine learning, data mining, pattern recognition, image analysis and bioinformatics. Clustering is the classification of a set of objects into different groups, or more precisely, the partitioning of a data set into subsets (clusters), so that the data in each subset share some common trait (often proximity), according to some defined distance measure.

The computed tomography perfusion imaging is a new technique, which appears to provide early diagnosis of major vessel occlusions in the brain. Computed Tomography perfusion (CT-perfusion) imaging also provides valuable information about the hemodynamic status of ischemic brain tissue (Tekşam, Çakır, & Coşkun, 2005), for example, CT perfusion imaging for childhood moyamoya disease before and after surgical revascularization (Sakamoto et al., 2006). The concept of developing Perfusion CT primarily as a procedure for functional imaging has proved especially advantageous for its practical clinical application, by using harmonised contrast medium and scan protocols and by implementing a series of postprocessing steps within the framework of image calculation (König, Klotz, & Heuser, 2000).

In this chapter fuzzy data partitional clustering method based on fuzzy relations is proposed to develop an image enhancement algorithm dedicated to CT-perfusion images. As a field of application, medical imagery was chosen. Medical imaging techniques such as X-ray, CT, Magnetic Resonance Imaging (MRI), Positron Emission Tomography (PET), Ultrasound (USG), and so forth, are indispensable for the precise analysis of various medical pathologies. Computer power and medical scanner data alone are not enough; we need the art to extract the necessary boundaries, surfaces, and segmented volumes of these organs in the spatial and temporal domains. This art of organ extraction is segmentation. Image segmentation is essentially a process of pixel classification, wherein the image pixels are segmented into subsets by assigning the individual pixels to classes. These segmented organs and their boundaries are very critical in the quantification process for physicians and medical surgeons in any branch of medicine which deals with imaging (Suri, Setarehdan, & Singh, 2002).

Colour image segmentation, viewed as the process of dividing the image into regions characterized by colour homogeneity, is one of the most widely used tools in image processing (Chamorro-Martinez et al., 2003). Many types of segmentation techniques have been proposed in the literature, for example those based on histogram analysis (Gillet, Macaire, Bone-Lococq, & Pastaire, 2001), clustering (Zhong & Yan, 2000), split and merge (Barges & Aldon, 2000), region growing (Moghaddamzadeh & Bourbakis, 1997), edge-based algorithms (Shiji & Hamada, 1999), and so forth. Most of the proposals that fall in the aforementioned categories provide a crisp segmentation of images, where each pixel has to belong to a unique region. However, the separation between regions is usually imprecise in natural images, so crisp techniques are not often appropriate. To solve this problem, some approaches propose the definition of region as a fuzzy subset of pixels, in such a way that every pixel of the image has a membership degree to that region. These regions form a fuzzy partition of the input set of pixels (Bezdek, 1981).

Recently, as it has been illustrated in numerous scientific publications, fuzzy techniques are often applied as complementary to existing techniques and can contribute to the development of better and more robust methods. It seems to be true that applications of fuzzy techniques are very successful in the area of image processing (Kerre & Nachtegael, 2000; Tizhoosh, 1998). Moreover, the field of medicine has become a very attractive domain for the application of fuzzy set theory. This is due to the large role that imprecision and uncertainty play in this field (Mordeson, Malik, & Cheng, 2000).

Another proposal for partition generation, considering the fuzzy concept, is to use a fuzzy relation, which represents the degree of dependency between the elements of a considered data set. An introductory research (Helgason, Jobe, Malik, & Mordeson, 1999; Helgason, Jobe, Mordeson, Malik, & Cheng, 1999) was related to medical diagnosis of patients (an analysis of data from patients for measuring the degree of casual efficacy or conditions reflecting the abnormalities of blood flow, coagulation, and vascular wall damage). Some other investigations were done relating to the possibility of using fuzzy equivalence relations to segment medical images (Tabakov, 2001) and remotely sensed images of urban environment (Li, Li, Dong, & Gao, 2002).

The predominant colour representation used in systems that deals with images is the RGB colour space representation. In this representation, the values of the red, green, and blue colour channels are stored separately. They can range from 0 to 255, with 0 being not present and 255 being maximal. A fourth channel, alpha, also provides a measure of transparency for the pixel. The alternative to the RGB colour space is the Hue-Saturation-Value (HSV) colour space. Instead of looking at each value of red, green and blue individually, a metric is defined which creates a different continuum of colours, in terms of the different hues each colour possesses. The hues are then differentiated based on the amount of saturation they have, that is, in terms of how little white they have mixed in, as well as on the magnitude, or value, of the hue. In the value range, large numbers denote bright colorations, and low numbers denote dim colorations. This space is usually depicted as a cylinder, with hue denoting the location along the circumference of the cylinder, value denoting depth within the cylinder along the central axis, and saturation denoting the distance from the central axis to the outer shell of the cylinder (Kamvysselis & Marina, 1999). This description of the HSV colour space can also be found in Smith (1997), also as a conical representation.

This chapter presents a method of colour image segmentation, using a measure of dependency between the image elements. The proposed measure is based on a pixel metric defined for HSV colour space. By this measure, a fuzzy similarity relation is defined, which next is used to introduce a clustering method that generates a partition and so a segmentation of the considered images. A formal comparison analysis between the proposed fuzzy clustering method and other clustering approaches has been discussed in Tabakov (2007). It has been presented a high quality, comparing the achieved results with respect to the corresponding fuzzy c-means algorithm segmentation results, for monochromatic computed tomography images. And this has been realised by introducing a new distance measure (i.e., metric space distance) and also a suitable definition of the notion of "segmentation accuracy" (the cluster pairs have been compared by using the well-known algebraic sum t-conorm: a more formal treatment is omitted). Thus, this chapter is a natural extension of the proposed concept (based on fuzzy relation) to colour images. The achieved segmentation results are used to develop an image enhancement algorithm, dedicated to CT-perfusion images.

BASIC NOTIONS

One can think of crisp relation as a tool for representing the presence or absence of association or interaction between the elements of two or more sets. A generalisation can be obtained by using the notation of fuzzy subsets to allow for various degrees or strength of dependency between elements. Let X and Y be finite sets. A fuzzy relation or a *binary fuzzy relation* $\rho(X,Y)$ (or in short: ρ) is a function from $X \times Y$ to the closed interval [0,1] (Zadeh, 1965). A convenient representation of ρ is the *membership matrix* of ρ, that is, $M_\rho =_{df} [\rho(x, y)] /_{(x, y) \in X \times Y}$, where any $\rho(x, y) \in [0,1]$. Some basic properties and operations over fuzzy relations, used in this chapter, are presented next.

Let $\otimes$: $[0, 1]^2 \rightarrow [0,1]$ be a binary operation over [0,1] which is commutative, associative, monotonic, and has 1 as unit element. Any such operation is called to be a *t-norm*. The t-norm operation provides the characterization of the AND operator. The dual t-*conorm* $\oplus$ (called also: *s-norm*), characterizing the OR operator, is defined in a similar way having 0 as unit element (a more formal treatment is omitted). In general any t-transitivity closure depends on the used triangular norms. Without loss of generality, Zadeh's t- and s- norms (corresponding to the logical operations minimum and maximum) are used below (Bronstein, Semendjajew, Musiol, & Mühlig, 2001; Schweizer & Sklar, 1963).

Let ρ(X,Y) and σ(Y,Z) be two fuzzy relations. The *composition* of ρ and σ, that is, $\rho \circ \sigma$ produces a new fuzzy relation τ(X,Z) defined as follows:

$\tau(x,z) =_{df} [\rho \circ \sigma](x,z) = \oplus_{y \in Y} \rho(x,y) \otimes \sigma(y,z)$, for any $x \in X$ and $z \in Z$.

It can be shown that any such composition is associative but not commutative. So it can be generalised for more than two (but a finite set) of fuzzy relations. Compositions of binary fuzzy relations can be performed conveniently in terms of the well-known membership matrices of the relations.

Next, we shall restrict our attention only to the case when ρ and σ are some binary fuzzy relations over X. The *t - union* of ρ and σ is defined as follows: $(\rho \cup \sigma)(x, y) =_{df} \rho(x, y) \oplus \sigma(x, y)$, where $x, y \in X$. The basic properties of the well known (crisp) binary relations over X are extended as follows. A binary fuzzy relation ρ is called *reflexive* if $\rho(x, x) = 1$, for any $x \in X$. We shall say that ρ is *symmetric* if $\rho(x, y) = \rho(y, x)$, for any $x, y \in X$ and also *t - transitive* if $\rho(x, z) \geq \rho(x, y) \otimes \rho(y, z)$, for any $x, y, z \in X$.

Without loss of generality the Zadeh's max-min transitivity is assumed below. The *transitive closure* of any binary fuzzy relation ρ which is not transitive can be realised by using the following simple algorithm (some more efficient algorithms can be also used, for e.g., see Mordeson, Malik, & Cheng, 2000). Here, the transitive closure of ρ is denoted by ρ^+.

Algorithm 1 (transitive closure)
Input: ρ
Output: ρ^+
Begin
1. Let $\rho^* =_{df} \rho \cup (\rho \circ \rho)$;
2. If $\rho^* \neq \rho$ Then set $\rho =_{df} \rho^*$ and go to (1) Else $\rho^+ =_{df} \rho^*$.

End. □

Let ρ be a binary fuzzy relation over X. We shall say that ρ is a *fuzzy equivalence relation* if ρ is reflexive, symmetric and t-transitive. If ρ satisfies only the first two properties (i.e., the reflexive and symmetric axioms) then ρ is known as *fuzzy similarity relation.*

Every fuzzy relation can be uniquely represented in terms of its α-*cuts*, that is,

$$\rho = \bigcup_{\alpha \in [0,1]} \alpha \cdot \rho_\alpha ,$$

where ρ_α is a crisp relation over X.

It can be shown that if ρ is an equivalence relation, then each α-cut ρ_α is a crisp equivalence relation on X. Let $\pi(\rho_\alpha)$ denote the partition corresponding to the equivalence relation ρ_α. Two elements *x* and *y* belong to the same class of this partition if and only if $\rho(x, y) \geq \alpha$. Analogically, a fuzzy similarity relation determines a crisp similarity relation for each of its α-cuts. And this relation can be used to generate a crisp partition by applying the above transitive closure algorithm.

THE SEGMENTATION METHOD

Image segmentation is essentially a process of pixel classification, wherein the image pixels are segmented into subsets by assigning the individual pixels to classes (clusters). Hence, any segmentation is a process of partitioning an image into some regions (or classes) such that each region is homogeneous and none of the union of two adjacent regions is homogeneous (Gonzalez & Woods, 2002). On the other hand, any such process can be considered as related to a construction of an equivalence relation over the set of all image pixels. The following definition is introduced below.

Definition 1

Let P be the set of all image pixels, $IS \subseteq P \times P$ be an equivalence on P, $Q \subseteq P$ be a subset and $H(\cdot)$ be one-argument homogeneity predicate such that: H(Q) = 'True' if and only if $Q \in P/IS$ (the quotient set with respect to IS). Then, image segmentation is a process of constructing IS. Any element of P/IS is said to be a cluster.

Let $\pi(IS)$ be the corresponding partition generated by IS. We have: $\pi(IS) = P/IS$. And hence, the accuracy of the segmentation process will depend on the accuracy of constructing IS.

The classical approach assumes some partition of the given set of pixels which is generalised to the notion of pseudopartition (or fuzzy partition: in accordance with the classical fuzzy c – means algorithm) (Bezdek, 1981). However, in this case an important problem is the need of information about the cardinality of the assumed initial partition.

The proposed approach is based on a fuzzy similarity relation and so it is independent on any such initial information. Any such relation determines a crisp similarity relation for each of its α-cuts. In accordance with Definition 1, this relation can be used to generate a crisp partition by the applying of a transitive closure algorithm. Thus, the fuzzy clustering problem can be thought as a problem of constructing an appropriate fuzzy relation on a given set of data. As in Kamvysselis and Marina (1999), an appropriately defined pixel metric in the *conical HSV colour space* is assumed below. The obtained fuzzy similarity relation is given in the next definition.

Definition 2

Let $P =_{df} \{p_1, p_2, \dots, p_n\} \subseteq HSV$ be a set of pixels, where $p_k =_{df} (h_k, s_k, v_k)$ (used notation: h_k – Hue(p_k), s_k – Saturation(p_k), v_k – Value(p_k), k=1,2,3…,n). Any element of P can be interpreted as a vector defined in the HSV colour space. The fuzzy relation ρ on P, that is, $\rho: P \times P \to [0,1]$, can be defined as follows:

$$\rho(p_i, p_j) =_{df} 1 - \sqrt{\frac{1}{2}\cdot(s_i^2 - 2\cdot s_i s_j \cos(\theta) + s_j^2 + (v_i - v_j)^2)}$$

(for any $p_i, p_j \in P$).

The used measure has an output value in the range of [0, 1], with 1 being most similar, and 0 being completely dissimilar. The weighting of the saturation results in darker colours being considered to be more similar than lighter colours, which is a known perceptual phenomenon. The multiplication by ½ is used only to map the distance values to the range of [0,1], and is determined by dividing 1 by the largest distance possible. In the equation, both saturation and value range from 0 to 1. Therefore, the largest distance possible is the diagonal of the circle of maximal saturation, which is 2.

The angle θ used in *Definition 2* is defined as follows:

$$\theta =_{df} 2\pi\Delta h\sqrt{\max(s_i, s_j)},$$

where Δh is computed as:

$$\Delta h =_{df} 2\cdot\begin{cases} 1 - |h_i - h_j|, & \text{if } |h_i - h_j| > \frac{1}{2} \\ |h_i - h_j|, & \text{else.} \end{cases}$$

Because the hue space is circular, the absolute distance between two hues is at most half the circumference of the hue space. The multiplication of Δh by $\sqrt{\max(s_i, s_j)}$ in the above equation decreases the magnitude of the angle between the two colours being compared, and therefore increases their similarity, in direct relation to the saturation of the two pixels. The square root is then taken so that the saturation weighting rises to 1 faster than linearly. As a result of the weighting, colours, which are saturated very little, and are close to the white-grey axis, are always similar regardless of hue. Colours which are very saturated, and which are very distinct from the colours in the range of white to grey, on the other hand, are always distinct. Because the max function is used, the comparison between two pixels is the same both ways. By weighting the saturation values used throughout this computation, a true distance in conical space is created. The only difference between a pure Euclidean distance and the distance being used is that the hue angle gets modified by the saturation of the pixels being compared. This results in the elongation of the matching space along the hue but not the saturation or the value axes of the space. This property is desirable due to the human visual clustering preferences (Kamvysselis & Marina, 1999).

In accordance with *Definition 2*, ρ is a fuzzy similarity relation, as it satisfies the reflexivity and symmetry axioms. Moreover, in order to achieve better clustering results with respect to the considered CT-perfusion images, the following two assumptions have been proposed:

1. It is assumed that the set of all hue values is divided into intervals in 60 degrees (see Figure 1 below).
2. The values of the defined fuzzy relation have been computed for each interval separately, which means that the degree of relationship for any two elements that belong to different intervals is assumed to be 0.

Definition 3

The α value, required to determine the appropriate α-cut of ρ, is defined as the fuzzy expected value (Schneider & Craig, 1992) of all co-domain values of the considered relation. These values can be interpreted as degrees of membership of some fuzzy set, as they range the number interval of [0; 1], and so a fuzzy expected value can be determined. Thus, the α value is considered as a fuzzy expected value:

$$\alpha =_{df} FEV(\,cod(\rho)\,).$$

The obtained value of α under Definition 3 was verified in experiments (using radiologists expert opinion for assessing).

THE IMAGE ENHACEMENT TECHNIQUE

A fuzzy similarity relation-based colour image segmentation algorithm is presented below. Provided there is no ambiguity and for convenience, here only the main steps of this method are included.

Algorithm 2 (image segmentation)

Input: P_o (*the input image*)

Output: P_s (*the segmented image*)

1. Input of the original image P_o: $P_o =_{df} \{p_1, p_2, \ldots, p_n\} \subseteq R^d$, where *n* is the number of pixels in P_o and $d = 3$ for colour images (using HSV colour space);
2. Define the appropriate fuzzy relation ρ under Definition 2. $\rho : P_o \times P_o \rightarrow [0,1]$;
3. According to Definition 3, chose the α-cut for ρ;
4. In accordance with the selected value for α, define the corresponding crisp similarity relation ρ_α;

Figure 1. The considered colour intervals

5. Apply the transitive closure Algorithm 1 to ρ_α (if ρ_α is not transitive) and so define ρ_α as a crisp equivalence relation;
6. Select the image clusters as the equivalent classes, obtained under the partition induced by ρ_α;
7. Generate the segmented image P_s with respect to the image clusters, obtained in step (6), which stops the algorithm. □

The corresponding process of image segmentation is always realised in a unique way (i.e., the process of constructing *IS*) and the algorithm converges in a finite time (Tabakov, 2003, 2006). Hence, for any P_0 there exists exactly one P_s obtained from P_0 under the above algorithm.

Next, the generated clusters (subsets of image pixels) are used to enhance the considered CT-perfusion images. For each cluster a dominant colour is chosen. The dominant colour is the colour of maximal number of pixels for a given cluster. Next, all pixels with colours that belong to this cluster take the value of the dominant colour. Thus, the effect of the proposed process is the clarification of the images. According to the opinion of radiologists and neurosurgery experts, the achieved image processing results allow processing with better understanding of the presented image information.

THE EXPERIMENTAL RESULTS

An illustration of the above approach concerning CT-perfusion images is given below. The aim of the performed image processing is the increasing of the clarity of the considered images, which allows better understanding of the brain perfusion process. The perfusion process is visualized using a colour palette that ranges from violet to red, concerning the corresponding brain perfusion characteristics (see Figure 2(a) below).

The proposed image processing technique enhances CT-perfusion images by reducing the unwanted colour fuzziness, as it is shown on Figure 2(b) above. Figure 3 shows a zoomed subimage of the considered CT-perfusion case, for better presentation of the image enhancement results.

Some other image enhancement result is shown next.

Figure 2. (a) CT-perfusion image; (b) The corresponding image enhancement result

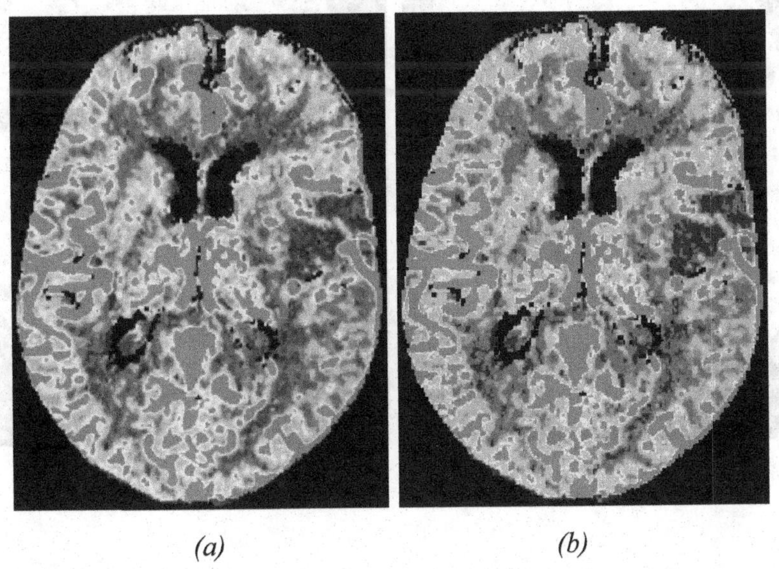

(a) *(b)*

Figure 3. The image on the left presents a subregion of the considered CT-perfusion image. The result achieved by the proposed technique is shown on the image of the right

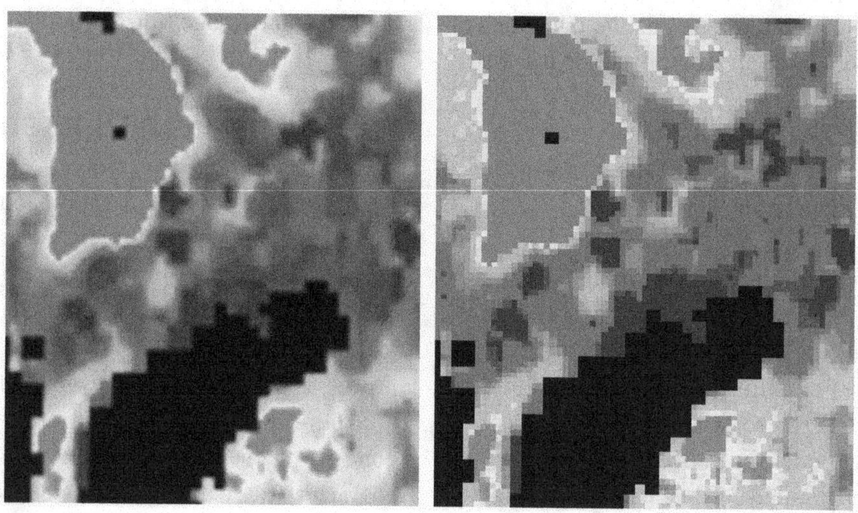

Figure 4. (a) CT-perfusion image; (b) The achieved result presents the regions of high activeness more precisely, reducing the unwanted colour fuzziness

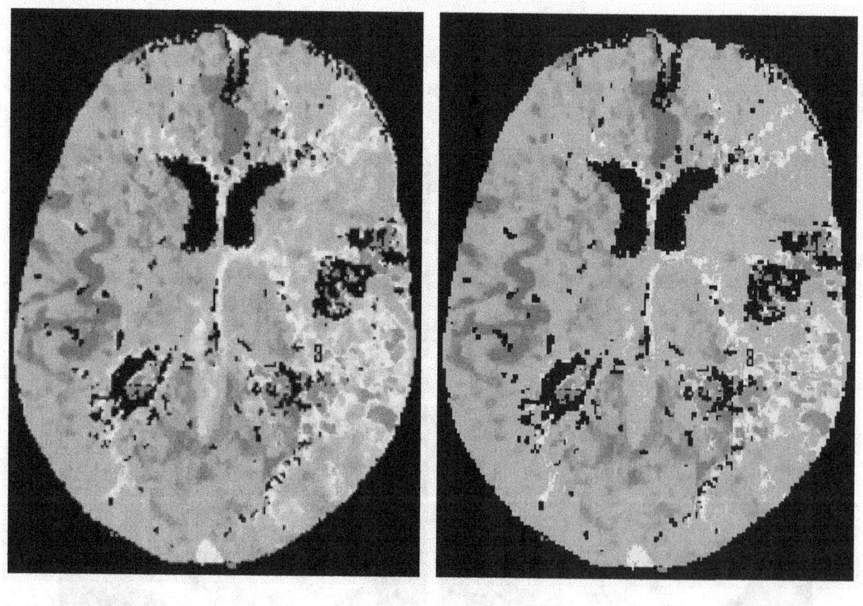

CONCLUSION

This chapter considers properties of fuzzy sets and relations and uses them as a foundation for the development of new methods for colour image enhancement, based on fuzzy segmentation. The major advantages of the proposed algorithm are its simplicity and speed, which allows it to run on large data sets. The corresponding process of image segmentation is always realised in a unique way and the algorithm converges in a finite time. The obtained image segmentation is realised as an automated computational process. The classical clustering algorithms require that the desired number of clusters be given in advance. This can be a problem when the clustering problem does not specify any desired number of clusters. The number of clusters should reflect the structure of the given data. The above-proposed method is based on using a properly defined fuzzy similarity relation and so it satisfies this need. In experiments, this algorithm demonstrated a high quality considering the expert's judgment. The segmentation results show appropriate structural separation, according to the corresponding radiological analysis, which gives the opportunity of making better decisions.

ACKNOWLEDGMENT

This work was supported by the Polish Ministry of Science and Higher Education grant No N518 022 31/1338, 2006 ÷ 2008.

REFERENCES

Bezdek, J.C. (1981). *Pattern recognition with fuzzy objective function algorithms* (p. 256). New York: Plenum Press.

Barges, G., & Aldon, M. (2000). A split-and-merge segmentation algorithm for line extraction in 2d range images. In *Proceedings of the 15th International Conference on Pattern Recognition*, (Vol. 1, pp. 441-444).

Bronstein, I.N., Semendjajew, K.A., Musiol, G., & Mühlig, H. (2001). Taschenbuch der Mathematik. *Verlag Harri Deutsch* (p.1258).

Chamorro-Martinez, J., Sanchez, D., Prados-Subrez, B., Galin-Perales, E., & Vila, M.A. (2003). A hierarchical approach to fuzzy segmentation of colour images. In *Proceedings of the 12th IEEE International Conference on Fuzzy Systems*, (Vol. 2, pp. 966-971).

Gillet, A., Macaire, L., Bone-Lococq, C., & Pastaire, J. (2001). Color image segmentation by fuzzy morphological transformation of the 3d color histogram. In *Proceedings of the l0th IEEE International Conference on Fuzzy Systems*, (Vol. 2, pp. 824-824).

Gonzalez, C.R., & Woods, E.R. (2002). *Digital image processing*. Upper Saddle River, NJ, USA: Prentice Hall.

Helgason, C.M., Jobe, T.H., Malik, D.S., & Mordeson, J.N. (1999). Analysis of stroke pathogenesis using hierarchical fuzzy clustering techniques. In *Proceedings of the 5th International Conference on Information Systems, Analysis and Synthesis: ISAS'99*, Orlando, FL, USA, (pp. 477-482).

Helgason, C.M., Jobe, T.H., Mordeson, J.N., Malik, D.S., & Cheng, S.C. (1999). Discarnation and interactive variables as conditions for disease. In *Proceedings of the 18th International Conference of the North American Fuzzy Information Society NAFIPS'99*, New York, (pp. 298-303).

Kamvysselis, M., & Marina, O. (1999). *Imagina: A cognitive abstraction approach to sketch-based image retrieval* (p. 157). Massachusetts Institute of Technology.

Kerre, E.E., & Nachtegael, M. (2000). *Fuzzy techniques in image processing*. Heidelberg, New York: Physica Verlag.

König, M., Klotz, E., & Heuser, L. (2000). Diagnosis of cerebral infarction using perfusion CT: State of the art. *Electromedica*, 9-12.

Li, Y., Li, J., Dong, H., & Gao, X. (2002). A fuzzy relation based algorithm for segmenting color aerial images of urban environment. In *Proceedings of the ISPRS Technical Commission II Mid-Term Symposium on Integrated System for Spatial Data Production, Custodian and Decision Support, Xi'an, P. R.* China, (pp.271-274).

Moghaddamzadeh, A., & Bourbakis, N. (1997). A fuzzy region growing approach for segmentation of color images. *Pattern Recognition, 30*(6), 867-881.

Mordeson, J.N., Malik, D.S., & Cheng, S.C. (2000). *Fuzzy mathematics in medicine*. Heidelberg, New York: Physica Verlag.

Sakamoto, S., Ohba, S., Shibukawa, M., Kiura, Y., Arita, K., & Kurisu, K. (2006). CT perfusion imaging for childhood moyamoya disease before and after surgical revascularization. *Acta Neurochirurgica, 148*(1), 77-81.

Schneider, M., & Craig, M. (1992). On the use of fuzzy sets in histogram equalization. *Fuzzy Sets and Systems, 45*, 271-278.

Schweizer, B., & Sklar, A. (1963). Associative functions and abstract semi-groups. *Publ. Math. Debrecen, 10*, 69-81.

Shiji, A., & Hamada, N. (1999). Color image segmentation method using watershed algorithm and contour information. In *Proceedings of the International Conference on Image Processing, 4*, 305-309.

Smith, R. J. (1997). *Integrated spatial and feature image systems: Retrieval, analysis and compression*. Doctoral thesis, Columbia University, Graduate School of Arts and Sciences.

Suri, J.S., Setarehdan, S.K., & Singh, S. (2002). *Advanced algorithmic approaches to medical image segmentation*. London: Springer-Verlag.

Tabakov, M. (2001). Using fuzzy set theory in medical image processing: Basic notions and definitions. *Reports of the Department of Computer Science of Wroclaw University of Technology PRE* (No.7, p. 35). Warsaw, Poland.

Tabakov, M. (2003). *Medical image segmentation algorithms using a fuzzy equivalence relation* (p. 80). Doctoral thesis, Wroclaw University of Technology, Wroclaw, Poland.

Tabakov, M. (2006). A fuzzy clustering technique for medical image segmentation. In *Proceedings of the 2006 IEEE International Symposium on Evolving Fuzzy Systems*, Ambleside, Lake District, UK, (pp.118-122).

Tabakov, M. (2007). A fuzzy segmentation method for Computed Tomography images. *International Journal of Intelligent Information and Database Systems, 1*(1), 79-89.

Tekşam, M., Çakır, B., & Coşkun, M. (2005). CT perfusion imaging in the early diagnosis of acute stroke. *Diagnostic and Interventional Radiology, 11*(4), 202-205.

Tizhoosh, H.R. (1998). Fuzzy image processing: Potentials and state of the art. In *Proceedings of IIZUKA'98, the 5th International Conference on Soft Computing*, Iizuka, Japan, (Vol. 1, pp. 321-324).

Zadeh, L. (1965). Fuzzy sets. *Information and Control, 8*, 338-353.

Zhong, D., & Yan, H. (2000). Color image segmentation using color space analysis and fuzzy clustering. In *Proceedings of the 2000 IEEE Signal Processing Society Workshop*, (Vol. 2, pp. 624-633).

Chapter XVI
Fuzzy Mediation in Shared Control and Online Learning

Giovanni Vincenti
Research and Development at Gruppo Vincenti, Italy

Goran Trajkovski
Algoco eLearning Consulting, USA

ABSTRACT

This chapter presents an innovative approach to the field of information fusion. Fuzzy mediation differentiates itself from other algorithms, as this approach is dynamic in nature. The experiments reported in this work analyze the interaction of two distinct controllers as they try to maneuver an artificial agent through a path. Fuzzy mediation functions as a fusion engine to integrate the two inputs to produce a single output. Results show that fuzzy mediation is a valid method to mediate between two distinct controllers. The work reported in this chapter lays the foundation for the creation of an effective tool that uses positive feedback systems instead of negative ones to train human and nonhuman agents in the performance of control tasks.

INTRODUCTION

Technology plays a dominant role as each new product is designed and placed on the market. When a technological solution oversees the cooking of rice, for example, one may or may not require much improvement. But, when we implement solutions that control a passenger jetliner, we need to be sure that the concepts and solutions are well founded.

The fly-by-wire airplane idea is similar to the one of an automobile that operates with little input from the driver, or a joystick-operated heavy-duty machine that lifts heavy loads. What these three examples have in common is the existence of mediation between signals from the operators and the actual sequence of actions taken by the machine in response to the operator. Sometimes, the response of the machine depends on the operator's input; there are, however, times when the controls are autonomous responses to situations at hand.

Who (and when) should have control over the machine: the human or the automatic operator? Should we share control, and if so, how? Norman (2005) points out that what we understand as *shared* control is not really shared. For example, in early stages of development, the machine used to supervise and control any part of the operations in fly-by-wire airplanes, and now they oversee operations and still control flight. But, when the conditions don't meet the standards of operations, the pilot is left alone. Therefore, it is either the pilot controlling the plane on his/her own, or the control system. There is *no interaction* between the two.

We need a system that allows for greater interaction between the human operator and the digital one. There are examples (e.g., Caterpillar machines) that allow operators to perform certain tasks only through automated systems (Grenoble O'Malley, 2005). Also, the two major airliner producers, Airbus and Boeing, are gearing toward advanced fly-by-wire technologies (Wallace, 2000), as the auto industry is attempting to infuse automation concepts in their products (Norman, 2005). These important initiatives stress that automation is becoming a predominant part of our everyday life. It is important, though, to realize that there is a need of some type of balance between controllers, humans, computers, or hybrids.

As the interaction between machines and digital controllers increases, we need to:

1. Find a better way to mediate control in dual control systems, when two (or more) operators are controlling the same machine;
2. Then, replace one of the human operators with a digital one and investigate the interaction between the two entities using a mediation system (such as the one introduced in this chapter). This can be done if we:
3. Create a framework for using simulation and virtual reality to test-drive solutions; and
4. Implement these systems as a part of the actual operations of the machines under scrutiny.

This chapter will review some of the current problems in this "new" Information Fusion problem area, which extends the classic, static and traditional approaches to add a dynamic component that we call Fuzzy Mediation.

This chapter is organized as follows. First, we give background information about Information Fusion, as well as other preliminary concepts relevant to the framework for Fuzzy Mediation. Then, we introduce our original concept of Fuzzy Mediation as a theoretical solution to shortcomings highlighted. Next, we give a detailed overview of our algorithm and a breakdown of the three functional units that create it, and report on our conceptual experiments. Then, we introduce our application of Fuzzy Mediation to a robotic line follower simulation and discuss other problem fields that may implement this algorithm. Finally, we conclude the chapter.

PRELIMINARIES

In this section, we overview preliminary concepts needed to follow the rest of the chapter.

Control is a concept that involves the interaction of multiple entities (by definition at least two). In a situation of control, one of the subjects is identified as the one who interacts directly with one object, that is, directing the object's every move (WordNET, n.a.).

It is the discipline of Information Fusion we turn to when we seek solutions for joint control in environments with more than one operator. Information Fusion is the process of taking multiple inputs and creating a single output (Kokar, Tomasik, & Weyman, 2004). As Kokar states, many Information Fusion algorithms are biased in their operations. As they are designed, their operations are set, and the fusion is carried out by simple execution of an algorithm. In most cases, fusion algorithms are predefined and static, and are not adaptable to the circumstances that influence the controlled object. They also reside at lower levels, such as the level of sensors. As an alternative, an adaptive Information Fusion system is able to offer shared control effectively.

A shared-control environment demands for applications of fusion to higher levels, in cases where supervised and collaborative learning is required, as in Trajkovski (2005). Supervised machine learning techniques represent a set of operations that "learn a task starting from a suite of examples" (Abad, Suarez, Gasca, & Ortega, 2002). Information Fusion provides for supervised machine learning algorithms that learn from examples as the expert (operator) is operating (controlling) the machine.

This approach represents a paradigmatical shift, as the learner becomes capable of interacting efficiently with the environment. It is an effective shift from the supervised learning domain to the collaborative learning domain (Gokhale, 1995; Hammonds, Jackson, DeGeorge, & Morris, 1997). It has been noted that many proposed solutions promote individual, instead of true collaborative learning (Dugan & Glinert, 2002). Kwek (1999) explored interaction between humans and computing when learning is performed by means of an apprentice model; however their focus is on the classification of items.

Another relevant concept here is the concept of agent. Agents attempt to solve issues of coordination, cooperation and learning (Arai & Ishida, 2004). Multiple agents' performance in an unknown environment is discussed by Trajkovski (2007), where online collaborative learning is applied to several agents, facilitating the learning process itself.

Positive Reinforcement

The research community widely accepts that learning in the brain resides in the plasticity associated with alterations in synaptic efficacy. The most widely accepted theory of learning at the cellular level is aligned with Hebb's reinforcement ideas (1949). This theory introduces the possibility that, as neurons fire, the synaptic connection between them is strengthened. The stronger synapses, in turn, lead to preferred pathways, which seem to then create memories. This process is called long-term potentiation (LTP) and has been widely investigated in Kandel's work, reported in Kandel, Schwartz, and Jessell (2000).

Much mathematical modeling has been drafted. Among the many, neural networks have been widely accepted as the closest representation of the concept of LTP. Among neural networks, we should mention the "artificial" neural networks, or ANNs, and various forms of feed-forward networks trained by back-propagation algorithms. ANNs were introduced by John Hopfield (1982) as a response to the then current faulty models that did not represent well the physiological basis of learning, especially true with back-propagation networks. This is because these models don't explain the interaction with another brain, and the inherent lack of self-organization.

In the reinforcement learning paradigm, the environment acts as a critic rather than a teacher. The downside of this approach is that the learning is slow and not portable to a new task. The subject will have to start all over when presented with another task. Chialvo and Bak (1999, p. 5) argue that this is true because of the lack of positive reinforcement rules. They say that these models thrive on the concept of LTP, but they also state that "long-term synaptic depression (LTD) in the mammalian brain is almost as prevalent as potentiation, but there appears to be little or no understanding of its functional role." Barnes et al. (1994, p. 81) argue that "although it is conceivable that LTP is the critical phenomenon used for storing information, and that LTD may exist simply to reset LTP, it must be noted that it is also conceivable for the converse to be true." Thus, Chialvo and Bak (1999) infer that the "depression" of synaptic efficacy is the fundamental dynamic mechanism in learning and adaptation, with LTP playing a secondary role. The collaboration of these two learning mechanisms produces the outcome that we are all familiar with, the acquisition of a new task. It is important though that LTP and LTD collaborate closely though, because it is much more likely that, as we learn, we perform some actions that are not successful, thus the need to "erase" that action from the set of successful ones through LTD, stimulating only the pathways that have given good results through LTP.

Learning by Imitation

The phenomenon of imitation was never seriously considered by scientists until the 1990s. It was mainly considered to be an unintelligent process in higher primates, and, as such, not worth researching. After its earliest consideration by Thorndyke (1898), it only appears sporadically in psychological literature. The only notable consideration was done by Piaget (1945) in his consideration about the developmental stages in children. Thorndyke imitation considered to be learning by observation, when one entity tries to mimic (copy) another entity's behavior.

With the discovery of mirror neurons by Rizzolatti and Arbib (1998) and Rizzolatti, Fadiga, Gallese, and Fogassi (1996), it seems likely that, after all, we are wired for imitation. The mirror neurons are located in Broca's F5 region in the frontal cortex, that has been found to be primary responsible for human linguistic expression. People with defects in Broca's region are usually not linguistically competent. In the observations of mirror neurons in primates, it has been noted that they fire in a monkey that is observing another monkey tearing or crumbling paper.

With the discovery of mirror neurons and the possible extent of their significance, imitation becomes a bona fide focus of research in learning. The indication for neonatal research is that it is by imitation that we start learning everything about the world and build our first conceptualization of it. Therefore, it seems that imitation could be the base of all learning that happens in the human and it is connected to our very basic biological self. As we know, in learning the more basic the motivation is, the easier it is to assimilate new knowledge toward the satisfaction of an active drive (Trajkovski, 2007). The main thesis of that work is that humans learn about the environment via interactions with it, based on their inborn Piagetian schemes, and from other humans via imitation conventions.

Mirror neurons, as hardware for imitation, were discovered by Rizzolatti et al. (1996), while performing experiments on primates. They discovered that a specific area of the brain, called F5, is involved with imitation of tasks. The neurons in this area are extremely active in macaque monkeys when they are performing movements that are goal-oriented. Such neurons were observed to be extremely active also when the monkey observed another monkey or a human perform those very movements. This finding, together with Kandel's LTP and the strengthening of synapses through repetitive stimulation, hints to the fact that observation and imitation are phenomena that lead to learning a task.

This past decade has been a great time for research in imitation. Heyes (2001) reports that advancements were made under various biological aspects. From an evolutionary standpoint, chimpanzees and birds were successfully studied for imitation patterns. Moving onto humans, the first point of view analyzed was the developmental one, where newborns were studied for the reflex of sticking out their tongue when an adult performs that action in front of them. Further studies were conducted on 18-month-old infants to study their imitation of movements that adults reported being intentional. When asked, the infants could not always motivate the performance of the imitated movements. Finally, Heyes reports that studies have shown that autism is linked to deficits the subjects have in the processes of imitating the performance of tasks they observe. Other studies were also performed on adults as well as at the organ level, with the study of the brain and patterns associated with task performance. The approach that we are presenting in this chapter is based on learning by imitation in adult humans.

The Fuzzy Paradigm

As a focal tool used in the proposed solution, in this section we briefly review the fuzzy sets paradigm. In classical, Aristotelian mathematics (and logic) everything is black and white, true or false, 1 or 0. The fuzzy paradigm introduces degrees of grayness, degrees of truth, the interval [0, 1] (Zadeh, 1965). In terms of sets, in Aristotelian terms, say, an element either belongs completely to a (crisp) set, or does not. In fuzzy terms, an element can belong to a set with some membership degree.

In addition to piece-wise linear (triangular, trapezoidal, etc.) membership functions can take any other shape, as long as the rank is [0, 1], depending on the problem at hand. In engineering applications, however, for the most part the explorations of the possible improvements based on the fuzzy paradigm start with piece-wise linear membership functions when providing proofs of concept, and deviate toward alternative, smoother functions later in the process.

Fuzzy Sets in Cognitive Processes in Humans

In the past, we have investigated the use of weighted outputs and fuzzy-rule based in a wide range of cognitive processes, relevant to this work (e.g., Trajkovski, 2004; Trajkovski et al., 1997), especially in the area of human learning, contingent negative variation (CNV) brain waves, and other EEG (Electroencephalogram brain potentials) analyses. In the Dynamic CNV experiment, for example, we use fuzzy sets and inference in substituting a CNV expert with a that automatically detects if a brain form contains negative variation, indicative of learning when to expect an imperative stimulus.

TOWARD A NEW TRAINING SOLUTION

This section introduces the ideas behind our original concept of Fuzzy Mediation and gives an overview of the contexts suitable for the implementation of our algorithm.

The problem of shared control is very relevant in a training setting. Here, we aim to create an efficient training framework for operating machinery using a computer simulation (virtual environment).

Initially, we focus on training car drivers, using a positive reinforcement approach, in a trainer-trainee shared control collaborative environment. This collaboration will be achieved greatly by devising a tool based on our Fuzzy Mediation approach. The setup involves two controllers that are trying to steer a car within a virtual environment. One of the controllers is an expert driver, and the second a novice. When both subjects are trying to control the car, whom should control go to? If the expert drives, the novice will passively observe the expert's driving, and not really learning how to drive; the novice is just being exposed to examples of driving situation cases. If the novice driver is (fully) in command, the expert can only interact with the novice by simply correcting his/her actions by negative reinforcement.

Our proposed approach shares the control between the expert and the novice by stimulating the involvement of the novice. Using a Fuzzy Mediation control, inputs from both participants in the process are evaluated and considered, and

Figure 1. Architecture for the training system

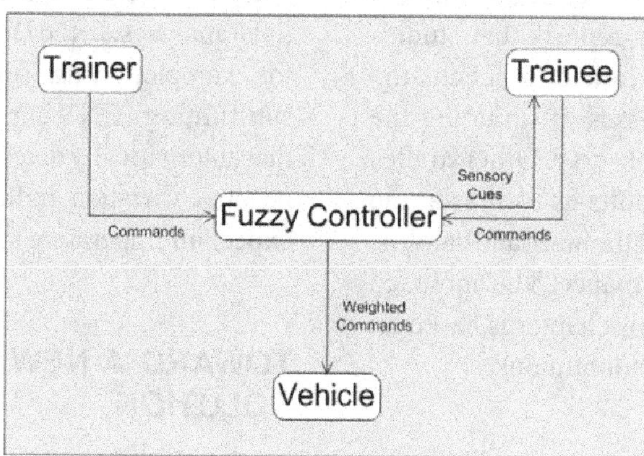

Figure 2. Fuzzy mediation system with two controllers

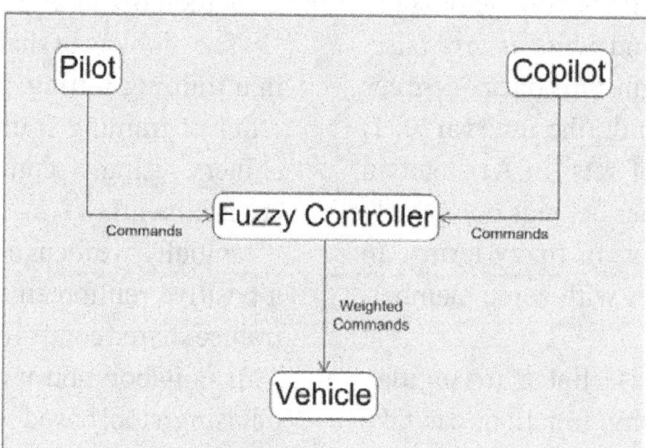

the overall control gives more or less weight to the novice based on his/her past performance. Our approach is a generalization of the negative reinforcement approach; the classical approach is just a marginal case of this one. The overview of our approach follows.

This solution involves both users at the same time. The expert and the novice are given a set of identical controls, and they both try to control the very same car within the same simulation. The inputs received by the controls of the car are analyzed and compared to the other participant's (novice to expert, expert to novice). Initially, the mediation system allows the vehicle to be controlled by the expert. As the novice's actions mimic/imitate more and more the correct ones of the expert, control shifts to the novice. Visual clues (indicators) are given for the novice to indicate progress (i.e., shifting control toward his/her side). If the novice's actions start deviating from the expert's, should control shift back to the expert? Figure 1 illustrates the architecture of such a system. Based on motivating results of success in positive reinforcement learning (Chialvo & Bak, 1999), we expect the novice to learn significantly faster than in a negative reinforcement environment.

An extension of the system architecture in Figure 1 is given in Figure 2, where the control mediates between a digital and a human controller, suitable for, for example, some fly-by-wire

Figure 3. Alternative system architectures with increased cardinality of novices: (a) One-on-one training; (b) One to many training, parallel stand-alone simulations; (c) Threaded training simulations architectures for one-to-many training sessions

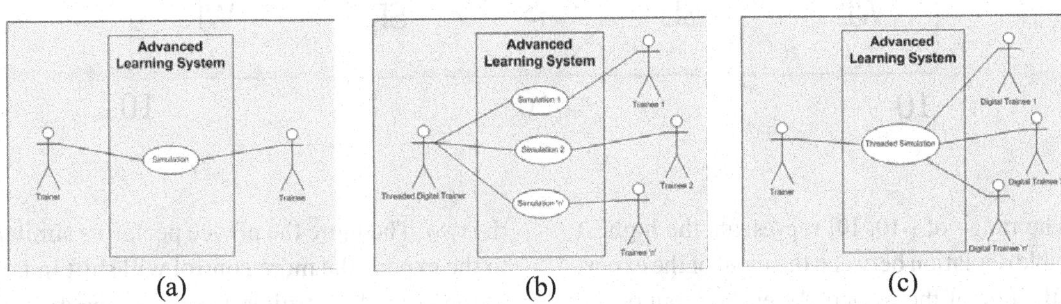

airplanes in service today. In this case, the pilot is the human controller, and the copilot can be either the human copilot or a digital smart box.

This work will also consider alternative architectures (Figure 3) in environments with one expert and multiple novices are being trained.

FUZZY MEDIATION

In this section, we elaborate on our concept of Fuzzy Mediation. We explore the three steps that allow us to successfully mediate between inputs coming from two controllers.

The concept of Fuzzy Mediation dynamicizes the static world of Information Fusion. It extends greatly the possibility of applications of this field of computing.

Fuzzy Mediation aims to solve problems that are innate with the concept of shared control among multiple agents. The birth of this concept was set in a training environment where two agents are interacting. Both agents can be human, nonhuman or a mix of the two. In the typical scenario, depicted in Figure 2, the first controller functions as a pilot, or expert user, and the second controller functions as a copilot, of novice user. It is our assumption for this work that the expert controller is the one that performs actions as expected, and the novice controller is at an early phase of learning the tasks.

As it would be impossible for a vehicle to be controlled simultaneously by two different controllers, we looked at concepts of Information Fusion for a solution. The fuzzy controller performs three distinct operations.

1. Analysis of the inputs to determine the closeness of control.
2. Performs a revision of the weight of control between the expert and the novice controller.
3. Computes the value of the single output.

Analysis of the Inputs

Classical set theory leaves little room for gray areas. For example, a room's temperature can be either hot, medium or cold. This classic framework does not take into consideration the possibility of the same room being perceived as comfortable or chilly by two different persons. Fuzzy sets provide a solution that takes into consideration values that fall within multiple sets.

These kinds of sets can be utilized within the field of Information Fusion applied to a situation of multiple controllers trying to interact with a vehicle by a means of comparison.

The analysis of the inputs coming from the two controllers aims at understanding the distance between the values. The input coming from the expert user is mapped to the center of the range [-10, 10], as shown in Figure 4.

Figure 4. Breakdown of the sets used for comparing controller inputs

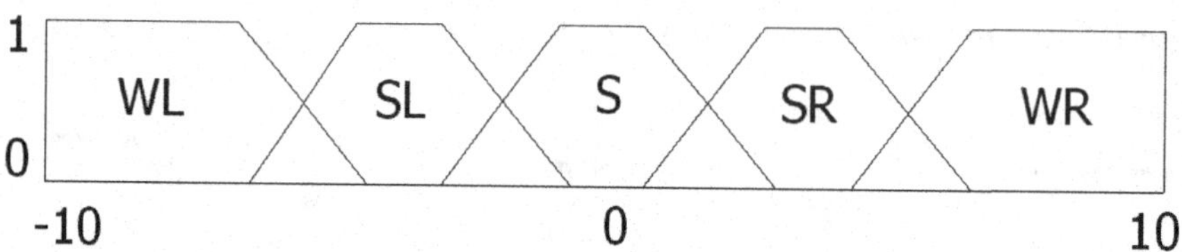

The range of [-10, 10] represents the highest possible deviation between the input of the expert and the one of the novice. After the input of the expert becomes the center of this domain, we calculate the difference between the value of the original input of the expert and the one of the novice. This value is then also mapped to the domain shown above.

When the distance between the inputs is mapped, it will fall within one or two sets that span over the range, as shown in Figure 4. The five sets we deal with are: WL (Wide deviation to the left), SL (Slight deviation to the left), S (Similar), SR (Slight deviation to the right), and WR (Wide deviation to the right). In fuzzy set theory, the value can belong to a set with a certain degree of belonging. Such degree of belonging is calculated by a membership function. In our case, we use a simple membership function, also shown in Figure 4.

The application of a linguistic modifier to the deviation between the controller inputs also keeps in consideration the degree of belonging–or membership value–of the difference to the different sets. It is important to note that a value may fall completely (degree of belonging = 1) within one set, or the value may belong mostly (.80) to the set of deviations deemed as Similar (S), and partly (.20) to the set of Slight deviations to the left (SL).

Revision of the Weight of Control

Fuzzy Mediation sees the fusion of the inputs of the expert and the novice as a balance between the two. The more the novice performs similarly to the expert, the more control will shift in favor of the second controller. Likewise, the more the control of the novice differs from the one of the expert, the more control will shift back toward the expert.

Given this preamble, this second part of the Fuzzy Mediation algorithm analyzes the linguistic modifiers applied to the deviation of the inputs during the first phase. Control is mapped to the range of [-1, 1], where a control weight of –1 identifies a control fully in the hands of the expert, a value of 1 instead refers to control managed by the novice. Figure 5 shows the visualization of this concept. As we apply the concept of fuzzy sets to this section of the algorithm, a value in the range (-1, 1) identifies a control that is mixed in a certain proportion between the expert and the novice. The arrow in Figure 5 shows a possible weight of the mediation of control between the trainer and the trainee. At the beginning of the simulation the weight has a value of –1.

When the classification of the distance between inputs is analyzed, there are several actions that can be taken, based on Mamdani rules (Mamdani & Assilian, 1975). If the inputs are classified as similar, control is given more to the novice. If the deviation between inputs is slight, then control stays unvaried. If instead the deviation is wide, more control is given to the novice. The shifting of the weight from one controller to the next occurs in a linear fashion, with increments or decrements of 0.2 points on the range [-1, 1] presented earlier. In the case of a distance between inputs that belongs to two sets, then we will multiply the degree of

Figure 5. Fuzzy sets that regulate the balance of control between the expert and the novice controller

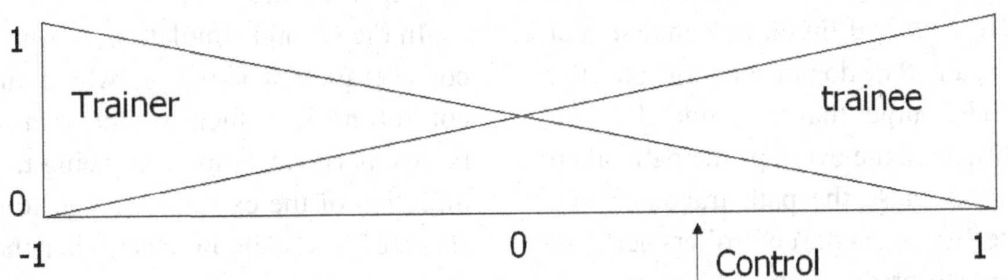

belonging to each of the sets to the action associated with that particular set. Using the example given earlier, a value that belongs to the set S with membership 0.8 will receive an increase in control of 0.2 (the standard increment) multiplied by the membership value, which means an increase of the weight of control of 0.16. The same value also belongs to the set SL with membership 0.2. The action associated with a deviation that is classified as slight is a movement of 0 of the balance between controllers, so the action for this set is calculated by multiplying 0.2, the membership value, to 0. The addition of these values, 0.16 and 0, shows the overall shift in control, which is of 0.16 in favor of the trainee.

Calculation of the Single Output

After the weight of control is updated, we need to calculate a value that will serve as a single input stemming from the original inputs of the two controllers. For this computation, we need to refer to the original values. The equation that regulates this third section of the algorithm is as follows:

Mediated output = μ_T*Expert's input + μ_t*Novice's input

where μ_T refers to the membership value of the weight of control to the Trainer set (Expert) and μ_t refers to the membership value of the weight of control to the trainee set (Novice).

In the case of a driving simulator, we may have an expert applying a turn of 15 degrees to the right and a novice applying a turn of 25 degrees. If the weight of control has a value of –1, the mediated output will have a value of 15 degrees to the right. Likewise, if the weight of control has a weight of 1, the mediated output will be of 25 degrees to the right. If the weight is anywhere in between, for example, $\mu_T = 0.5$ and $\mu_t = 0.5$, the mediated output will be of a 20 degree turn to the right.

SIMULATIONS AND RESULTS

This section reviews our initial demonstration of concept that show and further illustrate the conceptual niche that Fuzzy Mediation satisfies.

The proposed framework is expected to improve the efficiency of the training process. This will be measured through trainees' progress records stored in the database throughout training sessions, as specified in previous session.

In this section, we focus on simulation studies as a proof of concept for the Fuzzy Mediation engine. In our simulations of the operations of the engine, we assume that no learning happens in the novice.

Demonstration of Concept

The first simulation that we performed focuses on two linear paths that intersect at some point, without any deviation on the side of the novice to try and resemble the actions performed by the expert (Figure 6). As we can see, the lines that represent the expert and the novice intersect at one point only, and they do not converge past the intersection. The output that is produced by the Fuzzy Mediation engine overlaps the path taken by the expert, because the path traced by the novice is quite distant. As the controllers perform actions that are closer, the mediated output starts shifting toward the novice's line. After the intersection, the output of the fuzzy controller shows that control is shared somewhat equally between the expert and the novice for a few steps, but then it leans toward the expert's output once again. This is due to the fact that the novice's controls deviate widely from the ones of the expert, thus taking control away from the first and giving it more to the second. The output created by the fuzzy mediator does not adhere much to the line of control carried out by the novice because the time during which the output of the novice and the ones of the expert were somewhat similar is rather short. In cases where the lines would not intersect, the control will never move away from the expert's line.

In the second simulation, we have taken into consideration a situation where the expert is controlling the vehicle in a linear fashion, and the novice performs a motion trying to align to the direction of the expert in a logarithmic manner (Figure 7). Also in this case, when the simulation starts, the output of the fuzzy controller is overlapping the control of the expert. As the novice's output gets closer to the expert's, the output of the Fuzzy Mediation shifts. This time, the two values are close enough for quite a few cycles of comparison within the Fuzzy Mediation system, where each time the two values are considered similar, giving more control to the novice. Figure 8 shows a closer look at the area of the plot where the output of the Fuzzy Mediation engine starts moving toward the novice's controls. We should note that the difference between values, at that point, is only of 0.4 units. At this point, control is in the hands of the novice completely. After the lines intersect and then diverge, control will

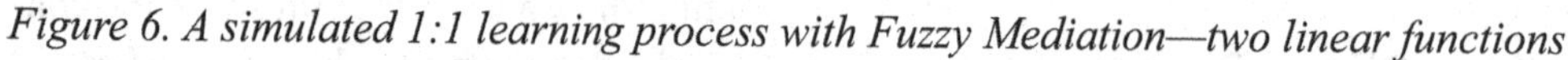
Figure 6. A simulated 1:1 learning process with Fuzzy Mediation—two linear functions

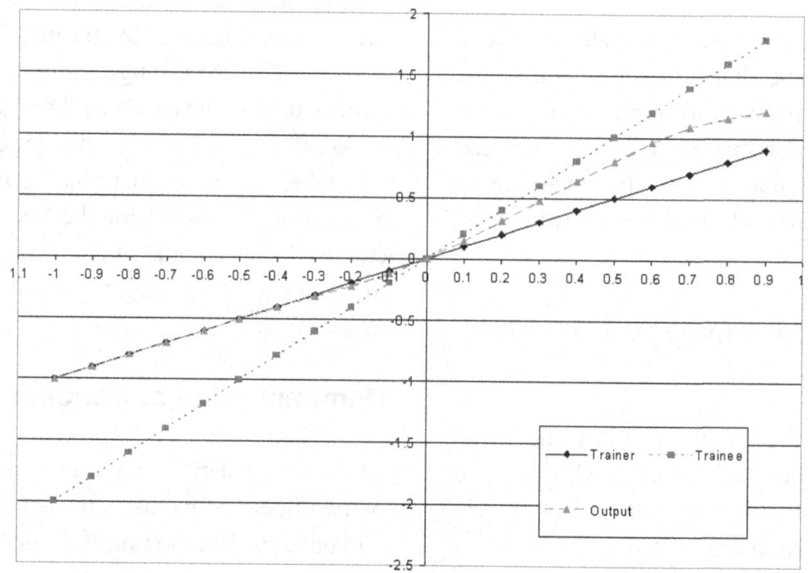

Figure 7. A simulated 1:1 learning process with Fuzzy Mediation—one linear function (expert) and one logarithmic function (novice)

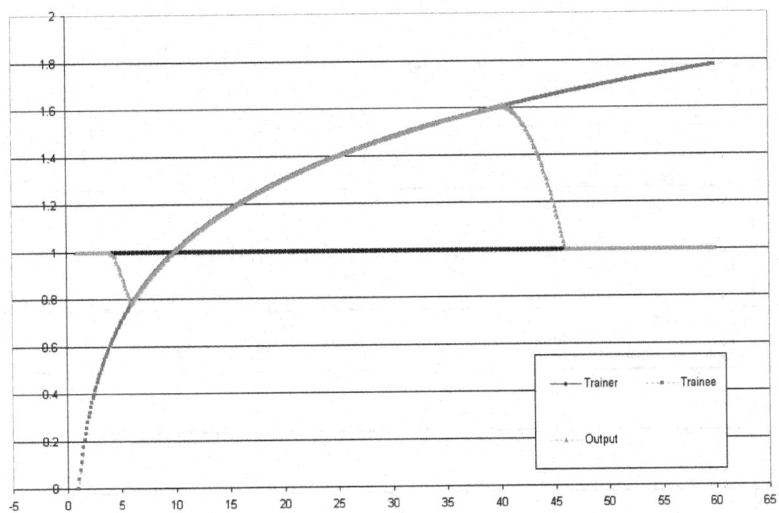

Figure 8. Magnification of the area where control shifts from the linear function (expert) to the logarithmic one (novice)

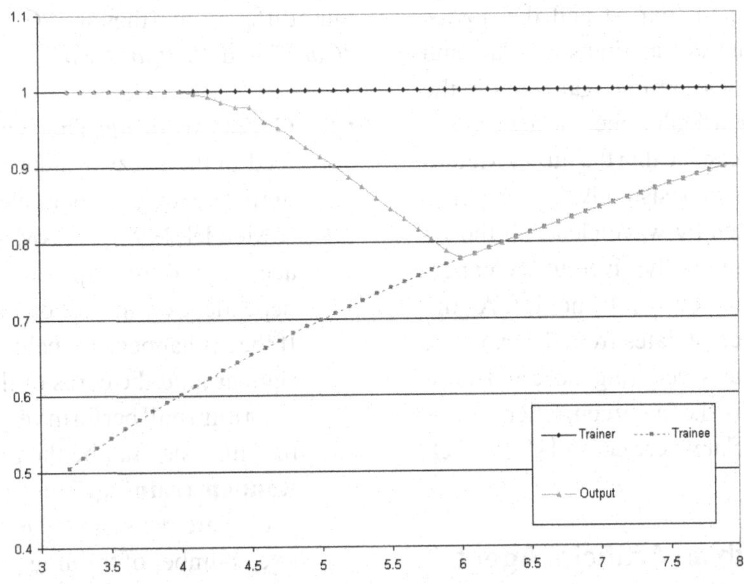

be retained by the novice, at least until the fuzzy mediator perceives a difference between the two inputs that is significant enough to start shifting control back in favor of the expert. This happens when the two values diverge by more than 0.6 in this simulation. The area where control starts shifting back to the expert is highlighted in Figure 9.

Figure 9. Magnification of the area where control shifts back from the logarithmic function (novice) to the linear one (expert)

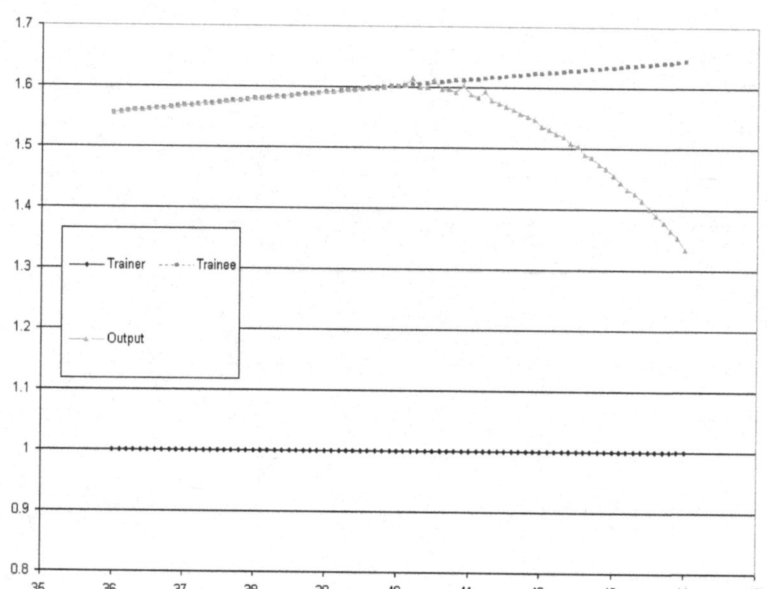

Figure 10 gives results of yet another simulation. In this case, the expert and the novice perform actions that are at times similar, and then diverge to become similar again later in the simulation. In this particular one, the actions are periodic, so that we can see that the Fuzzy Mediation engine performs consistently. The expert's control is represented by wave closer to the X-axis. The control is initially given to the expert and is slowly shifting toward the novice. As the novice's performance deviates from the expert's, the expert's input starts weighing more and more in the overall output of the fuzzy controller. Figure 11 shows the particular where control shifts back toward the expert.

Simulations with an Artificial Agent

In order to carry out a simulation that is realistic, we have analyzed several training scenarios that would mimic an autonomous agent based on the Wang-Mendel algorithm (1992) with imprecise knowledge at first that would improve as the operations continue. The training algorithms are quite different, and they are: "*Greedy,*" "*Random,*" "*Core,*" and "*Left to Right.*"

- **Greedy training:** This type of training will feed to the fuzzy system random numbers until the Fuzzy Associative Memory (Wang & Mendel, 1992), or FAM, is just filled. The accuracy of this type of learning is highly dependent on the set of training numbers. If the set happens to include values that are right around the cores of the sets, then this algorithm will perform very well; otherwise the rules may not be the most suitable.
- **Random training:** This training style is the one where the memory is presented with a fixed number of training pairs. If a rule has not been found for a particular association of sets, then the system will output the default value stored in the FAM. There is no assurance that the best example will be used for a certain association.

Figure 10. A simulated 1:1 learning process with Fuzzy Mediation—two sine waves

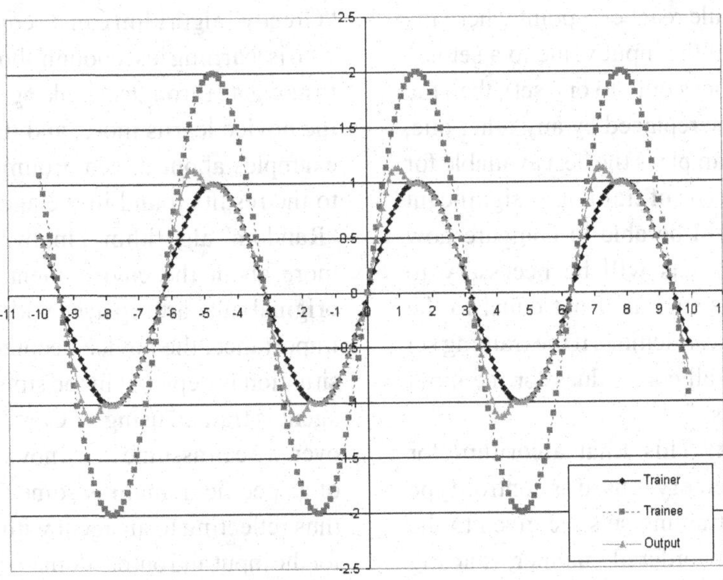

*Figure 11. Magnification of the area where control shifts back from the wave 2*sin(x) (novice) to the wave sin(x) (expert)*

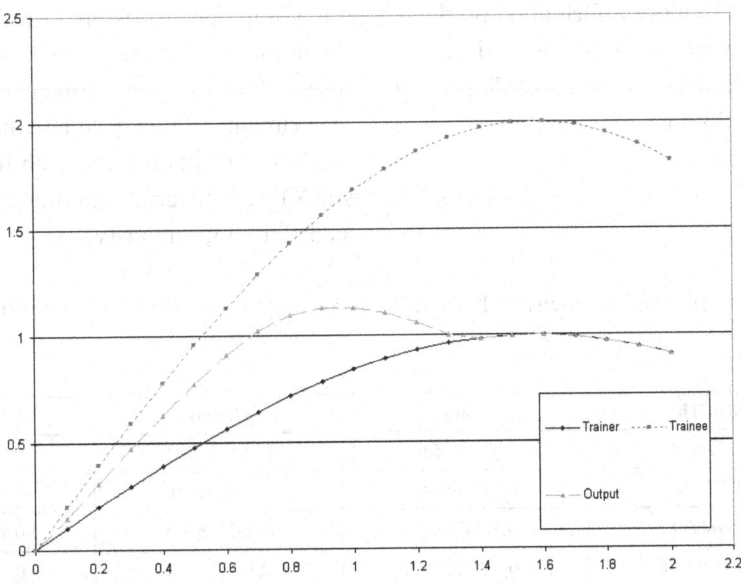

- **Core training:** This type of training feeds to the system the values of the cores. Theoretically, once a rule reaches a point where the membership of the input value to a set is 1 (the value belongs only to one set), then the rule will not be replaced by any other one, because that sample is the best available for that set. This type of training is significant because we will be able to compare how many training sets will be necessary to achieve results similar if not equal to the results that we can achieve if the training set was the set of all core values for the input and output sets.
- **Left to right:** This final algorithm for training the memory is used as control-type training. The training sets are given to the memory in an ordered manner, starting from the lower boundary of the range of the training set and ending at the high boundary. This type of training ensures that the FAM is given as many training pairs as possible, covering the entire range of the simulated environment. For more information on the evaluation of agents, a more detailed discussion is available in our recent work (Vincenti & Trajkovski, 2006).

The set of simulations discussed above show that a situation where an agent is trained using the "Greedy" algorithm can be compared to a novice who is learning just enough about an environment to navigate through it, making some mistakes. As the novice learns more, and thus observes more examples about the environment, we can relate to the results found in the agent trained using a "Random" algorithm. Finally, as a novice learns more about the environment and many of the original rules are replaced with others with higher importance, the novice becomes an expert. This situation is depicted in the simulations where the agent is trained using a "Core" algorithm. Moreover, we can assume that a novice's understanding of a specific domain is somewhat broad at first, thus reflecting to an agent with a low granularity for the input and output domains. As the agent goes from novice to expert, we can only assume that the understanding of the environment increases, with an increase in the granularity of the input and output ranges. This can be simulated by having an agent perform simulations first using a low number of sets, and gradually increasing the number of sets to reach the highest possible foreseen in the environment created for these experiments. Table 1 shows the different error levels for agents trained with the equation Y = sin(X) with different training algorithms as well as different granularity sets.

Table 1. Average error for behavior using Y = sin(X) across agents of different granularity and training approaches

	Left-to-right	Core	Greedy	Random
7	0,627506065	0,675936524	0,650292194	0,650422003
9	0,237706944	0,237706944	0,400731338	0,230864971
11	0,197668687	0,171135865	0,252355939	0,241180776
13	0,139887356	0,138705757	0,246697204	0,154017493
15	0,108492532	0,115646522	0,249215771	0,107077011
17	0,074813892	0,064634848	0,141544446	0,096659755
19	0,112382012	0,061053242	0,167742751	0,075847614
21	0,060825138	0,04926832	0,194218968	0,074670394

Table 2. Boundaries of trapezoidal fuzzy sets used to compute the linguistic modifier describing the difference between the inputs of the trainer and the trainee

Set name	Outer left boundary	Inner left boundary	Inner right boundary	Outer right boundary
Wide variation to the left	-∞	-∞	-4	-3
Slight variation to the left	-4	-3	-2	-1
Similar	-2	-1	1	2
Slight variation to the right	1	2	3	4
Wide variation to the right	3	4	∞	∞

This scenario will represent the sequence of events that we will simulate in this final round of experiments. When we introduced the concept of Fuzzy Mediation, we also addressed the analysis of the difference between the input of the trainer and the one of the trainee. For this simulation, the difference between the inputs was normalized to [-10, 10], and then the range was broken down into the sets reported in Table 2, which also specifies the boundaries of each set.

The fuzzy sets used map a value to a membership of 1 when the value falls between the two inner boundaries. If a value falls between an inner boundary and outer boundary on the same side, then it will have a partial membership to the set in question in according to the membership equation of choice. In this case, the equation chosen is a simple line equation. If a value falls outside of the outer boundaries, then the membership to the set in question is 0. For these simulations, the maximum absolute difference between the trainer's input and the trainee's is of one unit that is then mapped to the domain described above.

In the following simulations, we will use as an index the average value of the mediator weight. As we described above, the control of an ideal object must be mediated between two controllers, a trainer and a trainee. The average value for this index shows which controller has most of the weight. A weight between –1 and 0 shows that the trainer has more control, with total control to the trainer if the weight is –1. If the weight is between 0 and 1, then the trainee has the majority of control, with control completely assigned to the trainee if the value is 1. If the value is 0 then the control is shared equally between the trainer and the trainee.

The first aspect that we will discuss about the simulations performed is the influence of the number of fuzzy sets used for the input/output domain by the agents. Table 3 shows that, as the agent identifies the range in question with more fuzzy sets, also the weight of control shifts from a considerable average ownership of control to the trainer, as reported by the simulation with an agent based on seven fuzzy sets, to an almost complete ownership of control in favor of the trainee, as shown by the interaction with the agent based on 21 fuzzy sets.

Table 3. Core training for the sine environment with three different number of sets

Sets	Avg. mediator weight
7	-0.517277228
11	0.603871287
21	0.945544554

Figure 12. (a) Input and output values and (b) control mediation value for simulation using and agent with 7 sets and core training

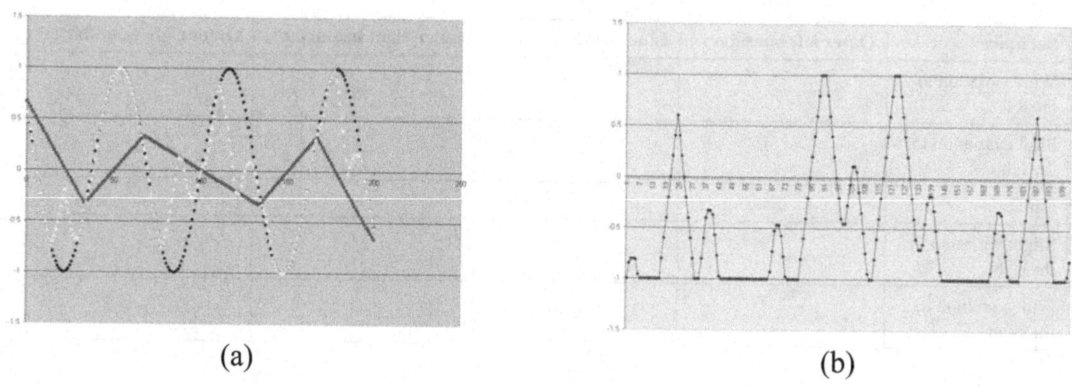

(a) (b)

Figure 13. (a) Input and output values and (b) control mediation value for simulation using and agent with 11 sets and core training

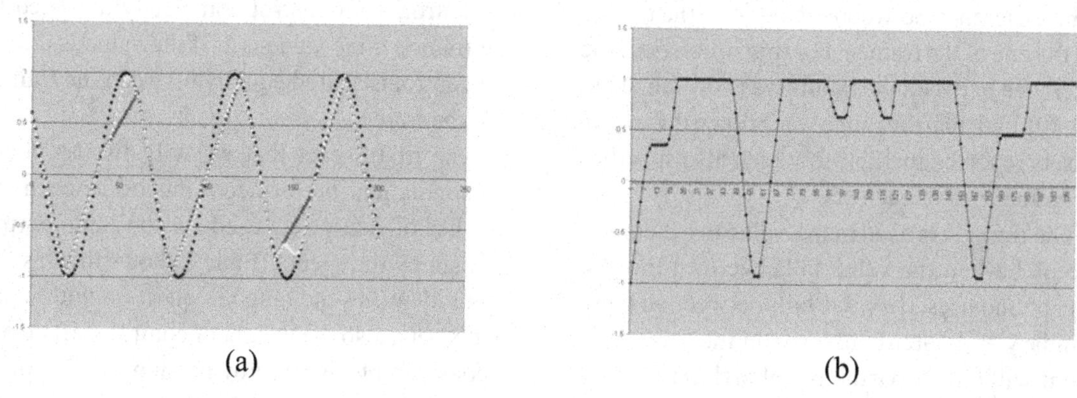

(a) (b)

Figure 14. (a) Input and output values and (b) control mediation value for simulation using and agent with 21 sets and core training

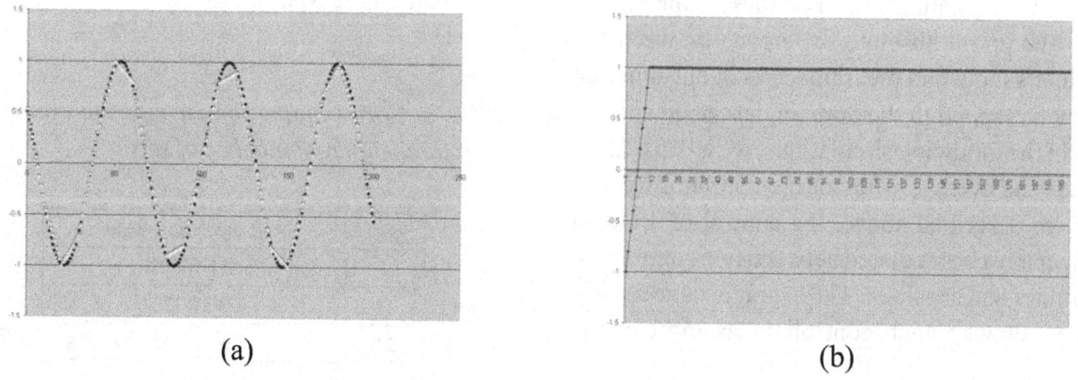

(a) (b)

Table 4. Different training algorithms for FAMs with the same amount of sets (11) within the sine environment

Training type	Avg. mediator weight
Greedy	0.149693069
Random	0.110108911
Core	0.603871287
L-to-R	0.300168317

Figures 12 through 14 show graphs reporting the input values of the trainer and the trainee, as well as the mediated output. Next to each simulation we also show graphs of the value of the mediator weight after each pair of trainer/trainee input are evaluated by the Fuzzy Mediation system. The graphs correspond to the simulations reported in Table 3. These figures show how dynamic this process of mediation between two controllers can be.

We also analyzed the performance of several training algorithms and the average weight of the mediator, reported in Table 4. For this particular example we chose an agent set with 11, and we can see that the agent that received "Core" training showed the best performance, keeping the average mediator value well over the side of the trainee for matters of control.

SELECTED APPLICATIONS

This section highlights our first "real-life" application of Fuzzy Mediation. We elaborate on other possible applications to which this concept is seamlessly applicable.

Fuzzy Mediation Applied to Robotics

The environment used for these experiments is a simple agent that follows a line. Figure 15 shows the pattern that was used. The pattern contains only the white background and the black line. The areas that have been highlighted show the segments of particular interest. Segment 1 was created to see the behavior of the agent in a mild turn to the right; Segment 2 instead simulates a sharp turn to the left followed by a moderate turn to the right. Segment 3 mimics a straight path. Segment 4 reveals a tight turn to the left, just like segment 5. Segment 6 is another tight turn to the right after a straightway, and finally, segment 7 shows quick changes in direction.

In these experiments we use different levels of control. When we want to create an environment when the two inputs need to be closer in order to shift the weight from the expert to the novice we simply need to set the boundaries of the sets shown in Figure 6 to tighter limits. The three levels of control we use are the following: *Tight* control, *Moderate* control and *Loose* control. Table 5 shows the values associated with each level of control. Each set carries four values, (OL, IL, IR, OR), where OL refers to the outer left boundary, IL to the inner left, IR to the inner right and OR to the outer right.

The simulated agent is composed of a central unit that contains sensors. The sensors check the terrain in front of the agent for color. The sensors can either pick up white, which is the background, or black, which is the line. The sensors are ar-

Figure 15. Path of the simulation

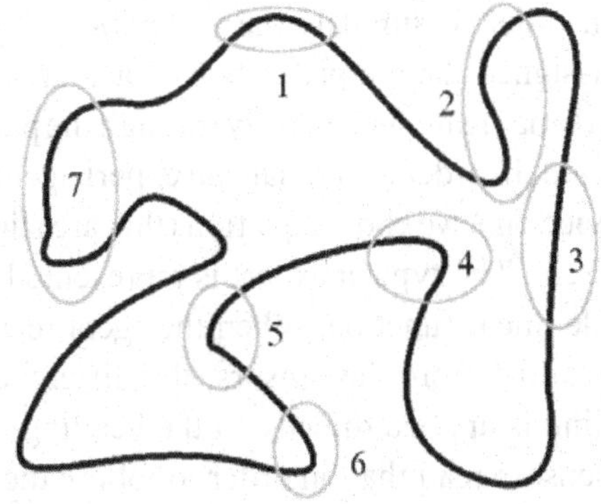

Table 5. Description of fuzzy sets used in classifying the difference between the expert and the novice inputs

	WL	SL	S	SR	WR
Tight	(-∞, -∞, -5, -3)	(-5, -3, -2, -1)	(-2, -1, 1, 2)	(1, 2, 3, 5)	(3, 5, +∞, +∞)
Moderate	(-∞, -∞, -6, -4)	(-6, -4, -2, -1)	(-2, -1, 1, 2)	(1, 2, 4, 6)	(4, 6, +∞, +∞)
Loose	(-∞, -∞, -8, -6)	(-8, -6, -4, -2)	(-4, -2, 2, 4)	(2, 4, 6, 8)	(6, 8, +∞, +∞)

Figure 16. Diagram of the simulated agent with its sensors

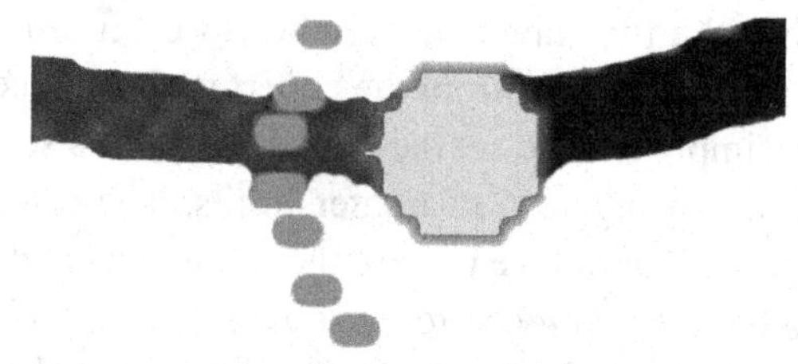

ranged on a probe that scans the range [-45, 45] in front of the agent at 5-degree intervals. Figure 16 shows an image of the agent that is following a line. The light gray is the body of the agent while the dark gray spots in front of it represent the range of action of the sensors.

The sensors communicate to the agent the color of the terrain at each angle. Then, the agent will group together the angles that recorded a reading of a line and calculate the average. Such average will be analyzed by the agent, which will select the new heading. An agent can perceive changes in direction up to ± 45 degrees.

In order to simulate the behavior of agents we assigned them a preset behavior that allows them to navigate successfully through the pattern selected. In order to simulate an expert agent and a novice one, we chose equations that are slightly different. The typical expert is represented by a simple linear function. When the agent receives the reading from the sensors, the difference in heading is applied directly to the heading, so if the sensors read that, in order to follow the line the agent needs to apply a 15-degree turn to the right, the agent will perform a 15-degree turn to the right.

The simulation that represents the case of a novice is powered by an agent that relies on the cube of the difference normalized to the range [-90, 90]. The two equations that are used to drive the agents are reported in Figure 17. The dotted line represents the behavior of the expert and the solid one the behavior of the novice.

This interpretation of the agents shows an expert that acts as expected, with a linear response to the situation. The novice instead reacts more slowly only to overcompensate as the deviation required in order to remain on track increases. We also carry out other simulations where the difference between the inputs of the expert and the one of the novice are very different. The

Figure 17. Control functions associated with the expert (dotted) and the novice (solid) controllers

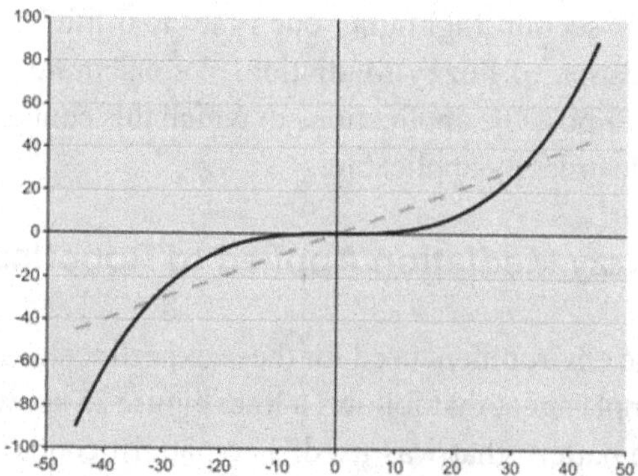

equations reported in Table 6 show the driving engines we used.

The first experiment we performed involves an agent navigating through the pattern using Fuzzy Mediation to blend the controls of the expert powered by equation 1 in Table 6 and the novice simulated by equation 2. The elements we studied were the differences in behavior of the agent based on different levels of control, as described above. The element we monitored is the value of the mediator's weight. Table 7 shows the average value for the three levels of control using the very same equations to simulate the controllers' inputs.

We can see that a tighter control reports an average weight of the controller that is lower when compared to the other indices. This means that, when we use the sets that correspond to a tight control of the novice's inputs, the expert retains control for more sections of the patters than in simulations performed using the moderately loose or loose controls.

In all the experimental runs we noticed that the area that consistently showed the most shift in control was number 2 in Figure 15. Figures 18, 19 and 20 show the shift in control weight as the agent goes through the turn. It is important to note that the agent was following the pattern in a clockwise motion.

These figures show that, in the case of a tight control, the expert will regain significantly the control of the agent and will perform the steeper section of the turn, as shown in Figure 18. Figure 19 instead shows that the Fuzzy Mediation that uses a moderately loose control still allows the expert to retain quite a bit of control, but overall the input of the novice is evaluated with a higher importance. Finally, Figure 20 shows that a loose control allows the novice to take care of the majority of the control in this situation. The different nature of the equations, as described earlier, does not allow the novice to be in control through the entire turn, as the difference in controllers' inputs are quite different.

Table 6. Equations used for the simulations

Num	Equation	Simulation
1	$Y = X$	Expert agent
2	$Y = X^3$	Simple novice agent
3	If $X > 0$ $Y = X$ Else $Y = 0$	Agent that turns to the right, but can't turn left. Instead it goes straight.
4	If $X > 0$ $Y = X^3$ Else $Y = 0$	Agent that turns to the right, but can't turn left. Instead it goes straight.
5	$Y = \|X\|$	Relatively novice agent that only turns right
6	$Y = \|X^3\|$	Novice agent that only turns right

Table 7. Mediation weight average for different control levels

	Tight control	Moderate control	Loose control
Average mediation weight	0.59	0.62	0.82

Figure 18. Average mediator weight during segment 2 of the simulation with tight control

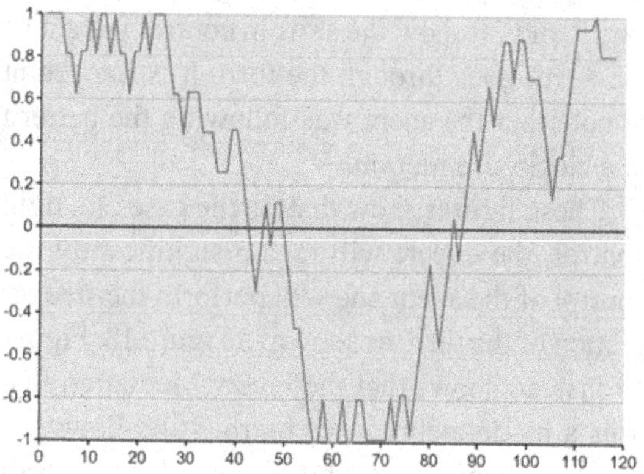

Figure 19. Average mediator weight during segment 2 of the simulation with moderate control

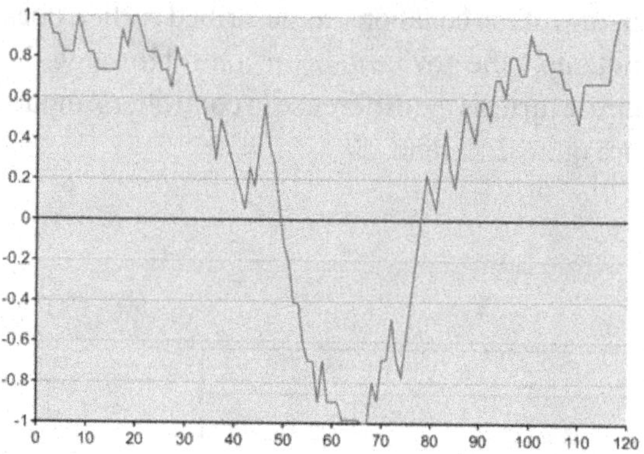

Figure 20. Average mediator weight during segment 2 of the simulation with loose control

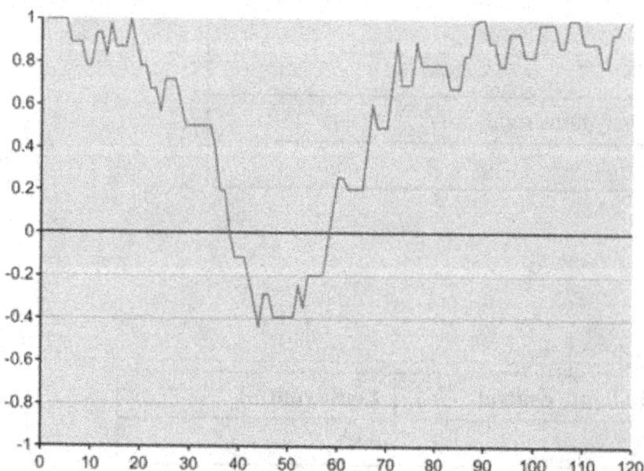

Table 8 shows the average weight of the controller's mediator weight for the values plotted in Figures 18, 19 and 20.

The following experiments were performed in order to study the interaction between an expert and a novice controller when the novice behaves quite differently from the expert. The first simulation was performed with the expert controller based on equation 1 in Table 6, while the second controller was powered by equation 3. In the case of a turn to the right the agent showed no problems. We recorded that in this case the agent was leaving the pattern completely in the case of a slight turn to the left. In the case of a sharp turn to the left the agent showed some problems at the beginning of the turn, but no problem after that. This is probably due to the fact that, in the case of a slight deviation control stays unaltered, thus leaving control for a longer period of time to the novice. The second setting involved a novice controller powered by equation 4 in Table 6. This simulation performed very similarly to the previous one, because both simulations for the novice controller show the same response to a left turn.

The third set of experiments focused on controllers that behaved completely different in the case of a left turn. In order to simulate this situation we used equations 5 and 6 to simulate the novice controllers. We were unable to record extensive data for these experiments because the agent was not able to complete a full loop of the pattern. Our observations show that in the case of equation 5, the agent would initially follow the line, but at the first left turn it would lose control, turn completely around and then get stuck looping around itself. The same behavior was observed when the equation for the novice controller was powered by equation 6. We then increased the value by which the weight value of the control is shifted from 0.2 to 0.5 At this point, the agent with the novice controller with equation 6 performed almost a full trip around the pattern. It showed problems when it was presented with a

Table 8. Mediation weight average for different control levels during a hard turn

	Tight control	Moderate control	Loose control
Average mediation weight	0.21	0.31	0.57

slight turn to the left. At that point it would also start looping around itself.

"Ready to Implement" Applications

The possible applications of this research are quite varied, given the possible applications in training, stressing those that involve machinery. This system may be used as a training environment as well as a monitoring one. Several types of machines require "check rides," where the user needs to perform certain operations under scrutiny of evaluators. If the operator fails to perform a certain number of tasks correctly, the license to operate that type of machine will be revoked. With this framework, the operator and the check ride officer may use this approach for evaluation. We may even push the technology a bit further, and allow the operator to be checked by a machine, supervising the operator's performance. Moreover, this framework will create a system that learns from the driver's everyday performance. Upon reaching a certain level of confidence for the "autopilot" system, this system will then become the monitor when the driver seems to be impaired in any way and correct the actions of the driver. Should the actions of the driver lead to dangerous driving, the car may calculate the best route to stop and do so to ensure the safety of the passengers as well as other drivers.

Such framework will be extensible to an embedded system that will allow an autopilot type of controller to take over when the machine (automobile/airplane) is capable of producing a path of motion similar to the one followed by the human operator. Such autopilot will also be able to learn from the human controller, so that, the next time similar conditions arise, the machine will be able to foresee the path required to avoid the problem.

This research will be extensible to other fields, such as the creation of an autopilot system that uses "drives" as parameters to control an automated robot, as explained in Trajkovski et al. (2005). Such robot will also be able to detect and avoid obstacles in an unfriendly environment using the structure described in Trajkovski et al. (2005). Many of the applications within this section are highly dependent on elements that are not in place yet. For example, a digital system that can drive a car efficiently within a regular urban environment is not existent. The authors envision this happening within a few years though, and believe that, when the times are ready, this research will show its true power in its applications. In the meantime, we will create simulations that will allow us to recreate some of the high-end sensors still unavailable to us.

CONCLUSION

This chapter introduced a dynamic concept of Information Fusion. We have discussed in detail the algorithm that powers the core of the mediation system as well as some architectures that are relevant to several training and shared control type settings. We have also shown how the algorithm behaves both in conceptual experiments as well as simulations of robots. We believe the future of Fuzzy Mediation is bright, as it is a technology that is applicable to several settings. Current research activities are exploring the relationship between the closeness in control of the controllers and the speed of mediation. Moreover, the experimentation with humans has been started with the construction of simulation environments where an expert human interacts with a novice in order to teach concepts of navigation in a moving fluid.

REFERENCES

Abad, P. J., Suárez, A. J., Gasca, R. M., & Ortega, J. A. (2002). Using supervised learning techniques for diagnosis of dynamic systems. In *Proceedings of the International Workshop on Principles of Diagnosis* (DX-02).

Arai, S., & Ishida, T. (2004). Learning for human-agent collaboration on the Semantic Web. In *Proceedings of the 12th International Conference on Informatics Research for Development of Knowledge Society Infrastructure (ICKS '04),* (pp. 132-139).

Barnes, C. A., Baranyi, A., Bindman, L. J., Dudal, Y., Fregnac, Y., Ito, M., et al. (1994). Group report: Relating activity-dependent modifications of neuronal function to changes in neural systems and behavior. In A. I. Selverston & P. Ascher (Eds.), *Cellular and molecular mechanisms underlying higher neural functions*. New York: John Wiley & Sons.

Chialvo, D. R., & Bak, P. (1999). Learning from mistakes. *Neuroscience, 90*(4), 1137-1148.

Dugan, B., & Glinert, E. (2002). Task division in collaborative simulations. In *Proceedings of the 35th Annual Hawaii International Conference on System Sciences (HICSS-35'02),* (p. 31).

Gokhale, A. (1995). Collaborative learning enhances critical thinking. *Journal of Technology Education, 7*(1), 22-30.

Grenoble O'Malley, P. (2005, March-April). Troubleshooting machine control: Contractors talk about making the most of 3D GPS. *Grading & Excavator Contractor, 7*(2).

Hammonds, K. H., Jackson, S., DeGeorge, G., & Morris, K. (1997). The new university—a tough market is reshaping colleges. *Business Week, 22,* 96-102.

Hebb, D. (1949). *The organization of behavior.* New York: John Wiley & Sons.

Heyes, C. (2001). Causes and consequences of imitation. *TRENDS in Cognitive Sciences, 5*(6), 253-261.

Kandel, E. R., Schwartz, J. H., & Jessell, T. M. (2000). *Principles of neural sciences*. New York: McGraw-Hill.

Kokar, M. M., Tomasik, J. A., & Weyman, J. (2004). Formalizing classes of information fusion systems. *Information Fusion, 5*(3), 189-202.

Kwek, S. S. (1999). An apprentice learning model. In *Proceedings of the Twelfth Annual Conference on Computational Learning Theory (COLT '99),* (pp. 63-74).

Mamdani, E. H., & Assilian, S. (1975). An experiment in linguistic synthesis with a fuzzy logic controller. *International Journal of Man-Machine Studies, 7*(1), 1-13.

Norman, D. A. (2005, November-December). There's an automobile in HCI's future. *Interactions,* 53-54.

Piaget, J. (1945). *Play, dreams, and imitation in childhood.* New York: Norton.

Rizzolatti, G., & Arbib, M. A. (1998). Language within our grasp. *Trends in Neurosciences, 21*(5), 188-194.

Rizzolatti, G., Fadiga, L., Gallese, V., & Fogassi, L. (1996). Premotor cortex and the recognition of motor actions. *Cognitive Brain Research, 3*(2), 131-141.

Thorndike, E. (1911). *Animal intelligence.* New York: Macmillan.

Trajkovski, G. (2004). Fuzzy sets in investigation of human cognition processes. In A. Abraham, L. Jain, & B. Van der Zwaag (Eds.), *Innovations in intelligent systems* (pp. 361-380). Springer-Verlag.

Trajkovski, G. (2005). E-POPSICLE: An online environment for studying context learning in human and artificial agents. In *Proceedings of the 16th Midwest AI and Cognitive Science Conference (MAICS 2005),* (pp. 61-66).

Trajkovski, G., Stojanov, G., & Vincenti, G. (2005). Extending MASIVE: The impact of stress on imitation-based learning. In *Proceedings of the IEEE 6th ACIS International Conference on Software Engineering, Artificial Intelligence, Networking and Parallel/Distributed Computing* (pp. 360-365).

Trajkovski, G. (2007). *An imitation-based approach to modeling homogenous agents societies.* Hershey, PA: Idea Group.

Trajkovski, G., Stojanov, G., Bozinovski, S., Bozinovska, L., & Janeva, B. (1997). Fuzzy sets and neural networks in CNV detection. In *Proceedings of the ITI'97,* Pula, Croatia, (pp. 153-158).

Vincenti, G., & Trajkovski, G. (2006, October 12-15). Fuzzy mediation for online learning in autonomous agents. In *Proceedings of the 2006 Fall AAAI Symposium*, Arlington, VA, USA, (pp. 127-133).

Wallace, J. (2000, March 20). Unlike Airbus, Boeing lets aviator override fly-by-wire technology. *Seattle Post-Intelligencer.*

Wang, L., & Mendel, J. (1992). Generating fuzzy rules by learning from examples. *IEEE Transactions on Systems, Man and Cybernetics, 22*(6), 1414-1427.

WordNET Search. (n.a.). *Control.* Retrieved April 3, 2008, from http://wordnet.princeton.edu/perl/webwn?s=control

Zadeh, L. (1965). *Fuzzy sets, information and control* (No. 8, pp. 338-353).

Chapter XVII
Utilizing Past Web for Knowledge Discovery

Adam Jatowt
Kyoto University, Japan

Yukiko Kawai
Kyoto Sangyo University, Japan

Katsumi Tanaka
Kyoto University, Japan

ABSTRACT

The Web is a useful data source for knowledge extraction, as it provides diverse content virtually on any possible topic. Hence, a lot of research has been recently done for improving mining in the Web. However, relatively little research has been done taking directly into account the temporal aspects of the Web. In this chapter, we analyze data stored in Web archives, which preserve content of the Web, and investigate the methodology required for successful knowledge discovery from this data. We call the collection of such Web archives past Web; a temporal structure composed of the past copies of Web pages. First, we discuss the character of the data and explain some concepts related to utilizing the past Web, such as data collection, analysis and processing. Next, we introduce examples of two applications, temporal summarization and a browser for the past Web.

INTRODUCTION

As the Web changes continuously, it is necessary to preserve the past content of pages for a future reuse. The Internet Archive[1] is the best-known and largest public Web archive containing data crawled since 1996. Other Web archives exist, for example, ones containing Web pages from particular countries (e.g., Arvidson, Persson, & Mannerheim, 2000; Hallgrimsson & Bang,

2003). Besides, there are also numerous repositories of past copies of pages such as caches, site archives, personal page repositories or search engine caches.

Web archives provide a view on the history of the Web reflecting past societal states. Past content of pages can reveal the histories of underlying elements represented by these pages, such as institutions, companies, people or other entities. For example, one could approximately detect when a particular member left some laboratory by detecting the time point at which her or his name was removed from the list of laboratory's personnel. In general, the use of Web archives can greatly benefit researchers and practitioners in many areas, such as history, sociology or marketing.

Furthermore, analyzing information from the past can help not only in better understanding the history of our society but also understanding its present state. This is because Web archives can provide contextual information about Web pages and the objects or concepts discussed on them as well as their inter-relations. For example, we can analyze information from Web archives concerning a given company in order to use it as a context for better understanding the present information about this company. In general, mining past Web content has a potential to stimulate and improve the traditional Web mining process in the sense that it provides contextual information and sheds new light on present data.

Past Web is considered here as a part of the WWW space where pages no longer have any change potential; they are "frozen" past snapshots of pages. The live Web, on the other hand, is the present Web, containing pages that we can currently view online. These pages may be changed or updated and they usually provide full interaction capabilities.

In the past Web each page has its history and lifetime. Links between the old content of pages can be reactivated again. In this way, a temporal structure can be obtained reflecting connectivity between pages in the past. Another aspect of the past Web is missing data. A given content after its deletion from a page may never be reproduced if it has not been preserved in any repository. Besides, due to the rapid growth of the Web, selective type archiving often needs to be done.

In this chapter, we approach the problem of discovering knowledge from the past Web. First, we discuss the character of data that is used and methods for acquiring and processing it. We propose techniques for analyzing and selecting candidate Web pages for mining. This approach is based on analyzing long-term characteristics of pages with a special focus on their content changes as they are most interesting from the viewpoint of pages' evolution. Next, we introduce temporal summarization, which is an adaptation of a traditional text mining task into the past Web scenario. We propose summarizing histories of Web pages to generate abstraction of events and salient concepts described in selected portions of the past Web. We also discuss the possibility of discovering object histories in past content of Web documents. Finally, we describe an application for browsing and navigating the past Web. We show an implementation that is similar to those of traditional browsers for the live Web and of video players.

The rest of this chapter is organized as follows. In the next section, we discuss the related research and attempt to place this work in the wider context of text and Web mining. The following two sections describe the data accumulation, preparation and analysis. In the next section we discuss temporal summarization and investigate the possibility of object history detection from the past Web. The next section describes a browser for the past Web, while the last section concludes the chapter with a brief summary.

RELATED RESEARCH

Web Dynamics

The dynamics of the Web has been measured in many experiments (Brewington & Cybenko, 2000; Cho & Garcia-Molina, 2000; Fetterly, Manasse, Najork, & Wiener, 2003; Ntoulas, Cho, & Olston, 2004) which demonstrated that the content and link structure of the Web continuously change. Although many pages on the Web are short-lived, meaning they are deleted shortly after being created (Ntoulas et al., 2004), many important Web documents persist over time. Popular and main, or top-ranked, pages usually belong to this category as it often takes a long time for a page or site to gain popularity and accumulate a high number of in-links.

The results of Web dynamics research indicate the level of volatility of the Web as a whole. On the other hand, the study of update patterns of individual pages has been carried out for prediction of their future changes (Cho & Garcia-Molina, 2000, 2003; Ntoulas et al., 2004). The frequencies and degree of changes are the most often used measures to set up crawling schedules for maintaining fresh indexes of search engines. In practice, however, it is usually difficult to predict content changes in pages although some Web documents, for example, newswire sources, change in a more or less periodical fashion. In this research, we go beyond the simple analysis of change statistics as we focus on the distribution of content and its context over time.

Text Mining

Text mining is defined as a nontrivial extraction of implicit, previously unknown and potentially useful information from textual data. Text mining evolved from data mining and is a promising field as much information nowadays is stored in the form of electronic text. We consider our approach to be similar to temporal text mining, because, to a certain extent it resembles efforts that were taken in analyzing and mining streams of text data. Generally, mining news articles or other text streams along the time dimension has been studied well (Allan, Gupta, & Khandelwal, 2001; Allan, 2002; Kleinberg, 2003; Li, Wang, Li, & Ma, 2005; Mei & Zhai, 2005; Papka, 1999; Swan & Allan, 2000; Wang & McCallum, 2006). For example, the well-known TDT (Topic Detection and Tracking) research initiative (Allan, 2002) was aimed at detecting, classifying, and tracking events in news corpora. Recently, Wang and McCallum (2006) identified topics persisting over dynamic collections of documents. Another work showed the development of topic patterns in news articles over time (Mei & Zhai, 2005). Li et al. (2005) proposed a probabilistic model for retrospective event detection in news corpora. An approach toward temporal summarization of news events was proposed in Allan (2001) where novelty and usefulness of sentences retrieved from newswire streams were calculated for the construction of a final summary. Another related work called TimeMines (Swan & Allan, 2000) was proposed for finding and grouping significant features in historical document collections based on applying chi-square test.

While news articles and, in general, any text streams are usually represented as transient text snapshots, the content of pages often persists over time. Duration of content has certain relation to its semantics and relative importance in a page. Thus, in contrast to typical text data streams, one has to consider three types of content in pages at every time point: static (persisting over time), deleted, and added. Additionally, pages have certain inherent topics that determine the context of their transitory content and that can enhance the mining process.

Web Mining

Web mining is often described as the application of data mining techniques for extracting knowledge from the Web. It is traditionally divided into usage, structure and content mining. Web usage mining identifies the behavior patterns of users visiting Web pages for the purpose of optimizing Web sites. It is usually based on historical data, which is collected during certain time periods for its subsequent analysis (Cooley, Srivastava, & Mobasher, 1997; Kosala & Blockeel, 2000). Web usage mining can show how the users' access to Web sites changes over time. Web structure mining focuses on the link structure and graphical representation of the live Web. There have been, however, several approaches proposed to analyze the evolution of links over time (Amitay, Carmel, Herscovici, Lempel, & Soffer, 2004; Chi et al., 1998; Toyoda & Kitsuregawa, 2003). For example, temporal link analysis was used for detecting trends in page collections (Amitay et al., 2004) or for visualizing evolutions of Web communities (Chi et al., 1998; Toyoda & Kitsuregawa, 2003).

Web content mining uses the content of Web pages for knowledge extraction. Blog related research is probably the most prominent example of Web content mining in which the temporal aspect of pages is considered (Gruhl, Guha, Liben-Nowell, & Tompkins, 2004; Kumar, Novak, Raghavan, & Tomkins, 2003). Blogs help to detect and analyze social structures and social relations as well as provide information on society opinions, hot topics or recent trends. Blogs, however, are a unique media type as they usually contain complete versions of their past content with explicit timestamps provided as well as they are highly personalized and subjective. We believe that a general framework for mining any page types in the past Web is required.

Although most approaches to Web content mining generally neglected the temporal dimension of pages (Cooley et al., 1997; Kosala & Blockeel, 2000), there were, however, several works that investigated the usefulness of data on page histories for knowledge discovery (Arms et al., 2006; Aschenbrenner & Rauber, 2006; Jatowt & Tanaka, 2007; Rauber, Ascenbrenner, & Witvoet, 2002; Yamamoto, Tezuka, Jatowt, & Tanaka, 2007). Rauber et al. (2002) discussed the possibility of analyzing past Web data for identifying changes in Web-related technologies, particularly in the features and characteristics of Web pages, such as a file format, language, size, and so forth. The objective was to create statistics describing Web changes over time. Aschenbrenner and Rauber (2006) surveyed the work that has been done toward mining large portions of Web content with consideration of its temporal aspect. They also provided a general outlook on the potential of mining Web archives. Arms et al. (2006) have reported on building a research library for facilitating study of the Web evolution. This is an ongoing project aiming to build an infrastructure for analysis of massive portions of the data that is stored in Internet Archive. Practical usage of the past Web has been recently demonstrated by Yamamoto et al. (2007), who have proposed an application similar to question answering systems for extracting and combining knowledge from the Web and Web archives. It uses Web archive data for detecting changes in opinions and user knowledge over time.

Mining the content of the past Web is different from the usual Web content mining in several aspects. First, the temporal dimension of content and links in page histories poses new challenges and opportunities for understanding their roles and interrelations in contrast to traditional Web content mining. Second, pages and Web sites should be treated as dynamic objects having certain age, histories, trends, patterns, and so forth. Thus, the notions of a page and its content need to be separated in a way in which the latter one is considered as a transient component occurring in a higher level object, that is, a page. Content has then its own duration of occurrence while the page history is considered as the composition of different content occurring throughout the page's lifetime. Third, there is an issue of missing and incomplete data. In order to obtain satisfactory results, multiple snapshots of the past content of pages have to be found and acquired as well as approximation methods need to be applied for an optimal page history reconstruction.

DATA ACQUISITION AND PREPARATION

Data acquisition and preparation are important steps in the knowledge discovery process. In the mining of the past content of the Web these steps mean the retrieval of data from Web archives and the reconstruction of Web document histories (Jatowt & Tanaka, 2007). The following issues are involved here. First, it is by definition an ex post facto process, as the data is the past content of pages. If one could predict beforehand which Web pages are going to be used, one could simply set up a crawler with a suitable crawling frequency so that page evolution would be captured with a desired precision. However, it is assumed that the user is unable to make such a prediction, and rather that she or he wishes to acquire knowledge in real time using the available, preserved data. Hence, past snapshots of Web pages are gathered in real time from available resources with the aim of reconstructing the past with the highest possible precision. Thus, when talking about crawling in the context of the past Web, we mean querying past Web repositories for the data they contain. Second, because data is scattered in different repositories, it has to be searched for and identified before being used. Therefore, it is necessary to use efficient search and download techniques to locate and gather multiple snapshots of past content with a minimal cost. Due to the large size of data, in practice, usually, only its small portion can be fetched and analyzed locally. Therefore, the focus of this research is on the analysis of the limited amount of data rather than on building a framework for examining the past Web from a macroscopic viewpoint. In addition, there is an issue of the trustworthiness of past content, which is directly related to the trustworthiness of past Web repositories. For example, data obtained from a personal Web repository would normally be less trustworthy than the data collected from a large Web archive containing millions of pages and having a professional maintenance and control. Finally, only fragmentary data can be obtained due to the unpredictable change pattern of the Web and limited resources of archival systems. This calls for employment of efficient techniques for estimation of actual content that pages had in the past.

Collecting Snapshots

***Definition 1*:** Past page snapshot is a copy of page content that was published in the Web at a given time point in the past. The timestamp of the snapshot indicates the date when it was captured.

As mentioned above, because of resource limitations, Web archives contain only fragmentary past data. As a general attempt to alleviate this problem a kind of meta-archive approach (Jatowt, Kawai, Nakamura, Kidawara, & Tanaka, 2006) can be used to maximize past Web coverage and consequently to increase the precision of history reconstruction. This approach presumes communication with several past Web repositories at the same time. An intermediary module is required between these repositories and the local system to translate queries into the format required for each repository. After receiving a request for a page history, the module queries the repositories about their data. The repositories should then send a list of stored page snapshots with their metadata so that a fetching policy can be determined.

The optimal strategy would be first to check the signatures (checksums) of snapshots, if they are provided, in order to detect the ones that actually contain content changes from among all data provided by the cooperating repositories. This would prevent downloading identical page snapshots from different repositories, thereby maximizing fetching efficiency[2]. However, currently, Web archives do not provide such information. Instead, some repositories, such as the Internet Archive, provide lists of page snapshots that have any changed content when compared to the neighboring snapshots. By utilizing this information, only the snapshots with content changes inside archives would be fetched. In general, the efficiency of the data collection would depend on the metadata that is provided in past Web repositories.

Such a meta-archive approach would provide a unified interface to the history of the Web, making the data acquisition process less dependent on the resources of single Web archives. However, as Web archive interfaces are diverse, different data acquisition methods would have to be used. In addition, we make an assumption here that the URLs of pages remain the same over time, although, in practice, they may change even though the content of pages remains almost the same.

McCown and Nelson (2006) and McCown Smith, and Nelson (2006) have recently measured the persistence and availability of page copies inside the repositories of major search engines and the Internet Archive. The objective was to estimate the possibility and to provide methodology for reproducing the latest versions of Web sites in case of the loss of Web data.

Reconstruction of Page Histories

Definition 2: Page history reconstruction is the process of reproducing the past content of a page using available snapshots for obtaining the continuous representation of page history.

Definition 3: Optimal page history reconstruction is a reconstruction which accurately reproduces page history; that is, the errors resulting from such a reconstruction are equal to zero. Having determined an optimal page history, it is possible to recreate page content for any time point in the past that shows the actual content the page had at that time.

However, unless the page was unchanging, it has been crawled continuously or the implicit information about its past changes is provided, there will be usually some error involved in the history reconstruction. Only for certain types of pages, for example wikis, complete past data is available as the preservation of their versions is usually automatically done. In case of such pages, the reconstruction error would be equal to zero as all past changes can be derived from available page versions. In addition, some pages may contain temporal annotations in their present content that can be used to enhance the history reconstruction. For example, blogs often provide timestamps of content insertion. Nevertheless, for the majority of hypertexts, usually, neither implicit version management nor temporal annotations are provided.

We propose a simple approach for the page history reconstruction (Jatowt & Tanaka, 2007). First, collected snapshots are chronologically ordered according to their timestamps. If past snapshots are not associated with any temporal metadata then they cannot be directly included in the ordered sequence of past snapshots without a prior determination of their timestamps. For example, Yahoo! search engine provides cached snapshots of Web pages but it does not attach any timestamps to them. Estimating a timestamp of a snapshot could be possibly done by comparing similarities between its content and the content of other snapshots with known timestamps.

Second, every previous page snapshot is considered to represent the actual state of page content for the time period until the next page snapshot in the sequence. For example, suppose that five snapshots have been collected, s_1, s_2, s_3, s_4 and s_5, with timestamps, t_1, t_2, t_3, t_4 and t_5, where $t_1<t_2<t_3<t_4<t_5$ (Figure 1). Let us also suppose that snapshots s_2 and s_3 are exactly same. After the simple approximation, the page content is assumed to be the same as that in s_1 during the period $[t_1, t_2)$, the same as that in s_2 during $[t_2, t_4)$ and equal to s_4 in $[t_4, t_5)$. The reconstructed page history is then represented as a minimal sequence of 2-tuples containing different page versions and

Figure 1. Example of page history reconstruction

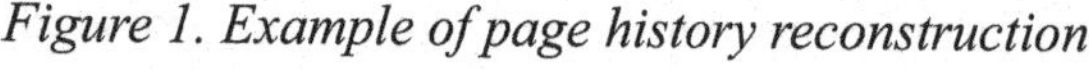

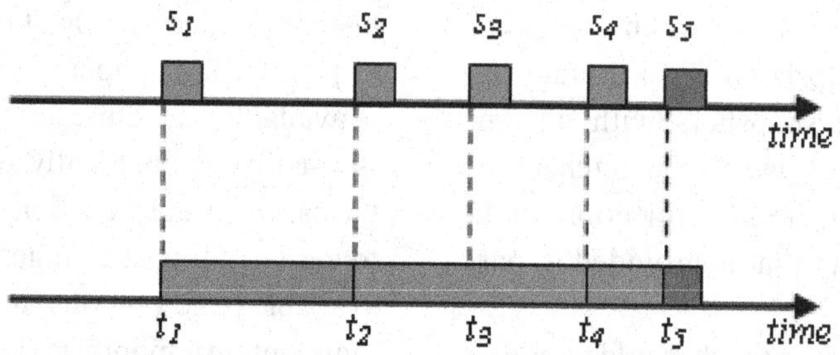

their starting dates ($\{(s_1, t_1), (s_2, t_2), (s_4, t_4)\}$ in the above case).

Page history reconstruction could be improved by considering additional information, for example, by analyzing changes in other pages belonging to the same site. Also, using the results of the temporal analysis of pages, especially their updating patterns, could make the reconstruction more accurate. Finally, historical snapshots of mirror pages, if there are any, could be utilized.

History Reconstruction Error

Usually, it is difficult to determine an accurate page history that would reflect the actual page content as it was at any arbitrary time point in the past unless the complete set of actual page versions is provided, for example, by a page author. Hence, mining the content of the past Web will typically be carried out using incomplete data with varying levels of precision and trust. It is thus necessary to consider the issue of missing data.

We can distinguish two types of errors in the page history reconstruction assuming that the page crawling was independent from the page update pattern (Jatowt & Tanaka, 2007). The first one, which we call a content error, is caused by uncertainty related to the content that appeared on a page in the past. Consider two retrieved past versions of the page (v_L and v_R) captured at time points t_L and t_R ($t_L<t_R$). The probability, $P(v_i)$, that there is any version v_i satisfying $t_L<t_i<t_R$ and containing any content different from that in v_L and v_R depends on many factors such as the length of the period from t_L to t_R, page type, content difference between v_L and v_R or the average change degree of the page. Intuitively, the shorter the time distance between the page snapshots and the more even their distribution over time are, the lower is the average probability of any transient, undetected content occurring in the page.

The second error type, which we call a timestamp error, is due to uncertainty in estimating the dates of content changes. The timestamp error, like the content one, depends also on the number of acquired past snapshots and their distribution over time. Figure 2 illustrates both error types. The top timeline shows available past snapshots of a page. For simplicity, let us assume that the page snapshots are empty (i.e., blank page) or they contain only one content element, be it a picture or a text snippet. Those snapshots that contain the element are marked by a green color, while the empty snapshots are marked by a grey color. After reconstructing the history of the page (the middle timeline) and comparing it with the bottom timeline that shows the actual page history, we can see that the reconstructed history contains both content and timestamp errors.

Site History Reconstruction

Pages usually belong to larger information units, or Web sites. Reconstructing histories of a Web site requires detecting the changes in site's topology over time and retrieving past content of pages

Figure 2. Content and timestamp errors in the reconstructed page history

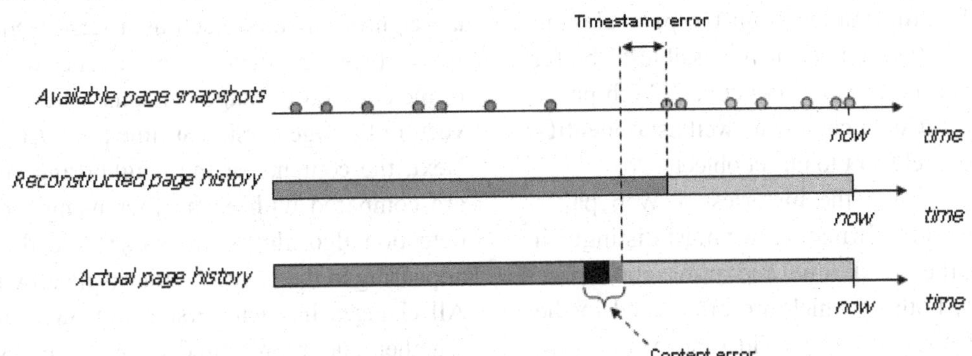

that belonged to the site. As an input the starting page (e.g., the top page of a site), time frame *T*, and the depth *D* (i.e., the number of hops from the starting page) need to be specified.

The data accumulation system collects all available snapshots of the starting page that have timestamps within *T*. It then searches their content for any links to other pages on the site (i.e., pages having the same domain name). For each such a link, it collects available, previous snapshots of the linked page that have timestamps within *T*. These snapshots are then searched in the same way for links to other pages on the same site. The entire process is repeated until the specified depth *D*. In general, a page is considered to belong to the site's history if, during the time frame *T*, it was linked from another page belonging to the site at that time and if it was located a smaller number of hops from the starting page than the specified depth *D*. Intuitively, the number of page snapshots collected at the initial steps of the crawl (few hops from the starting page) has an influence on detecting pages at later steps. This is because pages may remain undiscovered if the links pointing to them occurred only in the undetected, transient content of other pages in the site. We call the error caused by the missing links a page error.

A site history is represented as a set of reconstructed page histories that belonged to the site in the past. The precision of the site reconstruction can be enhanced by utilizing topological information preserved in the past content of site-map pages if they existed. Many Web sites include site-map pages designed to help users navigate sites. Utilizing the site-map page history could help to detect transient pages that have not been discovered by the above crawling approach and thus minimize the page error, as well as it could help to more precisely determine the actual time points of page creation and deletion within sites (timestamp error for whole pages).

PAGE TEMPORAL ANALYSIS

Page temporal analysis is the study of page content over time. Its results should be particularly useful if pages are associated with specific objects such as companies, institutions, persons or other entities. Understanding the temporal characteristics of a page over a long time frame can shed light on the associated objects or on other information appearing on the page. For example, if certain content occurred for a long time on a page which was updated frequently and regularly, then we can treat the content in a different way or with a different level of trust than if it occurred on a page that was generally static or even obsolete. A similar idea applies to a page devoted to a specific topic vs. a page that deals with many varying topics throughout its history. In other words, page temporal analysis can be used to find temporal context for information from the past. Having determined the context, it is possible to better understand the connection between Web pages and their transient content as well as to identify pages most relevant to target objects.

When mining the histories of Web pages for real-world information, we must distinguish between the valid time and transaction time of events, both of which are often used in the database research. The valid time of an event is considered as the time at which the event occurred in the real world. The transaction time is the time at which the information about the event was stored in a database or, in our case, added to a certain Web page. It can be estimated by searching the page history for the earliest occurrence of the content related to the event (Jatowt, Kawai, & Tanaka, 2007). The valid time, on the other hand, can be detected from temporal expressions appearing in the content of past page versions. This would require using special taggers and resolvers of temporal expressions in text. In addition, techniques such as the one described by Bar-Yossef, Broder, Kumar, and Tomkins (2004) could be applied for classifying page content as current or obsolete.

Next, we present a simple framework for analyzing page histories. After page history reconstruction, HTML tags, scripting code, and multimedia objects are removed from available page versions. Vector representation is then created for textual content of the past versions using a weighting method such as a term frequency. Let $V=(v_1,v_2,..,v_n)$ denote the sequence of vectors of the consecutive page versions, where v_j is the vector of a page version at time point t_j $(t_1 \le t_j \le t_n)$. Next, the contents of the neighboring versions are compared with each other using a change detection algorithm such as *diff*. Added content appearing in the page's history is thereby found. All changes in each version are then grouped together and represented as a change vector. Consequently, a sequence of change vectors is obtained, $C=(c_{(1,2)},c_{(2,3)},..,c_{(n-1,n)})$, where $c_{(j,j+1)}$ is a vector for an added-type change obtained by comparing page versions v_j and v_{j+1}.

The content of past versions can be compared against any query containing terms describing given topic of interest. In order to do so, at each selected time point, a query vector, q_j, is constructed by assigning uniform weights to all query terms. The sequence of query vectors is denoted as $Q=(q_1,q_2,..,q_n)$. Different values can be assigned to Q at different time points to reflect changes in the chosen topic of interest. Otherwise, the query vector is made static by having the same content at all times. To measure the relationship of past page content to the query topic, the similarity between V and Q is calculated using a cosine similarity measure. In result, the sequence of similarities is obtained: $sim(V,Q)=(sim(v_1,q_1),sim(v_2,q_2),..,sim(v_n,q_n))$,

Figure 3. Similarity calculation between the sequence of version vectors and the sequence of query vectors

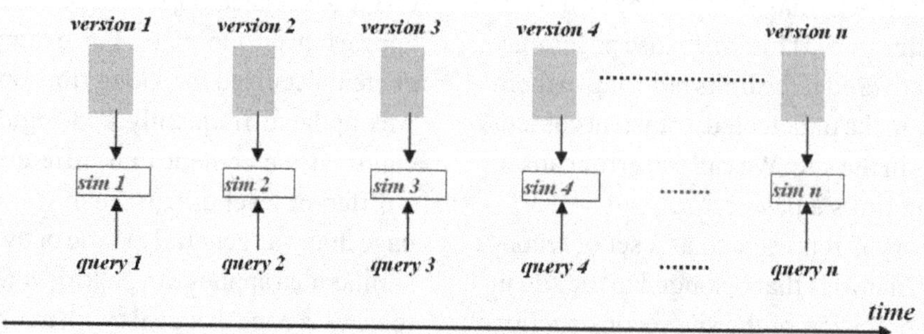

Figure 4. Similarity calculation between the sequences of change and query vectors; changes are depicted as small rectangles inside page versions

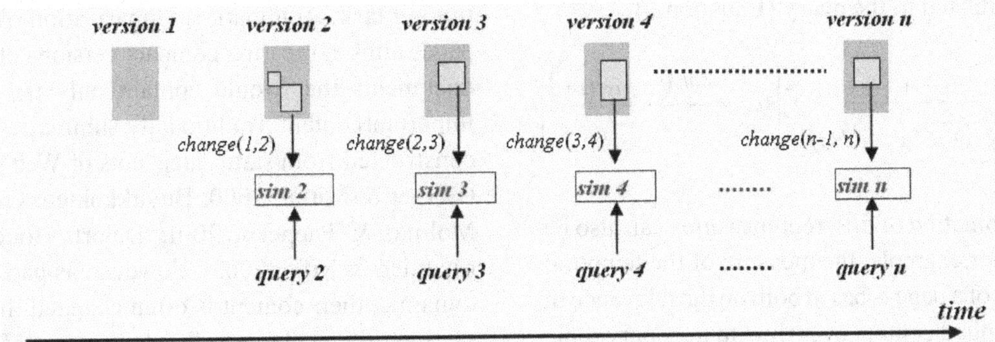

where $sim(v_j, q_j)$ is the cosine similarity between the vector of past version v_j and query vector q_j (Figure 3). Similarly, the sequence of similarities between the vectors of the changes and the query vectors is calculated, $sim(C,Q)=(sim(c_{(1,2)}, q_2), sim(c_{(2,3)}, q_3), \ldots, sim(c_{(n-1,n)}, q_n))$, where $sim(c_{(j,j+1)}, q_{j+1})$ is the cosine similarity between change vector $c_{(j,j+1)}$ and q_{j+1} (Figure 4).

First, a change frequency can be defined (Equation 1).

$$CF = \frac{fc}{n}$$

Here, *fc* is the number of non-zero elements in C. Another measure called a change degree indicates the average change size of a page (*size(a)* denotes the size of element *a*).

$$CD = \frac{\sum_{j=1}^{n} \frac{size\left(c_{(j,j+1)}\right)}{size\left(v_j\right)}}{n}$$

Besides these simple measures, the long-term relevance of a page to the query topic can be calculated. It is expressed as the weighted average of the elements of $sim(V,Q)$ by taking into account the duration of page content over time (Equation 3).

$$TR = \frac{1}{\sum_{j=1}^{n}\left(\alpha_{j+1} * \left(t_{j+1} - t_j\right)\right)} \sum_{j=1}^{n}\left[\alpha_{j+1} * sim\left(v_j, q_j\right) * \left(t_{j+1} - t_j\right)\right]$$

A page is considered relevant if its content overlaps with the sequence of query vectors during a large portion of a chosen time period. Using this approach, we can estimate the degree of page relevance to any topic within a given time frame. As the recent content is often likely to be more important, Equation 3 is adjusted by applying a weighting scheme depending on the age of page versions.

$$\alpha_j = e^{-\lambda\left(t_{now} - t_j\right)}$$

In addition, the long-term topic stability of a page can be computed by detecting the average similarity between consecutive past versions over time (Equation 5).

$$TS = \frac{1}{\sum_{j=1}^{n}\left(\alpha_{j+1} * \left(t_{j+1} - t_j\right)\right)} \sum_{j=1}^{n}\left[\alpha_{j+1} * sim\left(v_j, v_{j+1}\right) * \left(t_{j+1} - t_j\right)\right]$$

The long-term relevance and long-term topic stability are calculated considering the whole page content in the past, including the static content (the content that did not change between consecutive page versions). In contrast, we can compute a measure showing the degree of page updating based on the amount of changed content over time that is related to the query (Equation 6).

$$TRC = \frac{1}{\sum_{j=1}^{n}\left(\alpha_{j+1} * \left(t_{j+1} - t_j\right)\right)} \sum_{j=1}^{n}\left[\alpha_{j+1} * \frac{sim\left(c_{(j,j+1)}, q_{j+1}\right)}{t_{j+1} - t_j}\right]$$

A combination of different measures can also be used. For example, the measure of the temporal quality of a page is based both on the relevance of the changed content over time to the query topic and on the size of the changes:

$$TA = \frac{1}{\sum_{j=1}^{n}\alpha_{j+1}} \sum_{j=1}^{n}\left[\alpha_{j+1} * \frac{sim\left(c_{(j,j+1)}, q_{j+1}\right)}{\left(t_{j+1} - t_j\right)} * \frac{size\left(c_{(j,j+1)}\right)}{size\left(v_j\right)}\right]$$

According to this measure, a page is more attractive from the viewpoint of the query topic if its changes were relevant to that topic and if they were relatively large. Small changes are usually less likely to be attractive than large ones. Additionally, the temporal quality of the page is higher if the page was modified often in the past. In general, the greater the number and the larger the size of related changes that occurred within short time periods, the higher is the attractiveness of the page. The page temporal quality can be used to identify candidate pages for mining. Naturally, the precision of results depends on the amount and characteristics of the input data that is on the size of errors resulting from the history reconstruction process.

Finally, the trend of page relevance to query can be measured by fitting a regression line to the historical plot of the similarity between page content and query vectors. This allows for estimating the long-term change direction of the page relevance. A rising trend would mean that the page content becomes closer to the query topic.

TEMPORAL SUMMARIZATION

Document summarization is a well-known text mining task. Automatic summarization of Web pages aims at creating compact versions of Web documents that would contain only the most important content. Traditionally, summaries were constructed from static snapshots of Web pages (Berger & Mittal, 2000; Buyukkokten, Garcia-Molina, & Paepcke, 2001; Delort, Bouchon-Meunier, & Rifqi, 2003). However, as pages are dynamic, their content is often changed. In this section, we briefly describe the concept of temporal summarization which is the extension of the traditional summarization task into the time dimension (Jatowt & Ishizuka, 2004a, 2004b; Jatowt & Ishizuka, 2006). It is used to summarize temporal versions of Web documents in order to provide information on important content, hot topics or popular events described in pages over time. Web users are often overloaded with large amounts of data. Automatic temporal summarization would help them in discovering salient information from parts of the past Web such as histories of pages or their collections.

Following the classical division of document summarization research, two types of temporal summarization can be distinguished: single- and multi-page temporal summarization. Single-page temporal summarization attempts at capturing salient content that occurred on a page over a certain time period. The summary should thus reveal main page topics during a predefined time frame. On the other hand, in multi-page temporal summarization, multiple snapshots of a topical collection of pages are analyzed for changes over time. The summary should reveal important events or concepts that occurred in a given topical area over time. The key issue in this type of summarization is gathering pages which are up-to-date and related to the target topic so that a reliable and consistent topical collection can be synthesized. Below we discuss the multi-page temporal summarization in more detail.

Multi-Page Temporal Summarization

Web collection for multi-page temporal summarization can be obtained in several ways; for example, it can be created from a user-provided set of related Web documents that she or he usually revisits for fresh information or it could be downloaded from existing Web directories. While Web directories group topically related Web documents, they provide only a limited number of categories. In a more flexible way, the collection could be synthesized by filtering search engine results based on the analysis of their temporal characteristics such as long-term relevance or temporal quality. Naturally, duplicate pages should be discarded in this process. After the initial set of topically related pages is ready, it is extended in time by reconstructing page histories for a chosen time period.

In the following step, textual data is extracted from the accumulated past versions. Then, an extractive type summarization algorithm is used to detect useful sentences for constructing a summary. First, so-called long-term scores are calculated for all terms by comparing terms' distributions in documents over time. These scores are later used to identify important sentences to be included in the summary. We propose two approaches for the long-term score calculation. One uses a sliding window that is sequentially moving through the temporal collection to search for bursts of terms in added or deleted content in the collection (Jatowt & Ishizuka, 2004a). Any terms that were added to or deleted from many pages in the collection at around the same time have high values of the long-term scores. Another approach to the calculation of long-term scores is based on the analysis of term frequency plots. The parameters of term frequency plots such as variance, slope of a regression line and intercept are calculated and compared for identification of salient terms (Jatowt & Ishizuka, 2004b). The terms with outstanding features, such as the ones with upward trends or high variance would be then scored highly. More details on the both term scoring methods can be found in Jatowt and Ishizuka (2004a, 2004b) and Jatowt and Ishizuka (2006).

After the long-term scores of terms are computed, the summarization system searches for sentences suitable for constructing the summary. Sentence selection is based on analyzing plots of the terms that have the highest long-term scores. The plots are examined to identify intervals with the closest match to the shape of an ideal plot. For example, the system may search for a time period where the frequency plot of a term has a shape that most resembles the ideal shape in which the plot suddenly increases and remains at a relatively high level over a long time. Such a plot shape may indicate the onset of an important event represented by the term. Thus, sentences containing the term are extracted from the collection within the selected time period. The system tries here also to maximize the number of different terms with top long-term scores in the selected sentences. Lastly, after the predefined number of sentences is extracted, the system orders them based on their timestamps and relative locations in their original page versions. Each sentence may also have a link to its page version added to be used in case users wish to obtain more details. Furthermore, a number of additional heuristics may be used to increase the coherence and readability of the final summary, for example, by inserting explanatory content or by modifying or reordering the selected sentences.

Discovering Object Histories

Related to temporal summarization is object history reconstruction. Objects are defined here as higher level concepts and abstractions that represent persons, institutions, ideas, organizations, and so forth. Objects can be represented by groups of related words or n-grams. Thus, object histories could be modeled using the histories of the representative terms and their inter-relationships. Time points of changes and the durations of terms' occurrences on pages would provide clues about the timing of events related to objects represented by these terms.

Object's history should be most accurate if it has been derived from a source that directly represents the object (e.g., company homepage, personal blog). The relationship between the page and objects discussed on this page can help in understanding the content related to the objects. In general, contextual information about objects can be derived from the characteristics and topical scopes of analysed pages. Furthermore, the co-occurrence of similar information among different resources increases its trustworthiness as well as helps to better determine the starting and ending points of events. The larger is the number of different data sources devoted to an object, the more reliable and accurate the discovered knowledge should be.

A possible example of object history reconstruction is an automatic creation of personal bibliographies or their parts. There is much personal data published on the Web. For example, employment data is sometimes reported on company or personal Web pages (e.g., on blogs), and other personal information can be found. This information could be collected and processed to construct biography parts.

By analyzing semantic and temporal clues derived from past Web content it could be possible to improve the detection process by employment of various heuristics. For example, the temporal information derived from the chronological ordering of events reported on past pages might help in understanding the events and may provide hints for a further search. One such possible heuristic is the detection of person's employment dates. Suppose that at some time point a person's name was removed from the page of some laboratory. Then, the system could search for the page of another institution that reported hiring the person at around that time. Note, however, that there might be certain latency between the actual events and their reports in the Web (i.e., valid and transaction times).

BROWSING PAST WEB

Apart from mining the content of the past Web, it is important to have a tool that allows for viewing data in detail, for example, in order to manually inspect the data from the viewpoint of discovered results. Such a tool should be intuitive, easy to use and possibly resemble similar applications used for the current Web. In this section, we describe the framework for a past Web browser (Jatowt et al., 2006) that supports browsing and navigation in the past Web. A browser built using this framework would be a client-side system that downloads, in a real time, past page snapshots from Web archives for their customized presentation. Such a browser would enable viewing the evolution of pages and browsing the past structures of the Web.

The proposed browser integrates histories of Web pages with their present versions and has a standard functionality of a traditional browser for the live Web. Consequently, browsing the live and past Web can be done almost at the same time. Thanks to this, users browsing the live Web can access the histories of viewed pages in case they need to find some content from the past, observe the page evolution or, simply, to access the latest page snapshot if the present page cannot be properly viewed due to any reasons such as a server failure.

Browsing

Two basic types of browsing are distinguished here: vertical and horizontal. The former means browsing different pages around a certain point of time by following links, while the latter means viewing past snapshots of a single page along the time direction, that is, browsing the past Web in a horizontal direction. A mixture of both kinds of browsing enables users to traverse the past Web both in time and space dimensions.

To start the horizontal browsing, the URL of a page and a point of time have to be provided. The browser fetches a page snapshot whose timestamp is closest to the user-provided time point. Next, the browser automatically downloads the following page snapshots and displays them in a passive manner. This type of viewing results in a minimum user interaction, because page snapshots are presented to the user one by one, like in a slideshow, with a certain delay predefined for each snapshot. As when watching a video, the user can pause or stop the motion, enabling the detailed examination of the currently presented snapshot or following its links. Besides, the user may enter a new date or a different URL to make a jump to another snapshot. In addition, a timeline is automatically constructed and displayed above the page content (Figure 5). It shows the distribution of page snapshots indicating the points of time for which snapshots are available. The currently viewed snapshot is indicated in the timeline by a blue rectangle. The information provided by the timeline prevents users from being lost in the hyperspace of the past Web by informing them about the current time point of browsing and the overall distribution of snapshots. At the same time, it is also a navigation tool thanks to which users can make a jump to any page snapshot simply by clicking on any point on the timeline. The timeline can be also zoomed to provide the more detailed view. Besides the timeline, the clickable list of all page snapshots together with their timestamps is also displayed (Figure 5).

Horizontal browsing is enhanced by a page presentation in which content changes are detected and emphasized. Keeping in mind the large size of the past Web, with lots of static, redundant data, the most effective method for horizontal browsing seems to be the one using change visualization. We think that changed data is the most important in page histories and that enhancing horizontal browsing with the change indication can portray page evolution and, in addition, help reduce the amount of browsing needed, especially in the case of static (unchanging) pages. Both content additions and deletions between neighboring page snapshots are then detected using a change detection algorithm and emphasized to indicate the content variance in pages. This enables users to spot not only the added content in consecutive page snapshots but also to identify the removed one. However, effectively showing both change types in a combined view on a single page would be difficult, especially in the case of large and overlapping changes. Thus, we propose using animation effects in order to efficiently show both change types. The change presentation algorithm displays the changes gradually, in the form of animation. Content that was deleted in the page history first blinks for a certain time period and then disappears, followed by the inserted content that first appears on the page and then blinks for a short time. Page snapshots are processed in this way line by line from left to right and from top to bottom. Content that was static between consecutive snapshots remains displayed on the page. After the page transition between two consecutive page snapshots is completed, the browser waits a predefined time period with the latter page snapshot displayed and then it proceeds to analyze the following page snapshot. The user can control the speed of the presentation using a slider provided in the top-right corner of the browser (Figure 5). Besides, as sometimes page snapshots may be too large to be shown at once, a user can choose between the automatic scrolling option and the option of displaying only the top part of page content.

Animation of changed content results in a smooth transition between sequential page snapshots. By animating changes user's attention is drawn to the changed content. In addition, changes are also highlighted by different colors to increase their visibility. However, for simplicity, in the case when the amount of change in a page snapshot is higher than the predefined threshold, no animation is done and changes are emphasized using only different background colors.

The user can stop the horizontal browsing at any time by pressing stop or pause buttons in a similar way to video players. Next, she or he can view the currently displayed page snapshot in detail or follow any link. Upon clicking on a link, the browser loads the snapshot of the linked page that is closest in time to the one being currently viewed and, after a short time period, it automatically starts the horizontal browsing on the new page.

The browser is also equipped with two back and two forward buttons to enable navigation in the space as well as in time dimensions. Besides, there is an additional navigation mechanism provided (automatic jumping facility). It enables the browser to skip periods in the page history during which the content did not change or did not change much. When this functionality is switched on, the browser displays only those page snapshots that contain more than a certain amount of change. This enables faster viewing of page evolution by omitting changeless periods.

Finally, a search option enables users to specify queries for filtering changes. If a query is issued, only the changes that contain the query terms are animated. Other changes are treated as static content and thus are not animated. This browsing style results in the filtered view of page history. Users can thus observe page histories from the viewpoint of topics that they are interested in. For example, a newswire page history could be browsed for information about "Iraq" or "presidential election" over selected time periods.

Related Research and Future Work

Visual Knowledge Builder (VKB) (Shipman & Hsieh, 2000) was an early proposal of an application that provides a mechanism for enabling history navigation in private hypertexts. The objective was to allow users to play back the history of a hypertext for witnessing the authoring of hypertexts, understanding the context of their creation and authors' writing styles. The browser interface had some similarity to VCR players.

WERA[3] (Web ARchive Access) and Wayback Machine[4] are applications for accessing Web archives. WERA supports time and URL input for specifying a particular page snapshot. There is a timeline provided showing the available page snapshots and indicating the currently browsed one. Users can view the consecutive page snapshots by clicking arrows in the timeline.

Wayback Machine is a Web-based interface to the Internet Archive. After a user inputs a URL, optionally with a time period specified,

Figure 5. Past Web browser

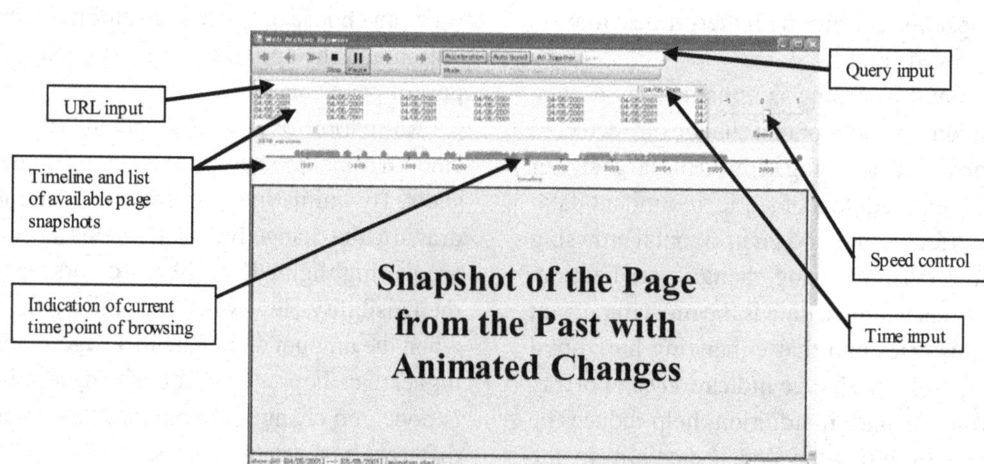

links to the available page snapshots are listed on the "directory" page. The user can then click on any snapshot to view its content or follow its links if the linked snapshots are also stored in the archive. The directory page indicates also page snapshots that contain changes by marking them with asterisks. Horizontal browsing using Wayback Machine is difficult, as users need to access the directory page each time if they wish to view other snapshots.

Both the Wayback Machine and WERA are server-side applications designed for single Web archives. Our proposed browser is a higher-level, client-side application that allows for the usage of multiple past Web repositories at the same time, thus, enabling browsing of the past Web rather than browsing single archives. Browsing the past Web is also facilitated by combining passive, automatic page viewing together with a change presentation. The framework has also functions that minimize the user effort and time required to find specific information in the past snapshots of pages. In addition, navigation mechanisms are provided to enable traversal of the link structure of the past Web. Testing the browser built on the proposed framework demonstrated its usefulness (Jatowt et al., 2006). Users were able to move freely in the past Web, find desired information and relatively easily obtain an overall view of pages' evolution.

In a multi-authoring area, an interesting application has been recently proposed for effective visualization of histories of wiki pages (Viégas, Wattenbeg, & Dave, 2004). It allows viewing contributions of different authors and their persistence over time as demonstrated on the example of Wikipedia pages[5].

There are several possible directions for expanding the proposed framework. For example, location-based browsing would allow a user to select a certain area on a page and then view its evolution over time, provided that the structure of the page did not change substantially. This would limit the presentation to only those changes that occurred in the selected area, for example, in the sports section of a newswire page. Next, links on visited snapshots could be annotated with timestamps of page snapshots that will be accessed when following these links. Thanks to it, a user would know how much time jump she or he is going to experience upon clicking on a certain link. Lastly, a comparative past web browser could enable comparison of histories of two or more pages highlighting their common or similar parts.

CONCLUSION

The Web has become nowadays a major means of communication and an important information repository. Due to its dynamic, ever evolving character, much of the content regularly disappears from the live Web and can only be accessed through Web archival repositories. Knowledge discovery from past Web is a challenging and promising research direction. Mining the content of the past Web differs from traditional Web content mining and thus requires a novel approach. In this chapter, we have described several issues related to mining data in Web archives. First, we provided the outlook on the data collection and preparation steps and emphasized their importance. Next, we demonstrated the methodology for determining page temporal characteristics as a source of contextual information for describing pages and their transient content. Then, data summarization and object history detection were described as examples of mining tasks on the past Web. Finally, we proposed the application for browsing and navigation in the past Web.

ACKNOWLEDGMENT

This research was supported by the MEXT Grant-in-Aid for Scientific Research in Priority Areas entitled: Content Fusion and Seamless

Search for Information Explosion (#18049041, Representative Katsumi Tanaka), and by the Informatics Education and Research Center for Knowledge-Circulating Society (Project Leader: Katsumi Tanaka, MEXT Global COE Program, Kyoto University) as well as by the MEXT Grant-in-Aid for Young Scientists B entitled: Information Retrieval and Mining in Web Archives (#18700111).

REFERENCES

Allan, J. (Ed.). (2002). *Topic detection and tracking: Event-based information organization.* Norwell, MA, USA: Kluwer Academic.

Allan, J., Gupta, R., & Khandelwal, V. (2001). Temporal summaries of news topics. In *Proceedings of the 24th Annual Conference on Research and Development in Information Retrieval*, (pp. 10-18). New Orleans, LA, USA: ACM Press.

Amitay, E., Carmel, D., Herscovici, M., Lempel, R., & Soffer A. (2004). Trend detection through temporal link analysis. *Journal of the American Society for Information Science and Technology, 55*(14), 1270-1281.

Arms, W.Y., Aya, S., Dmitriev, P., Kot, B. J., Mitchell, R., & Walle, L. (2006). Building a research library for the history of the Web. In *Proceedings of the Joint Conference on Digital Libraries*, (pp. 95-102). Chapel Hill, NC, USA: ACM Press.

Arvidson, A., Persson, K., & Mannerheim, J. (2000). The Kulturarw3 project—the Royal Swedish Web Archive—an example of "complete" collection of Web pages. In *Proceedings of the 66th IFLA Council and General Conference,* Jerusalem, Israel.

Aschenbrenner, A., & Rauber, A. (2006). Mining Web collections. In J. Masanes (Ed.), *Web archiving* (pp. 153-174). Berlin, Heidelberg, Germany: Springer-Verlag.

Bar-Yossef, Z., Broder, A. Z., Kumar, R., & Tomkins, A. (2004). Sic transit gloria telae: Towards an understanding of the Web's decay. In *Proceedings of the 13th International Conference on World Wide Web*, (pp. 328-337). New York: ACM Press.

Berger, A. L., & Mittal V. O. (2000). OCELOT: A system for summarizing Web pages. In *Proceedings of the 23rd Conference on Research and Development in Information Retrieval*, (pp. 144-151). Athens, Greece: ACM Press.

Brewington, E. B., & Cybenko, G. (2000). How dynamic is the Web? In *Proceedings of the 9th International World Wide Web Conference*, (pp. 257-276). Amsterdam, the Netherlands: ACM Press.

Buyukkokten, O., Garcia-Molina, H., & Paepcke, A. (2001). Seeing the whole in parts: Text summarization for Web browsing on handheld devices. In *Proceedings of the 10th International World Wide Web Conference,* (pp. 652-662). Hong Kong, SAR, China: ACM Press.

Chi, E. H., Pitkow, J., Mackinlay, J., Pirolli, P., Gossweiler, R., & Card, S. K. (1998). Visualizing the evolution of Web ecologies. In *Proceedings of Conference on Human Factors in Computing Systems*, (pp. 400-407), Los Angeles: ACM Press.

Cho, J., & Garcia-Molina, H. (2000). The evolution of the Web and implications for an incremental crawler. In *Proceedings of the 26th International Conference on Very Large Databases*, (pp. 200-209). Cairo, Egypt: ACM Press.

Cho, J., & Garcia-Molina, H. (2003). Estimating frequency of change. *Transactions on Internet Technology, 3*(3), 256-290.

Cooley, R. Srivastava, J., & Mobasher, B. (1997). Web mining: Information and pattern discovery on the World Wide Web. In *Proceedings of the 9th IEEE International Conference on Tools with Artificial Intelligence,* (p. 558). IEEE Press.

Delort, J.-Y., Bouchon-Meunier B., & Rifqi, M. (2003). Enhanced Web document summarization using hyperlinks. In *Proceedings of the 14th ACM Conference on Hypertext and Hypermedia*, (pp. 208-215). Nottingham, UK: ACM Press.

Fetterly, D., Manasse, M., Najork, M., & Wiener, J. (2003). A large-scale study of the evolution of Web pages. In *Proceedings of the 12th International World Wide Web Conference*, (pp. 669-678). Budapest, Hungary: ACM Press.

Gruhl, D., Guha, R., Liben-Nowell, D., & Tomkins, A. (2004). Information diffusion through blogspace. In *Proceedings of the 13th International World Wide Web Conference,* (pp. 491-501). New York: ACM Press.

Hallgrimsson, Þ., & Bang S. (2003). Nordic Web Archive. In *Proceedings of the 3rd Workshop on Web Archives in conjunction with the 7th European Conference on Research and Advanced Technologies in Digital Archives*, Trondheim, Norway. Springer-Verlag.

Jatowt, A., & Ishizuka, M. (2004a). Summarization of dynamic content in Web collections. In *Proceedings of the 8th European Conference on Principles and Practice of Knowledge Discovery in Databases*, (pp. 245-254). Pisa, Italy: Springer-Verlag.

Jatowt, A., & Ishizuka, M. (2004b). Temporal Web page summarization. In *Proceedings of the 5th Web Information Systems Engineering Conference,* (pp. 303-312). Brisbane, Australia: Springer-Verlag.

Jatowt, A., & Ishizuka, M. (2006). Temporal multi-page summarization. *Web Intelligence and Agent Systems: An International Journal, 4*(2), 163-180. IOS Press.

Jatowt, A., Kawai, Y., Nakamura, S., Kidawara, Y., & Tanaka, K. (2006). Journey to the past: Proposal for a past Web browser. In *Proceedings of the 17th Conference on Hypertext and Hypermedia*, (pp. 134-144). Odense, Denmark: ACM Press.

Jatowt, A., Kawai, Y., & Tanaka, K. (2007). Detecting age of page content. In *Proceedings of the 9th ACM International Workshop on Web Information and Data Management.* Lisbon, Portugal: ACM Press.

Jatowt, A., & Tanaka, K. (2007). Towards mining past content of Web pages. *New Review of Hypermedia and Multimedia, Special Issue on Web Archiving, 13*(1), 77-86. Taylor and Francis.

Kleinberg, J. M. (2003). Bursty and hierarchical structure in streams. *Data Mining Knowledge Discovery, 7*(4), 373-397.

Kosala, R., & Blockeel, H. (2000). Web mining research: A survey. *SIGKDD Explorations, 2*(1), 1-15.

Kumar, R., Novak, P., Raghavan, S., & Tomkins, A. (2003). On the bursty evolution of Blogspace. In *Proceedings of the 12th International World Wide Web Conference*. Budapest, Hungary: ACM Press.

Li, Z., Wang, B., Li, M., & Ma, W.-Y. (2005). A probabilistic model for retrospective news event detection. In *Proceedings of the 28th Annual International Conference on Research and Development in Information Retrieval,* (pp. 106-113). Salvador, Brazil: ACM Press.

McCown, F., & Nelson, M. (2006). Evaluation of crawling policies for a Web-repository crawler. In *Proceedings of the 17th Conference on Hypertext and Hypermedia*, (pp. 157-168). Odense, Denmark: ACM Press.

McCown, F., Smith, J.A., & Nelson, M.L. (2006). Lazy preservation: Reconstructing Web sites by crawling the crawlers. In *Proceedings of the 8th ACM International Workshop on Web Information and Data Management,* (pp. 67-74). Arlington, VA, USA: ACM Press.

Mei, Q., & Zhai, C-X. (2005). Discovering evolutionary theme patterns from text: An exploration of temporal text mining. In *Proceedings of the*

11th International Conference on Knowledge Discovery and Data Mining, (pp. 198-207). New York: ACM Press.

Ntoulas, A., Cho, J., & Olston, C. (2004). What's new on the Web? The evolution of the Web from a search engine perspective. In *Proceedings of the 13th International World Wide Web Conference,* (pp. 1-12). New York: ACM Press.

Papka, R. (1999). *Online new event detection, clustering and tracking.* Unpublished doctoral dissertation, Department of Computer Science, University of Massachusetts, USA.

Rauber, A., Aschenbrenner, A., & Witvoet, O. (2002). Austrian online archive processing: Analyzing archives of the World Wide Web. In *Proceedings of the 6th European Conference on Digital Libraries,* (pp. 16-31). Rome, Italy: Springer-Verlag.

Shipman, F. M., & Hsieh, H. (2000). Navigable history: A reader's view of writer's time: Time-based hypermedia. *New Review of Hypermedia and Multimedia, 6,* 147-167. Taylor and Francis.

Swan, R., & Allan, J. (2000). Automatic generation of overview timelines. In *Proceedings of the 23rd Conference on Research and Development in Information Retrieval,* (pp. 49-56). Athens, Greece: ACM Press.

Toyoda, M., & Kitsuregawa, M. (2003). Extracting evolution of Web communities from a series of Web archives. In *Proceedings the 14th Conference on Hypertext and Hypermedia,* (pp. 28-37). Nottingham, UK: ACM Press.

Viégas, F., Wattenberg, M., & Dave, K. (2004). Studying cooperation and conflict between authors with history flow visualizations. In *Proceedings of the CHI Conference,* (pp. 575-582). Vienna, Austria: ACM Press.

Wang, X., & McCallum, A. (2006). Topics over time: A non-Markov continuous-time model of topical trends. In *Proceedings of the 12th International Conference on Knowledge Discovery and Data Mining,* (pp. 424-433), Philadelphia, PA, USA: ACM Press.

Yamamoto, Y., Tezuka, T., Jatowt, A., & Tanaka, K. (2007). Honto? Search: Estimating trustworthiness of Web information by search results aggregation and temporal analysis. In *Proceedings of the APWeb/WAIM 2007 Conference,* (pp. 253-264). Hunagshan, China: Springer-Verlag.

ENDNOTES

1. Internet Archive: http://www.archive.org
2. This efficiency is important in case when stream data is required.
3. WERA: http://archive-access.sourceforge.net/projects/wera
4. Wayback Machine: http://www.archive.org
5. Wikipedia, the free encyclopedia: http://en.wikipedia.org/wiki/Wiki

Chapter XVIII
Example-Based Framework for Propagation of Tasks in Distributed Environments

Dariusz Król
Wroclaw University of Technology, Poland

ABSTRACT

In this chapter, we propose a generic framework in C# to distribute and compute tasks defined by users. Unlike the more popular models such as middleware technologies, our multinode framework is task-oriented desktop grid. In contrast with earlier proposals, our work provides simple architecture to define, distribute and compute applications. The results confirm and quantify the usefulness of such ad-hoc grids. Although significant additional experiments are needed to fully characterize the framework, the simplicity of how they work in tandem with the user is the most important advantage of our current proposal. The last section points out conclusions and future trends in distributed environments.

INTRODUCTION

The main goal of this project is to create an exemplary system that would allow a network of computers to serve as distributed computer, allowing a client to send computational tasks to this network. The task would be later split into smaller tasks and processed by computers in the network. The reason behind creating a network able to process tasks in distributed way is obvious (Lanunay & Pazat, 2001). Creating a network from many low-end computers in most cases gives us processing capability much greater than one high-end computer of the same cost as many slower computers (Haeuser et al., 2000; Laure, 2001; Matsuoka & Itou, 2001). Also, distributed computing allows for the more effective use of resources of many idle computers in the network (Mitschang, 2003). Creating a task-oriented network is also a good way to understand the basics of remoting and reflection, two mechanisms in C# that are crucial for such solution to work (Gybels, Wuyts, Ducasse, & Hondt, 2006).

On the one hand, to ensure communication between elements of the framework, .NET remoting technology is used. Remoting facilitates method invocation on remote objects works exactly the same way as on local objects. On the other hand, reflection provides necessary mechanisms that allow computer programs to modify themselves during runtime. C# implementation of reflection allows us to load assembly code, create objects, obtain information about assembly code, object, methods, properties and fields, and invoke a method of object. In our work, reflection is necessary to divide tasks into task portions. If a task cannot be divided into portions beforehand, this kind of task cannot be effectively distributed and our framework will be unable to facilitate parallel execution of that task.

The division of tasks into task portions is done during the coding of the task itself. The user needs to specify two classes for each task. One class serves as an "initiator" of task portions. The other class is a task portion, and needs a parameter and returns partial result. Every time a new task is run, an initiator class uses reflection to create instances of task portion class and puts instantiated objects into readyJobs array. This is done by a method called readyJobs.Add (i, Activator. CreateInstance (taskPortionType, parameters)). The first parameter to this method is task portion type and it needs to be declared inside of the "initiator" class. To create such a type of variable, reflection is needed. If the class, which type is set to the type of the task portion, is not loaded yet, reflection can be used to load appropriate assembly and then get the type of the class.

The rest of the chapter is organized as follows. The following section (Section "Background") introduces the cross-platforms for desktop grids. The next section (Section "Elements of the Multinode Framework") details the elements of the multinode framework. The fourth section (Section "Communication between Elements via .NET Remoting") proposes a communication schema between elements via .NET remoting. The next section (Section "Defining a Task to Distribute and Compute") describes how to define a task to distribute and compute. The next section (Section "Study of the Framework Performance") studies the performance. The last section (Section "Conclusion") points out conclusions and future trends in distributed environments.

BACKGROUND

There are many solutions for task propagation in distributed environments. The most known is the BOINC (Berkeley Open Infrastructure for Network Computing) project (Anderson, 2004). The system consists of a set of applications running in a Linux environment. They mostly have the form of independent daemons communicating with each other using a database or shared memory. Every project based on BOINC needs to have its own server. In order to take part in a project, community members need to install dedicated client software. This software connects to server and downloads data and executables necessary for carrying out tasks. The client may be connected to several projects at the same time. It is some form of reusability; however, it comes within client's administrator duties to connect manually to a new server. In this approach, there is no knowledge sharing or projects coordination. Each project runs independent. It may have some advantages when system failures are taken into consideration. In BOINC architecture, when one project is shut down, the others can continue to operate normally. The autonomy of projects increases the probability of error detection. Every project can define its own assertions to the data being the result of processing.

Recently, some .NET grids have been created. OGSI.net, developed at the University of Virginia (OGSI.net, 2007), is an implementation of the

OGSI specification on Microsoft's .NET platform. However, the OGSI.net project is committed to interoperability with other OGSI compliant frameworks (such as the Globus Toolkit 3) which run primarily on Unix systems and so represents a bridge between grid computing solutions on the two platforms. OGSI.NET provides tools and support for an attribute-based development model in which service logic is transformed into a grid service by annotating it with metadata. OGSI.NET also includes class libraries that perform common functions needed by both services and clients.

The Alchemi project from the University of Melbourne (Alchemi, 2007) is an open source software framework that allows you to painlessly aggregate the computing power of networked machines into a virtual supercomputer (desktop grid) and to develop applications to run on the grid. It has been designed with the primary goal of being easy to use without sacrificing power and flexibility. Alchemi includes the runtime machinery (Windows executables) to construct computational grids, a .NET API, and tools to develop .NET grid applications and grid-enable legacy applications.

The motivations of our work have similarities with BOINC, OGSI.net, and Alchemi, and differences from work of Poshtkohi, Abutalebi, and Hessabi (2007). This work proposes the DotGrid project to share, select and aggregate distributed resources in an integrated way based on Microsoft .NET in Windows and MONO .NET in Linux. This has come true via implementing a layer over the chosen operating systems. This approach eliminates the dependency of grid to the native system. The .NET programming environment includes features that are suitable for simplicity and efficiency computing: multithreading, remoting, and reflection. We use the last one to divide tasks into task portions.

Grid computing is one of the most innovative aspects in recent years. Multi-agent computing now becomes a promising solution in many domains. Hence, it is a natural choice to combine these two technologies together. Although grid technology heavily relies on efficient computation with interaction, most of the current systems or applications lack the vision of utilizing computer interaction. Recently, there has been a shift toward agent-based grid computing, with many researchers contributing to the field.

In the following section, we briefly present some of the main elements of the multinode framework, which addresses the problems described previously.

ELEMENTS OF THE MULTINODE FRAMEWORK

The framework that enables distributed code propagation (Deng, Han, & Mishra, 2005; Król, 2005; Król & Kukla, 2006; Lauzac & Chrysanthis, 2002; Lin & Kuo, 2000) is composed of the main elements (shown in Figure 1):

- The broker represents the central part of the system; it is the application which receives tasks, processes them and distributes parts of the tasks to remote nodes, then fetches the partial results and combines them into final result;
- The client is the application which connects to the broker and sends tasks with appropriate format; and
- The nodes are background applications running on many computers, constantly receiving task portions from the broker, processing them and finally returning the partial result to the broker.

Even though all elements of the system are connected with each other, all communication is channeled through the broker and there is no direct connection between client and any of the other nodes.

Figure 1. Elements of the project

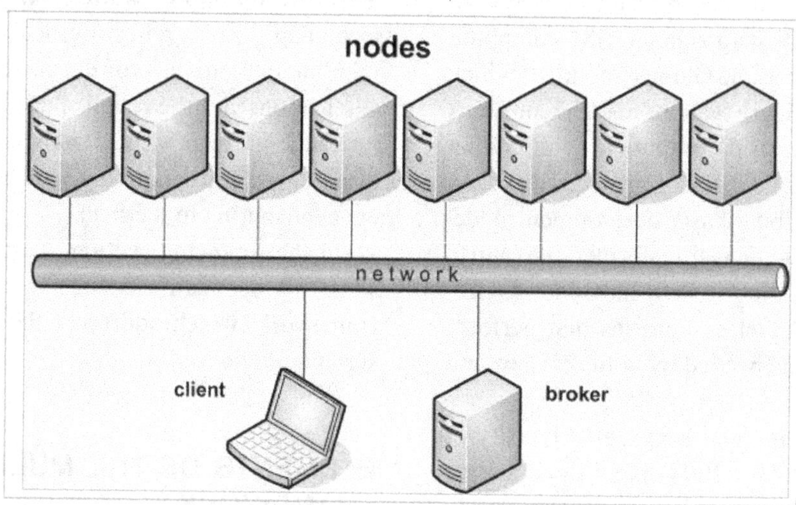

COMMUNICATION BETWEEN ELEMENTS VIA .NET REMOTING

Within our framework, communication is needed to:

- Send tasks from client to broker;
- Send task portions from broker to nodes;
- Send task results from nodes to broker; and
- Send other signals (login and logouts) between client, broker and nodes.

The communication between system elements can be described with an example of sending tasks from client to broker. Elements of the system that need to receive something become a server. The application that is sending something becomes a client. The broker creates a well-known service of MyTaskReceiver with method receive-Task (object Task). To pass a task instance, a client needs just to call remote method and give task as parameter. The method will return true of false value, depending on the result of the method. If a communication problem happens, a particular exception will be thrown. To allow two applications to communicate through remoting, one of them must be configured to be a server and another as a client.

Server needs to:

1. Register HTTP server port.
2. Register well-known service with remote object type.

Client needs to:

1. Register HTTP client port.
2. Create object of remote object type.
3. Activate remote object.
4. Run remote method.

Our project uses the following elements implemented as interfaces (shown in Figure 2).

As we can see in the Figures 3, 4, and 5, the interfaces define only what methods are available within the remote objects. The interfaces themselves do not define what is exactly executed by method. This is computed entirely on the server side. An application which is defined as server needs to have specific method definitions. In this system, this is done by:

Figure 2. Interfaces used in the framework

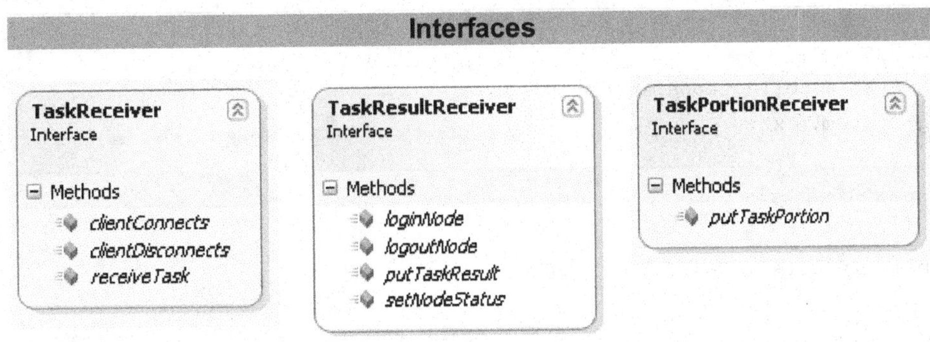

Figure 3. TaskReceiver interface

```
using System;
using System.Collections.Generic;
using System.Text;

namespace Nestor
{
  public interface TaskReceiver
  {
    bool clientConnects();
    bool clientDisconnects();
    string receiveTask(object task);
  }
}
```

- Class MyTaskReceiver on broker, extending TaskReceiver interface;
- Class MyTaskResultReceiver on broker, extending TaskResultReceiver interface; and
- Class MyTaskPortionReceiver on nodes, extending TaskPortionReceiver interface.

DEFINING A TASK TO DISTRIBUTE AND COMPUTE

To facilitate the process of sending tasks from a client to the broker, a generic task class needs to be defined. A task should be defined this way, so the broker will be able to divide a task into the smaller tasks (called here task portions), which will be later sent to the nodes. After gathering the

Figure 4. TaskResultReceiver interface

```
using System;
using System.Collections.Generic;
using System.Text;

namespace Nestor
{
  public interface TaskResultReceiver
  {
    /// <summary>
    /// Adds node to logged nodes list
    /// </summary>
    /// <returns>Returns assigned number of node (>=0) or error.</returns>
    int loginNode(string host);

    /// <summary>
    /// Removes node from logged node list
    /// </summary>
    /// <returns>True on success, false otherwise.</returns>
    bool logoutNode(int loggedNodeId);

    /// <summary>
    /// Passes to broker information about current status of node.
    /// This will be saved and displayed on connected nodes list.
    /// </summary>
    /// <returns>True on success, false otherwise.</returns>
    bool setNodeStatus(int loggedNodeId, string status);

    /// <summary>
    /// Passes to broker result of finished task.
    /// </summary>
    /// <returns>True on success, false otherwise.</returns>
    bool putTaskResult(int loggedNodeId, long portionNumber, string result);
  }
}
```

Figure 5. TaskPortionReceiver interface

```
using System;
using System.Collections.Generic;
using System.Text;

namespace Nestor
{
  public interface TaskPortionReceiver
  {
    /// <summary>
    /// Sends to a node a task portion
    /// </summary>
    /// <returns>true on success, false otherwise</returns>
    bool putTaskPortion(TaskPortion portions, long taskPortionNumber);

    /// <summary>
    /// Tests if a connection has been made
    /// </summary>
    /// <returns>Always true when connected, otherwise would return false
but without connection will throw an exception</returns>
    bool testConnection();

    /// <summary>
    /// Signals to a node that broker disconnects
    /// </summary>
    void brokerDisconnects();

  }
}
```

results from all the task portions, the final result is created by the broker and returned. In this project, a task is defined by two interface classes: Task and TaskPortion, with the following methods:

- Task.doBrokerJob: This method runs just after receiving the task. All initial instructions, which are preparing variables and job should be set here;
- Task.getResult: This method runs after all jobs return the results. This method aggregates partial results and displays result of the task; and
- TaskPortion.doNodeJob: This method runs on a node, which contains all the operations needed to get the partial results. This method is passed a parameter which is a reference to a table of results.

C# interfaces do not contain any information about the constructor, but when we are defining a task we need to add constructor for TaskPortion class. This constructor needs a parameter that will differentiate various task portions.

The body of the doBrokerJob method needs to create a set of tasks which will be put into readyJobs table (passed by reference). This is done using the reflection mechanism. First, we use reflection to create an object containing the type of TaskPortion class. Second, we use reflection to dynamically create a number of instances of Task-Portion classes with distinct parameters, and then to store those instances as ready jobs.

STUDY OF THE FRAMEWORK PERFORMANCE

The main aim of creating this framework is to utilize the power of more than one computer and create a simple way for parallel processing of various tasks. To apply for the success, the following study was made. The two defined tasks described use extensive mathematical calculations.

- **Task 1.** Checks which numbers from range 1000000001 ... 1000000100 are prime. This task is irregular, and there is no way to predict how long it might take to calculate one portion of the task.
- **Task 2.** Calculates 100 packs of 100 hexadecimal digits of the number π. This task is regular; all the portions take the same amount of processor time.

Normally, those tasks, depending on processor speed, require several minutes to complete. In our study, we have checked how the number of nodes influences the time of the task.

The results of the study are the following. Task 1, on a single node, was performed in 1 minute 11 seconds. After adding the second node, we have gained a decrease of 21 seconds. By adding the third node, we got an additional 26 seconds. Then, adding the next node did not result in speed gain. We can explain this with the nature of this task. When we send a task portion to a node, we don't know the size of this portion. If this number will be prime, it will be checked against all numbers smaller than itself (which is time consuming). If it will not be prime, the task will be stopped as soon as we find a number that is a divider of this number. From the result, we see that the complete loop of testing one number takes about 24-30 seconds. Also, from Figure 6 we see that in studied range we have only three prime numbers, because after adding the third node we do not gain any speed, which basically

Figure 6. Performance index for prime numbers checking on nodes

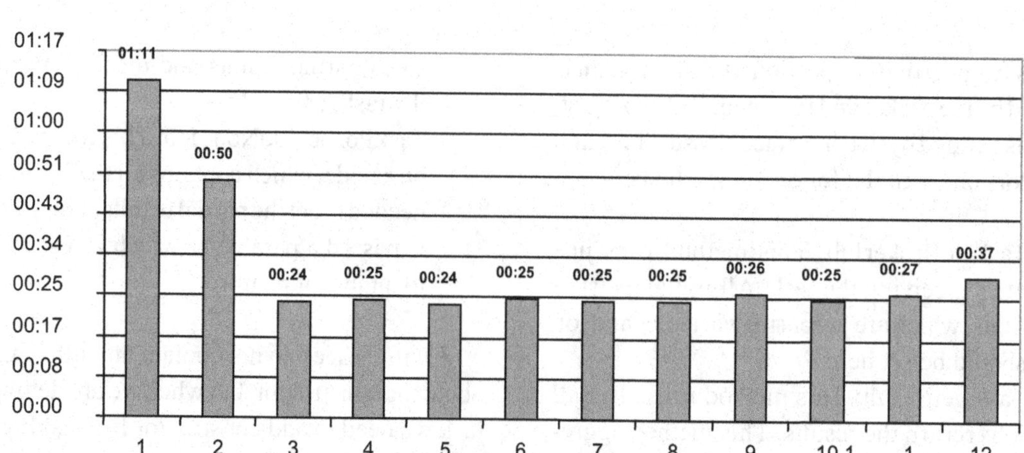

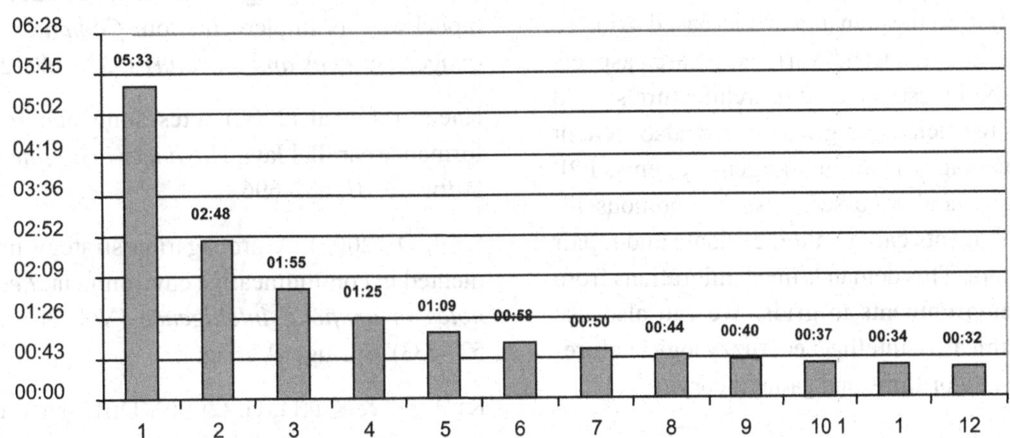

Figure 7. Performance index for digits of the number π calculation on nodes

means that we have three nodes processing, and the rest of the nodes are idle.

The second task proved the worth of building such networks. In this task, all the portions are even. From Figure 7, we can clearly see that with adding each new node, the total processing time was reduced. The relation is close to invert proportional (total time = total time for 1 node/ number of nodes).

CONCLUSION

This framework is prepared to only run a pair of predefined algorithms. To add more algorithms, the following actions should be made.

- Create two serializable task classes inheriting from Task and TaskPortion classes and implement all needed methods. Add those classes to application namespace, so that client application is able to create instances.
- The task class has to use the doBrokerJob method to fill the readyJobs table passed as a parameter. This table needs to be filled with instances of the TaskPortion class.
- The task class has to use the getResult method to aggregate results of the task portions (table jobResults) and return the single string containing the aggregated result.
- The taskPortion class can use the constructor parameters to pass parameters from the broker to a node.
- The taskPortion class has to use the doNodeJob method to perform task portion calculations and return the result of a task portion by reference to the table jobResult.

The only limitation of this framework is that algorithms used here must be easily divided into smaller tasks (which can require a lot of calculations). Task portions are distinguished by a parameter passed to a task portion. The type and number of those parameters are not limited, allowing the designer to make TaskPortion flexible to allow any kind of parameters.

In our approach, we can observe emergent simplicity while defining algorithms. In both algorithms, the first run of the doBrokerJob method creates all task portion instances that will be needed throughout the whole run of the task. While defining more parameters, we will need to add task portions conditioned on the results of previous task portions or other conditions. Modification of the doBrokerJob method will solve the issue noted here.

As computational framework evolves from network-oriented to task-oriented, our efforts are shifted to the semantic agent-based grid. To succeed into this trend, various research aspects should be investigated. Our architecture should not be restricted for grids. It can also benefit the advantages from multi-agent systems, P2P technique, and Web services. Autonomous intelligent agents can monitor, evaluate and repair the system. This demands many migrations from agent environments to grids. We can also use communicative intelligence, fuzzy logic, nature-inspired algorithms and game theory.

ACKNOWLEDGMENT

I wish to thank WUT (Wroclaw University of Technology), which gave me a grant. I am pleased to acknowledge the help of Marek Kowalczyk for his technical support and contribution. The author gratefully acknowledges helpful remarks and suggestions from Mariusz Nowostawski. Many thanks to anonymous reviewers who made some comments on the chapter. All errors and flaws are the sole responsibility of the author.

REFERENCES

Alchemi (2007). *.NET-based enterprise grid.* Retrieved April 3, 2008, from http://www.alchemi.net/

Anderson, D.P. (2004). BOINC: A system for public-resource computing and storage. In *Proceedings of the 5th IEEE/ACM International Workshop on Grid Computing,* (pp. 4-10).

Deng, J., Han, R., & Mishra, S. (2005). *Secure code distribution in dynamically programmable wireless sensor networks* (Tech. Rep. No. CU-CS-1000-05). Boulder, CO: University of Colorado.

Gybels, K., Wuyts, R., Ducasse, S., & Hondt, M. (2006). Inter-language reflection: A conceptual model and its implementation. *Computer Languages, Systems and Structures, 32,* 109-124.

Haeuser, J., et al. (2000). A test suite for high-performance parallel Java. *Advances in Engineering Software, 31,* 687-696.

Król, D. (2005). A propagation strategy implemented in communicative environment. *Lecture notes in artificial intelligence* (Vol. 3682, pp. 527-533). Springer-Verlag.

Król, D., & Kukla, G. (2006). Distributed class code propagation with Java. *Lecture notes in artificial intelligence* (Vol. 4252, pp. 259-266). Springer-Verlag.

Lanunay, P., & Pazat, J.L. (2001). Easing parallel programming for clusters with Java. *Future Generation Computer Systems, 18,* 253-263.

Laure, E. (2001). OpusJava: A Java framework for distributed high performance computing. *Future Generation Computer Systems, 18,* 235-251.

Lauzac, S.W., & Chrysanthis, P.K. (2002). View propagation and inconsistency detection for cooperative mobile agents. *Lecture Notes in Computer Science, 2519,* 107-124.

Lin, J.W., & Kuo, S.Y. (2000). Resolving error propagation in distributed systems. *Information Processing Letters, 74*(5-6), 257-262.

Matsuoka, S., & Itou, S. (2001). Towards performance evaluation on high-performance computing on multiple Java platforms. *Future Generation Computer Systems, 18,* 281-291.

Mitschang, B. (2003). Data propagation as an enabling technology for collaboration and cooperative information systems. *Computers in Industry, 52*(1), 59-69.

OGSI.net. (2007). *Main page.* Retrieved April 3, 2008, from http://www.cs.virginia.edu/~gsw2c/ogsi.net.html

Poshtkohi A., Abutalebi, A.H., & Hessabi, S. (2007). DotGrid: A .NET-based cross-platform software for desktop grids. *International Journal of Web and Grid Services, 3*(3), 313-332.

Chapter XIX
Survey on the Application of Economic and Market Theory for Grid Computing

Xia Xie
Huazhong University of Science and Technology, China

Jin Huang
Huazhong University of Science and Technology, China

Song Wu
Huazhong University of Science and Technology, China

Hai Jin
Huazhong University of Science and Technology, China

Melvin Koh
Asia Pacific Science & Technology Center, Sun Microsystems, Singapore

Jie Song
Asia Pacific Science & Technology Center, Sun Microsystems, Singapore

Simon See
Asia Pacific Science & Technology Center, Sun Microsystems, Singapore

ABSTRACT

In this chapter, we present a survey on some of the commercial players in the Grid industry, existing research done in the area of market-based Grid technology and some of the concepts of dynamic pricing model that we have investigated. In recent years, it has been observed that commercial companies are slowly shifting from owning their own IT assets in the form of computers, software and so forth, to purchasing services from utility providers. Technological advances, especially in the area of Grid computing, have been the main catalyst for this trend. The utility model may not be the most effective model and the price still needs to be determined at the point of usage. In general, market-based approaches are more efficient in resource allocations, as it depends on price adjustment to accommodate fluctuations in the supply and demand. Therefore, determining the price is vital to the overall success of the market.

INTRODUCTION

The term Grid computing was introduced for describing a new model for distributed computing. The basic concept refers to the sharing of distributed heterogeneous compute resources virtualized as a single resource pool (Foster & Kesselman, 1999; Foster, Kesselman, Nick & Tuecke, 2002). Typically, as grids are often used for running computational intensive applications, the common type of grid resource usually means compute cycle. However, the concept does not place any restrictions, as it can be all kinds of computing resources like network bandwidth, data storage, application licenses and even scientific devices.

Today, the practice of Grid computing is based on voluntary sharing of compute resources, which is sufficient for establishing small-scale private grid dedicated to a specific purpose. However, to build a global level generic grid, this is simply not sustainable. Organizations, especially from the industry, will find very little reason to share their resources for free, and will expect some gains from their participation. Therefore, in order for grid to be the mainstream computing model, an efficient supporting platform and mechanism should be designed for encouraging resource owners to offer their idle resources and customers to satisfy their resource needs. Therefore, the idea of using markets in Grid computing as a means for organizations to commercialize their grid resources was revitalized by many researchers.

A market is, as defined in economics, a social arrangement that allows buyers and sellers to discover information and carry out a voluntary exchange. Our definition of the Grid Market refers to a software platform with the business mechanisms to support trading between grid users. Its principle is similar to the conventional marketplace and the goods that are traded on are generic grid resources, including concrete computing/storage/network physical resources, grid services and complex workflows. The Grid Market provides the required business functions to support the business process to allow any customers to participate in the trading. Such functions have to cover the all the possible activities in a typical market such as registration of new customers, advertising the trade goods, searching and browsing the market, bartering, monitoring the prices and making or receiving payment.

An emergence of such a marketplace for grid brings the following advantages:

- Encourages more users to adopt Grid technology, especially in enterprises.
- Provides incentives for resource owners to provide their idle resources, which is helpful for establishing large-scale, mature grid systems.
- Enforces efficient utilization of grid resources in which buyers who value a resource most highly will buy from sellers most willing to sell. Provide access for even small businesses to temporary grid resources which may be too expensive to acquire on their own, or just to meet their short term peak demand.
- Customers, including both buyer and seller, can easily design their trading policies based on their current status so as to maximize resources' utilization and their benefits.

Currently, many works have been done on applying market-based economic paradigm to Grid computing. The objective of this chapter is to provide a review of the past and current efforts in commercialization of Grid computing, as well as some of the business and pricing models that have been considered for e-commerce and e-business which can be apply to the Grid Market.

OVERVIEW OF GRID COMMERCIALIZATION

Economic systems in human society can be broadly classified into two models: Central Planning Model and Free Market Model (Shetty, Padala, & Frank, 2003). In central planning model, a single institution has total authority and decides what to produce, how to produce and to whom. In a free market model, producers and consumers make these decisions suiting their benefits. Our Grid Market is a typical free market model.

Change in technology also brings change in our economic environment. The advent of computers and Internet brings new dimension to market trading. Online auctioning services like e-bay and Yahoo Auctions have major impact on how we buy various commodities (Shetty et al., 2003). Many works have been done to introduce market concept into grid systems. In this part, we will give a survey and detailed discussion on these research and application, including work done by the industry, standard organization and academia.

Grid from the Industry

Today, the industry has taken the first step in commercializing grid computing, such as the major Grid offerings like the Sun Grid Compute Utility, IBM's On-demand Computing (ODC), Platform Enterprise Grid Orchestrator (EGO) and HP Adaptive Enterprise.

Sun is changing the very nature of computing by delivering access to enterprise compute power over the Internet with its Sun Grid Compute Utility (http://www.sun.com/service/sungrid/). Sun Grid provides an easy and affordable access to an enormous computing resource for the predictable and all-inclusive price of $1/CPU-hr. Firstly, Sun Grid utility computing can provide zero barriers to entry and exit. Users can access the computing power they need, when they need it, with no hidden costs, without a long-term contractual obligation, and increase or decrease their usages as their demands require. Users only pay for what they use. And secondly, Sun Grid utility computing radically simplifies the way you select, acquire, and use next generation IT infrastructure. This utility model enables users to react quickly to business needs without investing in expensive infrastructure. In short, Sun Grid Compute Utility helps users reduce complexity, better utilize overbuilt infrastructures, and optimize IT resources.

Similar with Sun, in October 2002, IBM releases its "On-demand Computing" policy, which can provide IT resources dynamically based on user requirements (http://www-128.ibm.com/developerworks/ondemand/). ODC is a computing and communications infrastructure that facilitates flexible business service delivery and provides the basis for: (1) autonomic computing, (2) fast response to external business-affecting changes, (3) adaptive business processes to protect revenues and contain costs, (4) complex interactions inside and outside of organizational boundaries and (5) resilience against external threats such as viruses, intrusions, and power outages. IBM sees its ODC as being a way to help customers meet the market challenges of continuous changes, rigorous competition, unrelenting financial pressure, and unpredictable threats (e.g., to security and market dominance). All of this is happening in a market where customers need to become very responsive, able to focus on their business, avoid fixed costs where possible (to support that flexibility), and be resilient instead of vulnerable. In conjunction with Grid computing, it can be seen that ODC is conceptually similar to the global outsourcing phenomenon (Kourpas, 2006).

For the computing utility model in Sun and IBM, it is a kind of method to buy computing resource, software and management. Companies provide powerful, robust and security computing service, and users can buy the resources they need and pay money based on their usages with lower risks. The business models of user purchase are more fixed such as pay-per-use model.

Enterprise agility is the key to increasing competitive advantage and delivering customers with timely products and services. In order to support IT technology challenges faced by enterprise, Platform Enterprise Grid Orchestrator and HP Adaptive Enterprise projects give an efficient resolution.

Platform Enterprise Grid Orchestrator is a Grid platform that introduced by Platform Computing Inc., and delivers virtualization, automation and sharing of all IT resources to any enterprise application (http://www.platform.com/Products/Platform.Enterprise.Grid.Orchestrator). Businesses can improve performance, organizational efficiency and achieve accelerated results by using it. Platform EGO uses a single common agent on each server to orchestrate the sharing of resources specific to business and technical challenges that enterprise IT organizations face. The introduction of Platform EGO represents a paradigm shift in the Grid computing market, with enterprises now able to build a unified framework for Grid-enabled applications that allocates resources and responds to business needs in real time, allowing them to fully realize the benefits of utility, adaptive, and on-demand computing environments. So, Platform EGO offers the strength and reliability of grid computing to the enterprise. It provides the required infrastructure for deploying on-demand, scalable and utility computing solutions required for enterprise businesses ("Aligning IT", 2005).

For enterprise users, HP releases "Adaptive Enterprise" policy. With HP's Adaptive Enterprise strategy, companies are synchronizing business and IT to gain a competitive advantage (http://www.hp.com/go/adaptive). The main idea of adaptive Enterprise is to improve the agility of business so that IT environment can be suitable for dynamic business need (Hewlett-Packard, 2005).

In brief summary, the projects described so far are designed to share special resources and face IT technology challenges with simple, flexible, reliable and economic resolution for enterprise users. Enterprise only needs to pay relative acceptable expense to use resources or technologies freely without worrying about expensive infrastructure and relative IT technology construction and updating. EGO and HP provide a very good platform for enterprise.

All of the above can be regarded as successful applications for transactions between IT resources, technologies and services. Compared with our Grid Market concept, they just provide a kind of transaction model but not a platform. That is, resources, technologies and services are provided by special providers for trading, and clients can buy something based on their requirements.

Efforts by the Research Community

Along with the development of market-based grid research, several important standard organizations have done some related work. There is a group in OASIS (Organization for the Advancement of Structured Information Standards) working on concrete examples of business requirements for service-oriented architecture (SOA) implementations. Others are the Distributed Management Task Force, Inc., (DMTF) Utility Computing Working Group and Grid Economic Services Architecture Working Group (GESA-WG) under the Global Grid Forum (GGF).

OASIS SOA standardization effort focuses on workflows, translation coordination, orchestration, collaboration, loose coupling, business process modeling, and other concepts that support agile computing. As a result of the maturation of these standards, the on-demand computing model can enable a modular approach to infrastructure, including software design, development and execution (see http://www-128.ibm.com/developerworks/ibm/library/i-odoe2/). At present, some concrete examples of business requirements for SOA implementations have been identified, and because the on-demand computing model is based on industry standards, it can be used to define the business, applications and systems at various levels: within a department, across an entire enterprise or throughout an industry ecosystem. It enables true end-to-end business process integration.

DMTF Utility Computing Working Group aims to create interoperable and common object models for utility computing services within the DMTF's Common Information Model (CIM). The DMTF Utility Computing Working Group (http://xml.coverpages.org/DMTF-Utility.html) focuses on commercial enterprise Grid application use cases and requirements and defines how to assemble complete service definitions. This includes work focusing on the composition of the models in CIM, as well as business- and domain-specific functional interfaces. This working group also renders the utility computing classes of CIM in Unified Modeling Language.

The next related effort is the work of GESA-WG under GGF (https://forge.gridforum.org/projects/gesa-wg). The goal of this working group is to provide the supporting infrastructure to enable Computational and Data Grids operated by different organizations to "trade" services between each other. The main work of GESA-WG is to define the protocols and service interfaces needed to extensibility support a variety of economic models for the charging of grid services in the OGSA. The Grid economic services architecture (Newhouse, MacLaren, & Keahey, 2004) defined a Chargeable Grid Service (CGS), which wraps the grid service that is to be sold, that interacts with the Grid Banking Service (GBS) and the Resource Usage Service (RUS). Note that the GESA-WG is now officially closed and no further documents are planned.

For projects in the academic community, we have conducted a survey on such systems as GridBus, Compute Power Market, Computational Markets and Business Grid Computing. GridBus project (http://www.gridbus.org/intro.html) aimed at applying some economic rules for better Grid resource management. This project is a technically-oriented development project of fundamental, next-generation cluster and Grid technologies that support a true utility-driven service-oriented computing. The project consists of the following parts: GRid Architecture for Computational Economy (GRACE), Grid Resource Broker (GRB), GridBank and Grid Accounting Service Architecture (GASA). The GRACE-infrastructure (Buyya, Abramson, & Giddy, 2000b) supports generic interfaces (protocols and APIs) that can be used by the grid tools and applications programmers to develop software supporting the computational economy. Nimrod/G (Buyya, Abramson, & Giddy, 2000a) as a GRB is a grid application scheduler and it is responsible for resource discovery, selection, scheduling, and deployment of computations over them. Nimrod/G supports both deadline (soft real-time) based scheduling by keeping the cost of computation as low as possible and budget (computational economy) constraints in scheduling, and at the same time it can optimize execution time or budget expenses (Buyya, Murshed, & Abramson, 2002). The GRACE infrastructure will enable Nimrod/G to dynamically trade for grid resources in the open market environment and select resources that meet user requirements (deadline and cost). The GridBus project processes successfully task scheduling and resource allocation with user's constraints. It tries to optimize each kind of criteria to provide a more economical resource usage model, but the project does not touch how to setup resource market and trading.

The Business Grid Computing project was started to address the requirements of systems providing essential social services as part of the Focus 21 project of the Japanese government. The project is developing Business Grid technologies for building and operating business-oriented IT systems flexibly and inexpensively with high reliability. This project is developing the Business Grid middleware (Savva, Suzuki, & Kishimoto, 2004) based on the Open Grid Services Architecture (OGSA) and the features of it include: (1) all information relating to a business application can be described and retained in a defined format, (2) the IT resources used by business applications are virtualized as hosting environments, and (3) business applications can be deployed automatically on distributed IT resources.

The Compute Power Market (CPM) project (http://www.computepower.com) is a market-based resource management and job scheduling system for Grid computing. It allows application users to access computing power with ease and simplicity, and to choose computing power/resource providers that offer cost-effective service on demand. Thus, it aims at creating a competitive market approach to service-oriented Grid computing. The CPM project seeks to address complexities involved in developing a technology infrastructure that lets the users and resource providers operate under a computational economy over the Internet. The design of CPM comprises of three types of components: the Market, the Market Agent and the Market Broker. It supports three major economy models, Commodity Market, Tender/Contract-Net and Auction models (Buyya & Vazhkudai, 2001; Ling et al., 2003).

Another important research project is the Computational Markets project (http://www.lesc.ic.ac.uk/markets). It is funded under the DTI e-Science Core Technology program and is concerned with the development of mechanisms to support the trading of grid services. The project aims at designing and implementing facilities for pricing, accounting and charging for all types of grid resources (software, hardware, data and network capacity). These trading mechanisms are implemented as extensions to the OGSA and its reference implementation Globus Toolkit 3, and provide inputs to the standardization process through the GGF.

Building Blocks of the Grid Market

So far, the projects described are mainly focusing on expending grid system framework based on OGSA in existence, as well as providing middleware or framework for grid resource transaction. However, to build a Grid Market requires more than these. An independent, complete market platform with all the necessary mechanisms to support the entire trading process is required. In addition, it should be able to support different business and pricing models for different consumers and providers.

We have identified several fields that are vital to building the Grid Market.

- **Business model:** The business model refers to the method used for trading in market. It has been defined and categorized in many different ways. Any business model is hard to tackle all special requirements for different customers independently. A different business model has different application characters and use background. It provides various trading methods for customers (including buyers and sellers) in market with their particular requirements.
- **Pricing model:** The pricing model is the price formation model used for trading in market. Dynamic pricing is an important feature for trading in market (Narahari, Raju, Ravikumar, & Shah, 2005). Pricing model includes simple fixed price model, commodity market model, bargaining

model, tender/contract-net model and auction model. As buyers and sellers interact in grid market, the resulting dynamic prices more closely reflect the true market value of the products and services being traded.

- **Contract management:** All trade in the grid market will generate a contract between the consumer and the provider as the result of mutually accepted agreements on the terms of the grid service/resource purchase contract. The contract should include the detail information of the trade and the commitment conditions. Furthermore, the contract management system will save the contracts and track the status of the contract negotiation and execution thereof (Czajkowski, Foster, & Kesselman, 2005; Guth, Simon, & Zdun, 2003; Paschke, Dietrich & Kuhla, 2005; Paschke, Bichler & Dietrich, 2005). It provides the mechanisms to generate contracts, query the contract details for involved parties as well as tracking changes of the contract options.
- **Accounting and banking:** In market framework, accounting and banking systems are necessary (Frogner, Mandt, & Wethal, 2004). Accounting systems will enable resource owners to monitor the usage and utilization of their grid resources. There will be no restriction on the "types" of resource utilization that can be recorded and accounted for by the associated management tools. Banking systems will be implemented to provide a secure charging and payment mechanism for resource usage. The existing commercial electronic payment methods and their compatibility with Grid market for secure payment will be investigated.
- **Reputation management:** Credit problems exist in any market. Reputation management involves recording a person or agent's actions and the opinions of others about those actions (Resnick & Zeckhauser, 2001). Reputation management that is efficient and adapts market characters will provide the support on keeping markets safe and efficiently running and ensure customers' trading activities.
- Others: The other aspects of running and managing the grid market include marketplace security, propaganda and advertisement and property rights protection. The investigations of these problems will help to build the perfect function grid market.

Compared with the common market, the grid market has its own characteristics. We not only research commercial transaction platform in existence, but also consider the distributed and dynamic features of grid resources as trading contents and objects. In the process of designing and realizing market service mechanism, these features of grid resources should be supported by grid market. For customers in markets, a different business model and a pricing model can be combined flexibly for their requirements and benefits.

BUSINESS AND PRICING MODELS

In this section, we discuss some of the common business models that are applicable to the Grid context. The term business model is commonly used, but there is no single dominant definition. Here, what we are concerned with is how the buyers and sellers conduct business and operate in the Grid Market. Or, more specifically, the activities and interaction of the buyers and sellers for establishing the trade, as well as the pricing and charging model of the sellers. We expect that different businesses have different requirements, goals, policies and strategies, and therefore there is no single business model that can fit all.

To get a clearer understanding, we will first look at the possible interactions and activities of the buyers and sellers in the Grid Market. The group led by Dr. Rajkumar Buyya has done extensive research on different models for the Grid economy. In Buyya, Abramson, Giddy and Stockinger (2002), several models are proposed for adaptation to the Grid context:

- **Commodity market model:** In the commodity market model, resource owners specify their service price and charge users according to the amount of resource they consume. The pricing policy can be derived from various parameters and can be flat or variable depending on the resource supply and demand. In general, services are priced in such a way that supply and demand equilibrium is maintained. In the flat price model, once pricing is fixed for a certain period, it remains the same irrespective of service quality. It is not significantly influenced by the demand, whereas in a supply and demand model prices change very often based on supply and demand changes. In principle, when the demand increases or supply decreases, prices are increased until there exists equilibrium between supply and demand. Pricing schemes in a commodity market model can be based on flat fee, usage duration (time), subscription or supply and demand-based (McKnight & Boroumand, 2000). In the commodity market model, the consumer only considers the resource price specified by the provider as selection reference, and cannot negotiate with the resource owner for use price.
- **Posted price model:** The posted price model is similar to the commodity market model, except that it advertises special offers in order to attract consumers to establish market share or motivate users to consider using cheaper resources. In this case, consumers need not negotiate directly with providers for price, but use posted prices as they are generally cheaper compared to regular prices. The posted-price offers will have usage conditions, but they might be attractive for some users. By using the posted price model, the provider can formulate the flexible price strategies in terms of the utilization of resources, and make full use of the resources' capabilities. For example, during holiday periods, demand for resources is likely to be limited and providers can post tempting offers or prices aiming to attract users to increase resource utilization.
- **Bargaining model:** In the bargaining model, consumers bargain with resource providers for lower access price and higher usage duration. Both buyers and sellers have their own objective functions and they negotiate with each other as long as their objectives are met. The buyers might start with a very low price and sellers with a higher price. They both negotiate until they reach a mutually agreeable price or one of them is not willing to negotiate any further. This negotiation is guided by user requirements and buyers can take risk and negotiate for cheaper prices as much as possible and can discard expensive machines. This might lead to lower utilization of resources, so sellers might be willing to reduce the price instead of wasting resource capability. Buyers and sellers generally employ this model when market supply-and-demand and service prices are not clearly established. The users can negotiate for a lower price with promise of some kind favour or even using the provider's services in the future. It should be pointed out that the negotiation process will consume some resource and time, and if both sides cannot reach an agreement, the consumption of negotiation will not bring any profits.
- **Tendering model:** The tendering model is one of the most widely used models for service negotiation in a distributed problem-solving environment (Smith & Davis, 1980). It is modeled on the contracting mechanism used by businesses to govern the exchange

of goods and services. It helps in finding an appropriate service provider to work on a given task. The advantage of this model is that if the seller is unable to provide a satisfactory service or deliver a solution, the buyer can seek other sellers for the service. The tender model allows directed contracts to be issued without negotiation. The selected resource provider responds with an acceptance or refusal of award. This capability can simplify the protocol and improve the efficiency of certain services.

- **Auction model:** The auction model supports one-to-many negotiation, between a seller and many buyers, and reduces negotiation to a single value (i.e., price). The auctioneer sets the rules of auction, acceptable for the buyers and the sellers. Auctions basically use market forces to negotiate a clearing price for the service. In the real world, auctions are used extensively, particularly for selling goods/items within a set duration. The three key players involved in auctions are resource owners, auctioneers (mediators), and buyers. Many e-commerce portals such as Amazon.com and eBay.com are serving as mediators (auctioneers). Both buyers' and sellers' roles can also be automated. Depending on various parameters, auction models can be classified into several types (Sandholm, 2000), such as English auction, First-price sealed-bid auction, Vickrey auction (Vickrey, 1961) or Dutch auction.

Business models have been defined and categorized in many different ways. Internet business models continue to evolve. New and interesting variations can be expected in the future. The basic categories of business models discussed as follows.

According to the type of resource, business model can be classified into three classes: Service Model, Leasing Model, and Bartering Model. An agile trading activity can be provided for both business parties by using different models.

- **Service model:** Service model is the most popular resource providing and consuming model in OGSA currently. Seller or provider offers the available resources; and buyer or consumer purchases and uses the resources in terms of his requirements with paying the corresponding fees. In service model, the buyer will never own the resource. A provider offers a defined service for which the consumer pays a regular service fee. The service models differ regarding their clearing: (1) fix service model, in which the consumer pays a regular service fee for a specified service, and (2) consumption dependent service model, in which the fee is rated per unit and the consumer has to pay an amount calculated according to his consumption. Service model tries to provide resources for consumer in simple and agile way. Consumer can take into account the price wave sufficiently and grasp the ideal trading opportunity. To buy according to needs and to pay according to usage are the main features of the service model.
- **Leasing model:** The owner of the resource (the leaser) allows the customer (the leasee) to use the resource for a specified time in return for payment. The obvious difference between leasing model and service model is ownership may be transferred to the customer after the leasing period. In the leasing period, the lessee can use the leasing object freely. If permitted in the contract, the leasee even can resell or re-lease it. Leasing contract is used to formulate the detailed leasing matters, such as leasing period, leasing fee and payment mode and responsibility of resource maintenance in the leasing period. Comparing with service model, leasing model is often suitable for the case that the resource needs to be used for a long time or several times. And then, because in the leasing model what to purchase is the use of resource in specific period of time,

the consumer suffers little from the price wave.

- **Bartering model:** In the bartering model, several customers in the market form a resource sharing community. Once customers join this community, they need to contribute their resources to this shared resource pool, which can be used by other customers in this community. After the customers' resources are used by others, they also can use the resources from other customers in resource pool. The system may provide a virtual grid currency to measure the contribution and the capability of employing resource of each customer. The bartering model is suitable for the customers who are not only resource providers, but also resource consumers sometimes.

According to the time of resource providing and consuming, trading mode in grid market can be classified into three classes: Instant Mode, Subscription Mode and Agreement Mode. Different trading modes bring the convenience for providing or consuming resources.

- **Instant mode:** Instant mode is the most popular e-business trading mode. The customers can buy or sell the resources that are available currently. Sellers can publish their idle resources in market and buyers can purchase these resources and use them instantly. Instant mode is simple and suitable for requirements of the resources that need to be used right now. But the problem of this mode is that the wishes of both business sides are influenced by price wave in large measure. When the resource price is higher, sellers like to complete business quickly, but the buyers tend to find other sellers with lower price.
- **Subscription mode:** Grid resources as trading objects in grid market are capacity-type resources, or in the language of commodity markets, they are nonstorable commodities. Capacity not used yesterday is worthless today. It is necessary to reasonably foresee and arrange the use of resources in future. Subscription mode provides a mechanism that prearranges the use of resource in future periods. Sellers can publish resource information in the market, which can be used in the future. Buyers are able to subscribe the use of resources in a future period. "Buy now use later" is a characteristic of subscription mode. Subscription mode is more flexible than instant mode because both business sides not only arrange the use of resource ahead of time, but also have enough time to negotiate and try to satisfy both sides.
- **Agreement mode:** There is a great difference between agreement mode and the above two modes. Both business sides agree to put up business in a special time confirmed before through a certain way. Here, an agreement between two sides is made outside of the market. They just use the market platform to execute business, and the seller transacts with the specified buyer and buyer does likewise. The system does not need to search and choose trading objects for them, and just provides the basic functions for the execution of the trade, such as monitoring, accounting and banking. Agreement mode brings more convenience to over-the-counter business and transaction between long-term companions.

Besides the business models mentioned above, in Grid market customers have many ways to choose trading objects. Generally speaking, sellers and buyers can use three ways, including Commission Proxy, Confirmation Proxy and Non-Proxy, to search and match the trading objects.

- **Commission proxy:** The commission proxy way is used when customers provide business requirements and policies to the system and entrusts to proxy to deal with trading instead of oneself. The proxy chooses the proper objects according to the customer's demands. After determining the trading object, the proxy returns back the result to the customer. During the matchmaking process, proxy fully follows customer's requirements and policies, and maximizes his benefits. Using commission proxy, customer attaches itself littler to the matchmaking and the operation is relatively simple. But customer has to define requirements and policies in advance, which cannot be changed during the matchmaking process. So commission proxy way is a lack of agility.

- **Confirmation proxy:** Confirmation proxy is mainly used by the seller or resource provider. Compared with commission proxy, confirmation proxy needs a confirmation from customers after the system has selected a trading object according to the requirements. For sellers, business is successful if they have confirmed, and then customers can use the resource. Otherwise, the system will continue matching other trading objects. Moreover, customers can modify the original requirements and policies in the course of confirming and then carrying out the new search. This shows that confirmation proxy is more flexible than commission proxy for the customer's confirmation. However, the enhancement of the degree of customer's participant will increase the customer's overhead accordingly.
- **Non-proxy:** The non-proxy way is simpler and mainly regards the buyer or resource consumer. Customers only ask the system to provide a matching trading object list in which the proper object is chosen by oneself. For customers, it is easy to use this way to change trade requirements and policies in a flexible way according to the present dynamic change of the market during the process of choosing trading objects so as to ensure the best profits. However, this way needs customers to participate directly, and is unfavorable for a long time and requires large-scale resources matchmaking and negotiation.

It should be pointed out that no single model is suitable for all trade because the customer's demands and goals are varied. Both providers and consumers in the market need to select the proper business model according to their own demands and interests, and maximize their commercial profits.

Dynamic Pricing

The one most important thing of a market is price is the terms on which the trading objects (products or services) are exchanged. In an ideal market, the price is a reflection of the current state of the market, and therefore should be dynamic, that is, it varies when the market demand changes.

Dynamic pricing is the dynamic adjustment of prices to consumers depending upon the value these customers attribute to a product or service (Reinartz, 2001). Dynamic pricing includes two aspects as follows (Narahari et al., 2005).

Price dispersion—Price dispersion can be spatial or temporal. In spatial price dispersion, several sellers offer a given item at different prices. In temporal price dispersion, a given store varies its price for a given good over time, based on the time of sale and the supply-demand situation.

Price discrimination—Price discrimination describes the case that different prices are charged to different consumers for the same product. Varian (1996) describes three types:

- **First degree price differentiation:** This is also called perfect differentiation. A producer sells different units of output for different prices and these prices can differ from person to person. Here, each unit of the good is sold to the individual who values it most highly, at the maximum price that this individual is willing to pay for the item.

- **Second degree price differentiation:** This is also called nonlinear pricing and means that the producer sells different units of output for different prices, but every individual who buys the same amount of the product pays the same amount. Thus, prices depend on the amount of the product purchased, but not on who does the purchasing.
- **Third degree price differentiation:** This occurs when the producer sells products to different people for different prices, but every unit of the product sold to a given person sells for the same price. Price differentiation is achieved by exploiting differences in consumer valuations.

Elmaghraby and Keskinocak (2003) categorize dynamic pricing methods into two broad categories: posted price mechanisms and price discovery mechanisms. Under the first category, a product or service is sold at a take-it-or-leave-it price determined by the seller. The posted prices could be dynamic in the sense that the seller changes prices dynamically over time depending on the time of sale, demand information, and supply availability. In price discovery mechanisms, prices are determined through a bidding process. Auctions provide an immediate example.

Cost is perhaps the greatest factor precluding the widespread use of dynamic pricing, because in traditional markets, it is expensive to continuously re-price goods. But in digital markets, the costs associated with making frequent, instantaneous price changes are greatly diminished (Smith, Bailey, & Brynjolfsson, 2000).

A variety of mathematical models have been used in e-business dynamic pricing. Most of these models formulate the dynamic pricing problem as an optimization problem. Depending on the specific mathematical tool used and emphasized, dynamic pricing mainly includes five categories of models.

- **Inventory-based models:** These are models where pricing decisions are primarily based on inventory levels and customer service levels. Dynamic pricing in retail markets based on inventory considerations has been researched quite extensively. Elmaghraby and Keskinocak (2003) discuss three main characteristics of a market environment that influence the type of dynamic pricing problem a retailer faces: replenishment vs. nonreplenishment of inventory (R/NR), dependent vs. independent demand over time (D/I), and myopic vs. strategic customers (M/S). According to the authors, most existing markets can be classified under three categories: NR-I-M, NR-I-S, and R-I-M. Gallego and van Ryzin (1999) consider optimal dynamic pricing of inventories with stochastic demand over finite horizon. Federgruen and Heching (1999) consider the optimal inventory and pricing policy of a seller who faces an uncertain demand where prices are changed periodically over time. Bernstein and Federgruen (2003, 2005) consider inventory-based pricing in a two echelon supply chain with random demands. The approach used is based on game theory. Biller, Cha, Simchi-Levi, and Swann (2005) propose a strategy that incorporates dynamic pricing, direct-to-customer model, production scheduling, and inventory control under production capacity limits in a multiperiod horizon to improve the revenue and supply chain performance in automotive industry. Besides, a comprehensive review of models of traditional retail markets, where inventories are used as the main consideration for determining optimal prices, can be found in Elmaghraby and Keskinocak (2003), Swann (1999), and Chan, Shen, Simchi-Levi and Swann (2005).

- **Data-driven models:** These models use statistical or similar techniques for utilizing data available about customer preferences and buying patterns to compute optimal dynamic prices. Availability of customer data through e-business Web sites has opened up enormous opportunities for revenue enhancing measures. Some companies accumulate huge amounts of data about customers which they can leverage to improve their revenues and profits. In the real world, there are some examples of a data-driven approach for dynamic pricing. Boyd and Bilegan (2003) survey revenue management techniques to illustrate a successful e-commerce model of dynamic, automated sales enabled by central reservation and revenue optimization systems. Morris, Ree, and Maes (2000) examine the dynamic pricing strategies in the airlines industry by discovering patterns in customer preferences. By identifying product features for which consumers are willing to pay a premium, the Ford motor company has developed a pricing strategy that encourages consumers to purchase more expensive vehicles, resulting in a marked increase in revenue and profits (Coy, 2000). Rusmevichientong, Van Roy, and Glynn (2005) have developed a nonparametric, data-driven approach to determining optimal dynamic prices that uses online data on consumer preferences collected through a Auto Choice Adviser Web site developed by General Motors. Using the data available from the Web site, the authors formulate a revenue optimization problem. Once customer data becomes available through Web sites and customer relationship management software, a variety of techniques can be used for analyzing and using this data for determining better ways of pricing.
- **Auction-based models:** Auctions constitute a natural model for dynamic pricing. The outcome of an auction is determined by supply-demand characteristics and therefore the prices as determined by an auction can truly be based on market conditions, provided the bidders reveal their true valuations. Auction mechanisms can be designed to have truth revelation properties and the theory of auctions has a great deal to offer to the area of dynamic pricing. Auctions are now possibly the most popular mechanism for implementing price negotiations in B2B and B2C situations. Auctions can take several forms and each type of auction mechanism would implement a particular type of pricing outcome. Bichler et al. (2002) have described in detail the role of auctions in dynamic pricing, in the context of e-procurement, e-selling, bid preparation and reverse logistics. Narahari and Dayama (2005) discuss the combinatorial auctions, which represent an important class of auction mechanisms being employed in e-business situations. The paper by Elmaghraby (2005) is a focused survey on auctions and pricing in e-marketplaces. More surveys on general auctions can be found in the literature (McAfee & McMillan, 1987; Milgrom, 1989; Klemperer, 1999; Kagel, 1995; Kalagnanam & Parkes, 2005; Wolfstetter, 1996).
- **Game theory models:** Game theory models provide a natural tool to be used in modeling situations of conflict and cooperation arising in the interaction of rational and selfish agents. In a multiseller scenario, the sellers may compete for the same pool of customers and this induces a dynamic pricing game among the sellers. Game theory models lead to interesting ways of computing optimal dynamic prices in such situations. There are a few studies of using a game theoretic approach for dynamic pricing in e-business markets. Bernstein and Federgruen (2003, 2005) consider the dynamic pricing problem in a two echelon supply chain with one supplier servicing a network of competing

retailers under demand uncertainty. Game theory models have recently been used in the area of pricing of network/Internet resources (Cao, Shen, Milito, & Wirth, 2002). In network settings, dynamic pricing can be used as an effective means to recover cost, to increase competition among different service providers, to reduce congestion, and to control the traffic intensity. Game theory models which have been used in the context of Internet pricing (He & Walrand, 2005) and network pricing (La & Anantharam, 1999; Yaiche, Mazumdar, & Rosenberg, 2000) can be applied to e-business contexts in a fairly straightforward way.

- **Machine learning models:** Machine learning has recently emerged as a popular modeling tool for dynamic pricing in e-business. An e-business market provides a rich playground for online learning by buyers and sellers. Sellers can potentially learn buyer preferences and buying patterns and use algorithms to dynamically price their offerings so as to maximize revenues or profits. With learning-based models, one can put all available data into perspective and change the pricing strategy to adapt best to the environment. Machine learning models can be logically classified into single learning agent models and multiple learning agent models. A few studies of using single learning agent models are described in the papers by Brooks et al. (1999), Gupta, Ravikumar, and Kumar (2002), Carvalho and Puttman (2003), Leloup and Deveaux (2001), and Raju, Narahari, and Ravikumar (2003, 2006b). A few representative models that employ two or more learning agents can be found in the papers by Ravikumar, Batra, and Saluja (2002), Hu and Zhang (2002), Greenwald, Kephart, and Tesuaro (1999), Kephart and Tesauro (2000), Dasgupta and Das (2000), and Raju, Narahari, and Ravikumar (2006a).

The above way of categorizing dynamic pricing models is in no way a conclusive way. The categorization is neither mutually exclusive nor jointly exhaustive. A certain dynamic pricing scheme may include two or more of the above types. A given type of a model may use another type. For example, inventory-based models could be data-driven. Machine learning models may be data-driven. Machine learning models may use inventory levels in their learning algorithms.

FUTURE TRENDS

At present, a prime issue that is not yet resolved is how to organize and make efficient use of Grid infrastructure in a commercial context where several customers compete for the same Grid resources to support their computational tasks or the services they offer to their customers and business partners. The emergence of Grid market will solve these problems. It will support compute resource trading by enabling grid services to be registered, discovered, negotiated and paid for the usage.

By applying economic theories, we can gain insights to the problem by analyzing the characteristics of the Grid market. For example, the grid services, as the trading objects, are capacity-type resources and so are completely nonstorable commodities and unused capacity from yesterday is worthless today. These characteristics are expected to have a major impact on the business models. Prices are generally the combined result of supply and demand, so a price formation mechanism for matching these is required. A large body of research in auctions for a wide variety of goods is available, but for every new market a new mechanism is needed or an older one must be adapted to fit the idiosyncrasy of this market.

Another important issue is security, or more specifically, privacy. In order to convince commercial companies to participate in the Grid Market, they must be able to trust the system to protect their Intelligent Properties. For example, an animation company running rendering jobs for their new movie certainly would not want the movie to leak out to the public before it is released. Using encryption and some form of virtualization can alleviate the problem somewhat, but for stricter security requirements, especially from financial or medical institutes, would require innovation at the system or hardware level.

In the early part of the chapter, we also mentioned that the establishment and operation of the Grid Market, as a special market environment, also needs to support the basic functions that are considered in the conventional markets. So, a series of problems, such as arrangement, fulfillment and management of the contracts, monitoring of the transaction, payment and banking service, management of reputation of participants and security of market, need to be taken into consideration. Individually, each of these fields is well-researched, but to put them together into a real world system is not a trivial task. Already, we have seen many new projects taking on this challenge and we expect to see more in the future.

CONCLUSION

Today, many enterprises are working toward building agile businesses, where the buzzword in IT services are total cost of ownership and return of investment. It has already been observed that the trend in cost-cutting in the IT department is to outsource the management of their IT resources. However, now we are starting to see that companies are outsourcing their IT, that is, purchase compute services from external providers. Although this trend has been around for a long time (the term application service provider (ASP) refers to a company that offers application services over the Internet), the idea never really took off. However, with recent advances in Grid technology, we have observed that many big industry vendors have recognized this trend, the latest being Sun Microsystems, which has launched their Sun Grid Compute Utility.

Grids are often envisioned as a transfer of the deregulated electricity grid paradigm to high performance computing, but oddly enough, one distinct feature of electrical power has been mostly neglected. The term "utility" is rooted not only in the shared transportation network of electrical power, and in the plug-and-play user experience, but additionally in the fact that electrical power is a traded commodity. Commercial prices are set by markets, not policies.

On the other hand, Grid computing is recognized as a potential major platform for scientific computing as well as commercial computation in the future. However, despite the existing technical advances and commercial needs, up until now, almost all research efforts were focused on using Grids within the academic community. The adoption of Grid technology by commercial companies are still comparatively slower. We note that if there is a means for commercial companies to sell or buy extra resources using Grid technology, it will definitely speed up the adoption of Grid.

ACKNOWLEDGMENT

This chapter is supported by National Science Foundation of China under Grant Nos. 90412010, 60673174 and National High-Tech Research and Development Plan of China under Grant No. 2006AA01A115. It is also supported by the Program for New Century Excellent Talents at the University of China.

REFERENCES

A market for computational services: A proposal to the e-science core technology programme. (2002). Retrieved April 3, 2008, from Grid Market Project: http://www.lesc.imperial.ac.uk/markets

Aligning IT in real time with the speed of changing business demands. Platform Computing Inc. Retrieved April 3, 2008, from http://www.platform.com/Resources/Description/EGO-WP.htm

Bernstein, F., & Federgruen, A. (2003). Pricing and replenishment strategies in a distribution system with competing retailers. *Operations Research, 51*, 409-426.

Bernstein, F., & Federgruen, A. (2005). Decentralized supply chains with competing retailers under demand uncertainty. *Management Science, 51*(1), 18-29.

Bichler, M., Lawrence, R. D., Kalagnanam, J., Lee, H. S., Katircioglu, K., Lin, G. Y., et al. (2002). Applications of flexible pricing in business-to-business electronic commerce. *IBM Systems Journal, 41*(2), 287-302.

Biller, S., Chan, L. M. A., Simchi-Levi, D., & Swann, J. (2005). Dynamic pricing and the direct-to-customer model in the automotive industry. *Electronic Commerce Journal special issue on Dynamic Pricing, 5*(2), 309-334.

Boyd, E. A., & Bilegan, I. C. (2003). Revenue management and e-commerce. *Management Science, 49*(10), 1363-1386.

Brooks, C. H., Fay, R., Das, R., MacKie-Mason, J. K., Kephart, J. O., & Durfee, E. H. (1999). Automated strategy searches in an electronic goods market: Learning and complex price schedules. In *Proceedings of the 1st ACM Conference on Electronic Commerce,* (pp. 31-40). New York: ACM Press.

Buyya, R., Abramson, D., & Giddy, J (2000a, May). Nimrod/G: An architecture for a resource management and scheduling system in a global computational grid. In *Proceedings of the 4th International Conference on High Performance Computing in Asia-Pacific Region (HPC-Asia 2000).* IEEE Computer Society Press.

Buyya, R., Abramson, D., & Giddy, J. (2000b, June). An economy driven resource management architecture for global computational power grids. In H. R. Arabnia (Ed.), *The 2000 International Conference on Parallel and Distributed Processing Techniques and Applications (PDPTA 2000)*, Las Vegas, NV, USA.

Buyya, R., Abramson, D., Giddy, J., & Stockinger, H. (2002). Economic models for resource management and scheduling in grid computing. *Concurrency and Computation: Practice and Experience (CCPE), 14*(13-15), 1507-1542.

Buyya, R., Murshed, M., & Abramson, D. (2002). A deadline and budget constrained cost-time optimization algorithm for scheduling task farming applications on global grids. In H. R. Arabnia (Ed.), *Proceedings of the International Conference on Parallel and Distributed Processing Techniques and Applications,* (Vol. 3). CSREA Press.

Buyya, R., & Vazhkudai, S. (2001). Compute power market: Towards a market-oriented grid. In *Proceedings of the First IEEE/ACM International Symposium on Cluster Computing and the Grid,* (pp. 574-581). Washington, DC: IEEE Computer Society.

Cao, X., Shen, H., Milito, R., & Wirth, P. (2002). Internet pricing with a game theoretic approach: Concepts and examples. In *IEEE/ACM Transactions Networking: 10*(2), 208-216. Piscataway, NJ: IEEE Press.

Carvalho, A. X., & Puttman, M. L. (2003). Dynamic pricing and reinforcement learning. In *Proceedings of the International Joint Conference on Neural Networks*, (Vol. 4, pp. 2916-2921). IEEE Computer Society.

Chan, L. M. A., Shen, Z. J. M., Simchi-Levi, D., & Swann, J. (2005). Coordination of pricing and inventory decisions: A survey and classification. In D. Simchi-Levi, S. D. Wu, & Z. M. Shen (Eds.), *Handbook of supply chain analysis: Modelling in the e-business era,* (pp. 335-392). Norwell, MA: Kluwer Academic.

Coy, P. (2000). The power of smart pricing. *Business Week.* Retrieved April 3, 2008, from BusinessWeek Online: http://www.businessweek.com/2000/00_15/b3676133.htm

Czajkowski, K., Foster, I., & Kesselman, C. (2005). Agreement-based resource management. In *Proceedings of the IEEE* (Vol. 93, No. 3, pp. 631-643). IEEE Computer Society.

Dasgupta, P., & Das, R. (2000). Dynamic pricing with limited competitor information in a multi-agent economy. In Etzion & P. Scheuermann (Eds.), *Proceedings of the 7th International Conference on Cooperative Information Systems,* (pp. 299-310). London: Springer-Verlag.

Elmaghraby, W. (2005). Auctions and pricing in e-marketplaces. In D. Simchi-Levi, S. D. Wu, & Z. M. Shen (Eds.), *Handbook of supply chain analysis: Modelling in the e-business era* (pp. 213-246). Norwell, MA: Kluwer Academic.

Elmaghraby, W., & Keskinocak, P. (2003). Dynamic pricing: Research overview, current practices and future directions. *Management Science, 49*(10), 1287-1309.

Federgruen, A., & Heching, A. (1999). Combined pricing and inventory control under uncertainty. *Operations Research, 47,* 454-475.

Foster, I., & Kesselman, C. (Eds.). (1999). *The Grid: Blueprint for a future computing infrastructure.* Morgan Kaufmann.

Foster, I., Kesselman, C., Nick, J.M., & Tuecke, S. (2002). *The physiology of the Grid: An open grid services architecture for distributed systems integration.* Open Grid Service Infrastructure WG, Global Grid Forum.

Frogner, K., Mandt, T., & Wethal, S. E. (2004). *Cluster and grid computing: Accounting and banking systems.* Retrieved April 3, 2008, from the Cluster and Grid Computing Web site: http://hovedprosjekter.hig.no/v2004/data/gruppe05/?page=documents

Gallego, G., & Ryzin, G. V. (1994). Optimal dynamic pricing of inventories with stochastic demand over finite horizons. *Management Science, 40*(8), 999-1020.

Greenwald, A., Kephart, J. O., & Tesauro, G. J. (1999). Strategic pricebot dynamics. In *Proceedings of the 1st ACM conference on Electronic commerce,* (pp. 58-67). New York: ACM Press.

Gupta, M., Ravikumar, K., & Kumar, M. (2002). Adaptive strategies for price markdown in a multi-unit descending price auction: A comparative study. In *Proceedings of IEEE International Conference on Systems, Man, and Cybernetics,* (pp. 373-378).

Guth, S., Simon, B., & Zdun, U. (2003). A contract and rights management framework design for interacting brokers. In *Proceedings of the 36th Annual Hawaii International Conference on System Sciences: Track 9* (Vol. 9). Washington, DC: IEEE Computer Society.

He, L., & Walrand, J. (2005). Pricing of differentiated internet services. In *Proceedings of INFOCOM-2005,* (Vol. 1, pp. 195-204). New York: IEEE Computer Society.

Hewlett-Packard Development Company. (2005). *Adaptive enterprise: Business and IT synchronized to capitalize on change—an executive overview white paper.* Retrieved April 3, 2008, from http://www.hp.com/go/adaptive

Hu, J., & Zhang, Y. (2002). *Online reinformcenet learning in multiagent systems.*

Kagel, J. H. (1995). Auctions: A survey of experimental research. In J. H. Kagel & A. E. Roth (Eds.), *Handbook of experimental economics* (pp. 501-587). Princeton University Press.

Kalagnanam, J., & Parkes, D. (2005). Auctions, bidding and exchange design. In D. Simchi-Levi, S. D. Wu, & Z. M. Shen (Eds.), *Handbook of supply chain analysis: Modelling in the e-business era* (pp. 143-212). The Netherlands: Kluwer Academic.

Kephart, J. O., & Tesauro, G. J. (2000). Pseudo-convergent Q-learning by competitive pricebots. In *Proceedings of the 17th International Conference on Machine Learning,* (pp. 463-470). San Francisco: Morgan Kaufmann.

Klemperer, P. (1999). Auction theory: A guide to the literature. *Journal of Economic Surveys, 13*, 227-286.

Kourpas, E. (2006). *Grid computing: Past, present and future—an innovation perspective.* Retrieved April 3, 2008, from the IBM Grid Computing Web site: http://www-1.ibm.com/grid/grid_literature.shtml

La, R. J., & Anantharam, V. (1999). Network pricing with a game theoretic approach. In *Proceedings of the IEEE Conference on Decision and Control,* (pp. 4008-4013). New York: ACM Press.

Leloup, B., & Deveaux, L. (2001). Dynamic pricing on the Internet: Theory and simulations. *Journal of Electronic Commerce Research, 1*(3), 265-276. The Netherlands: Springer-Verlag.

Ling, A. W. K., Sun, L. C., Sodhy, G. C., Yong, C. H., Haron, F., & Buyya, R. (2003). Design framework of generic components for Compute Power Market. In S. Sahni (Ed.), *Proceedings of Computer Science and Technology.* Cancun, Mexico: ACTA Press.

McAfee, R. P., & McMillan, J. (1987). Auctions and bidding. *Journal of Economic Literature, 25*(2), 699-738. American Economic Association.

McKnight, L., & Boroumand, J. (2000). Pricing Internet services: Approaches and challenges. *Computer, 33*(2), 128-129. Los Alamitos, CA: IEEE Computer Society Press.

Milgrom, P. (1989). Auctions and bidding: A primer. *Journal of Economic Perspectives, 3*(3), 3-22. American Economic Association.

Morris, J., Ree, P., & Maes, P. (2000). Sardine: Dynamic seller strategies in an auction marketplace. In *Proceedings of the 2nd ACM Conference on Electronic Commerce,* (pp. 128-134). New York: ACM Press.

Narahari, Y., & Dayama, P. (2005). Combinatorial auctions for electronic business. *SADHANA - Academy Proceedings in Engineering Sciences, 30*(2), 179-212. India: Indian Academy of Science.

Narahari, Y., Raju, C. V. L., Ravikumar, K., & Shah, S. (2005). Dynamic pricing models for electronic business. *SADHANA - Academy Proceedings in Engineering Sciences, 30*(2&3), 231-256. India: Indian Academy of Science.

Newhouse, S., MacLaren, J., Keahey, K. (2004). Trading grid services within the UK e-Science Grid. In J. Nabrzyski, J. M. Schopf, & J. Weglarz (Eds.), *Grid resource management: State of the art and future trends* (pp. 479-490). Norwell, MA: Kluwer Academic.

Paschke, A., Bichler, M., & Dietrich, J. (2005). ContractLog: An approach to rule based monitoring and execution of service level agreements. In A. Adi, S. Stoutenburg, & S. Tabet (Eds.), *Proceedings for the 4th International Conference on Rules and Rule Markup Languages for the Semantic Web (RuleML 2005).* Galway, Ireland: Springer-Verlag.

Paschke, A., Dietrich, J., & Kuhla, K. (2005). A logic based SLA management framework. In *Proceedings of the 4th Semantic Web Conference (ISWC 2005).*

Raju, C. V. L., Narahari, Y., & Ravikumar, K. (2003). Applications of reinforcement learning to dynamic pricing of retail markets. In *Proceedings of the IEEE Conference on Electronic Commerce,* (pp. 339-346). Los Alanitos, CA: IEEE Computer Society.

Raju, C. V. L., Narahari, Y., & Ravikumar, K. (2006a, January). Learning dynamic prices in multi-seller electronic retail markets with price sensitive customers, stochastic demands, and inventory replenishments. *IEEE Transactions on Systems, Man, and Cybernetics, Part C: Applications and Reviews, 36*(1), 92-106.

Raju, C. V. L., Narahari, Y., & Ravikumar, K. (2006b, March). Learning dynamic prices in electronic retail markets with customer segmentation. *Annals of Operations Research, 143*(1), 59-75.

Ravikumar, K., Batra, G., & Saluja, R. (2002). Multi-agent learning for dynamic pricing games of service markets. *Communicated.*

Reinartz, W.J. (2001). Customizing prices in online markets. *European Business Forum, 6,* 35-41.

Resnick, P., & Zeckhauser, R. (2001). Trust among strangers in Internet transactions: Empirical analysis of eBay's reputation system. In M. R. Baye (Ed.), *Advances in Applied Microeconomics, 11*, 127-157.

Rusmevichientong, P., Van Roy, B., & Glynn, P. W. (2005). A non-parametric approach to multi-product pricing. *Operations Research, 54*(1), 82-98.

Sandholm, T. (2000). Distributed rational decision making. In G. Weiss (Ed.), *Multiagent systems: A modern approach to distributed artificial intelligence.* Cambridge, MA: The MIT Press.

Savva, A., Suzuki, T., & Kishimoto, H. (2004). Business Grid Computing Project activities. *Fujitsu Scientific & Technical Journal, 40*(2), 252-260.

Shetty, S., Padala, P., & Frank, M. P. (2003). *A survey of market-based approaches to distributed computing* (Tech. Rep. No. TR03-013). University of Florida: Computer & Information Science & Engineering.

Smith, M., Bailey, J., & Brynjolfsson, E. (2000). Understanding digital markets: Review and assessment. In E. Brynjolfsson & B. Kahin (Eds.), *Understanding the digital economy.* Cambridge, MA: MIT Press.

Smith, R., & Davis, R. (1980). The contract net protocol: High level communication and control in a distributed problem solver. *IEEE Transactions on Computers, C-29*(12), 1104-1113.

Swann, J. (1999). *Flexible pricing policies: Introduction and a survey of implementation in various industries* (Contract Rep. #CR-99/04/ESL). General Motors Corporation.

Varian, H. R. (1996). Differential pricing and efficiency. *First Monday: Peer-Reviewed Journal on the Internet.* Retrieved April 3, 2008, from http://www.firstmonday.org/issues/issue2/different

Vickrey, W. (1961). Counter-speculation, auctions, and competitive sealed tenders. *The Journal of Finance, 16*(1), 8-37.

Wolfstetter, E. (1996). Auctions: An introduction. *Journal of Economic Surveys, 10*(4), 367-420.

Yaiche, H., Mazumdar, R. R., & Rosenberg, C. (2000). A game theoretic framework for bandwidth allocation and pricing in broadband networks. *IEEE/ACM Transactions on Networking, 8*(5), 667-678.

Compilation of References

A market for computational services: A proposal to the e-science core technology programme. (2002). Retrieved April 3, 2008, from Grid Market Project: http://www.lesc.imperial.ac.uk/markets

Aarseth, E.J. (1997). *Cybertext: Perspectives on Ergodic literature.* Baltimore, MD: John Hopkins University Press.

Abad, P. J., Suárez, A. J., Gasca, R. M., & Ortega, J. A. (2002). Using supervised learning techniques for diagnosis of dynamic systems. In *Proceedings of the International Workshop on Principles of Diagnosis* (DX-02).

Abdul-Rahman, A., & Hailes, S. (2000). Supporting trust in virtual communities. In *Proceedings of the Hawaii International Conference on System Services* (Vol. 6, pp. 6007).

Acroname Robotics. (2007). *Garcia manual.* Retrieved April 3, 2008, from http://www.acroname.com/garcia/man/man.html

Adomavicius, G., & Tuzhilin, A. (2005). Toward the next generation of recommender systems: A survey of the state-of-the-art and possible extensions. *IEEE Transactions on Knowledge and Data Engineering, 17*(6), 734-749.

Alchemi, (2007). *.NET-based enterprise grid.* Retrieved April 3, 2008, from http://www.alchemi.net/

Aldewereld, H., Dignum, F., Garcia-Camino, A., Noriega, P., Rodriguez-Aguilar, J. A., & Sierra, C. (2006). Operationalisation of norms for usage in electronic institutions. In *Paper presented at the AAMAS*, Hakodate, Japan.

Aldewereld, H., Dignum, F., García-Camino, A., Noriega, P., Rodríguez-Aguilar, J., & Sierra, C. (2006). Operationalisation of norms for usage in electronic institutions. In *Proceedings of the Fifth International Joint Conference on Autonomous Agents and Multi-agent Systems*, (pp. 223-225).

Alferes, J. J., Damásio, C. V., & Pereira, L. M. (2003). Semantic Web logic programming tools. In F. Bry, N. Henze, & J. Maluszynski (Eds.), *Principles and practice of Semantic Web reasoning* (pp. 16-32). LNCS 2901. Springer-Verlag.

Aligning IT in real time with the speed of changing business demands. Platform Computing Inc. Retrieved April 3, 2008, from http://www.platform.com/Resources/Description/EGO-WP.htm

Allan, J. (Ed.). (2002). *Topic detection and tracking: Event-based information organization.* Norwell, MA, USA: Kluwer Academic.

Allan, J., Gupta, R., & Khandelwal, V. (2001). Temporal summaries of news topics. In *Proceedings of the 24th Annual Conference on Research and Development in Information Retrieval*, (pp. 10-18). New Orleans, LA, USA: ACM Press.

Allen, C. (Ed.). (2003). HR-XML recommendation. Competencies (Measurable Characteristics). *Recommendation, 2006 Feb 28.* Retrieved March 1, 2006, from http://www.hr-xml.org/

Amitay, E., Carmel, D., Herscovici, M., Lempel, R., & Soffer A. (2004). Trend detection through temporal link analysis. *Journal of the American Society for Information Science and Technology, 55*(14), 1270-1281.

Andersen, P.B. (1997). *A theory of computer semiotics.* Cambridge University Press.

Anderson, D.P. (2004). BOINC: A system for public-resource computing and storage. In *Proceedings of the 5th IEEE/ACM International Workshop on Grid Computing*, (pp. 4-10).

Anton, T. (2005). XPath-Wrapper Induction by generalizing tree traversal patterns. In C. Mathias Bădică, A. Bădică, E. Popescu, & A. Abraham (Eds.). (2007), L-wrappers: Concepts, properties and construction. A declarative approach to data extraction from Web sources.

Soft Computing—A Fusion of Foundations, Methodologies and Applications, 11(8), 753-772.

Arai, S., & Ishida, T. (2004). Learning for human-agent collaboration on the Semantic Web. In *Proceedings of the 12th International Conference on Informatics Research for Development of Knowledge Society Infrastructure (ICKS '04),* (pp. 132-139).

Arcos, J. L., Esteva, M., Noriega, P., Rodriguez-Aguilar, J. A., & Sierra, C. (2005). Engineering open environments with electronic institutions. *Engineering Applications of Artificial Intelligence, 18*(2), 191-204.

Argyris, C., & Schon, D. A. (1996). *Organizational learning II: Theory, method, and practice.* Reading, MA: Addison Wesley.

Arms, W.Y., Aya, S., Dmitriev, P., Kot, B. J., Mitchell, R., & Walle, L. (2006). Building a research library for the history of the Web. In *Proceedings of the Joint Conference on Digital Libraries,* (pp. 95-102). Chapel Hill, NC, USA: ACM Press.

Artikis, A., Pitt, J., & Sergot, M. (2002). Animated specifications of computational societies. In *Proceedings of the First International Joint Conference on Autonomous Agents and Multi-agent Systems* (Part 3, pp. 1053-1061).

Artz, D., & Gil, Y. (2006). *Survey of trust in computer science and the Semantic Web.* Submitted for publication, Information Sciences Institute, University of Southern California. Retrieved April 3, 2008, from http://www.isi.edu/~dono/pdf/ artz06survey.pdf

Arvidson, A., Persson, K., & Mannerheim, J. (2000). The Kulturarw3 project—the Royal Swedish Web Archive—an example of "complete" collection of Web pages. In *Proceedings of the 66th IFLA Council and General Conference,* Jerusalem, Israel.

Aschenbrenner, A., & Rauber, A. (2006). Mining Web collections. In J. Masanes (Ed.), *Web archiving* (pp. 153-174). Berlin, Heidelberg, Germany: Springer-Verlag.

Axelrod, R. (1984). *The evolution of cooperation.* New York: Basic Books.

Axelrod, R. (1986). An evolutionary approach to norms. *The American Political Science Review, 80*(4), 1095-1111.

Baader, F., Calvanese, D., McGuinness, D., Nardi, D., & Patel-Schneider, P. (Eds.). (2003). *The description logic handbook.* Theory, implementation and applications. Cambridge.

Banaszak, Z., & Jozefowska, J. (Eds). (2003). *Project driven manufacturing.* Warsaw, Poland: WNT.

Barabási, A. L. (2002). *Linked: The new science of networks.* PA: Perseus Books Group.

Barges, G., & Aldon, M. (2000). A split-and-merge segmentation algorithm for line extraction in 2d range images. In *Proceedings of the 15th International Conference on Pattern Recognition,* (Vol. 1, pp. 441-444).

Barnes, C. A., Baranyi, A., Bindman, L. J., Dudal, Y., Fregnac, Y., Ito, M., et al. (1994). Group report: Relating activity-dependent modifications of neuronal function to changes in neural systems and behavior. In A. I. Selverston & P. Ascher (Eds.), *Cellular and molecular mechanisms underlying higher neural functions.* New York: John Wiley & Sons.

Barthes, R. (1987). *Mythologies.* New York: Hill and Wang.

Bar-Yossef, Z., Broder, A. Z., Kumar, R., & Tomkins, A. (2004). Sic transit gloria telae: Towards an understanding of the Web's decay. In *Proceedings of the 13th International Conference on World Wide Web,* (pp. 328-337). New York: ACM Press.

Baudrillard, J. (1995). The Gulf War did not take place (trans Patton P.). Indiana University Press.

Bauer, B.B., Fürnkranz, J., Grieser, G., Hotho, A., Jedlitschka, A., & Kröner, A. (Eds.). *Lernen, Wissensentdeckung und Adaptivität (LWA)* (pp. 126-133). DFKI.

Baumgartner, R., Flesca, S., & Gottlob, G. (2001). The Elog Web extraction language. In R. Nieuwenhuis, & A. Voronkov (Eds.), *Logic for programming, artificial intelligence, and reasoning* (pp. 548-560). LNAI 2250. Springer-Verlag.

Bellifemine, F., Poggi, A., & Rimassa, G. (2000). Developing multi-agent systems with JADE. In *Proceedings of the 7th International Workshop on Intelligent Agents VII: Agent Theories Architectures and Languages,* (pp. 89-103).

Ben-Gal, I. (2007). Bayesian networks. In F. Ruggeri, F. Faltin, & R. Kenett (Eds.), *Encyclopedia of statistics in quality and reliability.* John Wiley & Sons.

Berge, C. (1973). *Graphs and hypergraphs.* Amsterdam: North-Holland.

Berger, A. L., & Mittal V. O. (2000). OCELOT: A system for summarizing Web pages. In *Proceedings of the 23rd Conference on Research and Development in Information Retrieval,* (pp. 144-151). Athens, Greece: ACM Press.

Berman, E. B. (1964). Resource allocation in a PERT network under continuous activity time-cost functions. *Management Sciences, 10*(4).

Berners-Lee, T., Hendler, J., & Lassila, O. (2001). The semantic Web. *Scientific American, 284*(5), 34-43.

Bernstein, F., & Federgruen, A. (2003). Pricing and replenishment strategies in a distribution system with competing retailers. *Operations Research, 51*, 409-426.

Bernstein, F., & Federgruen, A. (2005). Decentralized supply chains with competing retailers under demand uncertainty. *Management Science, 51*(1), 18-29.

Berry, G., & Boudol, G. (1989). *The chemical abstract machine.* New York: ACM Press.

Bex, G.J., Maneth, S., & Neven, F. (2002). A formal model for an expressive fragment of XSLT. *Information Systems, 27*(1), 21-39.

Bezdek, J.C. (1981). *Pattern recognition with fuzzy objective function algorithms* (p. 256). New York: Plenum Press.

Bichler, M., Lawrence, R. D., Kalagnanam, J., Lee, H. S., Katircioglu, K., Lin, G. Y., et al. (2002). Applications of flexible pricing in business-to-business electronic commerce. *IBM Systems Journal, 41*(2), 287-302.

Biller, S., Chan, L. M. A., Simchi-Levi, D., & Swann, J. (2005). Dynamic pricing and the direct-to-customer model in the automotive industry. *Electronic Commerce Journal special issue on Dynamic Pricing, 5*(2), 309-334.

Blackmore, S. (1999). *The meme machine.* Oxford University Press.

Boella, G., & Torre, L. v. d. (2006). An architecture of a normative system: Counts-as conditionals, obligations and permissions. In *Paper presented at the AAMAS*, New York.

Boella, G., & van der Torre, L. (2004). Virtual permission and authorization in policies for virtual communities of agents. In G. Moro, S. Bergamaschi, & K. Aberer (Eds.), *Proceedings of the Agents and P2P Computing Workshop at the 3rd International Joint Conference on Autonomous Agents and Agent Systems,* (pp. 86-97). New York.

Boella, G., & Van Der Torre, L. (2005). Normative multiagent systems and trust dynamics. *Trusting agents for trusting electronic societies, LNAI 3577,* (pp. 1-17).

Boella, G., Van Der Torret, L., & Verhagen, H. (2006). Introduction to normative multiagent systems. *Computational & Mathematical Organisation Theory, 12*(2-3), 71-79.

Boman, M. (1999). Norms in artificial decision making. *Artificial Intelligence and Law, 7*(1), 17-35.

Boyd, E. A., & Bilegan, I. C. (2003). Revenue management and e-commerce. *Management Science, 49*(10), 1363-1386.

Boyd, R., & Richerson, P. J. (1985). *Culture and the evolutionary process.* Chicago: University of Chicago Press.

Bradshaw, J. M., Dutfield, S., Benoit, P., & Woolley, J. D. (1997). KAoS: Toward an industrial-strength open agent architecture. *Software Agents*, 375-418.

Brams, S. J., & Kilgour, D. M. (2001). Fallback bargaining. *Group Decision and Negotiation, 10*, 287-316.

Brewington, E. B., & Cybenko, G. (2000). How dynamic is the Web? In *Proceedings of the 9th International World Wide Web Conference*, (pp. 257-276). Amsterdam, the Netherlands: ACM Press.

Bringsjord, S., & Zenzen, M. (2003). Superminds: People harness hypercomputation, and more. *Studies in cognitive systems* (Vol. 29). Kluwer Academic. Cen BF311 B 4867.

Bronstein, I.N., Semendjajew, K.A., Musiol, G., & Mühlig, H. (2001). Taschenbuch der Mathematik. *Verlag Harri Deutsch* (p.1258).

Brooks, C. H., Fay, R., Das, R., MacKie-Mason, J. K., Kephart, J. O., & Durfee, E. H. (1999). Automated strategy searches in an electronic goods market: Learning and complex price schedules. In *Proceedings of the 1st ACM Conference on Electronic Commerce,* (pp. 31-40). New York: ACM Press.

Bubnicki, Z. (2001a). Uncertain variables and their applications for a class of uncertain systems. *International Journal of Systems Science, 32*(5), 651-659.

Bubnicki, Z. (2001b). Uncertain variables and their application to decision making. *IEEE Transactions on SMC, Part A: Systems and Humans, 31*(6), 587-596.

Bubnicki, Z. (2002). *Uncertain logics, variables and systems.* Springer-Verlag. LNICS No. 276.

Bubnicki, Z. (2002). Uncertain variables and their applications for control systems. *Kybernetes, 31*(9/10), 1260-1273.

Bubnicki, Z. (2003). Application of uncertain variables to a project management under uncertainty. *Systems Science, 29*(2), 65-79.

Bubnicki, Z. (2004). *Analysis and decision making in uncertain systems*. Berlin, London, New York: Springer-Verlag.

Bubnicki, Z. (2004). Quality of an operation system control based on uncertain variables. In M. H. Hamza (Ed.), *Modelling, identification and control* (pp. 148-153). Zurich: Acta Press.

Bubnicki, Z. (2005). Application of uncertain variables in learning algorithms for uncertain systems. In G. Lasker (Ed.), *Advances in Computer Cybernetics: Proceedings of InterSymp 2005,* (pp. 25-29), Tecumseh: IIAS.

Buyukkokten, O., Garcia-Molina, H., & Paepcke, A. (2001). Seeing the whole in parts: Text summarization for Web browsing on handheld devices. In *Proceedings of the 10th International World Wide Web Conference,* (pp. 652-662). Hong Kong, SAR, China: ACM Press.

Buyya, R., & Vazhkudai, S. (2001). Compute power market: Towards a market-oriented grid. In *Proceedings of the First IEEE/ACM International Symposium on Cluster Computing and the Grid,* (pp. 574-581). Washington, DC: IEEE Computer Society.

Buyya, R., Abramson, D., & Giddy, J (2000, May). Nimrod/G: An architecture for a resource management and scheduling system in a global computational grid. In *Proceedings of the 4th International Conference on High Performance Computing in Asia-Pacific Region (HPC-Asia 2000)*. IEEE Computer Society Press.

Buyya, R., Abramson, D., & Giddy, J. (2000, June). An economy driven resource management architecture for global computational power grids. In H. R. Arabnia (Ed.), *The 2000 International Conference on Parallel and Distributed Processing Techniques and Applications (PDPTA 2000)*, Las Vegas, NV, USA.

Buyya, R., Abramson, D., Giddy, J., & Stockinger, H. (2002). Economic models for resource management and scheduling in grid computing. *Concurrency and Computation: Practice and Experience (CCPE), 14*(13-15), 1507-1542.

Buyya, R., Murshed, M., & Abramson, D. (2002). A deadline and budget constrained cost-time optimization algorithm for scheduling task farming applications on global grids. In H. R. Arabnia (Ed.), *Proceedings of the International Conference on Parallel and Distributed Processing Techniques and Applications,* (Vol. 3). CSREA Press.

Camarinha-Matos, L.M., & Afsarmanesh, H. (1999). The virtual enterprise concept. In *Proceedings of the IFIP Working Conference on Infrastructures for Virtual Enterprises: Networking Industrial Enterprises,* (pp. 3-14). Deventer, The Netherlands: Kluwer Academic.

Camarinha-Matos, L.M., & Afsarmanesh, H. (2003). Elements of a base VE infrastructure. *Journal of Computers in Industry, 51*(2), 139-163.

Camarinha-Matos, L.M., & Afsarmanesh, H. (2005). Collaborative networks: A new scientific discipline. *Journal of Intelligent Manufacturing, 16*, 439-452.

Camarinha-Matos, L.M., & Afsarmanesh, H. (2006). A modeling framework for collaborative networked organizations. In L.M. Camarinha-Matos, H. Afsarmanesh, & M. Ollus (Eds.), *Network-centric collaboration and supporting frameworks* (pp. 3-14). Boston: Springer Science Business Media.

Cao, X., Shen, H., Milito, R., & Wirth, P. (2002). Internet pricing with a game theoretic approach: Concepts and examples. In *IEEE/ACM Transactions Networking: 10*(2), 208-216. Piscataway, NJ: IEEE Press.

Čapkovič, F. (1995). Using fuzzy logic for knowledge representation at control synthesis. *Busefal, 63,* 4-9.

Čapkovič, F. (1996/1997). Petri nets and oriented graphs in fuzzy knowledge representation for DEDS control purposes. *Busefal, 69,* 21-30.

Čapkovič, F. (1999). Knowledge-based control synthesis of discrete event dynamic systems. In S.G. Tzafestas (Ed.), *Advances in manufacturing systems: Decision, control and information technology* (pp. 195-206). London, UK: Springer-Verlag.

Čapkovič, F. (2002). Control synthesis of a class of DEDS. *Kybernetes, 31*(9/10), 1274-1281.

Čapkovič, F. (2003). The generalised method for solving problems of DEDS control synthesis. In P. W. H. Chung, C. Hinde, & M. Ali (Eds.), *Developments in applied artificial intelligence: Lecture notes in artificial intelligence* (Vol. 2718, pp. 702-711). New York-Heidelberg-London: Springer-Verlag.

Čapkovič, F. (2004). An application of the DEDS control synthesis method. In M. Bubak, G. D. van Albada, P. M. A. Sloot, & J. J. Dongarra (Eds.), *Computational science—ICCS 2004, Part III: Lecture notes in computer science* (Vol. 3038, pp. 528-536). New York-Heidelberg-London: Springer-Verlag.

Čapkovič, F. (2004). DEDS control synthesis problem solving. In R. Yager & V. Sgurev (Eds.), *Proceedings of the 2nd IEEE Conference on Intelligent Systems,* Varna, Bulgaria (pp. 299-304.). Piscataway, NJ, USA: IEEE Press.

Čapkovič, F. (2005). An application of the DEDS control synthesis method. *Journal of Universal Computer Science, 11*(2), 303-326.

Čapkovič, F. (2006). Modeling, analysing and control of interactions among agents in MAS. In V. N. Alexandrov, G. D. van Albada, P. M. A. Sloot, & J. J. Dongarra (Eds.), *Computational science - ICCS 2006, Part III: Lecture notes in computer science* (Vol. 3993, pp. 176-183). Berlin-Heidelberg: Springer-Verlag.

Čapkovič, F. (in press). Modelling, analysing and control of interactions among agents in MAS. *Computing and informatics.*

Čapkovič, F., & Čapkovič, P. (2003). Petri net-based automated control synthesis for a class of DEDS. In L. Gomes, R. Zurawski, C. Couro, & J. Fuertes (Eds.), *Proceedings of the 2003 IEEE Conference on Emerging Technologies and Factory Automation—ETFA 2003*, Lisbon, Portugal (Vol. 2, pp. 297-304). Piscataway, NJ, USA: IEEE Electronic Society Press.

Carabelea, C., Boissier, O., & Florea, A. (2003). Autonomy in multi-agent systems: A classification attempt. In Nickles et al. (Eds.), (pp. 103-113).

Cardoso, L.C., & Oliveira, E. (2005). Virtual enterprise normative framework within electronic institutions. In M.-P. Gleizes, A. Omicini, & F. Zambronelli (Eds.), *Engineering societies in the agent world, Lecture notes in artificial intelligence* (Vol. 3451, pp. 14-32). Berlin: Springer-Verlag.

Carlsson, B., Davidsson, P., Jacobsson, A., Johansson, S.J., & Persson, J.A. (2005). Security aspects on inter-organizational cooperation using wrapper agents. In K. Fischer, A. Berre, K. Elms, & J.P. Müller (Eds.), *Proceedings of the Workshop on Agent-based Technologies and Applications for Enterprise Interoperability at the 4th International Joint Conference on Autonomous Agents and Agent Systems,* (pp. 13-25). Utrecht, The Netherlands: University of Utrecht.

Carruthers, M. (1998). *The craft of thought: Meditation, rhetoric, and the making of images, 400-1200.* Cambridge University Press.

Carter, J., & Ghorbani, A. A. (2004). Value centric trust in multiagent systems. In *Proceedings of the IEEE/WIC International Conference on Web Intelligence*, (pp. 3-9).

Carvalho, A. X., & Puttman, M. L. (2003). Dynamic pricing and reinforcement learning. In *Proceedings of the International Joint Conference on Neural Networks*, (Vol. 4, pp. 2916-2921). IEEE Computer Society.

Cassandras, C.G., & Lafortune, S. (1999). *Introduction to discrete event systems.* Boston: Kluwer Academic.

Castelfranchi, C. (1995). Guarantees for autonomy in cognitive agent architecture. In *Proceedings of the Workshop on Agent Theories, Architectures, and Languages, ATAL'94*, (Vol. 890 of LNAI, pp. 56-70). New York: Springer-Verlag.

Castelfranchi, C., & Falcone, R. (1998). Social trust: Cognitive anatomy, social importance, quantification, and dynamics. In *Proceedings of the First International Workshop on Trust,* Paris, France, (pp. 72-79).

Castelfranchi, R. C. C. (1995). *Cognitive and social action.* London: UCL Press.

Cavaleri, S., & Reed, F. (2000). Designing knowledge creating processes. *Knowledge and Innovation, 1*(1).

Chamorro-Martinez, J., Sanchez, D., Prados-Subrez, B., Galin-Perales, E., & Vila, M.A. (2003). A hierarchical approach to fuzzy segmentation of colour images. In *Proceedings of the 12th IEEE International Conference on Fuzzy Systems*, (Vol. 2, pp. 966-971).

Chan, L. M. A., Shen, Z. J. M., Simchi-Levi, D., & Swann, J. (2005). Coordination of pricing and inventory decisions: A survey and classification. In D. Simchi-Levi, S. D. Wu, & Z. M. Shen (Eds.), *Handbook of supply chain analysis: Modelling in the e-business era,* (pp. 335-392). Norwell, MA: Kluwer Academic.

Chandler, D. (2002). *Semiotics: The basics.* Routledge.

Chang, C.-H., Kayed, M., Girgis, M.R., & Shaalan, K. (2006). A survey of Web information extraction systems. *IEEE Transactions on Knowledge and Data Engineering, 18*(10), 1411-1428.

Chi, E. H., Pitkow, J., Mackinlay, J., Pirolli, P., Gossweiler, R., & Card, S. K. (1998). Visualizing the evolution of Web ecologies. In *Proceedings of Conference on Human Factors in Computing Systems*, (pp. 400-407), Los Angeles: ACM Press.

Chialvo, D. R., & Bak, P. (1999). Learning from mistakes. *Neuroscience, 90*(4), 1137-1148.

Chidlovskii, B. (2003). Information extraction from tree documents by learning subtree delimiters. In *Proceedings of IJCAI-03: Workshop on Information Integration on the Web (IIWeb-03),* (pp. 3-8).

Chituc, C.-M., & Azevedo, A.L. (2005). Enablers and technologies supporting self-forming networked organizations. In H. Panetto (Ed.), *Interoperability of enterprise software and applications* (pp. 77-89). London: Hermes Science.

Cho, J., & Garcia-Molina, H. (2000). The evolution of the Web and implications for an incremental crawler. In *Proceedings of the 26th International Conference on Very Large Databases*, (pp. 200-209). Cairo, Egypt: ACM Press.

Cho, J., & Garcia-Molina, H. (2003). Estimating frequency of change. *Transactions on Internet Technology, 3*(3), 256-290.

Choi, J. W., Oh, D. I., et al. (2006). R-LIM: An affordable library search system based on RFID. In *Proceedings of the International Conference on Hybrid Information Technology (ICHIT'06)*, (pp. 103-108).

Church, D. (1999). Formal abstract design tools. *Games Developer Magazine*, (August).

Chute, C.G. (2005). Medical concept representation. In H. Chen et al. (Eds.), *Medical informatics: Knowledge management and data mining in biomedicine* (pp. 163-182). Springer-Verlag.

Clutton-Brock, T. H., & Parker, G. A. (1995). Punishment in animal societies. *Nature, 373*, 209-216.

Colombetti, M., Fornara, N., & Verdicchio, M. (2002). The role of institutions in multiagent systems. In *Proceedings of the Workshop on Knowledge-based and Reasoning Agents, VIII AIIA*, (pp. 67-75).

Conte, R., & Castelfranchi, C. (1999). From conventions to prescriptions—Towards an integrated view of norms. *Artificial Intelligence and Law, 7*(4), 323-340.

Conte, R., Falcone, R., & Sartor, G. (1999). Agents and norms: How to fill the gap?. *Artificial Intelligence and Law, 7*(1), 1-15.

Cooley, R. Srivastava, J., & Mobasher, B. (1997). Web mining: Information and pattern discovery on the World Wide Web. In *Proceedings of the 9th IEEE International Conference on Tools with Artificial Intelligence*, (p. 558). IEEE Press.

Cooper, A., & Ostyn, C. (Eds.). (2002). *IMS reusable definition of competency or educational objective: Information model, (version 1.0), final specification.* Retrieved March 1, 2006, from http://www.imsglobal.org/competencies/rdceovlp0/imsrdceo_infovlp0.html

Cooper, A., & Ostyn, C. (Eds.). (2002). *IMS reusable definition of competency or educational objective: XML binding, (version 1.0), final specification.* Retrieved March 1, 2006, from http://www.imsglobal.org/competencies/rdceovlp0/imsrdceo_bindvlp0.html

Cooper, A., & Ostyn, C. (Eds.). (2002). *IMS reusable definition of competency or educational objective: Best practice and implementation guide, (version 1.0), final specification.* Retrieved March 1, 2006, from http://www.imsglobal.org/competencies/rdceovlp0/imsrdceo_bestvlp0.html

Cormen, T.H., Leiserson, C.E., & Rivest, R.R. (1990). *Introduction to algorithms.* MIT Press.

COSIGN. Retrieved from www.cosignconference.org

Cost, R., Chen, Y., Finin, T., Labrou, Y., & Peng, Y. (2000). Using colored petri nets for conversation modeling. *Lecture Notes in Computer Science*, 178-192.

Coté, R.A. (Ed.). (1975). SNOMED: Systematized nomenclature of medicine. *Diseases.* ACP.

Coy, P. (2000). The power of smart pricing. *Business Week.* Retrieved April 3, 2008, from BusinessWeek Online: http://www.businessweek.com/2000/00_15/b3676133.htm

Cranefield, S., & Purvis, M. (2002). A UML profile and mapping for the generation of ontology-specific content languages. *Knowledge Engineering Review, 17*(1), 21-39.

Cranefield, S., Nowostawski, M., & Purvis, M. (2002). Implementing agent communication languages directly from UML specifications. In *Proceedings of the First International Joint Conference on Autonomous Agents and Multi-agent Systems* (Part 2, pp. 553-554).

Crossbow Technology, Inc. (2006). *Stargate developers guide.* Retrieved April 3, 2008, from http://www.xbow.com/Support/Support_pdf_files/Stargate_Manual.pdf

Czajkowski, K., Foster, I., & Kesselman, C. (2005). Agreement-based resource management. In *Proceedings of the IEEE* (Vol. 93, No. 3, pp. 631-643). IEEE Computer Society.

D'Atri, A., & Motro, A. (2007). VirtuE: A formal model of virtual enterprises for information markets. *Journal of Intelligent Information Systems.*

Damasio, A.R. (1994). *Descarte's error: Emotion, reason and the human brain.* Papermac.

Dasgupta, P., & Das, R. (2000). Dynamic pricing with limited competitor information in a multi-agent economy. In Etzion & P. Scheuermann (Eds.), *Proceedings of the 7th International Conference on Cooperative Information Systems*, (pp. 299-310). London: Springer-Verlag.

Davidsson, P., Hederstierna, A., Jacobsson, A., Persson, J.A., et al. (2006). The concept and technology of plug and play business. In Y. Manolopoulos, J. Filipe, P. Constantopoulos, & J. Cordeiro (Eds.), *Proceedings of the 8th International Conference on Enterprise Information Systems Databases and Information Systems Integration,* (pp. 213-217).

Davidsson, P., Ramstedt, L., & Törnquist, J. (2005). Inter-organization interoperability in transport chains using adapters based on open source freeware. In D. Konstantas, J.-P. Bourrières, M. Léonard, & N. Boudjlida (Eds.), *Interoperability of enterprise software and applications* (pp. 35-43). Berlin: Springer-Verlag.

Davis, P.J., & Hersh, R. (1983). *The mathematical experience.* Pelican Books.

De Oliveira, M., Purvis, M., Cranefield, S., & Nowostawski, M. (2004). Institutions and commitments in open multi-agent systems. In *Proceedings of IAT 2004: IEEE/WIC/ACM International Conference on Intelligent Agent Technology,* (pp. 500-503).

De Oliveira, M., Purvis, M., Cranefield, S., & Nowostawski, M. (2005). The role of ontologies when modelling open multi-agent systems as institutions. In R. P. Katarzyniak (Ed.), *Ontologies and soft methods in knowledge management* (Vol. 4, pp. 181-199). Adelide: Advanced Knowledge International.

Dellarocas, C. (2000). Immunizing online reputation reporting systems against unfair ratings and discriminatory behavior. In *Proceedings of the ACM Conference of Electronic Commerce,* (pp. 150-157).

Dellarocas, C. (2000, June). Contractual agent societies: Negotiated shared context and social control in open multi-agent systems. In *Proceedings of the Workshop on Norms and Institutions in Multi-agent Systems, 4th International Conference on Multi-agent Systems (Agents-2000),* Barcelona, Spain.

DeLoach, S. (2002). Modeling organizational rules in the multi-agent systems engineering methodology. In *Proceedings of the 15th Canadian Conference on Artificial Intelligence,* Calgary, Canada.

DeLoach, S.A. (2000). *Specifying agent behavior as concurrent tasks: Defining the behavior of social agents.* Technical Report, U.S. Air Force Institute of Technology, AFIT/EN-TR-00-03.

DeLoach, S.A., Matson, E.T., & Li, Y. (2003). Exploiting agent oriented software engineering in cooperative robotics search and rescue. *The International Journal of Pattern Recognition and Artificial Intelligence, 17*(5), 817-835.

DeLoach, S.A., Wood, M.F., & Sparkman, C.H. (2001). Multiagent systems engineering. *The International Journal of Software Engineering and Knowledge Engineering, 11*(3), 231-258.

Delort, J.-Y., Bouchon-Meunier B., & Rifqi, M. (2003). Enhanced Web document summarization using hyperlinks. In *Proceedings of the 14th ACM Conference on Hypertext and Hypermedia,* (pp. 208-215). Nottingham, UK: ACM Press.

Demazeau, Y. (2003, September 29-October 4). MAS methodology. In *Proceedings of the Tutorial at the 2nd French-Mexican School of the Repartee Cooperative Systems - ESRC 2003.* Rennes, France: IRISA.

Dempster, A. P. (1968). A generalisation of BaysianBayesian inference. *Journal of the Royal Statistical Society, Series B, 30,* 205-247.

Deng, J., Han, R., & Mishra, S. (2005). *Secure code distribution in dynamically programmable wireless sensor networks* (Tech. Rep. No. CU-CS-1000-05). Boulder, CO: University of Colorado.

Deriaz, M. (2006). *What is trust? My own point of view.* University of Geneva. Retrieved April 3, 2008, from http://cui.unige.ch/ASG/publications/TR2006/

DiLeo, J., Jacobs, T., & DeLoach, S.A. (2002). Integrating ontologies into multiagent systems engineering. In *Proceedings of the Fourth International Bi-Conference Workshop on Agent-Oriented Information Systems (AOIS 2002) at AAMAS '02,* Bologna, Italy, July 16. Available from *http://CEUR-WS.org/Vol-59*

Ding, L., Kolari., P., Ganjugunte, S., Finin, T., & Joshi, A. (2004). Modeling and evaluating trust network inference. In *Proceedings of the 7th International Workshop on Trust in Agent Societies at AAMAS 2004,* New York.

Doben-Henisch, G. (1999). Alan Mathew Turing, the Turing Machine, and the concept of sign. Retrieved from www.inm.de/kip/SEMIOTIC/DRESDEN_ FEBR99/CS_Turing_and_Sign_febr99.html

Doctorow, C. (2001). *Metacrap: Putting the torch to seven straw-men of the meta-utopia* [Electronic Version]. Version 1.3. Retrieved April 3, 2008, from http://www.well.com/~doctorow/metacrap.htm

Doran, J. E., Franklin, S., Jennings, N. R., & Norman, T. J. (1997). On cooperation in multi-agent systems. *The Knowledge Engineering Review, 12*(3), 309-314.

Dugan, B., & Glinert, E. (2002). Task division in collaborative simulations. In *Proceedings of the 35th Annual Hawaii International Conference on System Sciences (HICSS-35'02),* (p. 31).

Durfee, E. H., & Lesser, V. (1989). Negotiating task decomposition and allocation using partial global planning. In L. Gasser & M. Huhns (Eds.), *Distributed artificial intelligence* (Vol. 2, pp. 229-244). San Francisco: Morgan Kaufmann.

Duvigneau, M., Moldt, D., & Rolke, H. (2002). Concurrent architecture for a multi-agent platform. In *Proceedings of the 2002 Workshop on Agent-oriented Software Engineering (AOSEÕ02), 2585.*

Eco, U. (1977). *A theory of semiotics.* Macmillan Press.

Electronic Business using eXtensible Markup Language (ebXML). Retrieved April 2, 2008, from http://www.ebxml.org/

Elio, R., & Haddadi, A. (1999). On abstract task models and conversation policies. *Working Notes of the Workshop on Specifying and Implementing Conversation Policies* (pp. 89-98).

Elmaghraby, W. (2005). Auctions and pricing in e-marketplaces. In D. Simchi-Levi, S. D. Wu, & Z. M. Shen (Eds.), *Handbook of supply chain analysis: Modelling in the e-business era* (pp. 213-246). Norwell, MA: Kluwer Academic.

Elmaghraby, W., & Keskinocak, P. (2003). Dynamic pricing: Research overview, current practices and future directions. *Management Science, 49*(10), 1287-1309.

Elster, J. (1989). Social norms and economic Theory. *The Journal of Economic Perspectives, 3*(4), 99-117.

Epstein, J. M. (2001). Learning to be thoughtless: Social norms and individual computation. *Computer Economy, 18*(1), 9-24.

Eshuis, R., & Dehnert, J. (2003). Reactive Petri nets for workflow modelling. In W.M.P. van der Aalst, & E. Best (Eds.), *Applications and theory of Petri nets: Lecture notes in computer science* (Vol. 2679, pp. 296-315). New York-Heidelberg-London: Springer-Verlag.

Eshuis, R., & Wieringa, R. (2001). *A formal semantics for UML Activity Diagrams—normalising workflow models* (Tech. Rep. No. CTIT-01-04). University of Twente, The Netherlands: Department of Computer Science.

Eshuis, R., & Wieringa, R. (2001). A real time execution semantics for UML activity diagrams. In H. Hussmann (Ed.), *Fundamental approaches to software engineering: Lecture notes in computer science* (Vol. 2029, pp. 76-90). Berlin-Heidelberg: Springer-Verlag.

Eshuis, R., & Wieringa, R. (2002, May 19-25). Verification support for workflow design with UML activity graphs. In *Proceedings of the 24th International Conference on Software Engineering (ICSE'02)*, Orlando, FL, USA, (pp. 166-176). New York: ACM Press.

Eshuis, R., & Wieringa, R. (2003). Comparing Petri net and activity diagram variants for workflow modelling—a quest for reactive Petri nets. In H. Ehrig, W. Reisig, G. Rozenberg, & H. Weber (Eds.), *Petri net technology for communication-based systems: Lecture notes in computer science* (Vol. 2472, pp. 321-351). New York-Heidelberg-London: Springer-Verlag.

European collaborative networked organizations leadership initiative (ECOLEAD). Retrieved April 2, 2008, from http://www.ecolead.org

Fargier, H., Galvagnon, V., & Dubois, D. (2000). Fuzzy PERT in series-parallel graphs. In *Proceedings of the IEEE International Conference on Fuzzy Systems,* (pp. 717-722). San Antonio, TX: IEEE Press.

Federgruen, A., & Heching, A. (1999). Combined pricing and inventory control under uncertainty. *Operations Research, 47*, 454-475.

Fencott, C. (1999). Content and creativity in virtual environment design. *Proceedings of Virtual Systems and Multimedia '99,* University of Abertay Dundee, Scotland.

Fencott, C. (1999). Towards a design methodology for virtual environments. *Proceedings of the International Workshop on User Friendly Design of Virtual Environments,* York, UK.

Fencott, C. (2003). Virtual saltburn by the sea: Creative content design for virtual environments. *Creating and using virtual reality: A guide for the arts and humanities.* Oxbow Books, Arts and Humanities Data Service.

Fencott, C. (2003). *Perceptual opportunities: A content model for the analysis and design of virtual environments.* PhD thesis, University of Teesside, UK.

Fencott, C. (2004). *Game invaders: Computer game theories.* In preparation.

Fencott, P.C., Fleming, C., & Gerrard, C. (1992). Practical formal methods for process control engineers. *Proceedings of SAFECOMP '92*, Zurich, Switzerland, October. City: Pergamon Press.

Fencott, P.C., Galloway, A.J., Lockyer, M.A., O'Brien, S.J., & Pearson, S. (1994). Formalizing the semantics of Ward/Mellor SA/RT essential model using a process algebra. Proceedings of Formal Methods Europe

'94. *Lecture Notes in Computer Science, 873*. Berlin: Springer-Verlag.

Ferber, J. (1999). *Multi-agent systems: An introduction to distributed artificial intelligence*. Harlow, UK: Addison-Wesley Longman.

Fetterly, D., Manasse, M., Najork, M., & Wiener, J. (2003). A large-scale study of the evolution of Web pages. In *Proceedings of the 12th International World Wide Web Conference*, (pp. 669-678). Budapest, Hungary: ACM Press.

Finin, T., Labrou, Y., & Mayfield, J. (1997). KQML as an agent communication language. *Software Agents*, 291-316.

Fipa. (2002). *Fipa ACL Message Structure Specification*.

Fix, J., Scheve, C. v., & Moldt, D. (2006). Emotion-based norm enforcement and maintenance in multi-agent systems: Foundations and petri net modeling. In *Paper presented at the AAMAS*.

Fonseca, S., Griss, M., & Letsinger, R. (2001). *Agent behavior architectures—a MAS framework comparison* (HP Labs Tech. Rep. No. HPL-2001-332). Palo Alto, CA, USA: HP Laboratories. Retrieved April 3, 2008, from http://www.hpl.hp.com/techreports/2001/HPL-2001-332.pdf

Foster, I., & Kesselman, C. (Eds.). (1999). *The Grid: Blueprint for a future computing infrastructure*. Morgan Kaufmann.

Foster, I., Kesselman, C., Nick, J.M., & Tuecke, S. (2002). *The physiology of the Grid: An open grid services architecture for distributed systems integration*. Open Grid Service Infrastructure WG, Global Grid Forum.

Foundation for Intelligent Physical Agents (FIPA). (2007). Retrieved April 3, 2008, from http://www.fipa.org

Fowler, M., & Scott, K. (2000). *UML distilled: A brief guide to the standard object modeling language*. Indianapolis, IN: Addison-Wesley.

Franklin, S., & Graesser, A. (1996). Is it an agent, or just a program?: A taxonomy for autonomous agents. In *Proceedings of the Third International Workshop on Agent Theories, Architectures, and Languages*, (pp. 21-36).

Freitag, D. (1998). Information extraction from HTML: Application of a general machine learning approach. In *Proceedings of AAAI'98*, (pp. 517-523).

Frogner, K., Mandt, T., & Wethal, S. E. (2004). *Cluster and grid computing: Accounting and banking systems*. Retrieved April 3, 2008, from the Cluster and Grid Computing Web site: http://hovedprosjekter.hig.no/v2004/data/gruppe05/?page=documents

Fromm, J. (2005). *Types and forms of emergence* [Electronic Version]. Adaptation and self-organizing systems. Retrieved April 3, 2008, from http://www.citebase.org/abstract?id=oai:arXiv.org:nlin/0506028

Gallego, G., & Ryzin, G. V. (1994). Optimal dynamic pricing of inventories with stochastic demand over finite horizons. *Management Science, 40*(8), 999-1020.

Gammasutra. Retrieved from www.gamasutra.com

Garcia-Camino, A., Rodriguez-Aguilar, J. A., Sierra, C., & Vasconcelos, W. (2006). Norm-oriented programming of electronic institutions. In *Paper presented at the AAMAS*, New York.

Gillet, A., Macaire, L., Bone-Lococq, C., & Pastaire, J. (2001). Color image segmentation by fuzzy morphological transformation of the 3d color histogram. In *Proceedings of the 10th IEEE International Conference on Fuzzy Systems*, (Vol. 2, pp. 824-824).

Gintis, H. (2003). Solving the puzzle of prosociality. *Rationality and Society, 15*(2), 155-187.

Giorgini, P., Massacci, F., & Zannone, N. (2005). Security and trust requirements engineering. *Foundations of security analysis and design III—tutorial lectures, LNCS 3655* (pp. 237-272). Springer-Verlag.

Giorgini, P., Mouratidis, H., & Zannone, N. (2007). Modelling security and trust with secure TROPOS. *Integrating security and software engineering: Advances and future vision*. Hershey, PA: Idea Group.

Goguen, J. (1999). An introduction to algebraic semiotics, with application to user interface design. Computation for metaphor, analogy and agents. *Springer Lecture Notes in Artificial Intelligence, 1562*, 242-291.

Gokhale, A. (1995). Collaborative learning enhances critical thinking. *Journal of Technology Education, 7*(1), 22-30.

Goldin, D., Keil, D., & Wegner, P. (2001). An interactive viewpoint on the role of UML. *Unified Modelling Language: Systems analysis, design, and development issues*. Hershey, PA: Idea Group Publishing.

Goldin, D.Q., Smolka, S.A., Attie, P.C., & Wegner, P. (2001). Turing Machines, transition systems, and interaction. *Nordic Journal of Computing*.

Gonzalez, C.R., & Woods, E.R. (2002). *Digital image processing.* Upper Saddle River, NJ, USA: Prentice Hall.

Gottlob, G. (2005). Web data extraction for business intelligence: The Lixto approach. In G. Vossen, F. Leymann, P.C. Lockemann, & W. Stucky (Eds.), *Datenbanksysteme in business, technologie und Web, 11. Fachtagung des GI-Fachbereichs "Datenbanken und Informationssysteme" (DBIS)* (pp. 30-47). Lecture Notes in Informatics 65, GI.

Gottlob, G., & Koch, C. (2004). Monadic datalog and the expressive power of languages for Web information extraction. *Journal of the ACM, 51*(1), 74-113.

Gottlob, G., Koch, C., & Schulz, K.U. (2004). Conjunctive queries over trees. In *Proceedings of PODS' 2004,* (pp. 189-200). ACM Press.

Gouaich, A. (2003). Requirements for achieving software agents autonomy and defining their responsibility. In Nickles et al. (Eds.), (pp. 128-139).

Gould, S. J., & Vrba, E. (1982). Exaptation—a missing term in the science of form. *Paleobiology, 8,* 4-15.

Grandison, T., & Sloman, M. (2000). A survey of trust in Internet applications. *IEEE Communications Surveys and Tutorials, 4th Quarter, 3*(4), 2-16.

Greenwald, A., Kephart, J. O., & Tesauro, G. J. (1999). Strategic pricebot dynamics. In *Proceedings of the 1st ACM conference on Electronic commerce,* (pp. 58-67). New York: ACM Press.

Grenoble O'Malley, P. (2005, March-April). Troubleshooting machine control: Contractors talk about making the most of 3D GPS. *Grading & Excavator Contractor, 7*(2).

Grossi, D., Aldewereld, H., Vázquez-Salceda, J., & Dignum, F. (2006). Ontological aspects of the implementation of norms in agent-based electronic institutions. *Computational & Mathematical Organization Theory, 12*(2), 251-275.

Gruber T. (1995). Towards principles for the design of ontologies used for knowledge sharing. *International Journal of Human-Computer Studies, 43*(5/6), 907-928.

Gruhl, D., Guha, R., Liben-Nowell, D., & Tomkins, A. (2004). Information diffusion through blogspace. In *Proceedings of the 13th International World Wide Web Conference,* (pp. 491-501). New York: ACM Press.

Grzegorzewski, P., Hryniewicz, O., & Gil, M. A. (Eds.). (2002). *Soft methods in probability, statistics and data analysis.* Heidelberg, Germany: Physica Verlag.

Guha, R., Kumar, R., Raghavan, P., & Tomkins, A. (2004). Propagation of trust and distrust. *In Proceedings of the 13th International Conference on World Wide Web (WWW 2004),* New York, (pp. 403-412).

Gupta, M., Ravikumar, K., & Kumar, M. (2002). Adaptive strategies for price markdown in a multi-unit descending price auction: A comparative study. In *Proceedings of IEEE International Conference on Systems, Man, and Cybernetics,* (pp. 373-378).

Guth, S., Simon, B., & Zdun, U. (2003). A contract and rights management framework design for interacting brokers. In *Proceedings of the 36th Annual Hawaii International Conference on System Sciences: Track 9* (Vol. 9). Washington, DC: IEEE Computer Society.

Gybels, K., Wuyts, R., Ducasse, S., & Hondt, M. (2006). Inter-language reflection: A conceptual model and its implementation. *Computer Languages, Systems and Structures, 32,* 109-124.

Habermas, J. (1985). *The theory of communicative action: Reason and the rationalization of society* (Vol. 1). Beacon Press.

Haeuser, J., et al. (2000). A test suite for high-performance parallel Java. *Advances in Engineering Software, 31,* 687-696.

Haken, H. (1983). *Synergetics, an introduction: Non-equilibrium phase transitions and self-organization in physics, chemistry, and biology* (3rd rev. ed.). Berlin: Springer-Verlag.

Hales, D. (2003). *Evolving specialisation, altruism and group-level optimisation using tags* (Vol. 2581/2003, Lecture notes in computer science). Berlin/Heidelberg: Springer-Verlag.

Hales, D. (2003, September 16-19). Understanding tag systems by comparing tag models. In *Paper presented at the to-Model Workshop (M2M2) at the Second European Social Simulation AssociationConference (ESSA'04),* Valladolid, Spain.

Hales, D. (2004). Change your tags fast!--a necessary condition for cooperation?. In *Paper presented at the The Multi-agent-based Simulation (MABS 2004),* (pp. 89-98).

Hales, D. (2004). Self-organising, open and cooperative P2P Societies--from tags to networks. In *Paper presented at the Engineering Self-Organising Applications (ESOA 2004),* (pp. 123-137).

Hales, D., & Edmonds, B. (2004). Can tags build working systems?—From MABS to ESOA. *Lecture notes in computer science* (pp. 186-194).

Hales, D., & Patarin, S. (2005). *How to cheat BitTorrent and why nobody does. UBLCS* (Tech. Rep. No. UBLCS-2005-12).

Hallgrimsson, Þ., & Bang S. (2003). Nordic Web Archive. In *Proceedings of the 3rd Workshop on Web Archives in conjunction with the 7th European Conference on Research and Advanced Technologies in Digital Archives*, Trondheim, Norway. Springer-Verlag.

Hammonds, K. H., Jackson, S., DeGeorge, G., & Morris, K. (1997). The new university—a tough market is reshaping colleges. *Business Week, 22*, 96-102.

Hardin, G. (1968). The tragedy of the commons. *Science, 162*, 1243-1248).

Harel, D., & Pnueli, A. (1985). On the development of reactive systems. In K.R. Apt (Ed.), *Logics and models of concurrent systems, NATO ASI Series F: Computer and systems sciences* (pp. 477-498). New York: Springer-Verlag.

He, L., & Walrand, J. (2005). Pricing of differentiated internet services. In *Proceedings of INFOCOM-2005*, (Vol. 1, pp. 195-204). New York: IEEE Computer Society.

Hebb, D. (1949). *The organization of behavior.* New York: John Wiley & Sons.

Helgason, C.M., Jobe, T.H., Malik, D.S., & Mordeson, J.N. (1999). Analysis of stroke pathogenesis using hierarchical fuzzy clustering techniques. In *Proceedings of the 5th International Conference on Information Systems, Analysis and Synthesis: ISAS'99*, Orlando, FL, USA, (pp. 477-482).

Helgason, C.M., Jobe, T.H., Mordeson, J.N., Malik, D.S., & Cheng, S.C. (1999). Discarnation and interactive variables as conditions for disease. In *Proceedings of the 18th International Conference of the North American Fuzzy Information Society NAFIPS'99*, New York, (pp. 298-303).

Heljanko, K. (2006). *T79.4301. Parallel and distributed systems (4 ECTS), Lecture 5* (pp. 1-24). Department of Computer Science, University of Helsinki, Finland. Retrieved April 3, 2008, from http://www.tcs.hut.fi/Studies/T-79.4301/2006SPR/lecture5.pdf

Hewlett-Packard Development Company. (2005). *Adaptive enterprise: Business and IT synchronized to capitalize on change—an executive overview white paper.* Retrieved April 3, 2008, from http://www.hp.com/go/adaptive

Heyes, C. (2001). Causes and consequences of imitation. *TRENDS in Cognitive Sciences, 5*(6), 253-261.

Hoffmann, T. (1999). The meanings of competency. *Journal of European Industrial Training, 23*(6), 275-286.

Holland, J. H. (1993). *The effect of labels (tags) on social interactions* (Vol. SFI Working Paper 93-10-064). NM: Santa Fe Institute.

Holsapple, C. W., & Joshi, K. D. (2004). A formal knowledge management ontology: Conduct, activities, resources, and influences. *Journal of the American Society for Information Science and Technology, 55*(7), 593-612.

Holzmann, G.J. (1997). The model checker spin. *IEEE Transactions on Software Engineering, 23*(5), 279-295.

Hopcroft, J. E., & Ullman, J. D. (1979). *Introduction to automata theory, languages, and computation.* Reading, MA: Addison-Wesley.

Hu, J., & Zhang, Y. (2002). *Online reinformcenet learning in multiagent systems.*

Huberman, B. A. (2001). *The laws of the Web: Patterns in the ecology of information.* Cambridge: MIT Press.

Hung, P. C. K., & Mao, J. Y. (2002). Modeling e-negotiation activities with Petri nets. In R. H. Spraguer, Jr. (Ed.), *Proceedings of the 35th Hawaii International Conference on System Sciences HICSS 2002*, Big Island, HI, (Vol. 1, p. 26), CD ROM. Piscataway, NJ, USA: IEEE Computer Society Press.

Huynh, D., Jennings, N. R., & Shadbolt, N. R. (2006). An integrated trust and reputation model for open multi-agent systems. *Autonomous Agents and Multi-Agent Systems, 13*(2), 119-154.

Jacobsson, A., & Davidsson, P. (2006). An analysis of plug and play business software. In R. Suomi, R. Cabral, J.F. Hampe, A. Heikkilä, J. Järveläinen, & E. Koskivaara (Eds.), *Project e-society: Building bricks* (pp. 31-44). New York: Springer Science Business Media.

Jacobsson, A., & Davidsson, P. (2007). Security issues in the formation and operation of virtual enterprises. In L. Kutvonen, P. Linnington, J.-H. Morin, & S. Ruohomaa (Eds.), *Proceedings of the Second International Workshop on Interoperability Solutions to Trust, Security, Policies and QoS for Enhanced Enterprise Systems at the Third International Conference on Interoperability for Enterprise Applications and Software,* (pp. 55-66).

Jatowt, A., & Ishizuka, M. (2004). Summarization of dynamic content in Web collections. In *Proceedings of the 8th European Conference on Principles and Practice of Knowledge Discovery in Databases*, (pp. 245-254). Pisa, Italy: Springer-Verlag.

Jatowt, A., & Ishizuka, M. (2004). Temporal Web page summarization. In *Proceedings of the 5th Web Information Systems Engineering Conference*, (pp. 303-312). Brisbane, Australia: Springer-Verlag.

Jatowt, A., & Ishizuka, M. (2006). Temporal multi-page summarization. *Web Intelligence and Agent Systems: An International Journal, 4*(2), 163-180. IOS Press.

Jatowt, A., & Tanaka, K. (2007). Towards mining past content of Web pages. *New Review of Hypermedia and Multimedia, Special Issue on Web Archiving, 13*(1), 77-86. Taylor and Francis.

Jatowt, A., Kawai, Y., & Tanaka, K. (2007). Detecting age of page content. In *Proceedings of the 9th ACM International Workshop on Web Information and Data Management*. Lisbon, Portugal: ACM Press.

Jatowt, A., Kawai, Y., Nakamura, S., Kidawara, Y., & Tanaka, K. (2006). Journey to the past: Proposal for a past Web browser. In *Proceedings of the 17th Conference on Hypertext and Hypermedia*, (pp. 134-144). Odense, Denmark: ACM Press.

Jensen, K. (1997). Coloured petri nets—Basic concepts, analysis methods and practical use: Basic concepts. *Monographs in theoretical computer science* (Vol. 1).

Josang, A. (1999). Trust-based decision making for electronic transactions. In L. Yngstrm & T. Scensson (Eds.), *Proceedings of the 4th Nordic Workshop on Secure Computer Systems*. Stockholm, Sweden: Stockholm University Report 99-005.

Josang, A. (2001). A logic for uncertain probabilities. *International Journal of Uncertainty, Fuzziness and Knowledge-Based Systems, 9*(3), 279-311.

Josang, A., & Ismail, R. (2002). The Beta reputation system. In *Proceedings of the 15th Bled Conference on Electronic Commerce*, Slovenia, (pp. 324-337).

Josang, A., Roslam, I., & Colin, B. (2006). A survey of trust and reputation systems for online service provision. *Decision support systems*.

Kagel, J. H. (1995). Auctions: A survey of experimental research. In J. H. Kagel & A. E. Roth (Eds.), *Handbook of experimental economics* (pp. 501-587). Princeton University Press.

Kahn, M. L., & Cicalese, C. D. T. (2003). CoABS grid scalability experiments. *Autonomous Agents and Multiagent Systems, 7*(1), 171-178.

Kalagnanam, J., & Parkes, D. (2005). Auctions, bidding and exchange design. In D. Simchi-Levi, S. D. Wu, & Z. M. Shen (Eds.), *Handbook of supply chain analysis: Modelling in the e-business era* (pp. 143-212). The Netherlands: Kluwer Academic.

Kamvysselis, M., & Marina, O. (1999). *Imagina: A cognitive abstraction approach to sketch-based image retrieval* (p. 157). Massachusetts Institute of Technology.

Kandel, E. R., Schwartz, J. H., & Jessell, T. M. (2000). *Principles of neural sciences*. New York: McGraw-Hill.

Kantorovich, L.B., Vulich, B.Z., & Pinsker, A.G. (1950). *Functional analysis in semi-ordered spaces* (in Russian). Moscow: GITTL.

Kaur, K. (1998). *Designing virtual environments for usability*. PhD Thesis, City University, London.

Kendall, E. (1998). Agent roles and role models: New abstractions for multiagent system analysis and design. In *Proceedings of the International Workshop on Intelligent Agents in Information and Process Management*, Bremen, Germany, September.

Kephart, J. O., & Tesauro, G. J. (2000). Pseudo-convergent Q-learning by competitive pricebots. In *Proceedings of the 17th International Conference on Machine Learning*, (pp. 463-470). San Francisco: Morgan Kaufmann.

Kerre, E.E., & Nachtegael, M. (2000). *Fuzzy techniques in image processing*. Heidelberg, New York: Physica Verlag.

Kleinberg, J. M. (2003). Bursty and hierarchical structure in streams. *Data Mining Knowledge Discovery, 7*(4), 373-397.

Klemperer, P. (1999). Auction theory: A guide to the literature. *Journal of Economic Surveys, 13*, 227-286.

Klir, G. (2006). *Uncertainty and information: Foundations of generalized information theory*. Hoboken, NJ: Wiley Interscience.

Kokar, M. M., Tomasik, J. A., & Weyman, J. (2004). Formalizing classes of information fusion systems. *Information Fusion, 5*(3), 189-202.

König, M., Klotz, E., & Heuser, L. (2000). Diagnosis of cerebral infarction using perfusion CT: State of the art. *Electromedica*, 9-12.

Kosala, R., & Blockeel, H. (2000). Web mining research: A survey. *SIGKDD Explorations, 2*(1), 1-15.

Kourpas, E. (2006). *Grid computing: Past, present and future—an innovation perspective.* Retrieved April 3, 2008, from the IBM Grid Computing Web site: http://www-1.ibm.com/grid/grid_literature.shtml

Król, D. (2005). A propagation strategy implemented in communicative environment. *Lecture notes in artificial intelligence* (Vol. 3682, pp. 527-533). Springer-Verlag.

Król, D., & Kukla, G. (2006). Distributed class code propagation with Java. *Lecture notes in artificial intelligence* (Vol. 4252, pp. 259-266). Springer-Verlag.

Kronlof, C. (1993). *Methods integration: Concepts and case studies.* New York: John Wiley & Sons.

Kulikowski, J.L. (1972). *An algebraic approach to the recognition of patterns.* CISM Lecture Notes No. 85. Wien: Springer-Verlag.

Kulikowski, J.L. (1986). *Outline of the theory of graphs (*in Polish). Warsaw, Poland: PWN.

Kulikowski, J.L. (1992). Relational approach to structural analysis of images. *Machine Graphics and Vision, 1*(1/2), 299-309.

Kulikowski, J.L. (2006). Description of irregular composite objects by hyper-relations. In K. Wojciechowski et al. (Eds.), *Computer vision and graphics* (pp. 141-146). Springer-Verlag.

Kumar, R., Novak, P., Raghavan, S., & Tomkins, A. (2003). On the bursty evolution of Blogspace. In *Proceedings of the 12th International World Wide Web Conference.* Budapest, Hungary: ACM Press.

Kushmerick, N. (2000). Wrapper induction: Efficiency and expressiveness. *Artificial Intelligence, 118*(1-2), 15-68.

Kwek, S. S. (1999). An apprentice learning model. In *Proceedings of the Twelfth Annual Conference on Computational Learning Theory (COLT '99),* (pp. 63-74).

La, R. J., & Anantharam, V. (1999). Network pricing with a game theoretic approach. In *Proceedings of the IEEE Conference on Decision and Control,* (pp. 4008-4013). New York: ACM Press.

Lacey, T.H. & DeLoach, S.A. (2000). Automatic verification of multiagent conversations. In *Proceedings of the 11th Annual Midwest Artificial Intelligence and Cognitive Science Conference,* University of Arkansas, Fayetteville, April 15-16 (pp. 93-100). AAAI Press.

Laender, A.H.F., Ribeiro-Neto, B., & Silva, A.S. (2002). DEByE—data extraction by example. *Data & Knowledge Engineering, 40*(2), 121-154.

Lanunay, P., & Pazat, J.L. (2001). Easing parallel programming for clusters with Java. *Future Generation Computer Systems, 18,* 253-263.

Laudon, K.C., & Traver, C.G. (2004). *E-commerce, business, technology, society* (2nd ed.). Pearson Addison-Wesley.

Laure, E. (2001). OpusJava: A Java framework for distributed high performance computing. *Future Generation Computer Systems, 18,* 235-251.

Laurel, B. (1992). Placeholder. Retrieved from www.tauzero.com/Brenda_ Laurel/Placeholder/Placeholder.html

Lauria, R. (1997). Virtual reality as a metaphysical testbed. *Journal of Computer Mediated Communication, 3*(2). Retrieved from jcmc.huji.ac. il/vol3/issue2/

Lauzac, S.W., & Chrysanthis, P.K. (2002). View propagation and inconsistency detection for cooperative mobile agents. *Lecture Notes in Computer Science, 2519,* 107-124.

Lee, T. B., Hendler, J., & Lassila, O. (2001). The Semantic Web. *Scientific American, 5,* 28-37.

Legerstee, M. (1991). The role of person and object in eliciting early imitation. *Journal of Experimental Child Psychology, 51*(3), 423-433.

Leibenstein, H. (1968). Entrepreneurship and development. *The American Economic Review,* 58, 72-83.

Leloup, B., & Deveaux, L. (2001). Dynamic pricing on the Internet: Theory and simulations. *Journal of Electronic Commerce Research, 1*(3), 265-276. The Netherlands: Springer-Verlag.

Lenat, D. B. (1995). Cyc: A large-scale investment in knowledge infrastructure. *Communications of the ACM, 38*(11), 33-38.

Lenhert, W., & Sundheim, B. (1991). A performance evaluation of text-analysis technologies. *AI Magazine, 12*(3), 81-94.

Lenz, K., Oberweis, A., & Schneider, S. (2001). Trust based contracting in virtual organizations: A concept based on contract workflow management systems. In B. Schmid, K. Stanoevska-Slabeva, & V. Tschammer (Eds.), *Towards the e-society—E-commerce, e-business, and e-government* (pp. 3-16). Boston: Kluwer Academic.

Levin, L. A. (1973). Universal sequential search problems. *Problems of information transmission, 9*(3), 265-266.

Leymann, F., & Roller, D. (2000). *Production workflow—Concepts and techniques*. New York: Prentice-Hall.

Li, N., & Mitchell, J. C. (2003). RT: A role-based trust-management framework. In *Proceedings of the 3rd DARPA Information Survivability Conference and Exposition (DISCEX III)*.

Li, Y., Li, J., Dong, H., & Gao, X. (2002). A fuzzy relation based algorithm for segmenting color aerial images of urban environment. In *Proceedings of the ISPRS Technical Commission II Mid-Term Symposium on Integrated System for Spatial Data Production, Custodian and Decision Support, Xi'an, P. R.* China, (pp.271-274).

Li, Z., Wang, B., Li, M., & Ma, W.-Y. (2005). A probabilistic model for retrospective news event detection. In *Proceedings of the 28th Annual International Conference on Research and Development in Information Retrieval*, (pp. 106-113). Salvador, Brazil: ACM Press.

Lin, J.W., & Kuo, S.Y. (2000). Resolving error propagation in distributed systems. *Information Processing Letters, 74*(5-6), 257-262.

Lindley, C., Knack, F., Clark, A., Mitchel, G., & Fencott, C. (2001). New media semiotics—computation and aesthetic function. *Proceedings of COSIGN 2001*, Amsterdam. Retrieved from www.kinonet.com/conferences/cosign2001/

Lindsay, Y., Yu, W., Han, Z., & Ray Liu, K. J. (2006). Information theoretic framework of trust modelling and evaluation for the ad hoc networks. *IEEE International Journal on Selected Areas in Communications, 24*(2), 305-317.

Ling, A. W. K., Sun, L. C., Sodhy, G. C., Yong, C. H., Haron, F., & Buyya, R. (2003). Design framework of generic components for Compute Power Market. In S. Sahni (Ed.), *Proceedings of Computer Science and Technology*. Cancun, Mexico: ACTA Press.

Llorens, M., & Oliver, J. (2004). Structural and dynamic changes in concurrent systems: Reconfigurable Petri nets. *IEEE Transactions on Computers, 53*(9), 1147-1158.

Llorens, M., & Oliver, J. (2006). A basic tool for the modelling of Marked-Controlled Reconfigurable Petri Nets. In *Proceedings of the Workshop on Petri Nets and Graph Transformation (PNGT 2006): Electronic Communications of the EASST: Petri Nets and Graph Transformations* (Vol. 2, pp. 1-13). Retrieved April 3, 2008, from http://eceasst.cs.tu-berlin.de/index.php/eceasst/article/viewFile/23/28 or http://eceasst.cs.tu-berlin.de/index.php/eceasst/issue/view/10

Loftin, R.B., & Kenney, P.J. (1994). *The use of virtual environments for training the Hubble Space Telescope flight team*. Retrieved from www.vetl.uk/edu/Hubble/virtel.html

Lombard, M., & Ditton, initial. (1997). At the heart of it all: The concept of telepresence. *Journal of Computer Mediated Communication, 3*(2). Retrieved September 1997 from jcmc.huji.ac.il/vol3/issue2/

López y López, F., Luck, M., & d'Inverno, M. (2006). A normative framework for agent-based systems. *Computational and Mathematical Organization Theory, 12*, 227-250.

Lopez, F. L. Y., & Marquez, A. A. (2004). An architecture for autonomous normative agents. In *Paper presented at the Fifth Mexican International Conference in Computer Science (ENC'04)*, Los Alamitos, CA, USA.

Lopez, F., Luck, M., & Inverno, M. (2002). Constraining autonomy through norms. In *Paper presented at the Proceedings of the First International Joint Conference on Autonomous Agents and Multi Agent Systems AAMAS'02*.

Luck, M., & d'Inverno, M. (1995). A formal framework for agency and autonomy. In *Proceedings of the First International Conference on Multi-agent Systems (ICMAS)*, (pp. 254-260).

Lynch, N., & Tuttle, M. R. (1989). An introduction to input/output automata. *CWI Quarterly, 2*(3), 219-246.

Mamdani, E. H., & Assilian, S. (1975). An experiment in linguistic synthesis with a fuzzy logic controller. *International Journal of Man-Machine Studies, 7*(1), 1-13.

Mamei, M., & Zambonelli, F. (2005). Spreading pheromones in everyday environments through RFID technology. In *Proceedings of the 2nd IEEE Symposium on Swarm Intelligence, (pp. 281-288)*. IEEE Press.

Margulis, L. (1970). *Origin of eukaryotic cells*. New Haven, CT: University Press.

Margulis, L. (1981). *Symbiosis in cell evolution*. San Francisco: Freeman.

Matsuoka, S., & Itou, S. (2001). Towards performance evaluation on high-performance computing on multiple Java platforms. *Future Generation Computer Systems, 18*, 281-291.

McAfee, R. P., & McMillan, J. (1987). Auctions and

bidding. *Journal of Economic Literature, 25*(2), 699-738. American Economic Association.

McCool, R. (2005). Rethinking the Semantic Web, part 1. *IEEE Internet Computing, 9*(6), 86-88.

McCool, R. (2006). Rethinking the Semantic Web, part 2. *IEEE Internet Computing, 10*(11), 93-96.

McCown, F., & Nelson, M. (2006). Evaluation of crawling policies for a Web-repository crawler. In *Proceedings of the 17th Conference on Hypertext and Hypermedia*, (pp. 157-168). Odense, Denmark: ACM Press.

McCown, F., Smith, J.A., & Nelson, M.L. (2006). Lazy preservation: Reconstructing Web sites by crawling the crawlers. In *Proceedings of the 8th ACM International Workshop on Web Information and Data Management*, (pp. 67-74). Arlington, VA, USA: ACM Press.

McElroy, M. W. (1999, October). The second generation of knowledge management. *Knowledge Management Magazine*.

McElroy, M. W. (2003). *The new knowledge management—Complexity, learning, and sustainable innovation*. Boston: KMCI Press/Butterworth-Heinemann.

McElroy, M. W. (n.d.). *The knowledge life cycle (KLC)*. Retrieved March 1, 2006, from http://www.kmci.org/media/Knowledge_Life_Cycle.pdf

McIntosh, P. Course notes on UML/VRML. Retrieved from www.public. asu.edu/~galatin/

McKnight, D. H., & Chervany. (1996). *The meanings of trust* (Tech. Rep.). University of Minnessota, Management Information Systems Research Center. Retrieved April 3, 2008, from http://www.misrc.umn.edu/workingpapers/

McKnight, L., & Boroumand, J. (2000). Pricing Internet services: Approaches and challenges. *Computer, 33*(2), 128-129. Los Alamitos, CA: IEEE Computer Society Press.

Meena, H.K., Saha, I., Mondal, K.K., & Prabhakar, T.V. (2005, October 16-17). An approach to workflow modeling and analysis. In *Proceedings of the 20th Annual ACM SIGPLAN Conference on Object-oriented Programming, Systems, Languages, and Applications (OOPSLA'05): The OOPSLA Workshop on Eclipse technology eXchange*, San Diego, CA, (pp. 85-89). New York: ACM Press.

Mei, Q., & Zhai, C-X. (2005). Discovering evolutionary theme patterns from text: An exploration of temporal text mining. In *Proceedings of the 11th International Conference on Knowledge Discovery and Data Mining*, (pp. 198-207). New York: ACM Press.

Mekprasertvit, C. (2004). AgentMom user's manual. Kansas State University. Retrieved June 30, 2004, from *http://www.cis.ksu.edu/~sdeloach/ai/software/agentMom_2.0/*

Melaye, D., & Demazeau, Y. (2005). Bayesian dynamic trust model. In *Proceedings of Multi-agent Systems and Applications IV: 4th International Central and Eastern European Conference on Multi-agent Systems, CEEMAS 2005*, (pp. 480-489). Springer-Verlag, LNCS 3690.

Microsoft BizTalk Server (BizTalk). Retrieved April 2, 2008, from http://www.microsoft.com/biztalk/

Milgrom, P. (1989). Auctions and bidding: A primer. *Journal of Economic Perspectives, 3*(3), 3-22. American Economic Association.

Milner, R. (1989). *Communication and concurrency*. Englewood Cliffs, NJ: Prentice-Hall.

Milner, R. (1989). *Communication and concurrency*. Upper Saddle River, NJ: Prentice Hall.

Minsky, N., & Ungureanu, V. (2000). Law-governed interaction: A coordination and control mechanism for heterogeneous distributed systems. *ACM Transactions on Software Engineering and Methodology (TOSEM), 9*(3), 273-305.

Mitschang, B. (2003). Data propagation as an enabling technology for collaboration and cooperative information systems. *Computers in Industry, 52*(1), 59-69.

Moghaddamzadeh, A., & Bourbakis, N. (1997). A fuzzy region growing approach for segmentation of color images. *Pattern Recognition, 30*(6), 867-881.

Molnar, D., & Wagner, D. (2004). *Privacy and security in library RFID: Issues, practices, and architectures*. New York: ACM Press.

Mon, D.-L., Cheng, C.-H., & Lu, H.-C. (1995). Application of fuzzy distributions on project management. *Fuzzy Sets and Systems, 73*, 227-234.

Mordeson, J.N., Malik, D.S., & Cheng, S.C. (2000). *Fuzzy mathematics in medicine*. Heidelberg, New York: Physica Verlag.

Morris, J., Ree, P., & Maes, P. (2000). Sardine: Dynamic seller strategies in an auction marketplace. In *Proceedings of the 2nd ACM Conference on Electronic Commerce*, (pp. 128-134). New York: ACM Press.

Moser, M.A. (1996). *Immersed in technology*. Boston, MA: MIT Press.

Mui, L., Halberstadt, A., & Mohtashemi, M. (2002). Notions of reputation in multi-agent systems: A review. In *Proceedings of the 1st International Joint Conference on Autonomous Agents and Multiagent Systems,* Bologna, Italy, (pp. 280-287).

Mui, L., Mohtashemi, M., Ang, C., Szolovtis, P., & Halberstadt, A. (2001). Ratings in distributed systems: A bayesian approach. In *Proceedings of the Workshop on Information Technologies and Systems (WITS),* Miami, Fl.

Murata, T. (1989). Petri nets: Properties, analysis and applications. *Proceedings of the IEEE, 77*(4), 541-588.

Murray, M. (1997). *Hamlet on the Holodeck: The future of narrative in cyberspace.* New York: The Free Press.

Muslea, I., Minton, S., & Knoblock, C. (2001). Hierarchical wrapper induction for semistructured information sources. *Journal of Autonomous Agents and Multi-Agent Systems, 4*(1-2), 93-114.

Narahari, Y., & Dayama, P. (2005). Combinatorial auctions for electronic business. *SADHANA - Academy Proceedings in Engineering Sciences, 30*(2), 179-212. India: Indian Academy of Science.

Narahari, Y., Raju, C. V. L., Ravikumar, K., & Shah, S. (2005). Dynamic pricing models for electronic business. *SADHANA - Academy Proceedings in Engineering Sciences, 30*(2&3), 231-256. India: Indian Academy of Science.

Newhouse, S., MacLaren, J., Keahey, K. (2004). Trading grid services within the UK e-Science Grid. In J. Nabrzyski, J. M. Schopf, & J. Weglarz (Eds.), *Grid resource management: State of the art and future trends* (pp. 479-490). Norwell, MA: Kluwer Academic.

Nickles, M., Rovatsos, M., & Weiß, G. (Eds.). (2004). Agents and computational autonomy—potential, risks, and solutions. In *Postproceedings of the 1st International Workshop on Computational Autonomy—Potential, Risks, Solutions (AUTONOMY 2003), held at the 2nd International Joint Conference on Autonomous Agents and Multi-agent Systems (AAMAS 2003),* July 14, 2003, Melbourne, Australia. Lecture Notes in Computer Science (Vol. 2969). Springer-Verlag.

Nicola, R. D., & Hennessy, M. (1987). CCS without tau's. In *Proceedings of the International Joint Conference on Theory and Practice of Software Development: Advanced Seminar on Foundations of Innovative Software Development I and Colloquium on Trees in Algebra and Programming,* (Vol. 1, pp. 138-152).

Noriega, P. (2006). Fencing the open fields: Empirical concerns on electronic institutions. *LNAI: Coordination, organizations, institutions, and norms in multi-agent systems* (Vol. 3913, pp. 81-98).

Norman, D. A. (2005, November-December). There's an automobile in HCI's future. *Interactions,* 53-54.

Nowostawski, M. (2002). *JFern manual.*

Nowostawski, M., Epiney, L., & Purvis, M. (2005). Self-adaptation and dynamic environment experiments with evolvable virtual machines. In S. Brueckner, G. M. Serugendo, D. Hales, & F. Zambonelli (Eds.), *Proceedings of the Third International Workshop on Engineering Self-organizing Applications (ESOA 2005),* (pp. 46-60). Springer-Verlag.

Nowostawski, M., Epiney, L., & Purvis, M. (2005). Self-adaptation and dynamic environment experiments with evolvable virtual machines. In *Proceedings of the Third International Workshop on Engineering Self-organizing Applications (ESOA 2005),* (pages 46-60). Utrech, The Netherlands: Fourth International Joint Conference on Autonomous Agents & Multi Agent Systems.

Nowostawski, M., Purvis, M., & Cranefield, S. (2001). Kea—multi-level agent infrastructure. In *Proceedings of the 2nd International Workshop of Central and Eastern Europe on Multi-agent Systems (CEEMAS 2001),* (pp. 355-362). Krak´ow, Poland: Department of Computer Science, University of Mining and Metallurgy.

Nowostawski, M., Purvis, M., & Cranefield, S. (2001, May 28-June1). A layered approach for modelling agent conversations. In T. Wagner & O. F. Rana (Eds.), *Proceedings of 2nd International Workshop on Infrastructure for Agents, MAS, and Scalable MAS, 5th International Conference on Autonomous Agents - AA 2001,* Montreal, Canada, (pp. 163-170). Menlo Park, CA, USA: AAAI Press.

Nowostawski, M., Purvis, M., & Cranefield, S. (2002). OPAL: A multi-level infrastructure for agent-oriented development. In *Proceedings of the First International Joint Conference on Autonomous Agents and Multi-agent Systems (AAMAS 2002),* (pp. 88-89).

Nowostawski, M., Purvis, M., & Cranefield, S. (2004). An architecture for self-organising evolvable virtual machines. In S. Brueckner, G. D. M. Serugendo, A. Karageorgos, & R. Nagpal (Eds.), *Engineering self organising sytems: Methodologies and applications.* Lecture Notes in Artificial Intelligence (No. 3464). Springer-Verlag.

Ntoulas, A., Cho, J., & Olston, C. (2004). What's new

on the Web? The evolution of the Web from a search engine perspective. In *Proceedings of the 13th International World Wide Web Conference,* (pp. 1-12). New York: ACM Press.

Nwana, H. S. (1996). Software agents: An overview. *The Knowledge Engineering Review, 11*(3), 205-244.

OGSI.net. (2007). *Main page.* Retrieved April 3, 2008, from http://www.cs.virginia.edu/~gsw2c/ogsi.net.html

Opp, K.-D. (2001). How do norms emerge? An outline of a theory. *Mind and Society, 2*(1), 101-128.

Orski, D. (2005a). Quality of cascade operations control based on uncertain variables. *Artificial Life and Robotics, 9*(1), 32-35.

Orski, D. (2005b). Application of uncertain variables to planning resource allocation in a class of research projects. In *Proceedings of the 18th International Conference on Systems Engineering,* (pp. 238-243). Los Alamitos: IEEE CS Press.

Orski, D. (2006). Application of uncertain variables in decision problems for a complex of operations. *IIAS Transactions on Systems Research and Cybernetics, VI*(1), 19-24.

Orski, D. (2007). Knowledge-based decision making in a class of operation systems with three-level uncertainty. *Acta Systemica, VII*(1), 19-24.

Orski, D., & Hojda, M. (2007). Application of uncertain variables to decision making in a class of series-parallel production systems. In A. Grzech (Ed.), *Proceedings of the 16th International Conference on Systems Science,* (Vol. II, pp. 131-142). Oficyna Wydawnicza PWr: Wroclaw.

Orski, D., Sugisaka, M., & Graczyk, T. (2006). Neural networks in adaptive control for a complex of operations with uncertain parameters. *Systems Science, 32*(2), 19-35.

Osman, N. (2006). *Formal specification and verification of trust in multi-agent systems.* School of Informatics, University of Edinburgh. Retrieved April 3, 2008, from http://homepages.inf.ed.ac.uk/s0233771/trust.pdf

Oxygen XML Editor. (2007). Retrieved April 2, 2008, from http://www.oxygenxml.com/

Papka, R. (1999). *Online new event detection, clustering and tracking.* Unpublished doctoral dissertation, Department of Computer Science, University of Massachusetts, USA.

Paschke, A., Bichler, M., & Dietrich, J. (2005). ContractLog: An approach to rule based monitoring and execution of service level agreements. In A. Adi, S. Stoutenburg, & S. Tabet (Eds.), *Proceedings for the 4th International Conference on Rules and Rule Markup Languages for the Semantic Web (RuleML 2005).* Galway, Ireland: Springer-Verlag.

Paschke, A., Dietrich, J., & Kuhla, K. (2005). A logic based SLA management framework. In *Proceedings of the 4th Semantic Web Conference (ISWC 2005).*

Peterson, J.L. (1981). *Petri nets theory and the modelling of systems.* Englewood Cliffs, NJ, New York: Prentice Hall.

Piaget, J. (1945). *Play, dreams, and imitation in childhood.* New York: Norton.

Pisanelli, D.M. (Ed.). (2004). *Ontologies in medicine.* Amsterdam: IOS Press.

Polsani, P. R. (2003). Use and abuse of reusable learning objects. *Journal of Digital Information, 3*(4).

Poshtkohi A., Abutalebi, A.H., & Hessabi, S. (2007). DotGrid: A .NET-based cross-platform software for desktop grids. *International Journal of Web and Grid Services, 3*(3), 313-332.

Poslad, S., Calisti, M., & Charlton, P. (2002). Specifying standard security mechanisms in multi-agent systems. In *Proceedings of the Workshop on Deception, Fraud and Trust in Agent Societies, AAMAS 2002,* Bologna, Italy, (pp. 122-127).

Preparata, F.P., & Yeh, R.T. (1974). *Introduction to discrete structures.* Reading, MA, USA: Addison-Wesley.

Pujol, J. M., & Sanguesa, R. (2002). Reputation measures based on social networks metrics for multi agent systems. In *Proceedings of the 4th Catalan Conference on Artificial Intelligence CCIA-01,* Barcelona, Spain, (pp. 205-213).

Purvis, M. K., Savarimuthu, S., Oliveira, M. D., & Purvis, M. A. (2006). Mechanisms for cooperative behaviour in agent institutions. In *Paper presented at the Intelligent Agent Technology (IAT 2006),* Hong Kong.

Purvis, M., Cranefield, S., et al. (2002). Opal: A multi-level infrastructure for agent-oriented software development. *Autonomous agents and multi-agent systems.* Bologna, Italy: ACM Press.

Purvis, M., Cranefield, S., Nowostawski, M., Ward, R., Carter, D., & Oliveira, M. A. (2002). Agentcities interaction using the Opal platform. In B. Burg, J. Dale, T. Finin, H. Nakashima, L. Padgham, C. Sierra, & S. Willmott (Eds.), *Proceedings of Workshop on Agentcities:*

Research in Large-scale Open Agents Environments, First International Joint Conference on Autonomous Agents and Multi-agent Systems–AAMAS 2002, Bologna, Italy. CD/ROM. New York: ACM Press. Retrieved April 3, 2008, from http://www.agentcities.org/Challenge02/Proc/Papers/ch02_56_purvis.pdf

Purvis, M., Cranefield, S., Nowostawski, M., Ward, R., Carter, D., & Oliveira, M. (2002). Agent cities interaction using the opal platform. In *Paper presented at the Proceedings of the Workshop on Challenges in Open Agent Systems, AAMAS.*

Purvis, M., Purvis, M., Haidar, A., & Savarimuthu, B.T.R. (2005). A distributed workflow system with autonomous components. In M.W.B., & N. Kasabov (Eds.), *Intelligent agents and multi-agent systems: Lecture notes in computer science* (Vol. 3371, pp. 193-205). New York-Heidelberg-London: Springer-Verlag.

Raju, C. V. L., Narahari, Y., & Ravikumar, K. (2003). Applications of reinforcement learning to dynamic pricing of retail markets. In *Proceedings of the IEEE Conference on Electronic Commerce,* (pp. 339-346). Los Alanitos, CA: IEEE Computer Society.

Raju, C. V. L., Narahari, Y., & Ravikumar, K. (2006, January). Learning dynamic prices in multi-seller electronic retail markets with price sensitive customers, stochastic demands, and inventory replenishments. *IEEE Transactions on Systems, Man, and Cybernetics, Part C: Applications and Reviews, 36*(1), 92-106.

Raju, C. V. L., Narahari, Y., & Ravikumar, K. (2006, March). Learning dynamic prices in electronic retail markets with customer segmentation. *Annals of Operations Research, 143*(1), 59-75.

Ramchourn, S. D., Huynh, D., & Jennings, N. R. (2004). Trust in multi-agent systems. *The Knowledge Engineering Review, 19*(1), 1-25.

Rasiowa, H., & Sikorski, R. (1968). *The mathematics of metamathematics.* Warsaw: PWN.

Rasmusson, L., & Jansson, S. (1996). Simulated social control for secure Internet commerce. In C. Meadows (Ed.), *Proceedings of the 1996 New Security Paradigms Workshop,* (pp. 18-26).

Rauber, A., Aschenbrenner, A., & Witvoet, O. (2002). Austrian online archive processing: Analyzing archives of the World Wide Web. In *Proceedings of the 6th European Conference on Digital Libraries,* (pp. 16-31). Rome, Italy: Springer-Verlag.

Ravikumar, K., Batra, G., & Saluja, R. (2002). Multi-agent learning for dynamic pricing games of service markets. *Communicated.*

Ray, T. S. (1991). An approach to the synthesis of life. In C. Langton, C. Taylor, J. D. Farmer, & S. Rasmussen, (Eds.), *Artificial life II: Santa Fe Institute Studies in the Sciences of Complexity* (Vol. XI, pp. 371-408). Redwood City, CA: Addison-Wesley.

Reinartz, W.J. (2001). Customizing prices in online markets. *European Business Forum, 6,* 35-41.

Research Information. (2007). *Radio-tagged books.* Retrieved April 3, 2008, from http://www.researchinformation.info/rimayjun04radiotagged.html

Resnick, P., & Zeckhauser, R. (2001). Trust among strangers in Internet transactions: Empirical analysis of eBay's reputation system. In M. R. Baye (Ed.), *Advances in Applied Microeconomics, 11,* 127-157.

RFID Gazette. (2007). *RFID applications for libraries.* Retrieved April 3, 2008, from http://www.rfidgazette.org/libraries/

Ricci, A., & Omicini, A. (2003). Supporting coordination in open computational systems with Tucson. In *Proceedings of WET ICE 2003,* (pp. 365-370).

Riolo, R. L. (1997). *The effects of tag-mediated selection of partners in evolving populations playing the iterated prisoner's dilemma* (Paper No. 97-02-016). NM: Santa Fe Institute.

Riolo, R. L., Cohen, M. D., & Axelrod, R. (2001). Cooperation without reciprocity. *Nature, 414,* 441-443.

Rizzolatti, G., & Arbib, M. A. (1998). Language within our grasp. *Trends in Neurosciences, 21*(5), 188-194.

Rizzolatti, G., Fadiga, L., Gallese, V., & Fogassi, L. (1996). Premotor cortex and the recognition of motor actions. *Cognitive Brain Research, 3*(2), 131-141.

Robinson, D. (2000). A *component based approach to agent specification.* Master's Thesis. AFIT/GCS/ENG/00M-22. School of Engineering, Air Force Institute of Technology, Wright-Patterson AFB.

Rollings, A., & Adams E. (2003). *Andrew Rollings and Ernest Adams on games design.* New Riders.

Rosen, K. (2002). *Discrete mathematics and its applications.* New York: McGraw-Hill.

Rosenschein, J.S., & Zlotkin, G. (1994). *Rules of encounter: Designing conventions for automated negotiation among computers.* Cambridge, MA: MIT Press.

Rothwell, W. J. (n.d.). *A report on workplace learner competencies.* Retrieved March 1, 2006, from http://www.ilpi.wayne.edu/files/roth_present.pdf

Rubiera, J. C., Molina Lopez, M. J., & Muro D. J. (2001). A fuzzy model of reputation in multi-agent systems. In *Proceedings of the 5th International Conference on Autonomous Agents*, Montreal, Quebec, Canada, (pp. 25-26).

Rusmevichientong, P., Van Roy, B., & Glynn, P. W. (2005). A non-parametric approach to multi-product pricing. *Operations Research, 54*(1), 82-98.

Rutkowski, L., Tadeusiewicz, R., Zadeh, L.A., & Zurada, J. (Eds.). (2006). *Artificial intelligence and soft computing–ICAISC 2006*. Berlin: Springer-Verlag.

Ryan, T. (1999). Beginning level design. Retrieved from www.gamasutra.com

Sabater, J., & Sierra, C. (2002). Reputation and social network analysis in multi-agent systems. In *Proceedings of the 1st International Joint Conference on Autonomous Agents and Multiagent Systems*, Bologna, Italy, (pp. 475-482).

Saint-Voirin, D., Lang, C., & Zerhouni, N. (2003, July 16-20). Distributed cooperation modelling for maintenance using Petri nets and multi-agents systems. In *Proceedings of the 5th IEEE International Symposium on Computational Intelligence in Robotics and Automation, CIRA'03, Kobe, Japan*, (Vol. 1, pp. 366-371). Piscataway, NJ, USA: IEEE Press.

Sakamoto, S., Ohba, S., Shibukawa, M., Kiura, Y., Arita, K., & Kurisu, K. (2006). CT perfusion imaging for childhood moyamoya disease before and after surgical revascularization. *Acta Neurochirurgica, 148*(1), 77-81.

Salimifard, K., & Wright, M. (2001). Petri net-based modelling of workflow systems: An overview. *European Journal of Operational Research, 134*(3), 664-676.

Saltzman, M. (ed.). (1999). *Games design: Secrets of the sages*. Macmillan.

Sánchez-Alonso, S., & Frosch-Wilke, D. (2005). An ontological representation of learning objects and learning designs as codified knowledge. *The Learning Organization, 12*(5), 471-479.

Sánchez-Segura, M.I., Cuadrado, J.J., de Antonio, A., de Amescua, A., & García L. (2003). Adapting traditional software processes to virtual environments development. *Software Practice and Experience, 33*(11). Retrieved from www3.interscience.wiley.com/cgi-bin/jhome/1752

Sandholm, T. (2000). Distributed rational decision making. In G. Weiss (Ed.), *Multiagent systems: A modern approach to distributed artificial intelligence*. Cambridge, MA: The MIT Press.

Savarimuthu, B. T. R., Cranefield, S., Purvis, M., & Purvis, M. (2007). *Mechanisms for norm emergence in multi-agent societies*. In Paper presented at the Sixth International Joint Conference on Autonomous Agents and Multiagent Systems (AAMAS), Honolulu, HI, USA.

Savarimuthu, B. T. R., Cranefield, S., Purvis, M., & Purvis, M. (2007). Role model based mechanism for norm emergence in artificial agent societies. In *Paper presented at the International Workshop on Coordination, Organization, Institutions and Norms (COIN) at AAMAS 2007*, Honolulu, HI, USA.

Savva, A., Suzuki, T., & Kishimoto, H. (2004). Business Grid Computing Project activities. *Fujitsu Scientific & Technical Journal, 40*(2), 252-260.

Schillo, M., Funk, P., & Rovatsos, M. (2000). Using trust for detecting deceitful agents in artificial societies. *Applied Artificial Intelligence, Special Issue on Trust, Deception and Fraud in Agent Societies, 14*(8), 825-848.

Schmidhuber, J. (2004). Optimal ordered problem solver. *Machine Learning, 54*, 211-254.

Schneider, M., & Craig, M. (1992). On the use of fuzzy sets in histogram equalization. *Fuzzy Sets and Systems, 45*, 271-278.

Schweizer, B., & Sklar, A. (1963). Associative functions and abstract semi-groups. *Publ. Math. Debrecen, 10*, 69-81.

Shafer, G. (1976). *A mathematical theory of evidence*. Princeton University Press.

Shapiro, C., & Varian, H.R. (1999). *Information rules: A strategic guide to the network economy*. Boston: HBS Press.

Shetty, S., Padala, P., & Frank, M. P. (2003). *A survey of market-based approaches to distributed computing* (Tech. Rep. No. TR03-013). University of Florida: Computer & Information Science & Engineering.

Shiji, A., & Hamada, N. (1999). Color image segmentation method using watershed algorithm and contour information. In *Proceedings of the International Conference on Image Processing, 4*, 305-309.

Shipman, F. M., & Hsieh, H. (2000). Navigable history: A reader's view of writer's time: Time-based hypermedia. *New Review of Hypermedia and Multimedia, 6*, 147-167. Taylor and Francis.

Shirky, C. (2003). *Permanet, nearlynet, and wireless*

data [Electronic Version]. Economics & Culture, Media & Community, Open Source. Retrieved April 3, 2008, from http://www.shirky.com/writings/permanet.html

Shirky, C. (2003). *Power laws, Web logs, and inequality* [Electronic Version]. Economics & Culture, Media & Community, Open Source, Version 1.1. Retrieved April 3, 2008, from http://shirky.com/writings/powerlaw_weblog.html

Shoham, Y., & Tennenholtz, M. (1995). On social laws for artificial agent societies: Off-line design. *Artificial Intelligence, 73*(1-2), 231-252.

Sicilia, M. A. (2005). Ontology-based competency management: Infrastructures for the knowledge-intensive learning organization. In M. D. Lytras & A. Naeve (Eds.), *Intelligent learning infrastructures in knowledge intensive organizations: A semantic Web perspective* (pp. 302-324). Hershey, PA: Information Science Publishing.

Sicilia, M. A., Lytras, M., Rodríguez, E., & García, E. (2006). *Integrating descriptions of knowledge management learning activities into large ontological structures: A case study.* Data and Knowledge Engineering.

Sicilia, M.A., García, E., Sánchez-Alonso, S., & Rodríguez, E. (2004). Describing learning object types in ontological structures: Towards specialized pedagogical selection. In *Proceedings of ED-MEDIA 2004: World Conference on Educational Multimedia, Hypermedia and Telecommunications*, 2093-2097.

Singh, M. P. (1999). An ontology for commitments in multiagent systems. *Artificial Intelligence and Law, 7*(1), 97-113.

Slater, M. (1999). Co-presence as an amplifier of emotion. *Proceedings of the Second International Workshop on Presence,* University of Essex, UK. Retrieved from www.essex.ac.uk/psychology/tapestries/

Slembeck, T. (1999). *Reputations and fairness in bargaining--experimental evidence from a repeated ultimatum game with fixed opponents* (Experimental). EconWPA.

Smith, M., Bailey, J., & Brynjolfsson, E. (2000). Understanding digital markets: Review and assessment. In E. Brynjolfsson & B. Kahin (Eds.), *Understanding the digital economy.* Cambridge, MA: MIT Press.

Smith, R. J. (1997). *Integrated spatial and feature image systems: Retrieval, analysis and compression.* Doctoral thesis, Columbia University, Graduate School of Arts and Sciences.

Smith, R., & Davis, R. (1980). The contract net protocol: High level communication and control in a distributed problem solver. *IEEE Transactions on Computers, C-29*(12), 1104-1113.

Smith, R.G. (1980). The contract net protocol: High-level communication and control in a distributed problem solver. *IEEE Transactions on Computers, C-29*(12), 1104-1113.

Sparkman, C.H., DeLoach, S.A., & Self, A.L. (2001). Automated derivation of complex agent architectures from analysis specifications. In *Proceedings of the Second International Workshop on Agent-Oriented Software Engineering, (AOSE 2001), Springer Lecture Notes in Computer Science* (Vol. 2222, pp. 188-205).

Spinney, L. (1998). I had a hunch…. *New Scientist,* (September 5).

Staab, S., Werthner, H., Ricci, F., Zipf, A., Gretzel, U., Fesenmaier, D.R., et al. (2002). Intelligent systems for tourism. *IEEE Intelligent Systems, 6*(17), 53-64.

Sterling, L., & Shapiro, E. (1994). *The art of Prolog* (2nd ed.). The MIT Press.

Sunassee, N., & Sewry, D. (2002). A theoretical framework for knowledge management implementation. In *Proceedings of the 2002 Annual Research Conference of the South African Institute of Enablement Through Technology* (pp. 235-245).

Suri, J.S., Setarehdan, S.K., & Singh, S. (2002). *Advanced algorithmic approaches to medical image segmentation.* London: Springer-Verlag.

Swan, R., & Allan, J. (2000). Automatic generation of overview timelines. In *Proceedings of the 23rd Conference on Research and Development in Information Retrieval,* (pp. 49-56). Athens, Greece: ACM Press.

Swann, J. (1999). *Flexible pricing policies: Introduction and a survey of implementation in various industries* (Contract Rep. #CR-99/04/ESL). General Motors Corporation.

Sycara, K. P. (1998). Multiagent systems. *Artificial Intelligence Magazine, 19*(2), 79-92.

Tabakov, M. (2001). Using fuzzy set theory in medical image processing: Basic notions and definitions. *Reports of the Department of Computer Science of Wroclaw University of Technology PRE* (No.7, p. 35). Warsaw, Poland.

Tabakov, M. (2003). *Medical image segmentation algorithms using a fuzzy equivalence relation* (p. 80). Doctoral thesis, Wroclaw University of Technology,

Wroclaw, Poland.

Tabakov, M. (2006). A fuzzy clustering technique for medical image segmentation. In *Proceedings of the 2006 IEEE International Symposium on Evolving Fuzzy Systems,* Ambleside, Lake District, UK, (pp.118-122).

Tabakov, M. (2007). A fuzzy segmentation method for Computed Tomography images. *International Journal of Intelligent Information and Database Systems, 1*(1), 79-89.

Tekşam, M., Çakır, B., & Coşkun, M. (2005). CT perfusion imaging in the early diagnosis of acute stroke. *Diagnostic and Interventional Radiology, 11*(4), 202-205.

The Foundation for Intelligent Physical Agents (FIPA). (2007). *Agent communication language specification.* Retrieved April 3, 2008, from http://www.fipa.org

Thomas, B. (2000). Token-templates and logic programs for intelligent Web search. *Intelligent Information Systems, Special Issue: Methodologies for Intelligent Information Systems, 14*(2/3), 241-261.

Thompson, L. (1998). *The mind and heart of the negotiator.* Upper Saddle River, NJ: Prentice Hall.

Tidd, J., Bessant, J., & Pavitt, K. (2005). *Managing innovation–integrating technological, market and organizational change.* Chichester, West Sussex: John Wiley & Sons.

Tizhoosh, H.R. (1998). Fuzzy image processing: Potentials and state of the art. In *Proceedings of IIZUKA'98, the 5th International Conference on Soft Computing,* Iizuka, Japan, (Vol. 1, pp. 321-324).

Toyoda, M., & Kitsuregawa, M. (2003). Extracting evolution of Web communities from a series of Web archives. In *Proceedings the 14th Conference on Hypertext and Hypermedia,* (pp. 28-37). Nottingham, UK: ACM Press.

Trajkovski, G. (2004). Fuzzy sets in investigation of human cognition processes. In A. Abraham, L. Jain, & B. Van der Zwaag (Eds.), *Innovations in intelligent systems* (pp. 361-380). Springer-Verlag.

Trajkovski, G. (2005). E-POPSICLE: An online environment for studying context learning in human and artificial agents. In *Proceedings of the 16th Midwest AI and Cognitive Science Conference (MAICS 2005),* (pp. 61-66).

Trajkovski, G. (2007). *An imitation-based approach to modeling homogenous agents societies.* Hershey, PA: Idea Group.

Trajkovski, G., Stojanov, G., Bozinovski, S., Bozinovska, L., & Janeva, B. (1997). Fuzzy sets and neural networks in CNV detection. In *Proceedings of the ITI'97,* Pula, Croatia, (pp. 153-158).

Trondstad, R. (2001). Semiotic and nonsemiotic MUD performance. *Proceedings of COSIGN 2001,* CWI, Amsterdam.

Tuomela, R. (1995). *The importance of us: A philosophical study of basic social notions.* Stanford, CA: Stanford Series in Philosophy, Stanford University Press.

Turing, A. M. (1936-1937). On computable numbers with an application to the entscheidungsproblem. *Proceedings of the London Mathematical Society, 42*(2), 230-265, also *43,* 544-546.

Turkle, S. (1995). *Life on the screen: Identity in the age of the Internet.* Phoenix.

Tutte, W.T. (1984). *Graph theory.* Menlo Park, CA: Addison-Wesley.

Tzafestas, S.G., & Čapkovič, F. (1997). Petri net-based approach to synthesis of intelligent control systems for DEDS. In S.G. Tzafestas (Ed.), *Computer assisted management and control of manufacturing systems* (pp. 325-351). London, UK: Springer-Verlag.

UML Version 1.1 Summary. Retrieved from www.rational.com/uml/resources/documentation/summary/

van der Aalst, W.M.P. (1998). The application of Petri nets to workflow management. *The Journal of Circuits, Systems and Computers, 8*(1), 21-66.

van der Aalst, W.M.P. (2000). Workflow verification: Finding control-flow errors using Petri net-based techniques. In W. van der Aalst, J. Desel, & A. Oberweis (Eds.), *Business process management: Models, techniques, and empirical studies, Lecture notes in computer sciences* (Vol. 1806, pp. 161-183). New York-Heidelberg-London: Springer-Verlag.

van Der Aalst, W.M.P., ter Hofstede, A.H.M., Kiepuszewski, B., & Barros, A.P. (2000). *Workflow patterns, BETA working paper series, WP 47,* Eindhoven University of Technology, Eindhoven, The Netherlands.

van Der Aalst, W.M.P., ter Hofstede, A.H.M., Kiepuszewski, B., & Barros, A.P. (2003). Workflow patterns. *Distributed and Parallel Databases, 14*(1), 5-51.

van Lamsweerde, A. & Letier, E. (2000). Handling obstacles in goal-oriented requirements engineering. *IEEE Transactions on Software Engineering, 26*(10), 978-1005.

Varian, H. R. (1996). Differential pricing and efficiency. *First Monday: Peer-Reviewed Journal on the Internet*. Retrieved April 3, 2008, from http://www.firstmonday.org/issues/issue2/different

Verhagen, H. (2000). *Norm autonomous agents.* Department of Computer Science, Stockholm University.

Verhagen, H. (2001). Simulation of the learning of norms. *Social Science Computer Review, 19*(3), 296-306.

Vickrey, W. (1961). Counter-speculation, auctions, and competitive sealed tenders. *The Journal of Finance, 16*(1), 8-37.

Viégas, F., Wattenberg, M., & Dave, K. (2004). Studying cooperation and conflict between authors with history flow visualizations. In *Proceedings of the CHI Conference*, (pp. 575-582). Vienna, Austria: ACM Press.

Vincenti, G., & Trajkovski, G. (2006, October 12-15). Fuzzy mediation for online learning in autonomous agents. In *Proceedings of the 2006 Fall AAAI Symposium*, Arlington, VA, USA, (pp. 127-133).

W3C Extensible Markup Language (XML). (2007). Retrieved April 2, 2008, from http://www.w3.org/XML/

W3C Extensible Stylesheet Language Family (XSL). (2007). Retrieved April 2, 2008, from http://www.w3.org/Style/XSL/

W3C HTML. (2007). Retrieved April 2, 2008, from http://www.w3.org/html/

W3C Semantic Web Activity. (2007). Retrieved April 2, 2008, from http://www.w3.org/2001/sw/

Walker, A., & Wooldridge, M. (1995, June 12-14). Understanding the emergence of conventions in multi-agent systems. In V. R. Lesser & L. Gasser (Eds.), *Proceedings of the First International Conference on Multi-agent Systems,* San Francisco, (pp. 384-389). Boston: The MIT Press.

Wallace, J. (2000, March 20). Unlike Airbus, Boeing lets aviator override fly-by-wire technology. *Seattle Post-Intelligencer.*

Wang, L., & Mendel, J. (1992). Generating fuzzy rules by learning from examples. *IEEE Transations on Systems, Man and Cybernetics, 22*(6), 1414-1427.

Wang, X., & McCallum, A. (2006). Topics over time: A non-Markov continuous-time model of topical trends. In *Proceedings of the 12th International Conference on Knowledge Discovery and Data Mining,* (pp. 424-433), Philadelphia, PA, USA: ACM Press.

Wang, Y., & Vassileva, J. (2003). Bayesian network-based trust model. In *Proceedings of IEEE International Conference on Web Intelligence,* Hallifax, Canada.

Wang, Y., Hori, Y., & Sakurai, K. (2006). On securing open networks through trust and reputation–architecture, challenges and solutions. In *Proceedings of the 1st Joint Workshop on Information Security,* Seoul, Korea.

Web Services Activity. (2007). Retrieved April 2, 2008, from http://www.w3.org/2002/ws/

Weigand, H., & Dignum, V. (2003). I am autonomous, you are autonomous. In Nickles et al. (Eds.), (pp. 227-236).

Weigand, H., Dignum, M. V., Meyer, J.J., & Dignum, F. (2003, September 16-17). Specification by refinement and agreement: Designing agent interaction using landmarks and contracts. In P. Petta, R. Tolksdorf, & F. Zambonelli (Eds.), *Engineering Societies in the Agents World III: Third International Workshop, ESAW 2002: Lecture Notes in Computer Science,* Madrid, Spain, (Vol. 2577, pp. 257-269). Berlin-Heidelberg: Springer-Verlag.

Whitby, A., Josang, A., & Indulska, J. (2004). Filtering out unfair ratings in bayesian reputation systems. In *Proceedings of the AAMAS 2004*, New York.

White, S.A. (2004, March). *Process modelling notations and workflow patterns.* BP Trends (pp. 1-24). IBM Corporation. Retrieved April 3, 2008, from http://www.omg.org/bp-corner/bp-files/Process_Modeling_Notations.pdf

Whitelock, D., Brna, P., & Holland, S. (1996). *What is the value of virtual reality for conceptual learning? Towards a theoretical framework.* Retrieved from www.cbl.leeds.ac.uk/~paul/papers/vrpaper96/VRpaper.html

Wiederhold, G. (1992). Mediators in the architecture of future information systems. *IEEE Transactions on Computers, 25*(3), 38-49.

Witkowski, M., & Stathis, K. (2003). A dialectic architecture for computational autonomy. In Nickles et al. (Eds.), (pp. 261-274).

Wolfstetter, E. (1996). Auctions: An introduction. *Journal of Economic Surveys, 10*(4), 367-420.

Wong, H. C., & Sycara, K. (1999). Adding security and trust to multi-agent systems. In *Proceedings of Autonomous Agents '99 Workshop on Deception, Fraud, and Trust in Agent Societies,* (pp. 149-161).

Wong, H.-C., & Sycara, K.P. (2000). A taxonomy of middle-agents for the Internet. In *Proceedings of the 4th International Conference on Multi-Agent Systems,* (pp. 465-466).

Wooldridge, M. (2004). *An introduction to multiagent systems.* Chichester, West Sussex: John Wiley & Sons.

Wooldridge, M., Jennings, N., & Kinny, D. (2000). The Gaia methodology for agent-oriented analysis and design. *Autonomous Agents and Multi-agent Systems, 3*(3), 285-312.

WordNET Search. (n.a.). *Control.* Retrieved April 3, 2008, from http://wordnet.princeton.edu/perl/webwn?s=control

Workshop on Structured Design of Virtual Environments. (2001). *Proceedings of Web3D Conference*, Paderborn, Germany. Retrieved from www.c-lab.de/web3d/VE-Workshop/index.html

Workshop on Usability Evaluation for Virtual Environments. (1998). De Montforte University. Retrieved from www.crg.cs.nott.ac.uk/research/technologies/evaluation/workshop/workshop.html

Xiao, L., Wissmann, D., Brown, M., & Jablonski, S. (2001). Information extraction from HTML: Combining XML and standard techniques for IE from the Web. In L. Monostori, J. Vancza, & M. Ali (Eds.), *Engineering of intelligent systems: Proceedings of the 14th International Conference on Industrial and Engineering Applications of Artificial Intelligence and Expert Systems, IEA/AIE 2001,* (pp. 165-174). Lecture Notes in Artificial Intelligence 2070, Springer-Verlag.

Yaiche, H., Mazumdar, R. R., & Rosenberg, C. (2000). A game theoretic framework for bandwidth allocation and pricing in broadband networks. *IEEE/ACM Transactions on Networking, 8*(5), 667-678.

Yamamoto, L. (2004). Automated negotiation for on-demand inter-domain performance monitoring. In *Proceedings of the 2nd International Workshop on Inter-domain Performance and Simulation,* (pp. 159-169).

Yamamoto, Y., Tezuka, T., Jatowt, A., & Tanaka, K. (2007). Honto? Search: Estimating trustworthiness of Web information by search results aggregation and temporal analysis. In *Proceedings of the APWeb/WAIM 2007 Conference,* (pp. 253-264). Hunagshan, China: Springer-Verlag.

Yen, J., Yin, J., Ioerger, T. R., Miller, M. S., Xu, E., & Volz, R. A. (2001). CAST: Collaborative agents for simulating teamwork. In B. Nebel (Ed.), *Proceedings of 17th International Joint Conference on Artificial Intelligence—IJCAI' 2001,*Seattle, WA, USA, (Vol. 2, pp. 1135-1142). San Francisco: Morgan Kaufmann.

Yu, B., & Singh, P. M. (2002). An evidential model of distributed reputation management. In *Proceedings of the 1st International Joint Conference on Autonomous Agents and Multiagent Systems*, Bologna, Italy, (pp. 294-301).

Yu, B., & Singh, P. M. (2003). Detecting deception in reputation management. In *Proceedings of the 2nd International Joint Conference on Autonomous Agents and Multiagent Systems,* Melbourne, Australia, (pp. 73-80).

Zadeh, L. (1965). *Fuzzy sets, information and control* (No. 8, pp. 338-353).

Zadeh, L. (1965). Fuzzy sets. *Information and Control, 8*, 338-353.

Zadeh, L. A. (1989). Knowledge representation in fuzzy logic. *IEEE Transactions on Knowledge and Data Engineering, 1*(1), 89-100.

Zadeh, L.A. (1975). The concept of a linguistic variable and its application to approximate reasoning. Part I. *Information Science, 8*, 199-249.

Zadeh, L.A. (1975). The concept of a linguistic variable and its application to approximate reasoning. Part II. *Information Science, 8*, 301-357.

Zadeh, L.A. (1975). The concept of a linguistic variable and its application to approximate reasoning. Part III. *Information Science, 9*, 43-80.

Zambonelli, F., Jennings, N., & Wooldridge, M. (2001). Organisational abstractions for the analysis and design of multi-agent systems. *Agent-Oriented Software Engineering, LNCS*, 98-114.

Zambonelli, F., Jennings, N., & Wooldridge, M. (2003). Developing multiagent systems: The Gaia methodology. *ACM Transactions on Software Engineering and Methodology (TOSEM), 12*(3), 317-370.

Zhong, D., & Yan, H. (2000). Color image segmentation using color space analysis and fuzzy clustering. In *Proceedings of the 2000 IEEE Signal Processing Society Workshop*, (Vol. 2, pp. 624-633).

About the Contributors

Dariusz Król is an assistant professor at the Institute of Applied Informatics, Wroclaw University of Technology. He received MSc in computer science in 1990 and PhD in computer science in 2001, both from the Wroclaw University of Technology. His major research interests are open distributed systems, semantic collaboration and communication and intelligent evaluation. His teaching includes database systems, Java and Internet technologies. He is an author of over 60 scientific publications (including journal and conference papers) and a co-author of two books in C programming and business information systems. In 2005, he won the IBM Award for the work on Eclipse Didactic Distribution.

Ngoc Thanh Nguyen currently works as a professor of computer science in Wroclaw University of Technology, Poland. His scientific interests consist of knowledge integration methods, intelligent technologies for conflict resolution, multi-agent systems and e-learning methods. He has edited six special issues in international journals and two books. He is the author of four monographs, editor of seven volumes and author of about 150 other publications. He serves as Editor-in-Chief of *International Journal of Intelligent Information and Database Systems*, editor-in-chief of two book series for IGI Global (*Advances in Applied Intelligence Technologies* and *Computational Intelligence and its Applications*), associate editor of *International Journal of Computer Science & Applications*; *Journal of Information Knowledge System Management* and *KES Journal* and a member of editorial boards of several other international journals. He is the chair of KES symposium series on agent and multi-agent systems. He is a senior member of IEEE and ACM.

* * *

Amelia Badica currently works as an associate professor at the Department of Economic Informatics at the University of Craiova, Romania. She has a specialization in management information systems at Binghamton University, USA, and she holds a PhD in economics. Amelia Badica has authored and co-authored more than 50 articles in journals, conference proceedings, and book chapters, one monograph and five textbooks. The publications in the last 2 years are related to e-business and e-commerce, Internet-based applications and software engineering. Amelia Badica also participated in national and international research projects and as a PC member of several international conferences. She is currently involved as principal investigator in a research project concerning data extraction from the Web.

Costin Badica is professor in the Department of Software Engineering at the University of Craiova, Romania, and scientific secretary of the Faculty of Automation, Computers and Electronics, University of Craiova. He has authored and co-authored more than 80 articles in journals, conference proceedings, and book chapters, one monograph and five textbooks. The publications in the last 2 years are related to multi-agent systems, e-business and e-commerce and the Web. Costin Badica is a member of the editorial board of four international journals, he served as program committee member for more than 25 international conferences, he co-organized two international conferences and three international workshops in the area of multi-agent, intelligent and distributed systems, and he is also guest editor for two special issues of internationally recognized journals. Costin Badica has also experience in project management, as he was director and participant in six national and seven international research projects.

František Čapkovič received his master's degree in 1972 from the Faculty of Electrical Engineering of the Slovak Technical University, Bratislava, Slovakia. Since 1972, he has been working with the Slovak Academy of Sciences (SAS), Bratislava, namely, in 1972-1991 at the Institute of Technical Cybernetics, in 1991-2001 at the Institute of Control Theory and Robotics and in 2001 to the present at the Institute of Informatics. In 1980, he received the PhD from SAS. Since 1998, he has been associate professor. He works in the area of modeling, analysing and control of discrete-event systems (DES). Since 1991, he has been a head of six national projects and a head of the Slovak participation in 11 international projects. He is the author of more than 150 publications in journals, book chapters and conference proceedings.

Stephen Cranefield, PhD, is an associate professor of information science at the University of Otago, Dunedin, New Zealand. His research interests include distributed information systems, multi-agent systems and the Semantic Web.

Paul Davidsson received his PhD in computer science in 1996 from Lund University, Sweden. His research interests include the theory and application of multi-agent systems, autonomous agents, and machine learning. The results of this work have been reported in more than 90 peer-reviewed scientific articles published in international journals, conference proceedings, and books. He is a member of the editorial boards of two international journals and regular reviewer for more than 10 different scientific journals. Davidsson is the founder and manager of the Distributed and Intelligent Systems Laboratory, which currently consists of six senior researchers and eight PhD students.

Toktam Ebadi, BE, is a PhD student at the University of Otago, Dunedin, New Zealand. Her research interests are in the areas of distributed computing, collaborative decision-making and multi-agent robotic systems.

Stathes Hadjiefthymiades received his BSc, MSc, and PhD in informatics from the Department of Informatics and Telecommunications at the University of Athens (UoA). He also received a Joint Engineering-Economics MSc from the National Technical University of Athens. In 1992, he joined the Greek consulting firm Advanced Services Group. In 1995, he joined the Communication Networks Laboratory (CNL) of UoA. During the period of 2001-2002, he served as a visiting assistant professor at the University of Aegean in the Department of Information and Communication Systems Engineering.

In the summer of 2002, he joined the faculty of the Hellenic Open University, Patras, Greece, as an assistant professor. Since December 2003, he has been in the faculty of the Department of Informatics and Telecommunications, University of Athens, where he is presently an assistant professor. He has participated in numerous EU and National projects. His research interests are in the areas of Web engineering, mobile/pervasive computing and networked multimedia. He has contributed to over 100 publications in these areas. Since 2004, he has coordinated the Pervasive Computing Research Group of CNL.

Jin Huang is a PhD candidate of computer software and theory at the Huazhong University of Science and Technology (HUST) in China. His research interests include performance evaluation, modeling and simulation, and economic grid.

Andreas Jacobsson works as a teacher and a doctoral candidate in computer science at the Blekinge Institute of Technology in Sweden. He will defend his thesis on security and virtual enterprises in the beginning of 2008. The results of this work have been published in more than 15 peer-reviewed scientific articles published in international journals and conference proceedings. Moreover, Jacobsson is a member of the research group Distributed and Intelligent Systems Laboratory.

Adam Jatowt received the M.S. in electronics and telecommunications from the Technical University of Lodz, Poland, in 2001. In 2005, he received the PhD in information science and technology from the University of Tokyo, Japan. He worked as a research fellow at the National Institute of Information and Communications Technology in Japan during 2005. Since 2006, he has been an assistant professor at Kyoto University. He is a member of ACM and Web Intelligence Consortium (WIC). His research interests include Web mining, Web information retrieval and analyzing Web history.

Hai Jin is a professor of Computer Science and Engineering at the Huazhong University of Science and Technology (HUST) in China. He is now the Dean of School of Computer Science and Technology at HUST. He received his PhD in computer engineering from HUST in 1994. In 1996, he was awarded the German Academic Exchange Service (DAAD) fellowship for visiting the Technical University of Chemnitz in Germany. He worked for the University of Hong Kong between 1998 and 2000 and participated in the HKU Cluster project. He worked as a visiting scholar at the University of Southern California between 1999 and 2000. He is the chief scientist of the largest grid computing project, ChinaGrid, in China. His research interests include computer architecture, cluster computing and grid computing, virtualization technology, peer-to-peer computing, network storage, and network security.

Yukiko Kawai received the BS in information science and technology from Kyushu Institute of Technology in 1998. She received the MS and PhD in information science and technology from Nara Institute of Science and Technology, in 1999 and 2001, respectively. She worked as a research fellow at the National Institute of Information and Communications Technology from 2001 to 2006. Since 2006, she has been a lecturer at Kyoto Sangyo University. Her research interests include data mining, information analyzing and Web information retrieval.

Melvin Koh is currently working as a Solution Architect in Sun Microsystems Global System Practise HPC team and Asia Pacific Science & Technology Center. He received his degree in computer science from the National University of Singapore in 2002. During his work in Sun, Melvin has worked on numerous Grid projects in many different regions. He is also actively involved in many Grid research projects and has collaborated with many research institutes and universities worldwide. Melvin is heading the Grid research projects in APSTC and has a number of publications on grid computing under his name.

Kostas Kolomvatsos received his BSc in informatics from the Department of Informatics at the Athens University of Economics and Business in 1995, and his MSc in computer science-new technologies in informatics and telecommunications from the Department of Informatics and Telecommunications at the National and Kapodistrian University of Athens (UoA) in 2005. He is now a PhD candidate in the National and Kapodistrian University of Athens, in the Department of Informatics and Telecommunications, under the supervision of assistant professor Stathes Hadjiefthymiades. His research interests are in the areas of Semantic Web technologies, ontological engineering, agent technologies and pervasive computing.

Juliusz L. Kulikowski received his MSc in electronic engineering from the Warsaw Technical University in 1955, C.and Sc. from the Moscow Higher School of Technology in 1959, and DSc from the Warsaw Technical University in 1966. Since 1966, he was a scientific worker in several Institutes of the Polish Academy of Sciences, and since 1981 in the Institute of Biocybernetics and Biomedical Engineering PAS in Warsaw. He was nominated a professor in 1973, and for 25 years he headed the Department of Biomedical Information Processing Methods of BaBE PAS. He is the author of more than 200 papers in information sciences, image processing methods, artificial intelligence, and application of computers in medicine, and of eight books and monographs in these domains. He is also the editor in chief of a scientific quarterly "*Computer Graphics & Vision*." For many years, he has been a member of IFIP TC13 on "Human-Computer Interaction," of IFAC TC on "Stochastic Systems," a chairman of the Polish National Committee of Data for Science and Technology CODATA, a vice-chairman of the Board of the Polish Image Processing Association and a full member of the Scientific Society of Warsaw.

Mariusz Nowostawski received degrees in information systems and artificial intelligence from Wroclaw University of Technology, Poland, and the University of Birmingham, UK. His interests lie in computer architectures, artificial intelligence, soft-computing methods and philosophy. Combining both computer architectures and soft computing methods, in 2007 he received his PhD on evolvable virtual machines from the University of Otago, New Zealand. Since 1998, he has worked in the Information Science Department of University of Otago, working on smart stochastic heuristics, adaptable computing, reprogrammable computing architectures, massive hardware-based parallelism and massively parallel non-uniform computing architectures. He is also researching techniques that facilitate high-performance computing on emerging multicore hardware architectures, and works part time as an independent consultant for World45 and Sun Microsystems, USA.

Marcos de Oliveira is a PhD candidate in information science at the University of Otago, Dunedin, New Zealand. His research interest is in the area of multi-agent systems. He is conducting research in representation of communication protocols by Petri Nets and Institution of Agents.

D. Orski received the MSc in 1995 and the PhD in 2000, and has been an assistant professor in the Institute of Information Science and Engineering at Wroclaw University of Technology since 2000 and deputy director of the institute. In 2004, he completed a short stay in the Artificial Life and Robotics

Lab on postdoctoral fellowship from Oita University, Japan. His research areas include knowledge-based control and decision systems, intelligent and expert systems, uncertain systems, and in particular, uncertain operation systems. He teaches expert systems, knowledge engineering, artificial intelligence, data mining, neural networks, and computer control systems.

Elvira Popescu is a PhD student at the University of Craiova, Romania, and University of Technology of Compiègne, France, and teaching assistant at the Software Engineering Department at the University of Craiova. She has authored and co-authored more than 20 articles in journals and conference proceedings. Her research interests include data extraction from Web sources, adaptive educational hypermedia, and intelligent and distributed computing. Elvira Popescu participated in eight national and international research projects; she was a reviewer for several international conferences and was a member of the local organizing committee for two international conferences.

Martin Purvis is a professor of information science at the University of Otago, Dunedin, New Zealand. He is the director of Software Engineering and Telecommunications programs. His research interests are in the areas of distributed information systems, agent-based computing, wireless computing, group learning and social networking.

Maryam Purvis, PhD, is a lecturer in information science at the University of Otago, Dunedin, New Zealand. Purvis obtained her MA in mathematics from the University of Texas at Austin and her PhD in information science from the University of Otago, Dunedin. Her research interests are in active learning, collaborative learning, distributed information systems and software engineering.

Bastin Tony Roy Savarimuthu, ME, is a lecturer in information science at the University of Otago, Dunedin, New Zealand. His research interests are in the areas of distributed computing, agent-based workflow systems, emergence of norms in agent societies and active learning.

Sharmila Savarimuthu is a PhD candidate in information science at the University of Otago, Dunedin, New Zealand. Her research interest is in the area of multi-agent systems. She is doing research in achieving cooperation in agent societies.

Simon See is the director for Advance Computing Solution System Practice and Global Science & Technology Network, and also an adjunct associate professor for National University of Singapore and Nanyang Technological University. Simon is also the director for the Sun Asia Pacific Science and Technology Center. His research interest is in the area of high performance computing, computational science, applied mathematics and simulation methodology. He graduated from University of Salford (UK) with a PhD in electrical engineering and numerical analysis in 1993. Prior to joining Sun, Simon worked for SGI, DSO National Lab of Singapore, IBM and International Simulation Ltd (UK).

Jie Song is a research scientist in Asia Pacific Science & Technology Center, Sun Microsystems. Her research interest is in the area of Grid computing, Internet pricing, quality of service and network communication. She received the BA Sc and MA Sc in computer engineering from Xi'an Jiaotong University, P.R. China, in 1995 and 1998, respectively. In 2004, she obtained the PhD from the School of Computer Engineering, Nanyang Technological University, Singapore.

Martin Tabakov is an assistant professor of information technologies in the Institute of Applied Informatics, Wroclaw University of Technology. He earned his PhD in computer science (medical image processing and segmentation algorithms research) from the Wroclaw University of Technology. He teaches undergraduate courses in digital image processing, computer graphics, discrete mathematics and programming languages. His research areas of interest include fuzzy techniques in image processing, image segmentation algorithms and fuzzy systems.

Katsumi Tanaka received the BS, MS, and PhD in information science from Kyoto University, in 1974, 1976 and 1981, respectively. In 1986, he joined the Department of Instrumentation Engineering, faculty of Engineering at Kobe University, as an associate professor. In 1994, he became a full professor at the Department of Computer and Systems Engineering Department, Faculty of Engineering, Kobe University. Since 2001, he has been a full professor of the Department of Social Informatics, Graduate School of Informatics at Kyoto University. His research interests include database theory and systems, Web information retrieval, and multimedia content retrieval. Tanaka is a member of the ACM, IEEE, the Database Society of Japan (DBSJ) and the Information Processing Society of Japan (IPSJ). He is currently a vice president of DBSJ and the fellow of IPSJ.

Goran Trajkovski is the chair of information technologies (IT) programs at South University and associate professor of IT on its Savannah, Georgia, campus. His research interest lies in the area of emergent phenomena in multiagent systems. He is the author of over 200 publications, including 10 books. He reviews for CHOICE and ACM Reviews.

Giovanni Vincenti is in charge of research and development at Gruppo Vincenti, a family-owned company with interests across several fields. His main areas of research include fuzzy mediation, information fusion, emotionally-aware agent frameworks and robotics. He held several positions at Towson University, including a lecturership with the Department of Computer and Information Sciences. He also taught courses for the Center of Applied Information Technology, also at Towson University. He is the author of many publications, and the father of the concept of Fuzzy Mediation, as applied to the field of Information Fusion.

Song Wu is an associate professor of computer science and engineering at the Huazhong University of Science and Technology (HUST) in China. He received his PhD from HUST in 2003. He is now the Vice Head of Computer Engineering Department at HUST. He also served as the vice sirector of Service Computing Technology and System Lab (SCTS) and Cluster and Grid Computing Lab (CGCL) of HUST now. He has worked for ChinaGrid project for almost four years and takes charge of the development of ChinaGrid Support Platform (CGSP). In the China Nation Grid (CNGrid) project, he is responsible for the construction of the Grid node in Central China. In 2007, he was awarded the New Century Excellent Talents in University (NCET). His current research interests include Grid computing and virtualization technology.

Xia Xie is a teacher of computer software and theory at the Huazhong University of Science and Technology (HUST) in China. She received her masters in computer software from HUST in 1998. She received her PhD in computer engineering from HUST in 2006. Her research interests include performance evaluation, modeling and simulation, cluster computing, and economic grid.

Index

A

adaptive resource distribution system 76
agent-based library management system 171–181
agent classes 211, 216
agent roles 126
agents behaviour 226–252
agent structure 239
agent trust in multi-agent environments 132–153
application negotiation scenario example 247
automated construction of logic wrappers 31
autonomy, definition 155
autonomy and EVM interactions 166
autonomy in distributed computation 154–170
autonomy in multi-agent systems 154–170

B

Bayesian network trust models 146
Bayesian reputation systems 139
belief theory models 141
book scenario 176
browsing 298
browsing past Web 298
business models 322
business processing environment (BPE) 105

C

C-uncertain variables, application 80
cascade-parallel structure algorithms 73
cascade operations algorithms 68
cognitive processes in humans, fuzzy sets 267
collective feedback mechanism, for norm emergence 199
coloured petri nets (CPN) 120
coloured petri nets, a formalization tool 122
colour image enhancement 253–262
communication between agents 177
competency management 112
computational autonomy in MAS 157
computed tomography perfusion images 253–262
conceptualizing Web pages for data extraction 24
concurrent task model 210
conversation space 126
cooperation using tags 185

D

data acquisition and preparation 290
data extraction from Web pages 17–47
deployment design 213
design phase 211
discovering object histories 297
discrete-event systems (DES) xi, 226
discrete event dynamic systems (DEDS) 228
distributed environments 305–315
distributed organizational knowledge base (DOKB) 106
dynamic pricing 326

E

entropy-based trust model 145
EVM search process 163
evolvable virtual machines (EVM) 160
external autonomy 156
extraction paths 32
extraction paths to XPath 35

F

first degree price differentiation 326
flat relational conceptualization 24
flexible manufacturing systems (FMS) 229
fuzzy mediation 269
fuzzy mediation in online learning 263–285
fuzzy mediation in robotics 279
fuzzy models 141
fuzzy ontological models 12
fuzzy ontologies 1–16
fuzzy paradigm 267
fuzzy segmentation 253–262

G

goal hierarchy diagram 209
grid commercialization overview 318
grid computing 316–334
grid from the industry 318
grid market building blocks 321

H

hierarchical conceptualization 27
history reconstruction error 292
HR-XML 108
hue-saturation-value (HSV) colour space 255
hyper-graphs 10
hyper-relations 11

I

image enhancement technique 258
IMS-RDCEO 108, 109
IMS consortium 109
indeterminacy as autonomy 159
InformationBearingThing (or IBT) 114
information model 109
information resource-sharing 55
institutional abstraction and design 122
institutional acts 124
institutional environments 124
interaction machine 89
interactive digital environments (IDEs) 86
interactive digital systems (IDSs) 86
interactive system 85
internal autonomy 156
issues concerning trust 143

K

KEA project 160
KLC model 111
KMCI viii, 104
KMCI lifecycle model viii, 104
KM framework viii, 104
knowledge-based resource distribution 63
knowledge-based resource distribution, quality 75
knowledge life cycle (KLC) viii, 104
knowledge management (KM) viii, 104
knowledge management lifecycle 112
knowledge production (KP) 105
knowledge representation problems 65

L

L-wrappers in XSLT, examples 39
learning by imitation 266
life cycle 108
logic wrappers as directed graphs 29

M

mapping L-wrappers to XSLT 37
moral norms (m-norms) 196
multi-agent environment description 198
multi-agent research tool (MART) 213
multi-agent society, cooperative behaviour 183
multi-agent system KEA 160
multi-agent systems engineering 207–225
multi-agent systems engineering (MaSE) 207
multi-agent systems performance 182–194
multi-page temporal summarization 297

multilateral collaboration 55
multinode framework elements 307

N

new training solution 267
non-functional goal 209
nondeterministic ontologies 12
normative multi-agent systems 197
norm emergence in multi-agent societies 195–206
norms, types 196

O

ontological models 2
OpenCyc 115
OpenCyc knowledge base 105
openess considerations 120
openness in multi-agent systems 119–131

P

P2P file sharing 185
page temporal analysis 293
partitioned goal 209
PeopleSoft GmbH 110
perceptual opportunities (POs) 94
persistent turing machines (PTMs) 89
place/transition Petri nets (P/T PN) xi, 226
PN-based modeling agents 234
power laws in network behaviour 184
pricing models 322
probability-based trust model 145
process model 88
prudential norms (p-norms) 196

R

reachability of states 230
reconstruction of page histories 291
referrals 186
reputation-based trust models 146
reputation in MAS 134
resource distribution for parallel and cascade operations 67
resource distribution problems 65
RFID, use issues 179
RFID infrastructure 176
RFID technology 171–181
robotic agent 176
role-based reputation 142
role-based trust 147
rule norms (r-norms) 196

S

second degree price differentiation 327
segmentation method 256
semi-ordered ontological models 14
semiotically closed interaction machine (SCIM) 90
signified sign 90
signifier sign 90
simple grid experiments 164
simple mixed structure (k-3) algorithms 69
simple mixed structure (k-4) algorithms 71
simple trust models 145
site history reconstruction 292
social networks 139
social norms (s-norms) 196
software agents 133

T

tag and referral experiments 186
taxonomies 2
text mining 288
third degree price differentiation 327
three-level uncertainty 79
tragedy of the commons 183
trust engineering 148
trust in MAS 135
trust propagation 144

U

ultimatum game 198
underlying model 88
unified modelling language (UML) 92

V

valuable knowledge 106
virtual enterprise creation 51, 54
virtual enterprise creation and operation 48–62
virtual enterprise definition 51, 54
virtual enterprise operation 51, 55
virtual environments, methodology of design 85–103
visual knowledge builder (VKB) 300

W

Web dynamics 288
Web mining 288

X

XML binding 109

XML schemas 110
XML technologies 17–47
XML technologies for data extraction 19
XSLT transformation language 36